HUMAN
FACTORS
ENGINEERING

HUMAN
FACTORS
ENGINEERING

Third Edition

ERNEST J. McCORMICK

Professor of Psychology
Occupational Research Center
Department of Psychology
Purdue University

McGraw-Hill Book Company

New York
St. Louis
San Francisco
Düsseldorf
London
Mexico
Panama
Sydney
Toronto

HUMAN
FACTORS
ENGINEERING

Library of Congress Catalog Card Number 74-107449

ISBN 07-044884-1

9 10 11   KPKP   7 9 8 7 6 5 4

to
Emily, Wynne, and Jan,
and to the memory of
Colleen (1945-1967)

PREFACE

Many of the man-made products and environments of our civilization are created for use by people in their everyday lives or in carrying out their work activities. In many such instances the nature (i.e., the design features) of these products and environments directly influences the extent to which they serve their intended human use. This text deals with some of the problems and processes that are involved in man's efforts to so design these products and environments that they optimally serve their intended use by human beings. These objectives, of course, are as old as man; human beings have always endeavored to adapt the things they make and their environments to their own use. It is only in recent years, however, that systematic, concerted action has been directed toward these objectives. This general area of human endeavor (and its various facets) has come to be known as *human factors engineering*, or simply *human factors, biomechanics, engineering psychology*, or (in most European countries) *ergonomics*.

The attention to this endeavor in the very recent decades reflects a significant shift in emphasis; this shift has been toward placing greater reliance upon systematic research and reduced emphasis upon human experience, as the basis for developing principles and data to be applied in adapting equipment, work space, and environments for human use. The impetus of the current attention to human factors engineering originated from the pressing requirement in the development of certain new types of systems of ensuring their suitability for their intended use. Although the initial focus of such attention was on certain items of military, aircraft, and electronics equipment, and space vehicles, some inroads of human factors engineering have since been made in the design of other items such as automobiles, transportation equipment, machine tools, computers, agricultural equipment, and certain types of consumer products. It should be added, however, that such inroads generally have been fairly restricted, and that there are many products of our civilization (including some that you and I use in our daily lives) that could indeed have benefited from more systematic human factors attention. As a frame of reference (and one that will be followed in this text), it is suggested that virtually all the things people use (ranging from can openers to huge industrial facilities and transportation systems), and the environments in which they live and work (including their houses and total communities), be viewed in terms of their possible human factors implications.

Several professions are concerned with human factors engineering in one way or another. It is essentially an interdisciplinary activity rather than a separate profession as such. While the interested professions overlap in their roles in this field, they generally fall into two groups. On one side of the coin are those professions that are concerned with creating the facilities that human beings use, the procedures and manners in which such facilities are used, and the environments in which people live and work; working within this area are engineers of all sorts (mechanical, electrical, industrial, illuminating, air conditioning, acoustical, civil, construction, traffic, etc.), industrial designers, architects, and city planners, and probably others; these are the potential *appliers* of human factors data and principles. They have it within their hands to facilitate, or conversely, to complicate, the subsequent use by human beings of those facilities which they create. On the other side of the coin are the disciplines that are the *producers* of human factors information; working within this area are primarily the behavioral scientists (such as psychologists, sociologists, and anthropologists) and the biological scientists (biologists, physiologists, etc.). Their dominant roles in the human factors area are those of carrying out research to generate new information about human beings that might be relevant to the human factors field, and of serving as consultants or advisors on matters relating to their areas of specialization.

This text is intended as a survey of human factors engineering. In line with this objective, the text deals with several of the most important aspects of human factors. In the case of each such topic it has been the intent to delineate it, to characterize its major dimensions and related concepts, and to present some of the research that is relevant to it. In the selection of research material dealing with any given topic, it was of course necessary to choose from a wide variety of sources. In doing so a dominant consideration was that of including the results of research that is reasonably central to the topic, and that therefore might have relatively general utility. In addition, however, in the case of most topics, examples of other relevant research have been summarized, but the selection of such research has been made more to illustrate various aspects of the topic in question than to provide exhaustive coverage (a five-foot bookshelf would be only the beginning of such a compendium). Although practical requirements have made it necessary to be selective in the inclusions of illustrative material, usually material was selected that would be broadly relevant rather than highly situational; however, situational material is included in some instances, this being done in part to represent the varied spectrum of the human factors area. The materials are drawn from a number of disciplines that are concerned with the human factors area, but especially from the behavioral and biological sciences; in this connection the bulk of the material comes from the field of psychology. This predominance is the consequence of psychology having undoubtedly contributed more to the human factors engineering area than any other discipline, and is also due to its being the author's own field of work.

A book such as this probably should be dedicated to the many individuals whose research has formed its basis. Their names are referred to throughout the text and are listed in the chapter references and in the authors' index, and so they will not be repeated here. If their data and conclusions have been misinterpreted, the present author assumes the responsibility; the reader should not blame the investigator. Appreciation is expressed to the publishers who have granted permission to reproduce or to adapt the many tables and figures. While many others who have contributed in some way to this effort will have to remain anonymous, there are a few whose contributions I should like particularly to acknowledge. These include Dr. Charles R. Kelley (Dunlap and Associates, Santa Monica, California), who reviewed Chapter 9 and offered suggestions related to it; Dr. Sidney L. Smith (Mitre Corporation, Boston, Massachusetts), whose comments and suggestions based on a thorough review of the previous edition have generally been incorporated in this edition; and Mr. Mark S. Sanders (Purdue University, Lafayette, Indiana), whose help was invaluable in many ways during the writing and final editing.

Ernest J. McCormick

CONTENTS

OVERVIEW

APPENDIXES

HUMAN
FACTORS
ENGINEERING

part
one
INTRODUCTION

chapter

one

THE

THINGS

PEOPLE

USE

If you were to list all the "things" people use in their everyday lives, plus those that are used in various and sundry work activities, it would become apparent that—for most people—by far the majority of the physical things we use are man made. In other words, we live in very much of a man-made world. Even those engaged in activities close to nature—the fishermen, farmers, campers, and bird watchers—use primarily man-made devices. An abbreviated listing of man-made facilities could include hand tools, kitchen utensils, vehicles, highways, machinery, TV sets, telephones, and space capsules (plus the binoculars for the bird watchers). A major thesis of this text is that the *human use* of virtually any man-made thing can be *enhanced*, or, conversely, *degraded*, by its design. Human factors engineering addresses itself to this problem.

HUMAN FACTORS ENGINEERING DEFINED

Definitions tend to be rather treacherous exercises in semantics but, for better or for worse, probably are necessary exercises. We shall approach the definition of human factors engineering in two stages, as follows:

- The *objectives* of human factors engineering are to enhance the effectiveness of the use of the physical objects and facilities people use and to maintain or enhance certain desirable human values in this process (e.g., health, safety, satisfactions).
- The *approach* of human factors engineering is the systematic application of relevant information about human characteristics and behavior to the design of both the things people use and the methods for their use, and to the design of the environments in which people work and live.

In capsule form, human factors engineering (that we shall frequently refer to simply as *human factors*) can be considered the process of designing for human use.

A BIT OF BACKGROUND

Clearly, the objectives of human factors are not new; the history of man is filled with evidence of his efforts, both successful and unsuccessful, to create tools and equipment which satisfactorily serve his purposes and to control more adequately the environment within which he lives and works. But during most of the centuries of man's history, the development of tools and equipment depended in large part on the process of evolution, of trial and error. Through the use of a particular device— an axe, an oar, a bow and arrow—it was possible to identify its deficiencies and to modify it accordingly, so that the next "generation" of the device would better serve its purpose. Experience, then, was a major basis for improvement and adaptation of man-made devices.

The industrial revolution of the past one and one-half centuries, of course, brought about major changes in the tools, equipment, devices, and environments people used; although the evolutionary process still was important as a basis for improvement in terms of human considerations, it is probable that the increased tempo of technological developments may have strained the evolutionary process as the basis for adapting devices for human use. The tempo has continued to increase in more recent years. The time during and since World War II probably will be recorded in history as a period of rapid scientific development. It has brought about an epidemic of entirely new and markedly altered types of equipment and devices for human use. It has been found, often through unhappy experiences, however, that some of these devices were not designed appropriately for human use. It was found, for example, that some items of military equipment, such as high-speed aircraft, radar, and fire-control systems, could not be managed effectively by their operators, that human errors were excessive, and that many accidents occurred because of human mistakes which were attributed to design deficiencies. Similar deficiencies have been documented for certain types of civilian equipment. Such deficiencies probably can be attributed, in part, to two factors: (1) in complex systems, the human factors aspects are amplified as contrasted with earlier, simpler types of systems and (2) many systems had been (and still are being) designed virtually "from scratch," with no opportunity to benefit from previous experience in the use of earlier versions—since there have been no earlier versions.

The upshot of all of this is that it is becoming increasingly important in the design of equipment, facilities, etc., to consider human factors aspects early in the design game and in a systematic manner. A few questions that illustrate some of the types of considerations that might be taken into account during the design stage could include the following:

Should a mail-sorting system have an optical scanner that automatically activates mechanisms for sorting mail by zip code or should human operators scan the mail and activate keying devices for sorting by zip code? Should a particular warning signal be visual or auditory? How much "feel" should be built into the power-assisted-steering mechanism of a car? Would the information load of a given (tentative) assortment of visual displays be within reasonable human limits? How much illumination should be provided for a given operation?

The solutions to these and many, many other kinds of questions should be based either on the availability of pertinent information about human characteristics, including human abilities and limitations, or on the judgment of individuals who are experts in the specific domain in question. Much of the rest of this text will deal with this, but let us now turn our attention to the general nature of man-machine systems and of the roles of people in such systems.

CONCEPT OF SYSTEMS

The concept of a *system* is indeed a very popular one, but also a very elusive one, as, for example, Jones [1][1] points out. Our current interests, however, can be somewhat restricted to what are typically referred to as *man-machine systems*, or what more accurately might be called *man-machine-environment systems*, since we shall be primarily concerned with systems that are a combination of people and machines and the environments in which they function. We shall be less concerned with what Jones refers to as social systems, biological systems, and symbol systems per se, except as these may be involved in man-machine-environment systems. We can consider a man-machine-environment system a combination of one or more human beings and one or more physical components interacting to bring about, from given inputs, some desired output within the constraints of a given environment. As we use the term man-machine system, let us keep in mind two qualifications, namely, (1) that the common concept of "machine" is really too restricted for our use and, rather, we should consider a "machine" to consist of virtually any type of physical object, device, equipment, facility, thing, even personal apparel, or "what have you" that people "use" in carrying out some activity, and (2) that some environment typically is implied, even if it is a "normal" environment. For the sake of semantic simplicity, however, we shall generally use the term *system*.

In a relatively simple form, a system can be a man with some common tool or device, be it a hoe, a hammer, a hod, or hair clippers. Going up the

[1] Numbers in brackets refer to references at the end of the chapter.

scale of complexity, one can regard as systems the family automobile, an office machine, a lawn mower, and a roulette wheel, each equipped with its operator. More complex systems include aircraft, bottling machines, conveyor systems, telephone systems, and automated oil refineries, along with their personnel. Some systems are less delineated and more "amorphous" than these, such as the servicing system of a gasoline station, the operation of an amusement park, a highway and traffic system, and the rescue operations for locating an aircraft downed at sea.

Especially in fairly complex systems, one can identify systems within systems within systems. In this connection Linvill [2] suggests that some systems can be viewed as coming in layers. An exterior system consists of a set of component interior systems, which, in turn, may be composed of component interior systems that are inferior to them, etc. We could view a complete telephone system as an exterior system, with a given telephone exchange as an interior system to it, and a specific switchboard as an interior system to the telephone exchange, etc. Interior systems frequently are referred to as *subsystems* or *components* (a component being itself something of a system). Within a given complex system, one can also envision a sequence of subsystems, such as the various sequential production processes in an industrial plant. Recognizing the various possible relationships among systems or subsystems, in system development processes one may, operationally, consider a particular entity—big or little, exterior or interior—the system to be dealt with.

CHARACTERISTICS OF SYSTEMS

All systems, at whatever level of detail, have certain general characteristics, or "properties," in common.

Purposes of the System

First, every system has some purpose or objective. This should be clearly understood and in most situations should be made a matter of record, including a definitive statement of the specifications that are to be fulfilled. In contemplating the design of an electric car for home-to-office and around-the-town use, for example, one should set forth in as precise terms as possible such specifications as speed, range, and maneuverability and set about designing a vehicle that will fulfill these specifications. Some of the specifications may be of a strictly engineering nature, (e.g., the horsepower), whereas others may have strong human factors overtones (e.g., the physical dimensions for the passengers or the ratio of the vehicular turn to steering-wheel turn).

System Functions and Components

A system, as we have defined it, carries out certain functions which are directed toward the purposes in mind, these functions being performed individually or collectively, or both, by the human being(s) and the physical component(s) of the system. The functions of systems generally can be characterized in different ways, as, for example, in terms of basic types or classes of functions and in "operational" terms.

Basic functions of systems. For our purpose, four basic classes of functions will be mentioned. These include sensing (information receiving), information storage, information processing and decision, and action functions; they are depicted graphically in Figure 1-1. Since

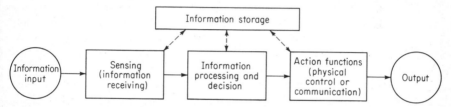

Figure 1-1 Types of functions performed by man or machine components of man-machine systems.

information storage interacts with all the other functions, it is shown above the others. The other three functions occur in sequence.

1. *Sensing (information receiving):* One of these functions is sensing, or information receiving. Some of the information entering a system is from outside the system, for example, airplanes entering the area of control of a control-tower operator, an order for the production of a product, the heat that sets off an automatic fire alarm, various cues regarding the presence of schools of fish, and telegraph communications. Some information, however, may originate from inside the system itself. Such information can be of a feedback nature (such as the reading on the speedometer of the action on the accelerator or the feel of a control lever), or it can be information that is stored in the system.

    The sensing, if by a human being, would be through the use of the various sense modalities, such as vision, audition, and touch. There are various types of machine sensing devices, such as electronic, photographic, and mechanical. Sensing by a machine in some cases is simply a substitute for the same sensing function by a man. The electronic device in an automated post office which identifies the location of a stamp on an envelope is simply doing the same thing that a man would otherwise do to place the envelope in proper position for canceling the stamp. The sonar used for detecting schools of fish, however, involves "sensing" the fish in a manner that a man is not capable of.

2. *Information storage:* For human beings, information storage is synonymous with memory of learned material. Information can be stored in physical components in many ways, as on punch cards, magnetic tapes, templates, records, and tables of data. Most of the information that is stored for later use is in a coded or symbolic form. Language and numerical systems are essentially coding systems in that they are used to represent the objects or concepts in question.

3. *Information processing and decision:* Information processing embraces various types of operations performed with information that is received (sensed) and information that is stored. The identity and nature of such operations as far as human mental processes are concerned are not yet well understood; in many respects the human mind must still be considered as a "black box." When human beings are involved in information processing, this process, simple or complex, typically results in a decision to act (or in some instances, a decision *not* to act). When mechanized or automated machine components are used, their information processing must be programmed in some way in order to cause the component to respond in some predetermined manner to each possible input. Such programming is, of course, readily understood if a computer is used. Other methods of programming involve the use of various types of schemes, such as gears, cams, electrical and electronic circuits, and levers.

4. *Action functions:* What we shall call the *action* functions of a system generally are the operations which occur as a consequence of the decisions that are made. These functions fall roughly into two classes. The first of these is some type of physical control action or process, such as the activation of certain control mechanisms or the handling, movement, modification, or alteration of materials or objects. The other is essentially a communication action, be it by voice (in human beings), signals, records, or other methods. Such functions also involve some physical actions, but these are in a sense incidental to the communication function.

Operational functions. Although most of the functions of a system can be viewed in terms of the above basic functions, for system design it is usually desirable, early in the game, to set down in black and white the operational functions that need to be performed by the system to accomplish its purpose. For example, in a postal operation it is necessary to perform such functions as stamp cancellation, mail sorting (actually a series of sorting operations), and delivery. In most situations the functions can be specified at various levels of specificity. In system-design processes it is sometimes the practice to specify these operational functions and to set them forth in blocks in a block diagram, with the tentative expectation that each function can be *allocated* to a corresponding physical component or to a human being. More will be said about this allocation process later. But it should be noted here, as Jones [1] points out, that this assumption of a one-to-one correspondence between functional blocks and separate physical (and presumably human) components is most applicable to *flow* systems, in which there is a sequence of clearly discriminable operations. But although all systems do involve the execution of functions by human or physical components, or by both, in some systems the functions are, as Jones [1] puts it, less "determinate" and may not be clearly discriminable from each other. In such instances the assumption of a one-to-one correspondence between functions and system components simply may not apply; Jones cites, as an example, the operation of an automobile by its driver, indicating that there is no quick and unambiguous way of knowing that any particular block diagram (of functions and components) represents the constantly changing interaction between the driver, the vehicle, and the environment. The moral of these observations seems to be that the function-component frame of reference may serve a useful purpose in the development of some systems but may be of less value for other systems.

Communication Links

Virtually every system involves some form of communications and communication links to make this possible. Some systems (such as telephone systems) exist for the sole purpose of sending communications. In some other systems the end action is a communication (as in computers). Although such systems have dominant communication objectives, there is some form of internal communication inherent in every system. In the most obvious form, it consists of voice or written communications. In less obvious form, it may be the transmission of a control signal (mechanical or otherwise) from an operator to the machine through the activation of his control device; in turn, the machine may "talk back" through, say, visual instruments that provide information or by direct cues such as sound and sight. By a bit of extrapolation, one can consider one component, whether man or machine, as "communicating" with another. Every such communication must be provided for by some type of communication link. In systems design, then, it is necessary to anticipate who (or what) is to "talk" to whom (or what) and to provide an appropriate link, human or mechanical, to make this possible.

Procedures

Another attribute of a system is the set of procedures or practices that are followed in its operation. These procedures may have been formally set forth as the system was developed, or they may have evolved with the use of the system. Whatever their origin—and whatever their status (whether formally set forth, or simply "ways of doing things")—they are part and parcel of the system and affect its operations for better or for worse.

Input and Output

Another essential feature of man-machine systems is their inputs and outputs. The input into a system consists of the ingredients that are necessary in order to achieve the desired outcome. The input may consist of physical objects or materials, such as lumber in a sawmill, crude petroleum in a refinery, or beef for the broth to be brewed. It may be information in some form, for example, account records, telegraphic messages, or the presence of objects such as aircraft in an area. In communication systems, the primary input would be information in some form. Further, the input may consist of energy, such as electric power, heat, or other types. In any given system, the input may consist of any or all of these. The output is of course the result or outcome of the system, such as a change in a product, a communication transmitted, or a service rendered. When the system in question has various components, the output of one component frequently serves as the input to another. And if one looks at systems within systems, the same relationships may exist, that is, the output of one system can serve as the input to another.

TYPES OF SYSTEMS

As indicated earlier, we are concerned primarily with systems that involve men, machines, and their environments. Before mentioning variations in such systems, let us clarify the distinction between closed- and open-loop systems. A closed-loop system is continuous, performing some process which requires continuous control (such as in vehicular operation), and requires continuous feedback for its successful operation. The feedback provides information about any error that should be taken into account in the continuing control process. An open-loop system is one which, when activated, needs no further control or at least cannot be further controlled. In this type of system the "die is cast" once the system has been put into operation, and no further control can be exercised, such as in firing a rocket that has no guidance system. Although feedback with such systems obviously cannot serve continuous control, feedback can, if provided, serve to improve subsequent operations of the system. It should also be noted that in most open-loop systems there is almost inevitably some internal feedback within the operator, even if not provided for outside the operator.

In a very gross sense, systems can be characterized by the degree of human versus machine control. In a fairly extensive classification of systems of various types, Jones [1] includes four types that he refers to as machine systems. Three of these are listed in Table 1-1 along with

Table 1-1 Certain Machine Systems Classified according to Mode of Operation and the Physical Nature of Their Components and Couplings

Kind of system and mode of operation	Components	Couplings between components	Examples
1. Manual system, operator-directed, flexible	Hand tools or aids	One human operator	Cook plus utensils Craftsman plus tools Singer plus amplifier
2. Mechanical system,* operator-controlled and inflexible	Highly interdependent physical parts forming indistinguishable components and couplings		Engine Automobile Machine tool
3. Automatic system, preset, programmed, or adaptive	Powered mechanical systems*	Cables, pipes, conduits, levers, etc., forming a control circuit	Process plant Automatic telephone exchange Digital computer

* The original source refers to these as subsystems. For our purpose, however, we shall refer to them as systems.
SOURCE: Adapted from Jones [1], Table 1.

their mode of operation and the nature of their physical components and couplings. Schematic illustrations of these are shown in Figure 1-2.

Manual Systems

Manual systems consist of hand tools and other aids which are coupled together by the human operator who controls the operation, using his own physical energy as a power source. The operator (usually a craftsman) transmits to, and receives from, his tools a great deal of information, typically operates at his own speed, and can readily exploit his ability to act as a "high variety" system.

Mechanical Systems

These systems (also referred to as *semiautomatic* systems) consist of well-integrated physical parts, such as various types of powered machine tools. They are generally so designed as to perform their functions with

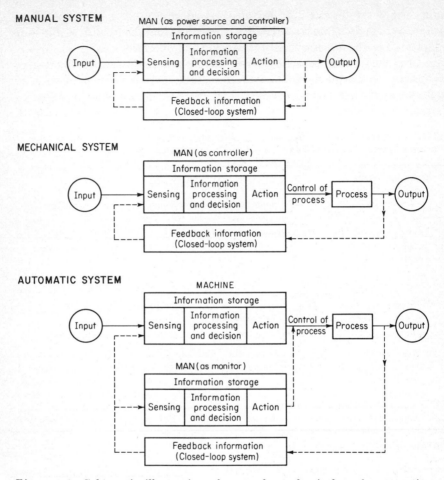

Figure 1-2 Schematic illustration of manual, mechanical, and automatic man-machine systems. In the case of closed-loop (continuous) systems, feedback information about conditions of the process is transmitted to the sensor (man or machine) for use in making necessary corrections in the control of the system.

little variation. The power typically is provided by the machine, and the operator's function is then essentially one of control, usually by the use of control devices. (Jones observes that these systems, or what he refers to as "mechanical subsystems," are the components of what he refers to as "mechanized systems"; the components typically are linked together by tracks, cables, conduits, etc., and by human operators to form the much larger ensemble of a mechanized system such as a railway or an assembly line.) The basic ingredients of mechanical systems are well represented by Taylor [3], as shown in Figure 1-3. In that figure the man receives information about the state of affairs of the system via displays, performs essentially an information-processing and decision function, and implements the decisions by the use of control devices—or as Taylor expresses it, the human operator is represented as an organic data-transmission and -processing link inserted between the mechanical

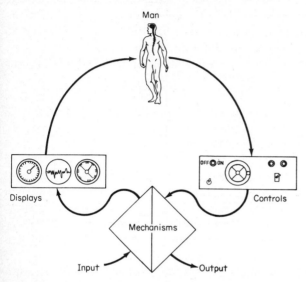

Figure 1-3 Schematic representation of a man-machine system as depicted by Taylor [3]. This represents essentially a mechanical or semiautomatic system.

or electronic displays and the controls of a machine. In some mechanical systems, however, the information about the state of affairs is sensed directly, not by the use of intervening displays.

Automatic Systems

When a system is fully automated, it performs all operational functions, including sensing, information processing and decision making, and action. Such a system needs to be fully programmed in order to take appropriate action for all possible contingencies that are sensed. Most automatic systems are of a closed-loop nature. A schematic illustration of automatic systems is given in Figure 1-2. If such a system were perfectly reliable, it conceivably could offer the possibility of taking over all functions and leaving men to twiddle their thumbs or take off for the golf course. No one would have to stay to tend the store. Since

perfectly reliable automated systems are not a likely possibility, however, at least in our lifetime, it is probable that certain primary human
functions in such systems will be those of monitoring and maintenance.

Discussion

The distinctions made between manual, mechanical, and automatic
systems are not really clear-cut. Rather, one should view these systems
as differing in the degree of manual versus automatic control. And, in
fact, within any given system, the different components (which can be
considered subsystems) can vary in the degree of their manual versus
automatic features.

If we were to try to view the development of systems historically,
it would seem that the period we refer to as the industrial revolution was
a period during which the trend was generally from strictly manual
systems to those of a mechanical nature. Much of the industrial and
military equipment today is somewhere within this segment of the
continuum since many of the sensing, decision-making, and control
functions in such systems are performed by men. The technological
developments of the last few decades, however, are causing further
shifts toward the automation end of the spectrum and undoubtedly
will bring about further changes in the nature of the activities of human
beings in carrying out the work of the world.

Aside from systems that might be characterized as manual, mechanical, or automatic (or some combination of these), there are some types
of physical equipment that defy such an ordering and that can only be
considered "systems" in the very loosest sense. It stretches one's credulity
a bit, for example, to regard as systems various items of personal equipment (glasses, girdles, apparel, etc.), furniture, buildings, printed material
(books, magazines, etc.), and beer mugs. But since people do "use"
such items, we might ask your (the reader's) indulgence in stretching
the common concept of systems enough to encompass these and many
other items that people use in their work, at home, or elsewhere.

SUMMARY

1. A man-machine-environment system is a combination of one or
   more human beings and physical components interacting to bring
   about, from given inputs, some desired output within the constraints
   of a given environment.
2. Most systems have certain general characteristics, such as a purpose,
   functions and components, communication links, procedures, and
   inputs and outputs.

3. Among various types of systems are manual systems (which involve human power and are operator directed and flexible), mechanical systems (which use external power sources and are operator controlled but generally inflexible), and automatic systems (which, by the use of automation, are programmed to perform all operational functions).

REFERENCES

1. Jones, J. C.: The designing of man-machine systems, *Ergonomics*, 1967, vol. 10, no. 2, pp. 101–111.
2. Linvill, W. K.: "Sensitivity of system design to system objectives," in *Proceedings of the National Symposium on Human Factors in Systems Engineering*, IRE Professional Group on Military Electronics, Dec. 3–4, 1957.
3. Taylor, F. V.: Psychology and the design of machines, *The American Psychologist*, 1957, vol. 12, pp. 249–258.

chapter

two

HUMAN

FACTORS

IN

SYSTEM

DEVELOPMENT

In recent years the term *system design and development* has become almost a byword in engineering-design stables. It refers to the various procedures and processes that are involved in designing and testing systems of all kinds. However, these procedures and processes do not consist of a single set of well-ordered, structured routines to be followed; rather, the term *system development* is simply something of an umbrella that can cover a broad conglomeration of operations that have to do with the basic objectives of designing and testing various kinds of systems.

This is not a book on system development.[1] As discussed in the first chapter, however, we do have a major stake in these processes in the sense that we are interested in the human factors aspects of the systems that are created, whether the system is a toothbrush, a pair of pliers, a tennis racquet, a typewriter, a highway, a radio, a tractor, an aircraft, a machine tool, an income tax form, or a cheese knife.

BASIC PROCESSES IN SYSTEM DEVELOPMENT

Granting differences in specific system development procedures, certain rather basic types of processes, however, are usually carried out, at least for complex systems. These processes require making various types of decisions at appropriate stages of the development processes. A graphic representation of these processes as formulated by Singleton [33] is given in Figure 2-1. In the development of some systems, especially rather simple ones, some of these processes may be very informal and in some

[1] Following are certain texts that deal with system development processes: N. N. Goode and R. E. Machol: *System engineering*, McGraw-Hill Book Company, New York, 1957; A. D. Hall: *A methodology for system engineering*, D. Van Nostrand Company, Inc., Princeton, N.J., 1962.

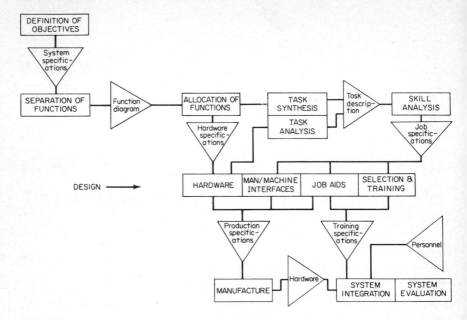

Figure 2-1   Graphic representation of the major types of processes (decisions) in the development of complex systems. For some systems certain of these processes may not be applicable. [From Singleton, 33.]

instances may be completely irrelevant. Certain of these processes will be discussed briefly.

Definition of Objectives

This phase typically includes the crystallization of the objectives of the system, including a statement of the inputs it will receive and of the outputs that the system will be expected to produce, including the performance requirements of the system (i.e., the performance standards which the system should achieve; actually the desired performance level of the output). In addition there should be a statement of all constraints under which the system will be operated—environmental, cost, social, etc. A very condensed example from the military field is given by Shapero and Bates [32], this including the objective and a listing of performance requirements.

· Objective (mission)

   The interception of manned bombers
· Performance requirements

   Capable of bomber kill

   Range, 200 miles

   Altitude, from 20,000 to 35,000 ft

The end result of this phase is the establishment of a set of system specifications for subsequent use during the later phases of system development.

Further along the processes, after the general nature of any subsystems and components become crystallized, it is usually in order to treat these parts also as systems and to set forth their objectives and performance requirements. In setting forth performance requirements (at whatever level—system, subsystem, or component) constraints include those imposed by the physical environment, the nature and availability of natural resources, the laws of nature, the stage of scientific and technological knowledge, human limitations, the values that have been accepted by the society in question, the policies of the organization(s) involved, funds and other economic considerations, and time, as well as others, including value judgments relating to trade-off advantages and disadvantages. Whatever their origins, if constraints imposed at any given level of an entire system result in modification of otherwise desired performance requirements, such modifications can have a chain-reaction effect on the performance requirements of preceding or succeeding systems, subsystems, or components.

Separation of System Functions

Once the objectives and performance requirements are pinned down, it usually behooves the designer to determine what functions need to be performed to achieve the desired objectives. As implied by Folley and his cohorts [13], at this juncture one usually should be concerned with *what* functions need to be provided for and not with the *way* in which they are to be performed.

Allocation of Functions

Given functions that have to be performed, in some instances there may be an option as to whether any particular function should be allocated to a human being or to some physical (machine) component(s). In this process, the allocation of certain functions to human beings, and of others to physical components, is virtually predetermined by certain manifest considerations, as obvious superiority of one over the other or economic considerations. Between these two extremes, however, may be a range of functions that are within the reasonable repertoire of both human beings and physical components. In this allocation process, however, we should keep in mind the qualms expressed by Jones [16] that there are circumstances in which the notion of a one-to-one correspondence between functions and components may be severely strained. In such instances one might view the functions (on one side of the fence) and the components (on the other side) in more of a "collective" frame of reference than a one-to-one relationship.

Because of the importance of these decisions for some systems, one would hope that there would be available some guidelines to aid the system designer in allocating specific functions to human beings versus physical machine components. The most common types of guidelines that heretofore have been proposed consist of general statements about the kinds of things human beings can do better than machines and vice versa. As Chapanis [8] points out, however, such comparisons serve a useful function in only the most elementary kind of way. He points out that such generalizations are useful primarily in directing one's thinking toward man-machine problems and in reminding one of some of the general characteristics that men and machines have as system components. Aside from the potential practical utility of such comparisons for use in the allocation process, however, Jordan [18] and others raise the more basic question of the appropriate role of human beings in systems, especially as we continue the trend toward automation. Before discussing such views, however, let us present a set of such generalizations, first to demonstrate what such "lists" are like, and second to serve what Chapanis refers to as the "elementary" purposes for which they may have some utility.

Relative capabilities of human beings and of machines. The following generalizations about the relative capabilities of human beings and of machine components are drawn from various sources [Chapanis, 7, p. 543; Fitts, 11 and 12; Meister, 22; Meister and Rabideau, 24; and others] plus some additional items and variations on previously expressed themes.

Humans are generally *better* in their abilities to:

- Sense very low levels of certain kinds of stimuli: visual, auditory, tactual, olfactory, and taste.
- Detect stimuli against high-"noise"-level background, such as blips on cathode-ray-tube (CRT) displays with poor reception.
- Recognize patterns of complex stimuli which may vary from situation to situation, such as objects in aerial photographs and speech sounds.
- Sense unusual and unexpected events in the environment.
- Store (remember) large amounts of information over long periods of time (better for remembering principles and strategies than masses of detailed information).
- Retrieve pertinent information from storage (recall), frequently retrieving many related items of information; but reliability of recall is low.
- Draw upon varied experience in making decisions; adapt decisions to situational requirements; act in emergencies. (Does not require previous "programming" for all situations.)
- Select alternative modes of operation, if certain modes fail.

- Reason inductively, generalizing from observations.
- Apply principles to solutions of varied problems.
- Make subjective estimates and evaluations.
- Develop entirely new solutions.
- Concentrate on most important activities, when overload conditions require.
- Adapt physical response (within reason) to variations in operational requirements.

Machines generally are *better* in their abilities to:

- Sense stimuli that are outside man's normal range of sensitivity, such as x-rays, radar wavelengths, and ultrasonic vibrations.
- Apply deductive reasoning, such as recognizing stimuli as belonging to a general class (but the characteristics of the class need to be specified).
- Monitor for prespecified events, especially when infrequent (but machines cannot improvise in case of unanticipated types of events).
- Store coded information quickly and in substantial quantity (for example, large sets of numerical values can be stored very quickly).
- Retrieve coded information quickly and accurately when specifically requested (although specific instructions need to be provided on the type of information that is to be recalled).
- Process quantitative information following specified programs.
- Make rapid and consistent responses to input signals.
- Perform repetitive activities reliably.
- Exert considerable physical force in a highly controlled manner.
- Maintain performance over extended periods of time (machines typically do not "fatigue" as rapidly as humans).
- Count or measure physical quantities.
- Perform several programmed activities simultaneously.
- Maintain efficient operations under conditions of heavy load (men have relatively limited channel capacity).
- Maintain efficient operations under distractions.

In discussing such "lists" of relative advantages of men and machines, Jordan [18] boils the assortment down to a nub, as follows: "Men are flexible but cannot be depended upon to perform in a consistent manner, whereas machines can be depended upon to perform consistently but have no flexibility whatsoever."

Limitations of man-machine comparisons. The previously implied limitations regarding the practical utility of general comparisons of

human and machine capabilities stem from various factors. Some of these
have been pointed up by Chapanis [8][2] and Corkindale [10] and are discussed below (with a fair portion of editorial license).

1.   General man-machine comparisons are not always applicable. Given
     some general superiority of men or machines, there are circumstances
     in which it would be inappropriate to apply the dictates of that gen-
     erality. For example, the amazing computational abilities of com-
     puters do not imply that one should use a computer whenever com-
     putations are required.
2.   Lack of adequate data on which to base function allocations. The
     gaps in, and limitations of, data on human performance in certain
     areas make it impossible, or at least treacherous, to set forth defini-
     tive statements about human capabilities in those areas. But, as
     Chapanis point out, these limitations also exist in certain areas of
     the machine side of the equation as well.
3.   Relative comparisons are subject to continual change. Any list stating
     the limitations of machines at any given point in time is subject to
     change because of continuous technological developments. To use a
     mundane example, there probably is no current model of automobile
     that has a crank for starting the engine (in the years gone by the
     crank was the *only* way to start a car). Today, machines are not very
     effective in pattern-recognition functions; but this may change in the
     years to come.
4.   It is not always important to provide for the "best" performance.
     Although men or machines may perform a function best, it may be
     that that relative superiority is not of major concern. To use an ex-
     ample given by Chapanis, although the use of men as toll collectors
     on superhighways offers some advantages over the use of machine
     toll collectors, the machine collectors do the job well enough to be
     acceptable, and they may offer enough other advantages to make up
     for their deficiencies.
5.   Function performance is not the only criterion. As Corkindale points
     out, from a system point of view the choice between man and ma-
     chine is not governed solely by considerations of the level of function
     performance of men versus machines. One also has to consider the
     trade-off values of other criteria such as availability, cost, weight,
     power, reliability, cost of maintenance, etc. For example, although
     remote-control devices for the family TV and garage are available,
     their cost probably has limited their widespread acceptance. At the

[2] Discussed by permission of the editor of *Occupational Psychology* (1965, vol. 39, no.
1, pp. 1–11), quarterly journal of the National Institute of Industrial Psychology,
14 Welbeck Street, London W1M, 8DR.

present stage of affairs, there are very few systematic guidelines to follow in figuring out the relative trade-off values of various criteria.

6. Function allocation should take social and related values into account. The process of allocation of functions to men versus machine components directly predetermines the role of human beings in systems and thereby raises important questions of a social, cultural, economic, and even political nature. The basic roles of human beings in the production of the goods and services of the economy have a direct bearing upon such factors as job satisfaction, human motivation, the value systems of individuals and of the culture, etc. Since our culture places a premium on certain human values, the system should not require human work activities that are incompatible with such values. In this vein, Jordan [18] postulates the premise that men and machines should not be considered comparable, but rather should be considered complementary. Whether one would agree with his conclusion that the allocation concept becomes entirely meaningless, the fact does remain that decisions (if not allocations) need to be made concerning the relative roles of men and machines. In this context, although the objectives of most systems are not the entertainment of the operators (pinball machines and gambling devices excepted), it would seem, for example, that, within reason, the human work activities that are generated by a system preferably should provide the opportunity for reasonable intrinsic satisfaction to those who perform them.

In this connection, it has been proposed that for jobs to be satisfying three conditions seem to be necessary and sufficient, namely [Jordan, 17]: (1) they must demand of the operator the utilization of skills; (2) they must be meaningful; and (3) the operator must have real responsibility. The point regarding the utilization of skills raises the question as to what it is in a job that requires skills. Several factors have been suggested [Jordan, 17]. In the first place the task, by definition, must be difficult, but within the range of potential competence of the incumbent. Second, it must permit degraded performance! If possible degradation can occur, the challenge to the incumbent is to develop, or select, those modes of performance that are better; there must be more than one degree of freedom in how the job can be done for "skill" to be developed. And third, there should be feedback to the individual about the consequences of his own job activities. Whitfield [37] shares Jordan's general view of the utilization of human skills in systems by making the assertion that the main justification for using human operators in advanced systems is the human being's capacity for skilled performance. Further support for this general theme comes from Bowen [4], who expresses the opinion that we cannot be successful in utilizing the resources of man in a system until we accept the fact that he contributes a qualitatively different form of operation in comparison with machine elements.

These and corresponding reflections cause one to view with a jaundiced eye the practice (however infrequent or frequent) of considering human beings as potential "components" in systems in the same mechanistic manner as one views various physical devices. Rather, they cause one to consider, in humanistic terms, the question of the optimum roles of human beings in the systems of which they are, or are to be, a part. The trend toward automation would seem to be at least partially in conflict with this idea, since it might relegate some human beings to the nominal role of monitors (a function which they really do not do well). This conflict, however, might be minimized. To do so would require that man *not* be viewed as subordinate to the system, but rather that his dominant role in the system be that of seeing that it works. As stated by Jordan, we have focused upon making jobs easy; let us now enlarge our focus and also try to make the jobs interesting for every human operator at every skill level—toward the dual ends of improving the lot of the human operator as a man and of improving overall system performance.

A strategy for allocating functions. The discussion above implies that there are no sets of clear-cut guidelines available for use in deciding what system functions should be performed by men and by machines. Rather, one needs to pursue a general strategy, bringing to bear at various phases the most adequate data that are available and exercising the best (well-informed) judgment possible. As proposed by Chapanis [8], this strategy should *not* be directed toward the allocation of functions as though each function were in vacuum-packed isolation from the others. Rather, the strategy should be directed toward making decisions about functions in such a manner as to enhance the operation of the system as a whole, and toward the creation of jobs that are interesting, motivating, and challenging to the human operator.

Design of Physical Components

Although the actual design of the physical components is dominantly an engineering chore, this phase represents a second stage at which human factors inputs usually are of considerable moment. The specific nature of the design decisions made during this phase can (if inappropriate in terms of human considerations) forever plague the user and cause decrement in system performance or, conversely (if appropriately designed), facilitate the user's use of the equipment and bring about better system performance. Some of the later chapters will include illustrations to fortify this point.

Personnel Aspects of Systems

The ultimate *users* of the items of physical equipment that are produced generally fall into two groups, namely, personnel working in the labor force who are producing goods and services for the economy and consumers who purchase things for their own use. In the case of personnel in the labor force, the design of the physical equipment used can have a very substantial impact on the nature of the jobs to be performed. This leads into a wide assortment of personnel-related affairs such as job design, task and job analysis, personnel incentives, personnel selection, training, and the development and use of training simulators and training aids. Unhappy experiences in the past have stimulated some organizations to start worrying about these personnel matters *during* the system development processes instead of waiting until *after* a system is developed. In fact, the military services (especially the Air Force) have developed systematic procedures for carrying out some of these personnel functions concurrently with the development of the physical system. This concept is sometimes referred to as the *personnel subsystem*. A particularly important facet of this procedure is that of carrying out task and activity analysis of the system on something of a continuous basis during its development. Such analyses, of course, are made on the basis of inferences from the tentative design of the system at any given point in time. In part the results of these analyses can serve the designers as the basis for modifying design characteristics that are found to be incompatible with human performance abilities. We shall touch on some of these matters in a later chapter, but it should be noted that some of these personnel matters need to be attacked during the development of physical equipment, especially in the case of complex systems. In the development of consumer products, the counterparts of some of these personnel aspects take on a somewhat different hue. For example, consumer products usually are much less complex than those procured by manufacturing and service organizations; further, manufacturing and service organizations generally can select their employees, whereas there is typically little control over which consumers buy what. Nonetheless, it probably still behooves the producer of at least some consumer products to be concerned about the nature of the "jobs" of consumers when using certain products, about the "training" of consumers (in a way, really self-instruction), and about the development of "job aids" for use by the consumer in using the products (such as in assembling Junior's new bicycle or operating a washing machine).

Evaluation in System Development

The evaluation of systems and their components is (or should be) virtually a continuous process through the entire system-development cycle including the systematic evaluation of completed systems. Our particular interest in evaluation is of course focused on the human factors aspects, some of which will be discussed later. For the moment, however, two points should be made. In the first place, simulation usually has a place in evaluation. Simulation, of course, has many different faces, covering the gamut of mathematical models, and prototypes, mock-ups, and other physical representations. In the second place, a prerequisite of evaluation is the availability of a criterion or standard as the basis of evaluation. This will be an item for our agenda in just a few more pages.

Production

Somewhere along the line, design of an item of equipment needs to be frozen, preferably, of course, after adequate testing and evaluation of actual items. But the "freezing" should not be forever. Subsequent testing and experience in the use of various items—hair curlers, automobiles, or submarines—should serve as the basis for retrofitting (if necessary) or the development of improved ("new generation") designs.

CRITERIA IN RESEARCH AND SYSTEM DEVELOPMENT

Passing references and inferences have been made above to such matters as system performance, the relative capabilities of human beings and of machines, and system evaluation. All of these matters imply some sort of a basis or standard for evaluation of that being evaluated (such as a system, component, or individual); such a standard is usually called a *criterion*. In experiments it is sometimes referred to as the *measured variable* or *dependent variable*. We might evaluate a traffic system against a criterion of average waiting time of cars at intersections, or typists in terms of speed or errors or both in typing. The primary uses of criteria in human factors engineering are in research activities (such as in studying the effects of vibration on manipulative performance of people) and in system and component evaluation (such as testing the performance of a dishwasher, an ICBM, or a grease gun). The selection of a specific criterion (whether for research, system evaluation, or other purpose) should depend upon the purposes at hand. The various criteria that might be used fall into two broad classes, namely, those associated with systems (and components) and those of a "human" nature.

System Criteria

Any reference to system criteria should, of course, be assumed to embrace criteria of subsystems and components. These generally fall into two subclasses.

    System performance criteria. Basically, system performance criteria are those which relate to system output, or in other words, reflect the degree to which a system achieves its intended purpose. Frequently there may be two or more relevant performance criteria, such as, for example, the number of telephone calls handled by a system and the number of erroneous connections. The gamut of system performance criteria is, of course, tremendous when one considers the wide range of systems and their components. A random assortment could include the following: percentage of hits (in bombing systems), output per unit of time (in certain industrial operations), speed (in vehicles), error rate in cards punched, computations per unit of time (in computers), customers served per unit of time, acres of ground plowed per day, cleanliness of clothes washed, and number of feet of pavement laid per day by a road-paving gang. Some such criteria are rather strictly mechanistic, in the sense that they reflect essentially engineering performance (e.g., the maximum rpm of an engine), whereas others reflect more the performance of the system as it is used by the people involved in it (such as errors in cards punched).

Nonfunctional criteria.   Aside from measures of the performance of systems or components, there are various types of *nonfunctional* criteria that are sometimes used, especially in system evaluation. These include *design* criteria, or what Meister and Rabideau [24] refer to as "representational" criteria, which describe desirable system characteristics. For example, certain standards might be (and actually have been) set forth regarding the desirable location of visual displays, the maximum allowable force requirements in operating a control device, and the minimum space to be allowed for hand access in using certain tools, etc. Given the basic validity of these criteria (standards), they can be applied not only during the design of systems but also in evaluating completed systems. Other types of nonfunctional criteria include economic factors, space required, and such considerations as practicality and trade-off values.

Human Criteria

Within the world of human affairs, there are four relatively distinct types of criteria which, in various contexts, may be pertinent indications of human "behavior." These are human performance measures, physiological indices, subjective responses, and accident frequency.

Human performance measures.   In a strict sense human performance must be considered in terms of various sensory, mental, and motor activities. In specific work situations, however, it is usually difficult, if not impossible, to measure human performance in strictly human activity terms, since such performance usually is inextricably intertwined with the performance characteristics of the physical equipment being used. Thus, the typing performance of a typist is not entirely a function of the typist but is also in part the consequence of the typewriter (its make, condition, etc.). Along these lines, Taylor [36] gives an interesting and preposterous example—but one which nicely illustrates the point. He postulates two types of systems with the same objective, namely, traversing ½ mile in the shortest span of time. One of the "systems" is a boy on a bicycle and the other is a boy on a pogo stick. The likelihood (perhaps the inevitability) that the boy-bicycle system would get there first tells us *only* which *system* is fastest (this would be a system performance criterion); it tells us nothing about *human* performance. It could well be that the boy on the pogo stick is doing a better job of pogo-stick jumping (in human motor coordination) than the bicycle rider is doing in his bicycle riding.

Physiological indices. For some purposes indices of various physiological conditions are pertinent criteria in connection with man-machine systems. Such possible indices include heart rate, blood pressure, composition of the blood, galvanic skin response, brain waves, respiration rate, skin temperature, blood sugar, and many other measures. Some of these and other physiological variables are used as indices of the physiological effects on people of various methods of work, of work performed with equipment of various designs, of work periods, and of work performed under various environmental situations (such as heat and cold).

Subjective response. On some occasions obtaining the subjective responses of people is in order. These may be of varied types and obtained by different methods. Although some of the later chapters will include examples, a few will be mentioned briefly here. Ratings provide systematic means of obtaining judgments by people; such ratings can range over a wide gamut, including ratings of people and their performance, of the quality of objects, of the desirability of different features of a system, and of the importance of different types of information for use in a system. The opinions of people—whether obtained through ratings or otherwise—may be useful criteria.

Accident frequency. In some systems the safety of human beings is an important criterion by which a system should be evaluated. Thus, the number of injuries or deaths would be useful in comparing one system with another. The number of injuries or deaths per million miles traveled, for example, gives a comparison (in terms of this criterion) of various types of transportation systems, such as commercial airlines, railways, buses, and automobiles.

Specific Types of Criteria

Because of the inextricable interaction of people and the physical devices they use, however, we probably should regard system and human performance criteria as comprising a continuum, rather than a neat dichotomy, ranging from strictly system (essentially mechanistic) criteria at one end, to strictly behavioral criteria at the other end. As we later discuss *human* performance, we shall be concerned primarily with the behavioral half of this imaginary continuum. In both system and human criteria, there are various specific types that can be developed. A few such types are listed and described briefly below, as based in part on Smode, Gruber, and Ely [34].

Recorded observations.   When certain "events" or behaviors are anticipated or planned in some operation, it is possible to have an observer, or incumbent, watch the operation and record the presence or absence of such events or behaviors or their sequence, typically using some form of checklist.

Errors, accidents, and injuries.   These may be observed, identified after the fact, or obtained from records.

Critical incidents.   Critical incidents are recorded events or behaviors that are deemed by the observer to be *critical*—in the sense that the event or behavior is particularly conducive to success or to failure.

Rating scales.   There are various methods of rating, the most typical providing for a judgment of *degree* of some attribute or performance.

Comparison systems.   With these systems, the personnel (or whatever is being rated) are compared with each other on a relative basis, such as by rank order, paired comparison (in which each is judged in comparison with every other), and forced distribution (in which specified percentages of cases are allocated to various categories).

Rating checklists.   Another system (used in rating personnel) consists of a list of "behaviors," usually oriented to the work in question. The rater checks the checklist statements he considers descriptive of the individual or the performance in question.

Performance tests.   Tests of human or system performance are occasionally used as criteria. Such tests typically are simulations of the activity, performed under specified, controlled conditions.

Quantity of output.   The output of a system or component (and in some cases of people) sometimes can be measured in quantitative terms, such as number, weight, volume, amount, length, and frequency, these usually having a specified time base.

Quality of output.   Such measures relate to some aspect of the qualitative nature of the output; depending upon the variables in question, quality can take the form of accuracy, number of errors, magnitude of errors, percentage meeting some standard, etc.

Time.   Time is generally the reciprocal of amount. When used, it characteristically consists of the time required to initiate or to complete some specified activity or the time required to reach some specified level of proficiency.

Requirements of Satisfactory Criteria

Criteria do not grow on trees; they usually are hard to come by, and one usually has to settle for something less than what is desirable. In the case of performance criteria, what is really desired is a measure of what is sometimes called *ultimate* performance of the system or personnel within it, in effect a perfect measure of the extent to which the performance requirements of the system (or component) are fulfilled. It is usually necessary to develop and use some criterion measure that falls short of that objective. These are sometimes referred to as *intermediate* and *immediate* criteria, depending generally on the time phasing of the measure. Generally speaking, criteria need to fulfill certain requirements, especially those given below.

Relevance. In the first place a criterion must be relevant to the intended use. In effect this means that it must be a reasonably appropriate measure of the performance in question. In the system-evaluation context, relevance must be weighed against the specified performance requirements.

Freedom from contamination. Since a criterion is intended to be a measure of performance (of a system or of people), it should not be influenced materially by extraneous, irrelevant variables. Contamination would occur, for example, if two alternate systems are being compared, one operated by experienced personnel and the other by inexperienced personnel.

Reliability of criterion measure. As the term reliability is used here, it refers to the *reliability* of the *measure* being used (as opposed to the concept of system or component reliability, to be discussed later). A measure is reliable to the extent that the criterion values will be about the same under similar circumstances, such as the heart rate measured twice under similar work conditions. If two "sets" of criterion values are available, reliability frequently is determined by computing a coefficient of correlation between the two.

Use of Two or More Criteria

When there are two or more relevant criteria, one may be faced with a situation akin to the "you can't have your cake and eat it too" problems in life. One cannot have both highest speed and greatest economy in a car. The use of a mechanical device to minimize human physical effort may be accompanied by mechanical breakdowns which disrupt a process. Speeding up an assembly line may result in more rejects. Such circumstances may require the trade-off practice of having to give up or reduce some desired objective in order to gain some other advantage. There usually are practical realities that preclude achieving the most of every desirable objective; thus, in a trade-off process, it may be desirable to attempt to achieve something of an overall optimum combination of the two or more criteria. This can be attempted in different ways. In the first place the problem can be resolved by judgment, perhaps by the collective judgment of several knowledgeable individuals. In the second place there are certain statistical procedures that can be used to achieve an optimum combination of criteria. Linear programming, multiple regression analysis, *minimax* solutions, and other methods can be used. Even these statistical approaches, however, usually depend upon someone's judgments which are injected into the statistical model. It should be added, however, that there may be purposes that can best be served by using different criteria individually, without combining them. If, for example, criteria of time and of accuracy of an operation are both available, it might be useful to analyze these two sets of data separately and to have data on both available when making any overall subjective evaluation of the system or personnel.

PERFORMANCE RELIABILITY

Akin to the question of performance criteria (either system or behavioral) is the matter of performance reliability. Having just discussed reliability of criterion *measures*, let us hasten to distinguish between that use of the term reliability and the use of the term as a measure of the *dependability* of *performance* of the system or individual in carrying out an intended function. This use of the term stems from engineering practice[3] in which it refers to quantitative values that characterize the dependability of system or component performance. Actually, there are different measures of system reliability, each being relevant to certain types of systems or situations. However, most of these measures of reliability can be applied to human performance as well as to system performance. Certain variations on the reliability theme include: (1) probability of successful perform ance (this is especially applicable when the performance consists of discrete events, such as detecting a defect or starting a car); (2) mean time to failure (abbreviated MTF; there are several possible variations on this, but they all relate to the amount of time a system or individual performs successfully, either until failure or between failures; this index is most applicable to continuous types of activities). There are other variations that could also be mentioned. For our present discussion, let us consider reliability in terms of the probability of successful performance. As an aside, an illustration of the application of this reliability con-

---

[3] For texts on reliability in engineering practice, the reader is referred to G. H. Sandler, *System reliability engineering*, Prentice-Hall, Inc., Englewood Cliffs, N.J., 1963, and W. G. Ireson, *Reliability handbook*, McGraw-Hill Book Company, New York, 1966.

cept to human beings comes from a study of the comparison of the probability of detection of auditory signals (which was .86) and of visual signals
(which was .76) [Buckner and McGrath, 5].

If a "system" includes two or more components (machine or human,
or both), the reliability of the composite system will depend on the reliability of the individual components and how they are combined within
the system. Components can be combined within a system either in *series*
or in *parallel*.

Components in Series

In many systems the components are arranged in series (or sequence) in
such a manner that successful performance of the total system depends
upon successful performance of each and every component, man or machine. By taking some semantic liberties, we could assume components
to be *in series* that may, in fact, be functioning concurrently and inter-
dependently, such as a human operator using some type of equipment.
In analyzing reliability data in such cases, two conditions must be fulfilled: (1) failure of any given component results in system failure, and
(2) the component failures are independent of each other. When these
assumptions are fulfilled, the reliability of the system for error-free operation would be the product of the reliabilities of the several components.
To estimate system reliability quantitatively then, it is necessary to express the reliability of the individual components quantitatively, specifically as the probability of error-free performance.

If, for example, there are two components in a system, each of which
has a reliability of .90 (meaning that it "works" 90 percent of the time),
the reliability of the system would be the product of these two, or .81.
The basic reliability formula for sequential situations as presented by
Lusser [20], but with changed notations, is

$$R_{system} = R_1 \times R_2 \times R_3 \cdots R_n$$

in which R_1, R_2, $R_3 \cdots R_n$ are the reliabilities of the individual components, expressed as percentages of successful functioning.

It can thus be seen that each component *decreases* system reliability
by its own factor. If a system consists of, say, 100 components, each with
a reliability R of .99, the system reliability would be only .365. If the
system consisted of 400 such components, the system reliability would
be only .03, which indicates almost certain failure. A graphic representation of some of these relationships is given in Figure 2-2. This shows
the overall system reliability as a function of complexity and of component reliabilities based on the same level of reliability of all components.
(The system reliability in the case of components with varying individual

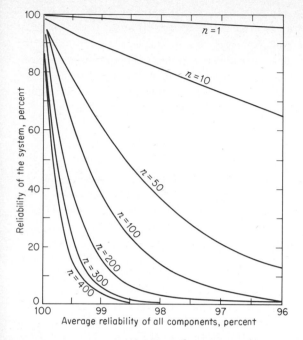

Figure 2-2 Overall system reliability as a function of complexity (number of components) and component reliability with components in series. [From Lusser, 20.]

reliabilities cannot easily be depicted graphically, but can be derived from the formula.)

When successful system performance depends upon successful performance of two components in series, for example, a physical component and a human operator, we can estimate the system reliability from the reliabilities of the two components, as shown in Figure 2-3. That figure is equally applicable for two physical components, or two human beings, operating independently, when system performance depends on performance of both.

Components in Parallel

The reliability of a system in which the components are in parallel is entirely different from the situation in which they are in a series. With parallel components there are two or more which in some way are performing the same function. This is sometimes referred to as a "back-up" or "redundancy" arrangement—one in which one component backs up another, increasing the probability that the function will be performed. In such a case the system reliability is estimated by combining the proba-

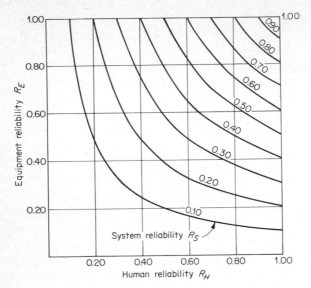

Figure 2-3 Effect of human and equipment reliabilities on system reliability. This same relationship
applies to two independent physical components or
human beings. [From Meister and Rabideau, 24,
p. 16.]

bilities of success (the reliabilities) of the several components, using the
following formula [Gordon, 14]:

$$R_{\text{system}} = [1 - (1 - r)^m]^n$$

where m = number of components in parallel for each function
n = number of functions
r = unit reliability

This formula applies only where the number of components in parallel
for each of the n functions is the same and the unit reliability of each
component is the same. Where these conditions do not apply, the derivation becomes somewhat more complex.[4]

[4] Where these conditions do not exist, the reliability can be built up by an iterative
technique, as follows: For the components in parallel, select the first. If the system
included only that one, the system reliability R_s would equal that of the component
R_1. The reliability of the system that included both components 1 and 2 (call it
$R_{s(1,2)}$) would be as follows:

$$R_{s(1,2)} = R_1 + (1 - R_2)R_2$$

and, in turn,

$$R_{s(1,2,3)} = R_{s(1,2)} + (1 - R_{s(1,2)})R_3$$

and

$$R_{s(1,2,3,\ldots,n)} = R_{s(1,2,\ldots,n-1)} + (1 - R_{s(1,2,\ldots,n-1)})R_n$$

In a strictly parallel arrangement, identical components would be used independently, either physical components or people as the case may be. If the reliabilities of individual components are, say, .90, the joint probability, for two components (which would be the system reliability) would be .99. Even with relatively low component reliabilities such as .70, system reliability of 99.2 can be achieved with four parallel components. Flights of bombers with a single target can be viewed as having parallel components. In some instances the parallel components can be of very different types, such as a man backing up a machine component.

Discussion

Lest one assumes that reliability—of physical components or men—is to be viewed only as the probability of successful performance, let us hasten to remind ourselves that there are other types of reliability, as mentioned before, such as MTF. Performance in some cases is measurable along a continuum from good to bad rather than on a *go* versus *no-go* basis. Such variations involve other statistical manipulations that will not be illustrated here.[5]

HUMAN PERFORMANCE

Although physiological measures and subjective judgments have their appropriate and important places as criteria in system development affairs, much of the raw data about human beings that should be used in this business deals with some aspect of human performance. For example, in various human factors contexts one might want to obtain answers (preferably hard, quantitative answers) to such questions as: What is the effect on speech intelligibility of various filters in a communication system? What are the effects on visual acuity of various types and levels of vibration? What is the probability of identification of a signal light at various angles from the operator? How many defects might occur in the task of soldering electric wires? The gamut of human performance information that might be required to answer these and scads of other questions ranges from very basic data about human beings (and their

---

[5] For the interested reader, where time considerations enter the reliability picture, as with MTF, the reliability of a component would decrease exponentially. The basic relationship in such a case may be described by the following function [Peters and Hussman, 28]:

$$P(t) = e^{-t/m}$$

in which $P(t)$ = reliability of the component (the probability of successful operation as a function of time)
  t = operating time
  m = mean life or mean time between failures

sensory, perceptual, mediation, and physical performance abilities) to data that are very task oriented in an operational sense and are specific to a very limited range of situations (such as the soldering task mentioned above). This broad spectrum of data has come to be referred to as *human performance data* and, at the next order of abstraction, *human performance theory*. Much of the illustrative data presented later will consist of an assortment of human performance data. Task-oriented performance data are particularly relevant to the above matter of system reliability and will warrant a bit of elaboration.

Human Performance of System Tasks

As human beings become involved in systems, their abilities and limitations are manifested in their performance of the tasks that are required. Considering the human being as an essential component of a system, for some purposes it may be necessary, as Meister [21] puts it, to measure his potentiality for failure (which, in behavioral terms, is error). This is, of course, a facet of the human reliability theme of a few paragraphs back. (Error is, of course, the complement of reliability.) If we wish to estimate the effect of human performance on system reliability, we need to do so in units of human performance that are system related, these units of behavior being *tasks*. The steps proposed by Swain [35] to do this are:

1. Define the system or subsystem failure which is to be evaluated.
2. Identify and list all the human operations performed and their relationships to system tasks and functions.
3. Predict error rates for each human operation or group of operations pertinent to the evaluation.
4. Determine the effect of human errors on the system.
5. Recommend changes necessary to reduce the system or subsystem failure rate as a consequence of the estimated effects of the recommended changes.

The particularly sticky wicket of these operations is item 3, the prediction of error rates for human operations (tasks), which would require something other than a Ouija board—it may be hoped, some reasonably quantitative data. Before discussing the collection of such data, however, let us scan the gamut of human errors in systems.

Types of human errors in system tasks. People are quite inventive in the kinds of bloopers they perpetrate, but it can serve some purposes (such as trying to reduce their frequency or severity) to figure out what kinds of mistakes people do make. As a first step in this direction, classification may be helpful. Rook [31] developed one scheme, to be used in the classification of errors in an operating system, that provides for a cross classification of errors in terms of two bases, as illustrated below:

Conscious level or intent in performing act	Behavior component		
	Input I	Mediation M	Output O
A Intentional	AI	AM	AO
B Unintentional	BI	BM	BO
C Omission	CI	CM	CO

In this formulation the intentional act obviously does not refer to an intentional error, but rather to an act that the individual performed intentionally, thinking he was doing the right thing when in fact he was not (such as pushing the "door close" button in an elevator instead of the "door open" button, thereby clobbering an entering passenger). The input-mediation-output model of this classification system corresponds to a common sequence of psychological functions that is basic to all behavior, namely, S (stimulus), O (organism), R (response). As Meister [22] points out, human error occurs when any element in this chain of events is broken, such as failure to perceive a stimulus, inability to discriminate among various stimuli, misinterpretation of meaning of stimuli, not knowing what response to make to a particular stimulus, physical inability to make a required response, and responding out of sequence. Identifying the *source* of errors in terms of input-mediation-output behaviors is a first step in developing inklings about how to reduce the likelihood of errors.

Rook's classification system above applies primarily to discrete (i.e., individual, separate, distinguishable) acts. Some variations in classifying errors in discrete acts, as well as in certain other tasks, have been proposed by Altman [1]. An abbreviated recap of his formulation is given below, with types of errors listed in each category:

Discrete acts	Continuous actions	Monitoring
Omissions	Failure to achieve end state in available time	False detections
Insertions		Failure to detect
Sequence	Displacement from target condition over time	
Unacceptable performance		

Besides defining the nature of the task itself, it is sometimes useful, for analytical purposes, to classify the task in terms of its degree of *revocability* as proposed by Altman [2], such as: immediate correction; correction only after intervening steps; no correction within a given "mission" (e.g., within the present operation); and irrevocable consequences.

Collection of Task Data

The types of task information that may be useful in at least some system-development operations cover primarily such parameters as reliability and time, although in some instances other parameters may be relevant, such as task criticality. In general there are two kinds of sources of such information, namely, empirical observations and human estimates.

Empirical task data.   Aside from sets of industrial engineering time-study data, there are two particular sources of essentially empirical task data that should be mentioned. One of these, the Data Store, was developed by The American Institute for Research [Munger, 25; Munger, Smith, and Payne, 26; and Payne and Altman, 27] particularly for tasks in the operation of electronic equipment. This Data Store was pulled together from many different sources, supplemented by special laboratory studies. Without going into the tale of how the data were developed, an example is given in Table 2-1 specifically related to the use of a joy stick.

Table 2-1   Time and Reliability of Operation of Joy Sticks of Various Lengths When Moved Various Distances

Dimension	Time to be added to base time,* sec	Reliability
Stick length, in.		
6–9	1.50	.9963
12–18	0.00	.9967
21–27	1.50	.9963
Stick movement, degrees		
5–20	0.00	.9981
30–40	0.20	.9975
40–60	0.50	.9960

* Base time, 1.93 sec.
Note: Other data (not shown here) are given for variations in other dimensions, specifically, control resistance, presence or absence of arm support, and time lag between movement of control and corresponding movement of display.
SOURCE: Munger [25].

Note that the data include performance time and reliability in performance under the variations of the illustrated dimensions. In using the Data Store, the time that would be estimated for a given instance would consist of the base time (in this case 1.93 sec) plus the times to be added for the particular dimension characteristics that would apply (such as 1.50 sec for a joy stick of 24 in., plus 0.50 sec if it is to be moved 50°, etc.).

The second source is referred to—in this day of acronyms—as SHERB (Sandia Human Error Rate Bank) [Rigby, 30], which is a compilation of error-rate data based on the THERP (Technique for Human Error Rate Prediction) [Swain, 35, and Rook, 31]. This body of data consists of human error rates for many industrial tasks based on large numbers of observations. Following are a few examples [Rook, 31]:

Type of error	Probability of error, P	Reliability, 1 − P
Two wires which can be transposed are transposed	.0006	.9994
A component is omitted	.00003	.99997
A component is wired backward	.001	.999

Judgmental task data. Where empirical data on tasks are not lying around loose, it may be useful to resort to human estimates about certain task parameters for use in a system development process. To illustrate such a process, in the estimation of the error rates of tasks, Irwin, Levitz, and Freed [15] asked experienced missile engineers, technicians, and mechanics to rate 60 tasks by sorting them into 10 piles ranging from those that would be performed with the least error to those with the greatest error. (As an aside, it might be noted that, in general, reading, inspecting, and installing tasks were judged more likely to produce errors than removal tasks.) By a subsequent manipulation (using empirical reliability data that were available for some of the tasks and were extrapolated to others) it was possible to derive quantitative estimates of the reliability of all 60 tasks. A few illustrations are listed below:

Task	Mean rating, 10-point scale (10 = greatest error)	Derived reliability
Read time (brush recorder)	8.2	.9904
Install gasket	6.0	.9945
Inspect reducing adapter	4.9	.9958
Open hand valves	3.8	.9968
Remove drain tube	2.6	.9976

Although data on human reliability in performing tasks would be useful in estimating total system reliability, it cannot be assumed that errors in performance of all tasks are necessarily equal in their effects on system performance. (To draw from Gilbert and Sullivan, it seems doubtful that to have "polished up the handle of the big front door" was a particularly critical task in the British Admiralty system.) Thus, another task parameter might be criticality, and it is this parameter that was used by Pickrel and McDonald [29] in one phase of their elaboration of human performance reliability. Specifically, they elicited ratings of task criticality, setting ranges of rating values for the following classification:

1. Safe (Error will not result in degradation, damage, hazard, or injury.)
2. Marginal
3. Critical
4. Catastrophic (Error will produce severe degradation—loss of system or death or multiple deaths or injuries.)

Using such data as a springboard, these investigators urge that a concentration of efforts for failure reduction be made for those errors which are most likely to occur and most likely to have the more serious consequences along the "criticality" scale.

Discussion

The above discussion of task performance, reliability, and error simply opens the door to a broad area of increasing importance to the whole human factors domain.[6] This door opener may be relevant at this time, however, to alert the reader to the importance of human performance and human error as they may relate to later chapters.

THE DEVELOPMENT OF HUMAN FACTORS DATA

The human factors information that is used in system development probably comes from two major sources, namely, "experience" or common sense, and research, with increasing dependence on research. It might, therefore, be useful for us to mention briefly a few facets of human research and certain of the problems that rise up to haunt both the experimenter and the user of the experimenter's data.[7]

Types of Variables Necessary in Research

There are two types of variables that must ordinarily be dealt with in most research projects. One of these is the factor that is being investigated, such as illumination, designs of instruments, information channels, environmental conditions, and gravitational forces. Such factors are referred to as the *independent variable*. The second type of variable, the

---

[6] For additional discussion the reader is referred to other sources such as Altman [2], Askren [3], Meister [21], Leuba [19], and Swain [35].
[7] For a more extensive discussion of research techniques applicable to human factors engineering, see Chapanis [6].

dependent variable, is a measure of the possible effects of the independent variable (frequently a measure of performance, such as reaction time). Actually the dependent variable is a criterion, as discussed earlier.

Laboratory versus Real-life Research

Frequently, if an investigator has a particular research purpose in mind, he may have some freedom to choose the situational context of his research—that is, whether to carry out his research in a laboratory setting or in the real world. There are other circumstances, however, in which the nature of the research problem virtually dictates the locale. For example, one cannot easily study in the laboratory the extent to which different types of speed-zone signs affect driver behavior; the laboratory would remove the spontaneous aspect of driver reactions to such signs. The experimenter in such a case obviously would set up shop along the highways and streets. On the other hand, if an investigator wished to study the differential thresholds for discriminating amplitude of vibration, he would have to do so in a laboratory, since there probably are no situations in the real world where such discriminations can be systematically observed as they take place.

Where the nature of the research problem virtually dictates the locale of the research, one simply has to accept such a constraint in his research and let it go at that—for better or for worse as the case may be. But where the investigator has the opportunity for choice, there are certain considerations that become pertinent in making such a decision, such as control of variables (which usually is easier within the pristine walls of the laboratory); the realism (if this is relevant), which is clearly greater in the real world; the differential motivation of subjects; the safety of subjects; and perhaps others.

Sampling

Practically every study or experiment includes a sample from some theoretical population. Depending on the nature of the research, the population from which a sample is drawn can consist of people, events, and even objects. Thus, a psychologist might consider the population in a study to be comprised of males with automobile driving permits in a given state, or of men undergoing a specific training course. A population of events could consist of the missile firings of a particular type of missile, of traffic at a particular location during rush hours, or of flights of commercial airlines. A population of objects could consist of rifle bullets, targets in aerial photographs, physical components of a given type (such as radio transmitters), or frozen meat pies.

In sampling, one wishes to select *part* of the people, events, or objects that comprise the population and to study the sample cases. The data from the *sample* then serve as the basis for extrapolating to, or drawing inferences about, the parent population. The sample should be of such a size and such a nature that it could be considered representing the population.

Depending somewhat on the nature of the research at hand, the selection of a sample may require very careful attention because of the possible effects upon the resulting data; in human factors research, probably one of the most common research sins is to use as subjects (in testing the use of items of equipment) engineers, technicians, and others "working on the project," instead of samples of typical users such as consumers or automobile drivers.

Statistical Analysis of Data

Once an experiment or study has been carried out and the data have been gathered, the experimenter is in a position to analyze the data to see what relationships there are between or among the independent and dependent variables that he has investigated. At this juncture it is common practice to summarize the data statistically and to carry out certain kinds of statistical analyses.

It is not intended in this text to deal extensively with statistics, or to discuss elaborate statistical methods.[8] Probably most readers are already familiar with most of the statistical methods and concepts that will be touched on later, especially frequency distributions, and different measures of central tendency (the mean, median, and mode). For those readers who are not familiar with the concepts of the standard deviation, correlations, and statistical significance, these will be described very briefly.

Standard deviation. The quantitative values of cases within a sample—values such as errors made, height of people, or scores on tests—naturally vary from each other. Certain statistical indexes can be used to quantify the degree of variability among the cases. One such measure is the *standard deviation*.[9] This is expressed in terms of the original numerical values of the data and reflects the variability in the distribution of the cases from the mean. In relatively normal distributions approximately two-thirds of all the cases fall within one standard deviation above or below the mean, and over 99 percent of the cases fall

---

[8] There are various statistical texts available to the reader who is interested in more intensive treatment of statistics, such as J. P. Guilford, *Fundamental statistics in psychology and education*, 4th ed., McGraw-Hill Book Company, New York, 1965.
[9] The standard deviation, mathematically, is the square root of the average of the squares of the deviations of the individual cases from the mean. Formulas for computing the standard deviation may be found in most elementary statistical texts.

within three standard deviations above or below the mean. If the standard deviation of the height of one group of boys is, say, 1 in., they would be more homogeneous (i.e., less variable) than another group whose standard deviation is 2 in.

Correlations. A *coefficient of correlation* is a measure of the degree of relationship between two variables. The magnitude of correlations may range from $+1.00$ (which is a perfect positive correlation) down through various intermediate positive values to zero (which is the absence of any relationship), down through various negative values to -1.00 (which is a perfect negative correlation).

Statistical significance. To evaluate his results, an experimenter should determine to what extent the results are *statistically significant*, such as the difference between two means or the size of a correlation. Statistical significance refers to the *probability* that the results (whatever they may be) could have occurred by *chance* (as opposed to being brought about by the experimental variables under investigation). It is common practice to use either the "5 percent" or the "1 percent" level as the acceptable level of statistical significance. If a difference is significant at the 1 percent level, this means that the obtained difference is of such a magnitude that it could have occurred *by chance* only 1 time out of 100. To express it another way, one could say that there are 99 chances out of 100 that there is some true difference between the groups in question. The 5 percent level can be interpreted in the same way, except that the probability is $95:5$, instead of $99:1$.

THE USE OF HUMAN FACTORS DATA

Although the consideration of the human factors aspects of systems usually is (or should be) an on-going process, there are two phases of system development in which the human factors aspects are especially pertinent (as mentioned before), namely, in the allocation of functions to men versus machines, and in the specific design of the components.

Types of Human Factors Data

In both these, and other phases of system development, the designer needs to have access to human factors *data* that can serve as the basis for resolving various design problems. The available data cover a wide span (incidentally, with many gaps yet to be filled), including some data of the following types:[10]

· Common sense and experience (such as the designer has in his "storage," some of which may be valid, and some not)

[10] Some illustrative data are given throughout this text; in addition, the reader is referred to published sources such as those referred to in this book and in Appendix C.

- Comparative quantitative data (such as relative accuracy in reading two types of visual instruments)
- Sets of quantitative data (such as anthropometric measures of samples of people and error rates in performing various tasks)
- Principles (based on substantial experience and research, that provide guidelines for design, such as the principle of avoiding or minimizing glare when possible)
- Mathematical functions and equations (that describe certain basic relationships with human performance, such as a human transfer function)
- Graphic representations (nomographs or other representations, such as tolerance to acceleration of various intensities and durations)
- Nonfunctional system criteria (such as specifications relating to design of displays and controls, some incorporated in check lists and in certain military specifications, called *Mil-Specs*)
- Judgments of experts

Evaluation of Human Factors Information

Umpteen years from now, it may be possible to crank human factors data into systems design by the use of computers (and, in fact, this is even now possible in very limited areas). Until that millenium arrives, however (and don't hold your breath until then), it probably will be necessary to inject a reasonable quota of judgment into the system design mix in *applying* information based on human factors research. In doing this, however, there are at least four considerations that are relevant in evaluating the potential applicability of the research information.

Practical significance of results. Although a research investigation may produce statistically significant results, the effects of the research if applied to a design problem might be of very limited practical significance. For example, although it might be found that the time in responding to signal A is significantly less than for signal B, the difference might be so slight, and of such nominal utility, that it would not be worthwhile going out of one's way to use signal A (especially if other factors argued against it).

The extension of research results. The second consideration in the evaluation of research results deals with the extent to which it is possible to take the results of one particular study and extend, or extrapolate, these results to other situations. In a very strict sense we should probably say that the results of a particular investigation should be considered applicable to the particular situation in which the investigation was carried out and that we should not extrapolate the results beyond this situation. Referring specifically to the relevance of laboratory studies, Chapanis [9] urges extreme caution, indicating that although results of such studies might suggest ideas and hunches that could be tried out in practical situations, one would be rash to generalize naively from such studies in the development of the solutions of real-world problems. These sobering reflections give one pause, since there may not be relevant data available from real-world studies and since it may not be feasible to try out one's ideas and hunches in practical situations.

Perhaps three points might be made about this dilemma. In the first place, despite Chapanis' reflections, there are some design problems for which available research data or experience are fully adequate to resolve the problems. In the second place, when fully adequate data are not available, one should, if possible, draw upon available research or carry out research that is as close to the real-world situation as possible; this is especially relevant when the *design decision is critical*. But in the third place (perhaps especially in noncritical situations), there are circumstances when fully confirmed data are not available and when it is not feasible to carry out research in the particular situation. In such circumstances one could argue that a judgment with regard to the applicability of research findings from somewhat comparable situations elsewhere would be better than a mere guess. This kind of generalization, of course, has its risks that depend on the variable that has been investigated and on the degree to which the situation in question is comparable with the one in which the investigation was initially carried out. Since there are no handy formulas for guiding one in making judgments about the extrapolation of research results to somewhat different situations, one must depend upon the background, knowledge, and experience of the individual making such judgments for his coming up with reasonably valid answers.

Consideration of risks. A third consideration in the evaluation of the applicability of research results is associated with the seriousness of the risks in making bad guesses. Bad guesses in designing a manned capsule for shipping an astronaut to the moon obviously would be a matter of greater concern than those in designing a hat rack.

Consideration of trade-off function. Still a fourth consideration is trade-off values. Since it frequently is not possible to achieve an optimum in all possible criteria for a given design of a man-machine system, some give-and-take must be accepted. The possible payoff of one feature (as suggested by the results of research) may have to be sacrificed, at least in part, for some other more desirable payoff.

DISCUSSION

The field of human factors engineering is still undergoing growing pains. Although the increasing pressures are recognized for the design of systems that in truth are reasonably optimum for human use, there is probably as yet no generally accepted *modus operandi* for systematically doing something about this during the development of a system. This is basically a problem of *applying* relevant human factors information at appropriate stages in system development activities. In this connection, Meister and Farr [23] carried out a systematic study of the way in which design engineers analyzed design problems when the nature of the problems demanded consideration of various operator (and other) factors. The analyses led to the dismal conclusions that the designers had little or no interest in human factors information or in the incorporation of human factors criteria in their designs, and that the degree of analysis performed was minimal and at a molecular level that was hardly conducive to the application of human factors principles to design. Add to this the implications woven into some of the above discussions that much of the human factors information available is highly situational and needs to be accepted and applied with caution, and we might assume that the prognosis for systematic consideration of the user of some systems looks bleak.

On the other side of the coin, however, are numerous success stories that indicate that, despite some of the difficulties, various types of systems and equipment have, in fact, been designed in such a manner as to facilitate their use. Perhaps the primary objectives for the further and more systematic application of human factors information in system development lie in increasing the awareness, by those concerned with system development, of the relevance of human factors considerations, in making human factors information more readily available to those who need to apply it, in so organizing and presenting such information that it can be more easily applied, and in developing (it may be hoped) more systematic procedures for the application of such information.

SUMMARY

1. The processes of system development include the definition of objectives, the separation of system functions, the allocation of functions (to men or machines), the design of physical components, personnel aspects, evaluation, and production.

2. When functions can be allocated to men or to machine components, the decisions in part can be based on the relative capabilities and limitations of men and of machines; however, such decisions need to take into account certain limitations related to such specific man-machine comparisons. The following more general comparison can provide some guidelines in making such allocations: Men are flexible but usually cannot be depended upon to perform in a consistent manner, whereas machines usually can be depended upon to perform consistently but have little flexibility.

3. Both in human factors research and in system development there are identified or implied standards for evaluation or testing, usually called *criteria*. (In research the criterion is also called the *measured variable* or the *dependent variable*.) Some such criteria relate to the system itself, such as system performance criteria and *nonfunctional* criteria (in particular *design* criteria); others relate more to human beings, such as human performance measures, physiological indices, subjective responses, and accident frequency.

4. Criteria used either in research or in evaluation need to be relevant, free from contamination, and reliable (in the sense of being a consistent or reproducible measure).

5. The reliability of performance of systems or of human beings (as contrasted with the reliability of measures, in item 4 above) refers to the degree of dependability of performance. The most common such indexes are (*a*) the probability of successful performance and (*b*) the mean time to failure (MTF).

6. In the case of components linked in a series, the reliability of the total system is the product of the reliability indexes of the individual components and thus is lower (in some cases much lower) than the reliability index of the least reliable component.

7. In the case of components in parallel (i.e., back-up or redundant components) the reliability of the total system is enhanced.

8. In designing systems in which human error is minimal (and human reliability is therefore maximal), it is useful to identify, if possible, the *sources* of human error. Such sources generally can be associated with information input (sensing), mediation processes, and output (response) processes.

9. Much of the human factors data used in system design comes from research. Such research involves the manipulation of the factor being investigated, such as illumination or type of control device; this is the *independent variable*. The *dependent variable* or criterion (mentioned above) is a measure of the effects of the independent variable.

10. Human factors research can be carried out in laboratories or in real-life situations. There are advantages and disadvantages to both types of studies.

11. Human factors *data* relevant to design problems cover a spectrum, including common sense and experience, comparative quantitative data, sets of quantitative data, principles, mathematical functions and equations, graphic representations, nonfunctional system criteria, and judgments of experts.

12. In considering the possible application of human factors research data to practical design problems, at least four considerations are

in order, as follows: (*a*) the practical significance of the results; (*b*) the degree of applicability of the results of research to the specific circumstance at hand; (*c*) the seriousness of the risks involved (i.e., the seriousness of the consequences of making bad guesses); and (*d*) the relative advantages and disadvantages of various trade-offs (i.e., of gaining certain advantages to the detriment of others).

REFERENCES

1. Altman, J. W.: Improvements needed in a central store of human performance data, *Human Factors*, 1964, vol. 6, no. 6, pp. 681–686.
2. Altman, J. W.: Classification of human error, in W. B. Askren (ed.), *Symposium on reliability of human performance in work*, AMRL, TR 67–88, May, 1967.
3. Askren, W. B. (ed.): *Symposium on reliability of human performance in work*, AMRL, TR 67–88, May, 1967.
4. Bowen, H. M.: The imp in the system, *Ergonomics*, 1967, vol. 10, no. 2, pp. 112–119.
5. Buckner, D. N., and J. J. McGrath: *A comparison of performances on single and dual sensory mode vigilance tasks*, Human Factors Research, Inc., Los Angeles, Calif., TR 8, ONR Contract Nonr 2649(00), NR 153–199, February, 1961.
6. Chapanis, A.: *Research techniques in human engineering*, The Johns Hopkins Press, Baltimore, 1959.
7. Chapanis, A.: "Human engineering," in C. D. Flagle, W. H. Huggins, and R. H. Roy (eds.), *Operations research and systems engineering*, chap. 19, pp. 534–582, The John Hopkins Press, Baltimore, 1960.
8. Chapanis, A.: On the allocation of functions between men and machines, *Occupational Psychology*, 1965, vol. 39, pp. 1–11.
9. Chapanis, A.: The relevance of laboratory studies to practical situations, *Ergonomics*, 1967, vol. 10, no. 5, pp. 557–577.
10. Corkindale, K. G.: Man-machine allocation in military systems, *Ergonomics*, 1967, vol. 10, no. 2, pp. 161–166.
11. Fitts, P. M. (ed.): *Human engineering for an effective air-navigation and traffic-control system*, NRC, Washington, D.C., 1951.
12. Fitts, P. M.: Functions of men in complex systems, *Aerospace Engineering*, 1962, vol. 21, no. 1, pp. 34–39.
13. Folley, J. D., Jr. (ed.): *Human factors methods for system design*, The American Institute for Research, Pittsburgh, 1960.
14. Gordon, R.: Optimum component redundancy for maximum system reliability, *Operations Research*, 1957, vol. 5, pp. 229–243.
15. Irwin, I. A., J. J. Levitz, and A. M. Freed: "Human reliability in the performance of maintenance," in *Proceedings, Symposium on quantification of human performance, Aug. 17–19, 1964, Albuquerque, New Mexico*, pp. 143–198, M–5.7 Subcommittee on Human Factors, Electronic Industries Association.

16. Jones, J. C.: The designing of man-machine systems, *Ergonomics*, 1967, vol. 10, no. 2, pp. 101–111.
17. Jordan, N.: Motivational problems in human-computer operations, *Human Factors*, 1962, vol. 4, pp. 171–175.
18. Jordan, N.: Allocation of functions between man and machines in automated systems, *Journal of Applied Psychology*, 1963, vol. 47, no. 3, pp. 161–165.
19. Leuba, H. R.: Quantification in man-machine systems, *Human Factors*, 1964, vol. 6, no. 6, pp. 555–583.
20. Lusser, R.: The notorious unreliability of complex equipment, *Astronautics*, Feb. 1958, vol. 3, no. 2, pp. 26, 74–78.
21. Meister, D.: Methods of predicting human reliability in man-machine systems, *Human Factors*, 1964, vol. 6, no. 6, pp. 621–646.
22. Meister, D.: "Human factors in reliability," in W. G. Ireson, *Reliability handbook*, sec. 12, McGraw-Hill Book Company, New York, 1966.
23. Meister, D., and D. E. Farr: The utilization of human factors in formation by designers, *Human Factors*, 1967, vol. 9, no. 1, pp. 71–87.
24. Meister, D., and G. F. Rabideau: *Human factors evaluation in system development*, John Wiley & Sons, Inc., New York, 1965.
25. Munger, S. J.: *An index of electronic equipment operability: evaluation booklet*, The American Institute for Research, Pittsburgh, 1962.
26. Munger, S. J., R. W. Smith, and D. Payne: *An index of electronic equipment operability: data store*, The American Institute for Research, Pittsburgh, 1962.
27. Payne, D., and J. W. Altman: *An index of electronic equipment operability: report of development*, The American Institute for Research, Pittsburgh, 1962.
28. Peters, G. A., and T. A. Hussman, Jr.: Human factors in systems reliability, *Human Factors*, 1959, vol. 1, no. 2, pp. 38–42.
29. Pickrel, E. W., and T. A. McDonald: Quantification of human performance in large, complex systems, *Human Factors*, 1964, vol. 6, no. 6, pp. 647–663.
30. Rigby, L. V.: *The Sandia human error rate bank (SHERB)*, paper presented at Symposium on Man-Machine Effectiveness Analysis: Techniques and Requirements, Human Factors Society, Los Angeles Chapter, Santa Monica, Calif., June 15, 1967.
31. Rook, L. W., Jr.: *Reduction of human error in industrial production*, Sandia Corporation, Albuquerque, N.M., SCTM 93–62(14), 1962.
32. Shapero, A., and C. J. Bates, Jr.: *A method for performing human engineering analysis of weapon systems*, USAF, WADC, TR 59–784, September, 1959.
33. Singleton, W. T.: The systems prototype and his design problems, *Ergonomics*, 1967, vol. 10, no. 2, pp. 120–124.
34. Smode, A. F., A. Gruber, and J. H. Ely: *The measurement of advanced flight vehicle crew proficiency in synthetic ground environments*, USAF, MRL, TDR 62–2, February, 1962.
35. Swain, A. D.: "THERP (Technique for human error rate prediction)," in *Proceedings, Symposium on quantification of human performance, Aug.*

17–19, 1964, *Albuquerque, New Mexico*, pp. 109–117, M–5.7 Subcommittee On Human Factors, Electronic Industries Association.

36.   Taylor, F. V.: Psychology and the design of machines, *The American Psychologist*, 1957, vol. 12, pp. 249–258.

37.   Whitfield, D.: Validating the application of ergonomics to equipment design: a case study, *Ergonomics*, 1964, vol. 7, no. 2, pp. 165–174.

part
two
INFORMATION
INPUT
ACTIVITIES

When, in the course of system development, the designer needs to develop plans for the transmission of information to a human being, he needs to so design the features of the system that appropriate stimuli (that convey information) can be reliably sensed by the individual through his sensory mechanisms. This requirement applies whether the system is a weather satellite, the family radio, or a page of Braille print. The primary senses are vision; hearing; the cutaneous senses of temperature, pain, and pressure (touch); the chemical senses of taste and smell; kinesthesis; and the orientation senses. In considering the use of information (really stimuli) received from our environment, certain conditions have to be fulfilled, these depending upon the sensory modality and the nature of the stimuli. In the first place, the sensory mechanism in question must be capable of sensing the relevant stimuli. Second, assuming the sensory mechanism is physically capable of receiving a given stimulus, the conscious awareness of the stimulus will depend on such situational factors as the intensity of the stimulus itself and the other stimuli present. With respect to other stimuli, some may be meaningful and thus require the attention of the individual in a "time-sharing" manner, whereas other stimuli may be irrelevant. The irrelevant stimuli are referred to as *noise*, whether they are actual auditory noise or visual noise (such as snow on the TV screen) or other types of stimuli. In the third place, reception will depend in part on the state of the individual himself, such as his interest, motivation, or momentary attention. Given a particular stimulus in a particular situation, it is at least theoretically possible to determine the probability of its being detected by an individual. This probability, expressed as a percentage, can be viewed in a reliability context, as discussed in Chapter 2.

The sensory reception of a stimulus, however, is only the first phase of using information received through the senses; it is also necessary that it be correctly recognized by the individual. This is essentially a perceptual process and involves the interaction of sensory processes and the cortex. These interactions are still not clearly understood. It should be noted, however, that it is usually necessary for there to be recall, from memory, of the nature or the meaning of the stimulus. For example, we have learned to recognize our neighbor's dog, to identify the sound of a dripping faucet, and to differentiate by touch the texture of fabrics. If the stimulus is a code symbol of some type, such as a red traffic light, it is necessary that its meaning (in this case, to stop) should be understood.

The information about objects and events in the environment that is received through the senses generally can be considered in two classes. In the first place, much of the information comes to us *directly* through our various sensory apparatuses. Thus, a person sees directly the traffic about him, hears the noise of an engine, and senses the air temperature, his body posture, odors in the air, and the taste of food. In the second place, some of the information man receives about his environment comes to him, usually in coded form, through displays that have been so designed that they convey the desired information *indirectly* rather than directly.

In the design of methods for the presentation of information (e.g., displays) to people, it is of course desirable to know how the various sensory mechanisms operate and what their relative capabilities and limitations are; we shall, therefore, discuss briefly the various sense modalities.[1]

VISION

Of the several senses, vision is undoubtedly the most important from the point of view of informing man about the world around him.

The Nature and Measurement of Light

Light is *visually evaluated radiant energy* [26, pp. 3–4]. The entire electromagnetic spectrum consists of waves of radiant energy that vary from about 1/1 billion of a millionth to about 100 million meters (m) in length. This tremendous range includes cosmic rays; gamma rays; x-rays; ultraviolet rays; the visible spectrum; infrared rays; radar; FM, TV, and radio broadcast waves; and power transmission—as illustrated in Figure 3-1.

The visible spectrum ranges from about 380 to 760 nanometers (nm). The nanometer (formerly referred to as a millimicron) is a unit of wavelength equal to 10^{-9} (one-billionth) m. Since light is radiant energy that is visible, it is then basically psychophysical in nature rather than purely physical or purely psychological. Variations in wavelength within the visible spectrum give rise to the perception of color, the violets being around 400 nm, blending into the blues (around 450 nm), the greens (around 500 nm), the yellow-oranges (around 600 nm), and the reds (around 700 nm and above).

---

[1] For more extensive treatment of this topic the reader is referred to such sources as Graham [6], Gregory [7], Tuttle and Schottelius [17], or Wyburn, Pickford, and Hirst [24].

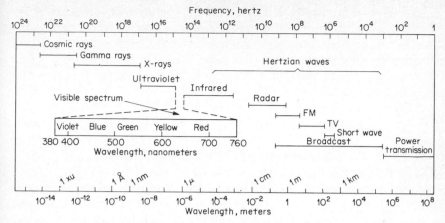

Figure 3-1 Spectrum of electromagnetic radiation, showing the visible spectrum. [Adapted from IES lighting handbook, 26, fig. 1-1.]

Light that strikes the eye may come directly from some luminous body, such as the sun, a light bulb, a flame, or a candle. Light also comes to the eyes by being reflected from an object or surface. In such a case the light comes from some luminous source and after striking the object or surface is reflected to the eyes. What we see in our environment is, of course, largely reflected light.

Reflected light. The process of reflection of light from a surface is called *subtraction*, since each surface absorbs some specific combination of wavelengths. The light reflected from a colored object is the effect of the interaction of the spectral characteristics of the originating light source with the spectral absorption characteristics of the object. If a colored object is viewed under white light (light that includes all wavelengths in about equal proportions), it will be seen in its "natural" color. If it is viewed under a light that has a concentration of energy in a limited segment of the spectrum, the reflected light may alter the apparent color of the object, such as a blue necktie appearing green when viewed under a yellow light.

The measurement of light: photometry. The Illuminating Engineering Society has established a standard nomenclature, with abbreviations, for the measurement of light [26]. These measurements include the candela, candlepower, footcandle, and footlambert.

The *candela* (cd), formerly called the *candle*, is a measure of luminous intensity.[2]

Candlepower (cp) is luminous intensity (light at its source) expressed in candelas.

---

[2] The candela is defined as $\frac{1}{60}$ of the luminous intensity of 1 cm² of projected area of a blackbody radiator operating at the temperature of solidification of platinum.

The *footcandle* (fc) is a measure of illumination at some given distance from a source when the foot is taken as the unit of length. This measure can best be characterized if one assumes that light comes from a *point source*. One footcandle is the amount of illumination from a 1-cp point source at a distance of 1 ft. The distribution of light follows the *inverse-square law*, as follows:

$$fc = \frac{cp}{D^2}$$

in which D is the distance in feet. At 2 ft a 1-cp source would produce $\frac{1}{4}$ fc, and at 3 ft it would produce $\frac{1}{9}$ fc.

The *footlambert* (fL) is a measure of photometric brightness (luminance) and is equal to $1/\pi$ cd/ft^2. It is typically used as a measure of light reflected from a surface. The *millilambert* (mL), equal to 0.929 fL, is another measure of brightness.

Color

The physical characteristics of the light reflected from objects that produce our sensations of color are the *dominant wavelength* of the physical (light) energy present, its *purity* (i.e., the predominance of a particular wavelength, as contrasted with an admixture of various wavelengths), and its *luminous reflectance*. The attributes of the perceived color of an object that correspond to these physical attributes are [26, pp. 3–5] *hue* (the attribute which determines whether it is red, yellow, green, etc.), *saturation* (the attribute used to describe its departure from gray of the same lightness), and *lightness* (the attribute by which it seems to transmit or reflect a greater or smaller fraction of the incident light). These three attributes are depicted in the color cone shown in Figure 3-2. In the color cone, hue is indicated by position around the circumference. The colors blend into one another from red to orange, yellow, green, blue, violet, and back to red. Saturation (sometimes called *chroma*) is shown in the color cone as the radius. A saturated color consists of a single hue and would be positioned on the circumference of the cone. Colors toward the center are mixtures of various hues, and while they may have a dominant hue, they do not appear to be pure. Lightness (sometimes referred to as *value* or *brightness*) is shown on the vertical dimension, the center of which ranges from white through varying levels of gray, to black. Any color of a given hue and saturation can be varied in its lightness. With the addition of black, various shades are produced; with the addition of white, various lighter tints. Although there is a general relationship between the luminance of light and the subjective response of lightness, all colors that reflect equal total amounts of light energy are not necessarily perceived as equal in lightness. This is due to the fact that the eye is differentially sensitive to various wavelengths.

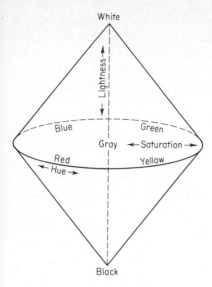

Figure 3-2 The color cone. Hue is shown on the circumference, lightness (from light to dark) on the vertical, and saturation on the radius from circumference to the center.

Color systems. Two types of color systems are used as standards of colors. First are those which consist of color plates or color chips for use as standards in characterizing colorants. The Munsell color system [28], the Ostwald color system [25], and the Maerz and Paul *Dictionary of color* [10] are of this type. Some of these (for example, the Munsell and Ostwald systems) correspond substantially to the color cone and provide standard nomenclature to identify selected colors of certain hues, saturations, and lightness levels. The Ostwald system, for example, identifies 680 color samples (28 color chips for each of 24 hues, plus 8 representing lightness levels on the white to black continuum).

The second type of system designates colors in terms of mixtures of theoretical colored lights. The Commission Internationale de l'Eclairage developed such a color system, designated as the *CIE system* [30]. This system is based on the fact that all colors can be reproduced by proper combinations of the three primary colors of light, namely, red, green, and blue.

The Process of Seeing

The eye is in many respects something like a camera. It has a lens through which the light rays are transmitted and focused, and a sensitive area (the retina) upon which the light rays fall. The retina corresponds to the film of the camera. Figure 3-3 illustrates the principal features of the eye in cross section. The lens of the eye is normally flexible, so that it can adjust itself to bring about proper focus on the retina. The image of the object upon the retina is reversed and inverted, just as it is in a camera, as illustrated in Figure 3-4. The retina consists of two types of sensitive

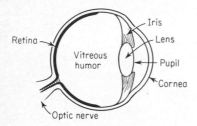

Figure 3-3   Principal features of the human eye in cross section. Light passes through the pupil, is refracted by the lens, and is brought to a focus on the retina. The retina receives the light stimulus and transmits an impulse to the brain through the optic nerve.

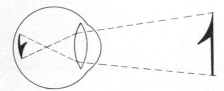

Figure 3-4   Illustration of the manner in which the image of an object is reproduced in inverted form on the retina of the eye.

areas, namely, rods and cones. The cones are primarily sensitive to variations in the wavelength of light, which gives rise to the subjective sensation of color. There are approximately 6 or 7 million cones in the eye; these generally predominate in the center section of the retina. The rods are primarily sensitive to the amount of light and are not particularly sensitive to differences in wavelength. There are about 130 million rods in the eye; they tend to predominate toward the outer reaches of the retina around the sides of the eyeball. The rods and cones, upon receiving light through the lens, set up nerve impulses which are transmitted through the optic nerve to the brain, where translation then takes place.

Visual Acuity

Visual acuity is the ability to resolve (distinguish) black and white detail. This is very largely controlled by the *accommodation* of the eyes. Accommodation is the adjustment of the lens of the eye to bring about proper focusing of the light rays on the retina. In normal accommodation, if one is looking at a far object, the lens flattens, and if one looks at a near object, the lens tends to bulge, in order to bring about proper focusing of the image on the retina. If the lens is accommodated properly to the focal distance of the object from the eye, the light rays that emanate from a specific point on the object and that strike the lens are redirected by the lens and are brought to a focus at a specific point on the retina. This is llustrated in Figure 3-5a for far objects, and 3-5b for near objects.

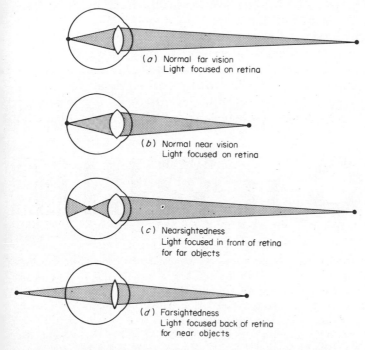

(*a*) Normal far vision
Light focused on retina

(*b*) Normal near vision
Light focused on retina

(*c*) Nearsightedness
Light focused in front of retina
for far objects

(*d*) Farsightedness
Light focused back of retina
for near objects

Figure 3-5 Illustration of normal accommodation at far and near distances, *a* and *b*, and of nearsightedness *c* and farsightedness *d*.

In some individuals the accommodation of the eyes is inadequate. This causes the conditions that we sometimes call *nearsightedness* and *farsightedness*. When a person is nearsighted, his lens tends to remain in a bulged condition, so that while he may achieve a proper focus of near objects, he cannot achieve a proper focus of far objects, as shown in

Figure 3-5c. Farsightedness, in turn, is a condition in which the lens tends to remain too flat. While such a person may see clearly at far distance, he encounters difficulty in seeing properly at near distance, as shown in Figure 3-5d. Such conditions can sometimes be corrected by appropriate glass lenses which change the direction of the light rays before they reach the lens of the eye and thereby bring about proper focusing on the retina.

Actually, there are various kinds of visual acuity, including the following: minimum-distinguishable acuity (this is the most commonly measured type of acuity, and refers to the smallest detail the eye can detect); minimum-perceptible acuity (this refers to the smallest target— such as a dot—that the eye can detect); vernier, or minimum-separable, acuity (this refers to the smallest offset or lateral displacement of two lines that can be detected, when the lines are positioned end to end); and dynamic visual acuity (DVA, the ability to make visual discriminations when the target object or the viewer or both are moving).

Visual acuity tests.   Of various vision tests, tests of minimum-distinguishable acuity are by all odds the most commonly used.[3] This type is usually defined as the reciprocal of the visual angle, in minutes of arc, that is subtended at the eye by the smallest visual detail that can be discriminated. Various kinds of visual targets are used for this purpose, some of them being letters and others being geometric figures. Figure 3-6 illustrates several.

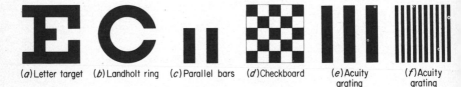

(a) Letter target   (b) Landholt ring   (c) Parallel bars   (d) Checkboard   (e) Acuity grating   (f) Acuity grating

Figure 3-6   Illustrations of various types of targets used in visual acuity tests and experiments. The features to be differentiated in targets a, b, c, d, and e are all the same size and would, therefore, subtend the same visual angle at the eye. With target a the subject is to identify each letter; with c, e, and f he is to identify the orientation (such as vertical or horizontal); and with b he is to identify any of four orientations. With target d he is to identify one checkerboard target from three others with smaller squares.

The different kinds of targets can be varied in size and distance, and the acuity of the subject is determined by the smallest targets that he can properly identify. Usually the reciprocal of a visual angle of 1 minute of arc is used as a standard in scoring. The reciprocal of 1 is, of course, 1,

---

[3] Tests of other visual skills can be measured with various types of testing devices, such as the Ortho-Rater produced by the Bausch and Lomb Optical Co.; this provides for measuring acuity, phorias, depth perception, and color discrimination.

and it provides a base of unity against which poorer or better levels of acuity can be compared. If, for example, one individual can identify only a detail that subtends an arc of 1.5 minutes, his acuity score would then be the reciprocal of 1.5 minutes, or 0.67. On the other hand, if an individual can identify a detail that subtends an arc of 0.8 minute, his score, the reciprocal of 0.8 minute, would be 1.25.

Convergence (Phoria)

As we direct our visual attention to a particular object, it is necessary that the two eyes converge on the object so that the images of the object on the two retinas are in corresponding positions; in this way we get an impression of a single object. The two images are said to be *fused* if they are in corresponding positions on the retinas and so give an impression of a single image. Convergence is controlled by muscles that surround the eyeball. Normally, as an individual looks at a particular object, these muscles operate automatically to bring about convergence. But some individuals tend to converge too much, and others tend not to converge enough. In still another circumstance an individual may have one eye that tends to point up relative to the other eye. These conditions are called *phorias*. Since the double images that occur when convergence does not take place are visually uncomfortable, such people usually have compensated for this by learning to bring about convergence. Where such an individual is required to use his eyes for continuous periods of time, however, muscular stresses and strains occur in overcoming these muscular imbalances.

Depth Perception

Depth perception, or *stereopsis*, is an impression of depth or distance due to the fact that the eyes see an object from slightly different angles. If you hold a matchbox in front of your eyes and close first one eye and then the other, you will see that you get slightly different views of it. Through experience we learn to translate these two slightly different views into distance or depth. A person who does not have depth perception, however, may judge distance or depth by various cues, such as size of objects relative to each other, apparent speed of moving objects in the distance, relative position of objects, and relative clarity of objects.

Color Discrimination

While people are still arguing about the specific process by which we differentiate between colors, color blindness basically is a deficiency in the ability of the cones to differentiate various wavelengths. True and complete color blindness is very rare. The various degrees and types of partial color blindness include cases in which there is difficulty in differentiating between certain colors, such as between red and green or between blue and yellow.

Dark Adaptation

The adaptation of the eye to different levels of light and darkness is brought about by two functions. In the first place, the pupil of the eye increases in size as we go into a darkened room, in order to admit more light to the eyes; it tends to contract in bright light, in order to limit the amount of light that enters the eye. This process takes a little while, and as we proceed from one condition of illumination to another, we may be partially blinded until this process is completed. Another function that affects how well we can see as we go from the light into darkness is a physiological process in the retina in which *visual purple* is built up. Under such circumstances the cones (which are color sensitive) lose much of their sensitivity. Since in the dark our vision depends very largely on the rods, color discrimination is limited in the dark. The time required for complete dark adaptation is usually 30 to 40 min. The reverse adaptation, from darkness to light, takes place in some seconds, or at most in a minute or two. In situations where someone needs to become dark-adapted, such as to go on shipboard watch at night, it is frequently the practice to wear red goggles for a period of time (say, half an hour) before going on watch. Since the rods are not generally sensitive to red light, the goggles facilitate the dark-adaptation process.

Conditions That Affect Visual Discriminations

The ability of individuals to make visual discriminations is of course dependent upon their visual skills, especially acuity. Aside from individual differences, however, there are certain variables (conditions), external to the individual, that affect visual discriminations. Some of these variables are listed or discussed briefly below.

Luminance contrast. Luminance contrast refers to the difference in luminance of the features of the object being viewed, in particular of the feature to be discriminated by contrast with its background (for example, an arrow on a direction sign against the background area of the sign). The luminance contrast is expressed by the following relationship:

$$\text{Contrast} = \frac{B_1 - B_2}{B_1} \times 100$$

in which B_1 = brighter of two contrasting areas
B_2 = darker of two contrasting areas

The contrast between the print on this page and its white background is considerable; if we assume that the paper has a reflectance of 80 percent and that the print has a reflectance of 10 percent, the contrast would be

$$\frac{80-10}{80} \times 100 = \frac{70}{80} \times 100 = 88 \text{ percent}$$

If the printing were on medium-gray paper rather than on white, the contrast would, of course, be much less. In general, the greater the contrast between an object (or its features) and the background, the greater the visual discriminability; with high contrast, smaller detail can be distinguished.

Amount of illumination. (It will be discussed in Chapter 14.)

Time. Within reasonable limits, the longer the viewing time, the greater is the discriminability.

Luminance ratio. The luminance ratio is the ratio between the luminance of any two areas in the visual field (usually the area of primary visual attention and the surrounding area).

Movement. As implied above, the movement of the target object or the observer (or both) brings into play a special type of visual acuity, specifically DVA. Such acuity generally deteriorates as a function of speed of movement, usually expressed in degrees of movement per second [Goodson and Miller, 5]. Dynamic visual acuity seems not to be strongly related to other visual skills [Burg and Hulbert, 2].

Glare. (It will be discussed in Chapter 14.)

Combinations of variables. Available evidence generally suggests that there may be interactions among certain variables, when two or more are used in combination. This was illustrated in a fairly classic study by Cobb and Moss [3] in which visual acuity was measured under varying combinations of contrast, time, and illumination. Acuity was expressed in terms of the visual angle of the smallest target that could be discriminated. The results are shown in Figure 3-7. To illustrate the use of this figure, let us take as an example the 100-mL level and the 0.300-sec exposure-time curve. We can see that for a very small test target (one of 0.7 minute of arc) a contrast of 50 percent is necessary to discriminate the target positions, whereas for a large target (one of 5 minutes of arc) a contrast of only 2 percent permits discrimination.

The interaction effects of contrast, viewing time, movement of objects, and luminance on the ability of people to make visual discriminations have been very thoroughly investigated by Blackwell [1] in connection with his studies of illumination. Some of his results will be presented in Chapter 14.

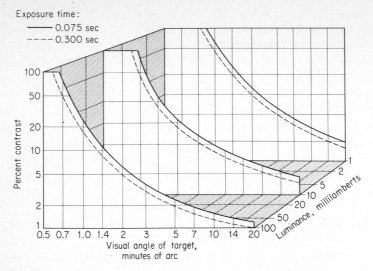

Exposure time:
——— 0.075 sec
– – – – 0.300 sec

Figure 3-7 Relation between contrast, luminance of background, exposure time, and visual acuity (actually the visual angle in minutes of arc of the smallest targets that can be discriminated). [After Cobb and Moss, 3.]

HEARING

In discussing the hearing process, let us first describe the physical stimuli to which the ear is sensitive, namely, sound vibrations.

The Nature and Measurement of Sound

Sound is originated by vibrations from some source. While such vibrations can be transmitted through various media, our primary concern is with those transmitted through the atmosphere to the ear. Two primary attributes of sound are *frequency* and *intensity*.

Frequency of sound waves. The frequency of sound waves can be visualized if we think of a simple sound-generating source such as a tuning fork. When it is struck, the tuning fork is caused to vibrate at its "natural" frequency. In so doing it causes the air particles to be moved back and forth. This alternation creates corresponding increases and decreases in the air pressure. The number of alternations per second is the frequency of the sound expressed in hertz (Hz) or cycles per second (cps). The frequency of a physical sound gives rise to the human sensation of pitch.

The vibrations of a simple sound-generating source such as a tuning fork form *sinusoidal,* or *sine,* waves that can be represented as the projection of the movement of a point around a circle that is revolving at a constant rate, as shown in Figure 3-8. As point P revolves around its center O, its vertical amplitude, as a function of time, will be that represented by the sine wave. The height of the wave above the midline at any given point in time represents the amount of above-normal air pressure at that point in time, the crest, of course, being the maximum. Positions below the midline, in turn, represent the reduction in air pressure below normal. On the musical scale, middle C has a frequency of 256 Hz. Any given octave has double the frequency of the one below it. In general terms, the human ear is sensitive to frequencies in the range

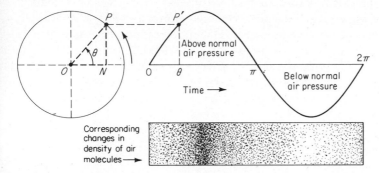

Figure 3-8   Reproduction of sinusoidal, or sine, wave.   The magnitude of the alternating changes in air pressure caused by a sound-generating source with a given frequency can be represented by a sine wave. A sine wave is represented as the projection of a point on the circumference of a circle as that point rotates about its center at a constant speed. The lower part of the figure depicts the changes in the density of molecules of the air caused by the vibrating source.

from about 20 to 20,000 Hz, although there are marked differences among individuals.

Intensity of sound.   Referring to Figure 3-8, the amplitude of sound waves reflects the *intensity* of the physical sound stimulus and determines the subjective *loudness* of the sound to the person hearing it. The absolute intensity is measured in power units (such as watts) or pressure units (such as newtons or microbars). In practice, however, it has been found most useful to express sound intensity as a ratio between two sounds rather than by absolute magnitude of power or pressure. The *bel* (named after Alexander Graham Bell) is the basic ratio used for this purpose. The number of bels is the logarithm (to the base 10) of the ratio of the two intensities. Actually the most convenient and most common measure of sound intensity is the *decibel* (dB). A decibel is $\frac{1}{10}$ of a bel;[4] like the bel, it expresses a ratio. Still another step in the conventional standardization of intensity measurement is that of using, for the lower of the two sounds, a standard reference level that represents zero decibels. This reference level can be characterized in various ways. For airborne sounds the reference level is generally 20 micronewtons per square meter (abbreviated 20 $\mu N/m^2$). In this expression, micro stands for a factor of 1/1 million with an abbreviation of the Greek letter μ (mu). The μN stands for 0.000001 newton; m^2 refers to square meters. For some purposes a reference level of one microbar has been used. One microbar (1 μbar) equals 0.1 newton per square meter ($0.1N/m^2$) or 1 dyne per square centimeter (1 $dyne/cm^2$), which is approximately one-millionth of the normal atmospheric pressure. The more commonly used reference level of 20 $\mu N/m^2$ can also be expressed as 0.0002 $dyne/cm^2$ or 0.0002 μbar.[5] Figure 3-9 shows the decibel scale with examples of several sounds that fall at varying positions along the scale. That figure also shows the power ratios with increasing decibel levels; an increase of 10 dB reflects a tenfold increase in relative sound power.

Complex sounds.   There are very few sounds that are pure tones. Even tones from musical instruments are not pure, but rather consist of a fundamental frequency in combination with certain others (especially harmonic frequencies that are multiples of the fundamental). Most complex sounds, however, are nonharmonic. Complex sounds can be depicted in two ways. One of these is a wave form which is the composite of the wave forms of the individual component sounds; such a wave form is illustrated in Figure 3-10. The other method of depicting complex sounds is by the use of a sound spectrum that shows the intensity of various frequency bands, as illustrated in Figure 3-11. The four curves illustrate spectral analyses of the noise of a rope-closing machine, using analyzers of varying bandwidths, namely, an octave, a half octave, a third of an octave, and a thirty-fifth of an octave. The narrower the bandwidth, the greater the detail of the spectrum and the lower the level of each band-

---

[4] The difference in the intensities of two sounds, given in decibels, is expressed as follows:

Number of decibels $= 20 \log \dfrac{P_1}{P_2}$

in which P_1 and P_2 represent the pressures of the two sounds.

[5] It should be noted that the United States of America Standards Institute (USASI) —formerly the American Standards Association—has established a standard to which sound-level meters should conform. This standard requires that three alternate *frequency-response* characteristics be provided in such instruments, these consisting of *weighting* networks (designated A, B, and C) which selectively discriminate against low and high frequencies in accordance with certain equal-loudness contours (to be discussed later). Recommended practice provides for using all three weightings, but in any event, designating any single one that is used in presenting data, such as the "A-weighted sound level is 45 dB."

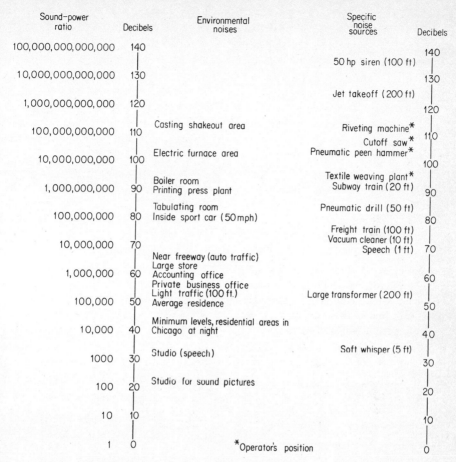

Sound–power ratio	Decibels	Environmental noises	Specific noise sources	Decibels
100,000,000,000,000	140			140
			50 hp siren (100 ft)	
10,000,000,000,000	130			130
			Jet takeoff (200 ft)	
1,000,000,000,000	120			120
100,000,000,000	110	Casting shakeout area	Riveting machine*	110
			Cutoff saw*	
10,000,000,000	100	Electric furnace area	Pneumatic peen hammer*	100
			Textile weaving plant*	
1,000,000,000	90	Boiler room Printing press plant	Subway train (20 ft)	90
100,000,000	80	Tabulating room Inside sport car (50 mph)	Pneumatic drill (50 ft)	80
			Freight train (100 ft) Vacuum cleaner (10 ft)	
10,000,000	70		Speech (1 ft)	70
		Near freeway (auto traffic) Large store		
1,000,000	60	Accounting office		60
		Private business office Light traffic (100 ft.)		
100,000	50	Average residence	Large transformer (200 ft)	50
		Minimum levels, residential areas in		
10,000	40	Chicago at night		40
1000	30	Studio (speech)	Soft whisper (5 ft)	30
100	20	Studio for sound pictures		20
10	10			10
1	0	*Operator's position		0

Figure 3-9 Decibel levels (dB) and sound power ratios for various sounds. Decibel levels are *A*-weighted sound levels measured with a sound-level meter. (See footnote 5 for discussion of *A*-weighting.) [Examples from Peterson and Gross, 12, p. 5.]

width. In characterizing octave bands, there have been varying practices in the division of the spectrum into octaves. The current preferred practice, as set forth by the USASI (United States of America Standards Institute), however, is that of dividing the audible range into 10 bands, with the *center* frequencies of these bands being 31.5, 63, 125, 250, 1000, 2000, 4000, 8000, and 16,000 Hz [USASI, S1.11–1966]. (Many existing sets of sound and hearing data, however, are based on other octave divisions, such as that in which the above frequencies define the *ends* of the class intervals instead of their *centers*.)

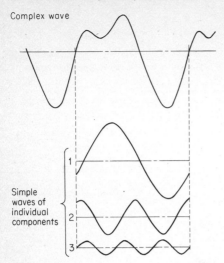

Complex wave

Simple
waves of
individual
components

1

2

3

Figure 3-10 Wave form of a violin
tone, including the simple sine waves
of the individual components that combine to form the complex wave.
[Adapted from Miller, 11.]

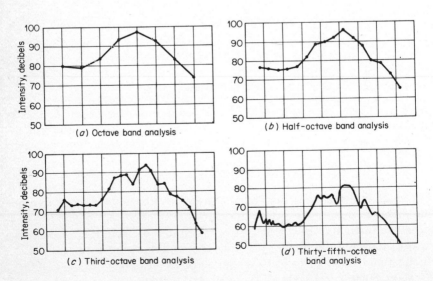

Intensity, decibels

(*a*) Octave band analysis

(*b*) Half-octave band analysis

Intensity, decibels

(*c*) Third-octave band analysis

(*d*) Thirty-fifth-octave
band analysis

Figure 3-11 Spectral analyses of noise of a rope-closing machine, using
analyzers of varying bandwidths. The narrower the bandwidth, the greater
the detail and the lower the level of any single bandwidth. [Adapted from
Industrial noise manual, 27, p. 25.]

The Anatomy of the Ear

The ear has three primary anatomical divisions, namely, the outer ear, the middle ear, and the inner ear. These are shown schematically in Figure 3-12.

The outer ear.   The outer ear consists of the external part (called the *pinna* or *concha*), the auditory canal (the *meatus*), which is a tube about an inch long that leads inward from the external part, and the eardrum (the *tympanic membrane*) at the end of the auditory canal.

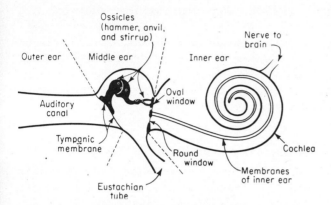

Figure 3-12   Schematic drawing of the ear, showing auditory canal through which sound waves travel to tympanic membrane, which in turn vibrates the ossicles of the middle ear. This vibration is transmitted through the membrane of the oval window to the cochlea, where the vibrations are transmitted by liquid through membranes to sensitive hair cells, which send nerve impulses to the brain.

The middle ear.   The middle ear is separated from the outer ear by the membranous eardrum. The middle ear includes a chain of three small bones called *ossicles*, the hammer, the anvil, and the stirrup (also called the *malleus*, the *incus*, and the *stapes*). These three ossicles, by their interconnections, transmit vibrations from the eardrum to the oval window of the inner ear. The stirrup acts something like a piston on the oval window, its action transmitting the changes in sound pressure to the fluid of the inner ear, on the other side of the oval window membrane. In this process, the extremely minute pressure changes on the eardrum are amplified about 22 times by the time they reach the fluid of the inner ear [von Békésy, 18].

The inner ear. The important part of the inner ear as far as hearing is concerned is the *cochlea*, a spiral-shaped tube that resembles a snail. The cochlea, if uncoiled, would be about 30 mm long, and at its widest point (nearest the oval window) is about 5 or 6 mm. There are two membranes that divide the cochlea throughout its curved length into three parallel canals. The outer two of these canals are filled with liquid, and there is a small connection between them, at their innermost end. The oval window, which is caused to vibrate by the stirrup, transmits its pressure through this closed hydraulic system. The termination of the induced vibration is completed at the round window, which is located below the oval window and is also separated from the middle ear by a membrane.

The hydraulic pressure transmitted through these canals, however, exerts itself also on the membranes that separate these canals from the one between them. This in-between canal, which is also filled with fluid, contains hair cells that are sensitive to very slight changes in pressure. These hair cells in turn transmit nerve impulses to the brain, where they are interpreted.

The Conversion of Sound Waves into Sensations

Thus, we see that the ear receives and funnels sound waves in the air, converts these vibrations into mechanical vibrations of the three ossicles, which convert the vibrations into hydraulic pressure in the cochlea, which in turn has sensitive areas that pick up these minute changes in pressure and transmit nerve impulses to the brain. While these mechanical processes have been understood for some time, the procedures by which the vibrations are *heard* and differentiated are still not entirely known, although various theories have been proposed to explain this phenomenon. In this connection, the work of von Békésy [18] has led to some greater insight. He explains the discrimination of loudness and pitch by two processes. In the low-frequency range (up to about 60 Hz) the vibration of the basilar membrane produces, in the auditory nerves, volleys of electric nerve impulses in the form of "spikes" which are synchronous with the rhythm of the sound. With increases in sound pressure (intensity) the number of spikes packed into each period of the rhythm increases. The number of spikes and their rhythm convey the loudness and pitch of the sound.

Starting with frequencies of about 60 Hz, however, a different phenomenon begins to occur. The basilar membrane now begins to vibrate unequally over its entire area, but each tone produces a maximum vibration in a different area of that membrane. This selectivity of different areas of the membrane to different vibrations gradually (through higher frequencies) takes over the determination of pitch, since the rhythm of the spikes (which indicates pitch at low frequencies) becomes irregular at higher frequencies. Above about 4000 Hz, pitch presumably is determined entirely by the location of the maximal vibration amplitude along the basilar membrane. There presumably is some inhibitory mechanism that is not yet understood, which suppresses weaker stimuli and sharpens the sensation around the maximum vibration amplitude.

Loudness Sensations

The subjective sensation of loudness is in large part a function of intensity, but it is also influenced in part by sound frequency. In this connection there are two measures of subjective sensation that we should mention, namely, *phons* and *sones*.

Loudness level in phons.   Some years ago, Fletcher and Munson [4] developed what have become known as *equal-loudness* contours. They obtained judgments of subjects with regard to sounds of different combinations of frequencies and intensities to find out which ones were judged to have equal loudness. For this purpose they used 1000-Hz tones of various intensities as reference tones and had subjects compare other tones of varying frequencies and intensities with the 1000-Hz reference tones. The result was the identification of the intensities of tones of other frequencies that were judged equal in loudness with each of several 1000-Hz tones such as 10, 20, and 30 dB.

More recently Robinson and Dadson [13] in Great Britain have developed somewhat more refined equal-loudness curves, these being shown in Figure 3-13. Each contour shows the decibel intensities of different frequencies that were judged to be equal in loudness to that of a 1000-Hz tone of the specified decibel level of intensity. To illustrate the curves, one can see that a tone of 50 Hz of about 62 dB is judged equal in loudness to a 1000-Hz tone of only 40 dB.

The unit *phon* (of German origin) is used to indicate the loudness level of sounds. Thus, any point along a given contour in Figure 3-13 represents sounds of the same number of phons. The loudness level in phons, then, is numerically equal to the decibel level of a tone of 1000 Hz, which is judged equivalent in loudness.

Loudness in sones.   The phon tells us only about subjective *equality* of various sounds, but it tells us nothing of the *relative subjective loudness* of different sounds. For such comparative purposes we need still another yardstick. Fletcher and Munson [4] developed such a scale—a ratio scale of loudness. Stevens [14], in turn, labeled the scale, using the term *sone*. In developing this scale (as with the phon) a reference sound was used. One sone is defined as the loudness of a 1000-Hz tone of 40 dB. A sound that is judged to be twice as loud as the reference sound has a loudness of 2 sones, a sound that is judged to be three times as loud

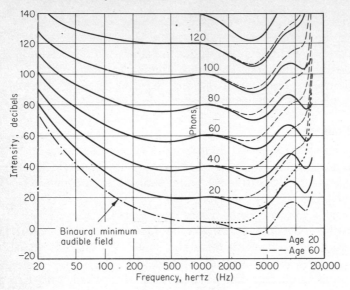

Figure 3-13 Equal-loudness curves of pure tones. Each curve represents intensity levels of various frequencies that are judged to be equally loud. The lowest curve shows the minimum intensities of various frequencies that typically can be heard. [From Robinson and Dadson, 13. Crown copyright reserved. Courtesy National Physical Laboratory, Teddington, Middlesex, England.]

as the reference sound has a loudness of 3 sones, etc. In turn, a sound that is judged to be half as loud has a loudness of ½ sone. To provide some basis for relating sones to our own experiences, the following examples are given [Bonvallet, 29, p. 43]:

Noise source	Decibels	Loudness, sones
Residential inside, quiet	42	1
Household ventilating fan	56	7
Automobile, 50 ft	68	14
"Quiet" factory area	76	54
18-in. automatic lathe	89	127
Punch press, 3 ft	103	350
Nail-making machine, 6 ft	111	800
Pneumatic riveter, 4 ft	128	3000

A procedure for estimating the loudness of complex sounds has been developed and revised by Stevens [15, 16]. This procedure uses a table of data for deriving a loudness index for each octave band. Excerpts from this table are presented in Table 3-1 [Peterson and Gross, 12, pp. 50–51]. First, the measurement of the band level is taken, in decibels, of each octave. Given these values, one proceeds as follows:

1. From the table (Table 3-1), find the proper loudness index for each band level (S).
2. Add all the loudness indexes $\left(\sum S\right)$.
3. Multiply this sum by 0.3.
4. Add this product to 0.7 of the index that has the largest index (S_{max}). The total loudness in sones then is $(0.3 \sum S + 0.7S_{max})$.
5. This total loudness (sones) can be converted to loudness level (phons) by using the two columns at the right of Table 3-1.

Table 3-1 Excerpts from Table for Use in Calculating Loudness (Sones) of Complex Sounds

Band level, dB	Band loudness index (S) for octave bands Midpoint of octave band									Loudness, sones	Loudness level, phons
	31.5	63	125	250	500	1000	2000	4000	8000		
20						0.18	0.30	0.45	0.61	0.25	20
30				0.16	0.49	0.67	0.87	1.10	1.35	0.50	30
40		0.07	0.37	0.77	1.18	1.44	1.75	2.11	2.53	1.00	40
50	0.26	0.62	1.13	1.82	2.24	2.68	3.2	3.8	4.6	2.00	50
60	0.94	1.56	2.44	3.4	4.1	4.9	5.8	7.0	8.3	4.00	60
70	2.11	3.2	5.0	6.2	7.4	8.8	10.5	12.6	15.3	8.00	70
80	4.3	6.7	9.3	11.1	13.5	16.4	20.0	24.7	30.5	16.0	80
90	8.8	13.6	17.5	21.4	26.5	32.9	41	52	66	32.0	90
100	18.7	28.5	35.3	44	56	71	90	113	139	64.0	100
110	44	61	77	97	121	149	184	226	278	128	110
120	105	130	160	197	242	298	367			256	120

SOURCE: Peterson and Gross [12, Table 3-1, pp. 50–51].

Masking

When listening to complex sounds, the ear usually can analyze the sound, differentiating its various components; this makes it possible sometimes to give attention to one component of the total sound environment and to ignore others. The ability to do this, however, is far from perfect. Under some circumstances one component may mask another. Masking is the opposite of analysis. It is a condition in which the hearing process fails to differentiate certain sound components from others. Masking, however, is no discriminator between desirable and undesirable sounds.

Operationally defined, *masking* is the amount (usually expressed in decibels) by which the threshold of audibility of a sound is raised by the presence of another (masking) sound [Harris, 8, pp. 1–11]. In studying the effects of masking, an experimenter typically measures the absolute threshold (the minimum audible level) of a sound (the sound to be masked) when presented by itself and then measures its threshold in the presence of the masking sound. The difference is attributed to the masking effect.

The effects of masking vary with the type of masking sound and of the masked sound itself—whether pure tones, complex sound, white noise, speech, etc. Two examples of masking effects of pure tones are shown in Figure 3-14. In the masking effects of noise (*a*) we can see that

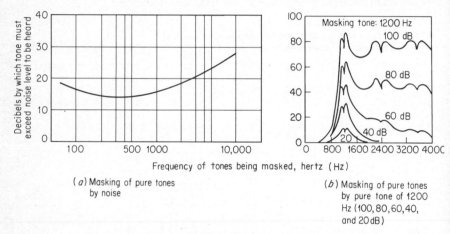

(*a*) Masking of pure tones by noise

(*b*) Masking of pure tones by pure tone of 1200 Hz (100, 80, 60, 40, and 20 dB)

Figure 3-14 The effects of masking pure tones *a* by noise and *b* by other tones. The curves show the number of decibels by which tones of various frequencies are masked by noise or other tones. [*a* Adapted from W. A. Munson in C. M. Harris, 8, pp. 5–16; *b* from Wegel and Lane, 20.]

the threshold of pure tones is increased by 15 to nearly 30 dB, the increase being greatest in the case of the tones of higher frequencies. In the masking effects of pure tones on other pure tones (*b*) we can see that the masking effects are generally greatest on tones around the frequency of the masking tone, that they are generally greater for tones above the masking tone than for those below, and that they increase somewhat around the harmonics of the masking tone. In our everyday lives we frequently experience the effects of masking, such as when an airplane sound drowns out the audio portion of a TV program (in the case of some shows and commercials this would be a good thing!) or when the noise of office machines makes it difficult to hear others talk.

THE CUTANEOUS SENSES

The *cutaneous* senses have their origins in the nerve endings in the skin. The nerve terminations in the skin come in assorted varieties and occur in the various skin layers. Each type of cutaneous receptor is capable of being most efficiently stimulated by one form of stimulus and thus gives rise to a specific sensation. Although various cutaneous senses have been postulated, there is still some question about the number and character of distinct cutaneous senses. There are, however, certain senses which have come to be generally recognized as being different and having distinct neural receptors in the skin. These are temperature (actually there are distinct receptors for heat and cold), pain, and pressure, or touch [Wyburn, Pickford, and Hirst, 24].

KINESTHETIC SENSE

Kinesthesis is the awareness of position or movement of parts of the body such as the arms, fingers, or legs. The kinesthetic receptors are nerves in the muscles, tendons, and the covering of the bones, particularly around the joints. Such nerve receptors are called *proprioceptors* and are stimulated primarily by actions of the body members.[6] The sensations transmitted by these receptors enable a person to control the movement of muscles through the motor nerves. This is essentially a feedback process; as a machine operator reaches for a control lever that he does not see, the continuation of the motion produces the changing sensations of position that, in turn, help to control the continuation of the movement until the lever is reached.

ORIENTATION SENSES

The awareness of body orientation and movement comes, in part, from various sensory mechanisms. There are, however, two mechanisms whose functions are primarily concerned with orientation, the semicircular canals and the vestibular sacs. Other senses that have some relationship to orientation include vision, the sense of touch, and kinesthesis.

Sensory Mechanisms in Orientation

The semicircular canals and vestibular sacs are located in the inner ear, but they have no relationship with hearing as such. They are shown in Figure 3-15.

[6] It might be added here that proprioceptors also occur in the musculature of the body organs, such as the stomach. In these locations they keep us posted on internal conditions, such as hunger, as opposed to movement of body members.

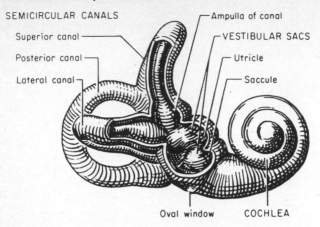

SEMICIRCULAR CANALS — Ampulla of canal

Superior canal — VESTIBULAR SACS

Posterior canal — Utricle

Lateral canal — Saccule

Oval window   COCHLEA

Figure 3-15   The body orientation organs. The semicircular canals form roughly a three-coordinate system that provides information about body movement. The vestibular sacs (the utricle and saccule) respond to the forces of gravity and provide information about body position in relation to the vertical.

Semicircular canals.   The three semicircular canals in each ear are interconnected doughnut-shaped tubes that form, roughly, a three-coordinate system as shown in Figure 3-15. Each one has a small tube filled with a liquid (endolymph). With changes in acceleration or deceleration, the fluid shifts its position in these tubes, which stimulates nerve endings that then transmit nerve impulses to the brain. It should be pointed out that movement of the body at a constant rate does not cause any stimulation of these canals. Rather, they are sensitive only to a *change* in rate (acceleration or deceleration). *Dynamic equilibrium*, the ability to react to body motions and to adjust body movements under conditions that operate against equilibrium, depends primarily on information from the semicircular canals.

Vestibular sacs: utricle and saccule.   The vestibular sacs (also called the *otolith organs*) are two organs with interior hair cells and containing a gelatinous substance. The *utricle* is generally positioned in a horizontal plane, and the *saccule* is more in a vertical plane. As the body changes position, the gelatinous substance is affected by gravity, setting up nerve impulses via the hair cells. The utricle apparently is the more important of the two organs. The primary function of these organs is that of sensing body posture in relation to the vertical and thus serving as something of a gyroscope that helps to keep us on an even keel. While their dominant role is that of aiding in sensing postural conditions of the body, they are also somewhat sensitive to acceleration and deceleration and presumably supplement the semicircular canals in sensing such changes. The ability to maintain a relatively fixed posture is called *static equilibrium* and depends primarily on information from the vestibular sacs.

Interdependence of Senses in Orientation

The ability of (some) people to "fly through the air on the flying trapeze" or to maintain balance on an I beam of a building under construction and of most people to maintain body orientation in more mundane functions is based on the interaction of stimuli received through the semicircular canals, the vestibular sacs, the kinesthetic receptors, the cutaneous senses (especially pressure, such as on the soles of the feet), vision, and occasionally audition. Although these interactions are not well understood, the role of vision seems to be a relatively important one. This has been illustrated, for example, in experiments with people in a "tilting room," in some cases with a chair that could be tilted within the room [Witkin, 21, 22; Wapner and Witkin, 19, 23]. Subjects seated in the room were tilted to various angles and asked to indicate what direction they considered vertical. When blindfolded, they were able to indicate the vertical more accurately than when they could see the inside of the room; when they could see, they tended more to indicate that the ceiling of the room was upright. The implication of such investigations is that misperceptions of the *true* upright direction may occur when there is a *conflict* between the sensations of gravity and visual perceptions; in such a case one's visual perceptions may dominate, even when they are erroneous. In this connection, however, there is evidence to suggest that some people tend to be *field-dependent* (they tend to be dominated more by their visual field) and others tend to be *field-independent* [Witkin, 22].

PERCEPTION

As indicated earlier, perception is the interaction of sensory processes and the cortex of the brain. It can be considered as the process of interpretation of stimuli that are sensed, or as the formation of impressions of the outside world from those stimuli. In considering perceptual processes, let us back up to the sensory processes. The sensory mechanisms (the eyes, ears, etc.) are not in *direct* contact with the objects or events that are being sensed, but rather receive some form of energy from them, such as light energy or vibrations. The objects are sometimes referred to as the *distal* stimuli, and the energy that impinges upon the sensory mechanisms as the *proximal* stimuli. Thus, we can "know" the objects and events about us only through the proximal stimuli that are picked up by the sensory mechanisms. Our perception of objects and events, based on our proximal stimuli, is then not a carbon copy of the real world. In many situations, however, our perception sufficiently corresponds with the real world to provide a reasonably valid basis for interpreting reality. In other situations, however, there is less correspondence between the real world and our perceptions. This is especially the case with various types of common illusions, some of which are almost universally experienced. In addition, there are some situations in which stimuli are rather consistently misperceived. This frequently occurs, for example, because of the context of stimuli; for instance, a given color usually appears to be darker against the background of a light color than when it is against a background of a darker color, and straight lines appear to be curved when against the background of certain curved or radiating configurations.

It should also be noted that aside from such contextual aspects of physical stimuli as they influence our perceptions, there are individual factors that influence perceptions. Thus, certain stimuli may be interpreted (perceived) by different people in different ways or may be interpreted differently by the same individual at different times. These variations depend upon many complex variables such as experience, the motivation of the individual, or his "set" or "expectation."

The many varied aspects of perception cannot be discussed here.[7] It should be pointed out, however, that certain aspects of perception have a very important bearing on human engineering. This is particularly so in the case of variations in perceptions caused by the distal stimuli of displays of various types that, via the proximal stimuli as sensed by the sensory mechanisms, can be perceived incorrectly. Many human errors can be directly traced to stimuli that are commonly mistaken for others. For example, certain visual displays are more commonly misread than others, certain colors are frequently confused with other colors, and certain sound stimuli cannot reliably be discriminated from others.

The moral of all this for the designer is that in the development of displays for presenting information to people, he needs to take into account their sensory abilities and perceptual behavior, toward the end of creating those displays that will ensure the most reliable interpretation of the stimuli that are presented by the displays.

SUMMARY

1.   The sense organs of man are the avenues through which he receives information from his environment. Such information may be received directly or in coded form through the use of some type of display.

[7] For a discussion of perception the reader is referred to such sources as Hochberg [9].

2. Where information is to be presented by a display, the designer may have some choice of the sensory modality to be used, the form of the display, and the manner of coding information.

3. Of the several senses of man, vision is undoubtedly the most important from the point of view of informing man about the world around him.

4. Light is *visually evaluated radiant energy*. The visible portion of the total spectrum includes those wavelengths to which the retina of the eye is sensitive.

5. Light comes to the eye in two ways: (*a*) *directly* from some luminous body, such as the sun (a "hot" source), and (*b*) *indirectly* by being reflected from some object (a "cold" source).

6. Light is measured in different units, as follows: (*a*) the *candela* (cd) is a measure of luminous intensity; (*b*) *candlepower* (cp) is luminous intensity expressed in candelas; (*c*) light falling upon a surface is measured in *footcandles* (fc); and (*d*) light reflected from a surface is measured in *footlamberts* (fL).

7. The sensitivity of the eye to differences in wavelengths of light gives rise to color perception.

8. The physical characteristics of reflected light are (*a*) *dominant wavelength*, (*b*) *purity*, and (*c*) *luminous reflectance*. The corresponding attributes of color as perceived by people are (*a*) hue, (*b*) saturation, and (*c*) lightness.

9. Surfaces of objects absorb certain wavelengths and reflect others. The "color" of a surface depends on the wavelengths reflected from it.

10. There are two types of color systems for identifying colors: (*a*) those which consist of color samples for use as standards in characterizing colorants, such as the Ostwald color system, and (*b*) the CIE system, which designates colors in terms of mixtures of theoretical colored lights.

11. The lens of the eye focuses light upon the retina. The retina consists of sensitive receptors called *rods* and *cones*, which transmit nerve impulses to the brain through the optic nerve.

12. There are several different visual skills, including the following: (*a*) *visual acuity*, which is the ability to perceive black-and-white detail at specific distances; (*b*) *convergence*, which is the ability to bring the two eyes into symmetrical focus on the object; (*c*) *depth perception*; (*d*) *color discrimination*; and (*e*) *dark adaptation*.

13. The ability to make visual discriminations is influenced by various conditions external to the individual, including (*a*) *luminance contrast* (the difference in luminance of the features of the object being viewed), (*b*) *amount of illumination*, (*c*) *time*, (*d*) *luminance ratio* (the

ratio of luminance of the object being viewed in relation to that surrounding the object), (e) *movement*, and (f) *glare*.

14. Sound consists of vibrations transmitted through the air or other medium. The *frequency* of the sound waves in hertz (Hz), or cycles per second, produces the sensation of *pitch*. The *intensity* of the sound waves, usually measured in decibels (dB), produces the sensation of *loudness*.

15. Sound intensity is usually expressed as a ratio between two sounds. The number of *bels* is the logarithm (to the base 10) of the ratio of two intensities, and the *decibel* is $\frac{1}{10}$ of a bel.

16. Most sounds are complex in that they include various frequencies. The most useful representation of sound is a *sound spectrum*, which shows the intensities (usually in decibels) of frequencies or frequency ranges.

17. The ear has three primary divisions, as follows: (a) the outer ear, which channels sound waves to the eardrum; (b) the middle ear, which consists of three small bones, called *ossicles*, that receive vibrations from the eardrum and transmit them to the oval window of the inner ear; and (c) the inner ear, which includes the cochlea, which transmits nerve impulses to the brain.

18. Sounds of various frequencies but of the same intensity do not necessarily create sensations of equal loudness. The *phon* is a measure of loudness level. Tones of 1000-Hz are used as reference tones. The loudness level in phons is numerically equal to the decibel level of a tone of 1000 Hz which is judged to be equivalent in loudness.

19. Relative subjective loudness of sounds is measured in *sones*. One sone is defined as the loudness of a 1000-Hz tone of 40 dB. The relative loudnesses of other sounds in comparison with this reference tone are expressed in sones.

20. *Masking* is a condition in which certain sounds are not differentiated from others. The degree of masking of a sound is determined by the shift in the threshold of that sound in the presence, versus the absence, of the masking sound.

21. In the masking of pure tones by white noise, the masking is least in the intermediate frequencies (300 to 500 Hz), is greater in the lower frequencies and is still greater in the higher frequencies.

22. The cutaneous senses have their origins in nerve endings in the skin. There are three senses that are generally recognized as being different, namely, temperature, pain, and pressure, or touch.

23. The *kinesthetic receptors* (*proprioceptors*) in the muscles, tendons, and bone coverings and around the joints provide feedback information regarding movement and position of body members.

24. The semicircular canals of the inner ear consist of a three-coordinate system filled with fluid that is sensitive to acceleration and deceleration; they generally bring about dynamic equilibrium (i.e., the ability to sense motion and to maintain equilibrium in a state of motion).

25. The vestibular sacs (the utricle and saccule) of the inner ear contain a gelatinous substance that is affected by gravity; they generally bring about static equilibrium (i.e., the ability to sense the vertical and to maintain a relatively fixed posture).

26. Aside from the sensory mechanisms that are directly related to orientation, other senses can also have an effect on orientation. Vision is especially important in perception of the upright; where there is a conflict between the sensations of gravity and visual perception, the visual perception will usually tend to dominate.

REFERENCES

1. Blackwell, H. R.: Development and use of a quantitative method for specification of interior illumination levels on the basis of performance data, *Illuminating Engineering*, 1959, vol. 54, pp. 317–353.
2. Burg, A., and S. F. Hulbert: Dynamic visual acuity and other measures of vision, *Perceptual and motor skills*, 1959, vol. 9, p. 334.
3. Cobb, P. W., and F. K. Moss: "Four fundamental factors in vision," in M. Luckiesh and F. K. Moss, *Interpreting the science of seeing into lighting practice*, vol. I, 1927–1932, General Electric Co., Cleveland.
4. Fletcher, H., and W. A. Munson: Loudness, its definition, measurement, and calculation, *Journal of the Acoustical Society of America*, 1933, vol. 5, pp. 82–108.
5. Goodson, J. E., and J. W. Miller: *Dynamic visual acuity in an appplied setting*, USN School of Aviation Medicine, Pensacola, Fla., Report 16, Project NM 17 01 99 Subtask 2, May 25, 1959.
6. Graham, C. H.: *Vision and visual perception*, John Wiley & Sons, Inc., New York, 1965.
7. Gregory, R. L.: *Eye and brain; the psychology of seeing*, World University Library, McGraw-Hill Book Company, New York, 1966.
8. Harris, C. M. (ed.): *Handbook of noise control*, McGraw-Hill Book Company, New York, 1957.
9. Hochberg, J. E.: *Perception*, Prentice-Hall, Inc., Englewood Cliffs, N.J., 1964.
10. Maerz, A., and M. R. Paul: *Dictionary of color*, 2d ed., McGraw-Hill Book Company, New York, 1950.
11. Miller, D. C.: *The science of musical sounds*, The Macmillan Company, New York, 1926.
12. Peterson, A. P. G., and E. E. Gross, Jr.: *Handbook of noise measurement*, 6th ed., General Radio Co., New Concord, Mass., 1967.

13.   Robinson, D. W., and R. S. Dadson: Threshold of hearing and equal-loudness relations for pure tones, and the loudness function, *Journal of the Acoustical Society of America*, 1957, vol. 29, no. 12, pp. 1284–1288.

14.   Stevens, S. S.: A scale for the measurement of a psychological magnitude: loudness, *Psychological Review*, 1936, vol. 43, pp. 405–416.

15.   Stevens, S. S.: Calculation of the loudness of complex noise, *Journal of the Acoustical Society of America*, 1956, vol. 28, pp. 807–832.

16.   Stevens, S. S.: Procedure for calculating loudness: Mark VI, *Journal of the Acoustical Society of America*, 1961, vol. 33, no. 11, pp. 1577–1585.

17.   Tuttle, W. W., and B. A. Schottelius: *Textbook of physiology*, 15th ed., The C. V. Mosby Company, St. Louis, 1965.

18.   von Békésy, Georg: The ear, *Scientific American*, August, 1957.

19.   Wapner, S., and H. A. Witkin: The role of visual factors in the maintenance of body-balance, *American Journal of Psychology*, 1950, vol. 63, pp. 385–408.

20.   Wegel, R. L., and C. E. Lane: The auditory masking of one pure tone by another and its probable relation to the dynamics of the inner ear, *Physiological Review*, 1924, vol. 23, pp. 266–285.

21.   Witkin, H. A.: Perception of body position and the position of the vertical field, *Psychological Monographs*, 1949, vol. 63, no. 7, pp. 1–46.

22.   Witkin, H. A.: The perception of the upright, *Scientific American*, February, 1959.

23.   Witkin, H. A., and S. Wapner: Visual factors in the maintenance of upright posture, *American Journal of Psychology*, 1950, vol. 63, pp. 31–50.

24.   Wyburn, G. M., R. W. Pickford, and R. J. Hirst: *Human senses and perception*, University of Toronto Press, Toronto, 1964.

25.   *Color harmony manual*, Container Corporation of America, Chicago, 1942.

26.   *IES lighting handbook*, 4th ed., IES, New York, 1966.

27.   *Industrial noise manual*, 2d ed., American Industrial Hygiene Association, Detroit, 1966.

28.   *Munsell book of color*, Munsell Color Co., Baltimore, 1929.

29.   *Noise*, lectures presented at the Inservice Training Course on the Acoustical Spectrum, Feb. 5–8, 1952, sponsored by the University of Michigan, School of Public Health and Institute of Industrial Health, University of Michigan Press, Ann Arbor, Mich.

30.   Optical Society of America, Committee on Colorimetry: *The science of color*, Thomas Y. Crowell Company, New York, 1953.

31.   *USA Standard specification for octave, half-octave, and third-octave band filter sets*, USASI, S1.11–1966.

Our common notion of *information* can cover a fairly wide span of examples, such as what we read in the newspapers or hear on TV, the bill for automobile repairs, the gossip over the backyard fence, and the directions given on road signs. We can, however, conceive of information in a much broader sense. Since we really are talking about stimuli that convey meaning, we could even think of information as the transfer of energy that has some meaningful implications in the given situation, such as a driver "communicating" with his car via the control mechanisms; mechanical, hydraulic, and servo linkages in various types of equipment; and a Geiger counter. As man interacts with physical objects, with other people, and with the environment about him, it is reasonable to consider information flowing along two-way streets between all elements of a situation, such as

Man ↔ man
Man ↔ machine
Man ↔ environment
Machine ↔ machine
Machine ↔ environment

INFORMATION IN SYSTEMS

In a systems frame of reference, information can serve either of two roles, or functions. In the first place, it can contribute to the physical output of a system (or component), such as the baker's recipe (either in his head or on a sheet of paper) that is used to mix a batch of bread dough, or the computer program of a computer-controlled machine tool, which controls the mechanism to produce the desired output. In the second place, a system (or component) may exist solely for the purpose of transmitting or processing information; information may be the exclusive grist for the mill, both as input and output. Then the output information (actually stimuli that convey information) usually becomes the input to an individual or to another system. Some such systems, as the telephone company, exist simply to facilitate the transfer of information; their performance is evaluated by the degree to which the output corresponds with the input. Other information systems may involve some processing or modification of the information, such as computers, whose output is of course information, but usually in some consolidated or reorganized form.

Collectively, various types of systems perform a number of information-processing functions, such as sensing, filtering, storing, retrieving, "using" (as in decision functions), transforming, and transmitting. In given situations, certain information-processing functions may most appropriately be the functions of machine components; others, the functions of human beings. In system design, it is of course necessary to determine whether a particular information process should be allocated to a machine component or to a human being in the system. This determination, of course, needs to be made in part by a comparison of the relative abilities of machine components and of human beings to perform the function in question. In making such determinations it is helpful to understand something of the abilities and limitations of human beings for performing information-processing functions. In this chapter we shall discuss generally certain such functions, particularly those concerned with the reception (i.e., input) of information. Chapter 8 will deal with other aspects, such as learning and decision processes. It should be kept in mind, however, that the reception of information by human beings and its subsequent processing (e.g., storage, retrieval, use in decision making, and use in making responses) are not mutually exclusive functions, but rather form a continuum, the various phases of which cannot be clearly separated. Before dealing with human information-receiving functions, however, let us sidetrack ourselves to discuss briefly the topic of information theory.

INFORMATION THEORY

Information theory is a field of scientific investigation that is concerned with the transmission of information in various contexts. The crystallization of some of the concepts involved generally can be attributed to Wiener [56], and Shannon was responsible for the development of some of the mathematical formulations [Shannon and Weaver, 47]. While the original possible applications of information theory were viewed as being within the field of engineering, the theory was soon realized to have a wide range of possible applications, such as in the fields of psychology and the biological sciences.

The Bit: A Measure of Information

For scientific inquiry to become possible, it is necessary to measure, or to identify, in relatively quantitative or objective terms, the variables with which the inquiry is concerned. In this connection, probably the major contribution of information theory is the development of a measure of information, namely, the *bit*, this term being a boiled-down version of *binary unit*.

The measurement of information in bits. The bit has been defined as the amount of information we obtain when one of two equally likely alternatives is specified [Abramson, 1, p. 12]. When in fact the various alternatives are equally probable, the total amount of information (usually symbolized by the letter H) is derived from the following formula:

$$H = \log_2 n$$

where n is the number of equally probable alternatives. This, in turn, can be expressed in terms of the probabilities of each alternative, that probability being the reciprocal of n. Thus,

$$H = \log_2 \frac{1}{p}$$

where p is the probability of each such alternative.

In the simplest case, then, where the probabilities of various alternatives are equal, the amount of information in bits is measured by the logarithm, to the base 2, of the number of such alternatives. With only 2 alternatives, the information, in bits, is equal to the logarithm of 2 to the base 2, which is then 1. When Paul Revere was to receive a signal from the Old North Church, he was to see 1 signal (1 lantern) if the enemy came by land and 2 if by sea. Assuming that these 2 alternatives were equally probable, the amount of information available would be 1 bit; a discrimination was to be made between 2 alternatives.

Let us now take 4 alternatives, such as 4 lights on a panel, only 1 of which may be on at a time. In this case we should have 2 bits of information ($\log_2 4 = 2$).[1] If we had 8 such lights, we should have 3 bits of information ($\log_2 8 = 3$), etc.

If we were playing Twenty Questions, it would be possible to determine the correct answer out of 1,048,576 possible alternatives if the questions were properly framed and if all the information were used. To accomplish this, it would be necessary so to phrase each question that the number of remaining alternatives would be reduced by $\frac{1}{2}$. This is illustrated in Figure 4-1 with 4 questions. If, with each question, the remaining alternatives are reduced by $\frac{1}{2}$, then with 4 questions, the unerring reduction will be to 1 of 16 subclasses; this will produce 4 bits of information ($\log_2 16 = 4$). The amount of information that would be obtained with 20 questions would be 20 bits ($\log_2 1,048,576 = 20$).

We have here been assuming equal probabilities of each event or pair of alternatives. This, of course, does not exist in all circumstances. For ex-

[1] $\log_2 n$ = the exponent that must be applied to 2 to obtain n.

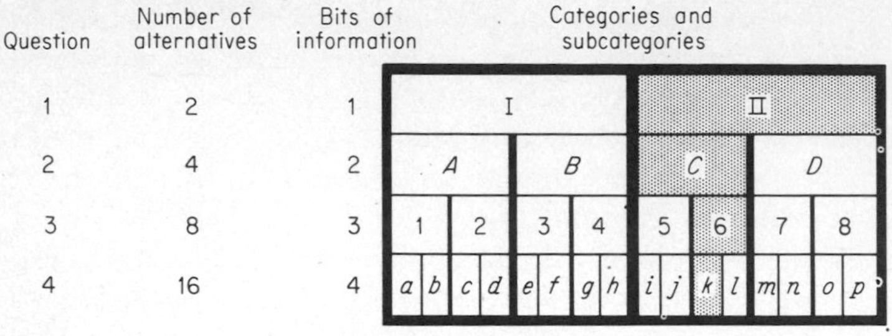

Question	Number of alternatives	Bits of information	Categories and subcategories							
1	2	1	I				II			
2	4	2	A		B		C		D	
3	8	3	1	2	3	4	5	6	7	8
4	16	4	a b	c d	e f	g h	i j	k l	m n	o p

Figure 4-1 Illustration of information theory applied to sequential elimination as in Twenty·Questions. Assuming equal probabilities, the number of possibilities is reduced by half at each question; so in the 4 questions it is possible to select 1 of 16 alternatives. as shown by the shaded areas.

ample, let us take the case of a bent coin that lands heads up $\frac{9}{10}$ of the time, $p = .90$, and tails up $\frac{1}{10}$ of the time, $p = .10$. We can compute the amount of information *for each alternative* h_i as follows:

h (heads) $= \log_2 1/.90 = \log 1.11 = 0.15$ bits
h (tails) $= \log_2 1/.10 = \log 10 = 3.32$ bits

We see from these computations that we get more information from a throw of tails (it is less expected and more of a surprise) than from a throw of heads (which we should expect from our knowledge of the probabilities). But in deriving an estimate of the *total* amount of information H that one would receive across *all* possible alternatives, we must weight the amount of information from *individual* alternatives h_i by their *respective* probabilities, as follows:

$H = p$ (heads) $\times h$ (heads) $+ p$ (tails) $\times h$ (tails)
$\quad = .90 \times 0.15$ bits $+ .10 \times 3.32$ bits
$\quad = 0.47$ bits

This value, in turn, can be compared with the amount of information H from a situation where the probabilities would be equal, $p = .50$, which would turn out to be 1 bit ($H = \log_2 2 = 1$ bit).[2]

As the probabilities of two alternatives become further and further from being equal, the amount of information H becomes less and less. When the probabilities become 1.0 and .0 respectively, no new information is gained; there is no uncertainty at all. Where there are various

[2] This derivation can be confirmed by the same process illustrated above in which the weighted values are summed, as follows:

H (equal probabilities) $= p$ (heads) $\times h$ (heads) $+ p$ (tails) $\times h$ (tails)
$\qquad = .50 \times \log_2 1/.50 + .50 \times \log_2 1/.50$
$\qquad = .50 \times \log_2 2 + .50 \times \log_2 2$
$\qquad = .50 \times 1 + .50 \times 1 = 1$ bit

possible alternatives, with their own individual probabilities, the *average* information associated with the various possible events is equal to the sum of the information values of all the individual alternatives multiplied by their individual probabilities.[3]

The Nature of Information

In the application of the concept of information theory in such fields as electrical engineering, the measure of information (i.e., the bit) is based on the objectively determined probabilities of occurrence of various symbols, codes, signals, or other essentially physical stimuli or events. As we adapt the notion of information theory to situations with human beings, however, we usually become involved with the *meaningfulness* of the signals or events. In this context of information theory, *information* has very much the same meaning as the term has in everyday life, namely, that it refers to knowledge that was not previously known. In this sense information can be considered the reduction of ignorance, and one bit is that measure of information which reduces one's ignorance of some area of knowledge by one-half, as in the case of Paul Revere's "one if by land, two if by sea" message.

Consideration of the individual's own state of knowledge, however, complicates the problem, since for any content area, some individuals already know more than others. In the actual application of information-theory principles to situations with human receivers, however, it usually is not practical to take into account the level of ignorance (or knowledge) of individuals. Rather, it is the more common practice to deal quantitatively with the information that is available in different situations (i.e., the information that is implied by the symbols, signals, or other stimuli that are present) or with the information which is, or can be, received by people. Let us consider, for example, a single TV picture, assuming it to consist of an array of 30,000 black, white, and gray dots (500 rows and 600 columns), with 10 distinguishable brightness levels. With

---

[3] In the case of n independent symbols or messages, the amount of information in bits is determined from the following equation [Shannon and Weaver, 47, p. 105]:

$$H = -(p_1 \log p_1 + p_2 \log p_2 + \cdots + p_n \log p_n)$$

or

$$H = -\sum p_i \log p_i$$

in which

H = amount of information in bits

$\sum p_i \log p_i$ = sum of all items like the typical one, $p_i \log p_i$

$p_1, p_2, \ldots, p_n$ = probabilities of individual symbols or messages

n = number of independent symbols or messages

The minus sign occurs simply because any probability is a number less than, or equal to, 1, and the logarithms of numbers less than 1 are themselves negative. Thus, the minus sign is necessary in order that H itself be positive.

equal probabilities, the amount of information so presented would be H = 300,000 log 10 ≈ 10 million bits [Abramson 1, p. 13]. Since the variations in the stimuli dots are not equal in their probabilities, the actual amount of information available is very substantially less than the above, but even so, our ability to perceive the many individual features of a television picture (or other complex configuration) is markedly limited. For certain practical purposes, then, we need to be concerned with the effective amount of information people can actually receive, rather than with that which is available to them.

Sources and Pathways of Information

We shall return to information theory as related to the human *use* of information, but first let us pause to consider the general sources of information (stimuli) that may bombard us in our hurly-burly world, the *pathways* of such information, and the variations in the form of the information that may occur in between the original source and the receiver. Perhaps most typically the original source (the distal stimuli if we want to use the long-haired term referred to in the last chapter) is some object, event, or environmental condition; in some instances this source (at least as far as the human receiver is concerned) consists of displays that represent something else (such as numbers representing dollars in a banking account). Information from these original sources may come to us *directly* (such as by *direct* observation of an airplane), or it may come to us *indirectly* through some intervening mechanism or device (such as radar or telescope). In either case, the distal stimuli are sensed by the individual only through the energy that they generate (directly or indirectly) through proximal stimuli (light, sound, mechanical energy, etc.). In the case of *indirect* sensing, the *new* distal stimuli may be of two types. In the first place, they may be *coded* stimuli, such as visual or auditory displays. In the second place, they may be *reproduced* stimuli, such as those presented by TV, radio, or photographs or through such devices as microscopes, microfilm viewers, binoculars, and hearing aids; in such cases the reproduction may be intentionally or unintentionally modified in some way, as by enlargement, miniaturization, amplification, filtering, or enhancement. With either coded or reproduced stimuli, the new, or converted, stimuli become the actual distal stimuli to the human sensory receptors. Figure 4-2 illustrates in a schematic way these various pathways of information reception to the individual, both direct and indirect.

Limitations in Information Measurement

Some forms of information that are received by human beings, whether directly or in coded or reproduced form, actually can be quantified in terms of bits. Other forms of information, however, do not yet lend themselves to such measurement. For example, it is not as yet possible to do

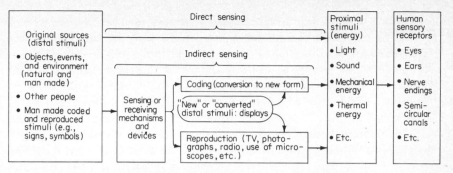

Figure 4-2 Schematic illustration of pathways of information from original sources to sensory receptors. (Although typically the original, basic source is an object, event, condition, the environment, etc., in some situations the effective original source to an individual consists of some man-made coded or reproduced stimuli; office personnel, for example, usually deal with recorded symbols which, for practical purposes, are their "original" distal stimuli.)

so in the case of continuously changing stimuli, such as targets in a tracking task, continuous changes in instrument readings, complex traffic patterns, or football games. It has become apparent over the last several years that information theory is not an open sesame to all the locked doors of knowledge regarding human behavior. But the concept of information transmission which has become thus crystallized may be of some value in the analysis of man's functions in man-machine systems, even where it is not possible to assign a quantitative number to the information that actually is involved. It may be relevant, however, to take a quick look at a few types of situations in the human factors context in which rather fuzzy concepts of information have been pinned down in quantitative terms.

The Information in Stimulus Dimensions

In the transmission of information through the various sensory modalities, variations along some *stimulus dimension* frequently are used as the basis for judgmental determinations regarding the meaning of the stimulus. Let us cite a few possible examples. The loudness of a train whistle may be used to judge the distance of a train; the perceived position of the hands of a clock, to tell time (when one may not be able to see the numbers); the colors of electric wires, to identify them; the number and time spacing of Morse code signals, to characterize various letters; and the kinesthetic feedback of the arm, to tell where it is located as it is positioned without visual guidance.

For any given sensory modality (vision, audition, touch, etc.), there can of course be different *kinds* or *classes* of stimuli, such as brightness, color, and shape (in vision) and loudness, duration, and pitch (in audition). Each of these (and other classes of stimuli for these senses, as well as various classes of stimuli for other senses) may be thought of as a separate stimulus dimension. For any such stimulus dimension, it is usually possible for people to make judgmental discriminations based on stimulus differences along the dimension (such as loudness). To the extent that a designer of a system has some latitude in the selection of the sensory modality and the stimulus dimension to use in transmitting information to human beings, he can take into account the variations in human abilities to make discriminations of stimuli of each of the various stimulus dimensions. He can also take advantage of any variation in appropriateness of various possible stimulus dimensions to the purpose at hand.

Absolute versus relative judgments. The discriminations in, or judgments of, stimuli of some dimension may be made on a *relative* or on an *absolute* basis. A relative judgment is one which is made when there is an opportunity to compare two or more stimuli; thus, one might compare two or more sounds in terms of loudness or two or more lights in terms of brightness. In absolute judgments, there is no opportunity to make comparisons, such as identifying a given note on the piano (say, middle C) without being able to compare it with any others, or identifying a given color out of several possible colors when it is presented by itself.

As one might expect, people are generally able to make fewer discriminations on an absolute basis than on a relative basis. For example, it has been estimated that most people can differentiate as many as 100,000 to 300,000 different colors on a relative basis when comparing two at a time (taking into account variations in hue, lightness, and saturation). On the other hand, the number of colors that can be identified on an absolute basis is limited to no more than a dozen or two.

Emphasis on absolute judgments. In operational situations absolute discriminations of stimuli along some dimensions are much more common than relative discriminations are. Absolute judgments can be required in either of two types of circumstances. In the first place, several discrete positions (levels or values) along a stimulus dimension might be used as codes, each position representing a different item of information. If the stimuli consist of tones of different frequencies, the receiver is supposed to identify the *particular* tone. In the second place, the stimulus may be of any value along the stimulus dimension, and the individual needs to make some judgment regarding its value or position along the dimension. Such judgments might be used in various operational ways. A radio operator, for example, might use his judgment of loudness of a radio signal to adjust the gain (volume) up or down to some subjective standard. On the other hand an inspector might use his judgment to classify all items as either "pass" or "fail," or possibly by grades, such as A, B, and C.

The amount of information in absolute judgments. The accuracy with which individuals can, on the average, make absolute discriminations of various levels of any given stimulus dimension can be quantified, with the bit as the measure of this accuracy. Thus, one can evaluate any given stimulus dimension by the amount of information it can convey. Let us suppose that for a given stimulus dimension (say, loudness of sounds), individuals can only differentiate two levels of loudness, namely, loud and soft (actually they can discriminate more than this). Given equal probabilities, then, the amount of information that could be transmitted by using these would be 1 bit ($\log_2 n = \log_2 2 = 1$ bit). If, however, people could generally discriminate 4 levels of loudness, the amount of information that could be transmitted would be 2 bits ($\log_2 4 = 2$ bits). It should be added, however, that the number of possible degrees or levels along some dimension that can be discriminated on an absolute basis is, itself, not an absolute number, since it varies somewhat among individuals and even with the same individual from time to time. It also varies somewhat with the total range of stimuli used.

But granting that such a measure is an estimate (rather than a highly precise value), let us take a look at some examples. A number of such examples are given in Table 4-1. This table shows, for each of various dimensions, the number of absolute judgments which people can make and the number of bits of information, H, represented by those judgments.

An examination of the data in this table and elsewhere points up the fact that, in general, the number of possible discriminations that people can make along typical *individual* stimulus dimensions is really not very high, being generally in the range from 4 to 9 or 10 (in some cases higher), with corresponding bits from about 2.0 to 3.0 or 3.3. In this connection Miller [33] refers to the "magical number seven, plus or minus two," meaning that the range of such discriminations is somewhere around 7 ± 2 (5 to 9); for some the number is greater, and for some less, than this specific range. Seven discriminations would transmit 2.8 bits.

In *combinations* of dimensions, however, the information that can be transmitted sometimes is noticeably greater, ranging from 3.5 bits to as high as 7.2 bits. The 7.2 bits characterizes a situation in which people made absolute discriminations of as many as 150 combinations of 6 auditory dimensions.

The Information in Language

A language consists of a finite number of symbols, used in varying combinations but following certain principles and rules (along with many exceptions). The frequencies (probabilities) of individual symbols and combinations and of words and their combinations conceivably could always be determined, and in certain circumstances they have been de-

Table 4-1 Amount of Information in Absolute Judgments of Various Stimulus Dimensions

Sensory modality and stimulus dimension	*No. of levels which can be discriminated on absolute basis*	*No. of bits of information transmitted, H**	*Source*
Vision, single dimensions:			
Pointer position on linear scale	9	3.1	*a*
Pointer position on linear scale:			
Short exposure	10	3.2	*b*
Long exposure	15	3.9	*b*
Visual size	5–7	2.3–2.8	*c, j*
Hue	9	3.1	*c*
Brightness	3–5	1.7–2.3	*c, j*
Vision, combinations of dimensions:			
Size, brightness, and hue†	17	4.1	*c*
Hue and saturation	11–15	3.5–3.9	*d*
Audition, single dimensions:			
Pure tones	5	2.3	*e*
Loudness	4–5	1.7–2.3	*f, g*
Audition, combination of dimensions:			
Combination of six variables‡	150	7.2	*h*
Odor, single dimension	4	2.0	*i*
Odor, combination of dimensions:			
Kind, intensity, and number	16	4.0	*i*
Taste:			
Saltiness	4	1.9	*j*
Sweetness	3	1.7	*j*

* Since the number of levels is rounded to the nearest whole number, the number of bits does not necessarily correspond exactly.

† Size, brightness, and hue were varied concomitantly, rather than combined in the various possible combinations.

‡ The combination of six auditory variables was frequency, intensity, rate of interruption, on-time fraction, total duration, and spatial location.

SOURCES:

  a. Hake and Garner [25]
  b. Coonan and Klemmer [as reported by Miller, 33]
  c. Eriksen [19]
  d. Halsey and Chapanis [26]
  e. Pollack [43]
  f. Morgan et al. [34]
  g. Garner [22]
  h. Pollack and Ficks [44]
  i. Engen and Pfaffmann [18]
  j. Beebe-Center, Rogers, and O'Connell [7]

termined. Thus, we have the makings of an information analysis of messages in any given language.

If we were to assume that the 26 letters of .the English language were equally frequent in use and we were to throw in spaces as another symbol, making 27 symbols in all, the upper limit of the amount of information per letter would be $H = \log_2 n = \log_2 27 = 4.75$ bits. In practice, however, the symbols are not equal in frequency. Further, the combinations of various letters are not random. The letter u after q in spelling is not useful in differentiating between words; the letter u is completely redundant. Taking these and several other factors into account, in practice, the information content of ordinary English has been estimated [Raisbeck, 46] to be about 1 bit per letter. But the 26 letters form words by the thousands. Even allowing for variations in the frequencies of words as used in speech and writing, the information content of language material can be seen to be quite great.

While some rough estimates could thus be made, the systematic analysis of information from language can perhaps better be illustrated with a fairly well-controlled study, such as the one by Pierce and Karlin [42]. In that study, various vocabularies were used, the vocabularies varying in the number of words in them (2, 4, 8, 16, 32, 64, 128, 256, 2500, and 5000), the number of bits per word ranging from 1 bit (2-word vocabulary) to 8 bits (256-word vocabulary) to 12.3 bits (5000-word vocabulary). Words from each of these were listed randomly on sheets of paper and were read aloud. The amount of *information* transmitted would be the product of speed of reading times bits per word. Thus a 4-word vocabulary (2 bits per word) read at a rate of about 3.5 words per second would transmit about 7 bits (2 bits $\times$ 3.5 words per second), whereas a 256-word vocabulary (8 bits per word) read at the same rate would transmit about 28 bits. For comparative purposes, the amounts of information transmitted under various experimental situations in this study, including prose reading, are given below (the ranges illustrate differences from slow to fast readers).

Type of speech material	Range, bits per second
4-word vocabulary	7.0 (average)
256-word vocabulary	24–30
2500-word vocabulary	34–42
5000-word vocabulary	26–33
Prose (5 bits per word)	19–24
Prose (10 bits per word)	39–48

Pierce and Karlin estimated the human channel capacity (presumably for language material at least) to be of the order of 40 to 50 bits. In this connection, they have contrasted this approximate human capacity with the transmission capacity of telephone and television channels (about 50,000 and 50 million bits per second, respectively). The fairly obvious implication of this discouraging comparison is that we mortals miss a lot of information!

Information in Other Contexts

Aside from the measurement of information as an input to human beings through their sensory channels, the measuring of information in subsequent human processes (as in memory and making motor responses) is also conceivable. Since we are discussing information theory, we might as well jump the gun a bit and bring in here at least brief reference to information theory in those contexts.

Information in human memory.   While the specific processes of human learning are not entirely understood, it is generally recognized that it is based on certain changes within the nerve cells (the neurons) of the brain. It has been estimated that there are something like 10 billion of these. Assuming the efficient utilization thereof, it has been estimated that the overall storage capacity of the human memory is somewhere between 10^8 and 10^{15} bits (i.e., 100 million to 1 million billion bits) [Geyer and Johnson, 23]. This range, of course, is far greater than the storage capacity of any computer now in existence or likely to be developed within any reasonable time. To extend the analogy between human information storage and that of computers, it has been pointed out that storage components of electronic systems are of two types, namely, static (consisting of special patterns of binary data unchanging in time) and dynamic (information in the form of electrical or mechanical impulses). In turn, there is some evidence that the brain also combines both of these schemes [Geyer and Johnson, 23]. One of the schemes is provision for storage of "old" information; the other has been described as a "circulatory" conception, accounting for the recording of current or recent information (short-term memory).

Information in motor activities.   Information analysis has also been used in analyzing human motor activities. In a relatively straightforward illustration in one experiment, subjects were presented with Arabic numerals (2, 4, or 8 under a number of experimental conditions) at specified rates (1, 2, or 3 per second) and were asked to repeat them *verbally* or to respond by pressing keys which corresponded to the numerals to be presented [Alluisi, Muller, and Fitts, 3]. The maximum rate for verbal responses was 7.9 bits per second, and for the motor (key-pressing) responses it was 2.8 bits per second. While this comparison is of course of some interest, the more pertinent point at the moment is that the information in motor responses can be measured, at least in such simple situations.

In a somewhat more complex task, an analysis was made of the information of placing pegs in holes with varying degrees of tolerance (the difference in diameter of the peg and of the hole) and varying amplitudes (distance of movement) [Fitts, 21]. Information in bits was measured by a special formula, the number of bits for the various experimental conditions ranging from 3 to 10. This type of analysis was. extended to a study of elements of the same type of task [Annett, Golby, and Kay, 5].

Discussion

These illustrations of information analysis are, of course, only some examples, and should not be considered as representing the complete range of present or possible applications in the human factors context. (We shall cite other illustrations in later chapters.) In taking an overview of the application of information techniques to psychological problems generally, however, it has been pointed out that their applications in some situations were successful and illuminative, some were pointless, and some were downright bizarre [Attneave, 6, p. v]. According to Attneave, two generalizations may be stated with considerable confidence: (1) Information theory is not going to provide a ready-made solution to all psycho·logical problems. (2) Employed with intelligence, flexibility, and critical insight, information theory can have great value both in the formulation of certain psychological problems and in the analysis of certain psychological data.

It is with this understanding that we should view the implications of information theory for human factors problems.

CHANNEL CAPACITY

Brief mention was made above of the notion of *channel capacity* of human beings. There are, however, two ways in which this term has been used. In the first place, it is applied to any *given* channel of information, or what we have referred to as a stimulus dimension. Although a channel usually refers to a particular sense organ, two or more input channels can be presented through the same sensory modality. Considering any particular channel (or stimulus dimension), there is substantial evidence to indicate that there is some upper limit of the amount of information per unit of time that can be received through that channel. As discussed above, for example, there is an upper limit to the number of discrete steps or levels of stimuli of any given dimension that can be discriminated on an absolute basis.

The second, and broader, concept of channel capacity relates to the possibility of some upper limit of information that can be received and processed by people, usually within a given time, taking into account the various sensory modalities and the entire complex of stimuli that impinge upon the individual. This theory assumes some bottleneck concept—that some given volume can "get through" within a given time, but no more. Although there are some differences of opinion regarding this question, some of the research of Broadbent [9] and others has clearly indicated that, in any event, the multiple inputs of information by separate individual channels typically result in some loss, or squeezing out, of some of the information being presented, especially if the rate of information transmission is fairly high. Thus, it is reasonable to believe that there is some overall channel capacity of the human being that imposes a ceiling on the total amount of information that can be received and processed. In this connection, as indicated above, Pierce and Karlin [42] from their study suggest that the maximum they found (about 43 bits per second) may approximate a real human information capacity of around 40 to 50 bits per second.

We should note, however, that we have gradually oozed from an initial concern with sensory processes into the domain of comprehension, or, as Abramson puts it [1, p. 139], an interior point of the human information-processing system. In turn, we can very easily slide on into the response phase of the whole interaction sequence. In this connection it may be useful to take a look at a conceptual formulation [Welford, 55] relating to the chain of postulated mechanisms in sensorimotor performance. In this formulation (shown in Figure 4-3), the several sensory in-

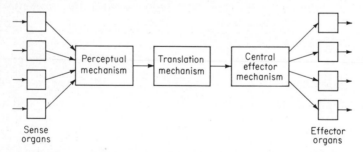

Figure 4-3 Schematic illustration of postulated mechanisms used in sensorimotor processes. Following this formulation, sensory inputs come through one or more sense organs, and are then channeled through a central perceptual mechanism, a translation mechanism, and a central effector mechanism which is then responsible for the coordination of motor responses. [From Welford, 55.]

puts, through a single sensory modality or through various senses, are channeled into a central, integrative perceptual mechanism, then through a central translation mechanism, and subsequently through a central effector mechanism. This last mechanism coordinates and controls the several subsequent motor processes that are executed by the different effector organs (fingers, hands, etc.). These mechanisms should be thought of more as different functions of the central nervous system than as different parts of the brain. But—if this formulation is basically sound—it suggests something of a funneling process with some possible limitation of the information input into the perceptual mechanism and with resulting competition among the various sensory inputs, which supports the general theory of an overall channel capacity. Although we cannot neatly separate the different sequential phases of information processing (i.e., sensation, perception, decision, response, etc.), it should be noted that the primary bottleneck in the system probably is the brain rather than the human sensory mechanisms themselves. In this connection, Steinbuch [51] has summarized estimates of the information reduction that occurs from the initial reception by the sense organs through the intermediate processes to permanent storage (memory) and presents the following estimates:

Process	Maximum flow of information, bits/sec
Sensory reception	1,000,000,000
Nerve connections	3,000,000
Consciousness	16
Permanent storage	0.7

He postulates certain as yet unexplained intermediate reduction processes between the neural connections of the sensory organs and the conscious perception of the stimuli. Granting that the above estimates are rough, we can indeed see that the central-nervous-system processes of consciousness and storage are capable of handling only a fraction of the potentially tremendous information input to the sensory receptors. If we consider the brain, there is some evidence to suggest that it operates like a time-shared data-processing system with a time cycle of 50 msec [Kristofferson, 31]. If this is so, we can accept inputs into the central processor (i.e., the cortex of the brain) from only one of the several or many input channels at a time.

Entirely aside from possible theoretical aspects, we probably need to accept both of these concepts of channel capacity, the one relating to the information-capacity limits of individual stimulus dimensions and the other relating to the overall information-processing capacity of the human being. However, accepting these concepts as relating to theoretical maximums, let's face it—we humans do not operate at perfect efficiency. There are certain kinds of situational variables that have some bearing upon one's performance, some for the worse (for example, noise), others for the better. Thus, it might be useful to pin down the situational features that are related to total-information-handling abilities of people. If we could do this, we might, in the design of systems, capitalize on those features or schemes which facilitate information handling, ranging from the more strictly sensory processes through the perceptual, mediational, and response processes.

If we consider now primarily the input end of this continuum, some of the variables that are related to human information-processing abilities are the coding of information stimuli, the organizing of information to be presented, and the presence of multiple-sensory inputs.

CODING OF SENSORY INPUTS

Any type of indirect presentation of information that has a display implies some selection of the particular sensory modality and the stimulus dimension to be used. For any given stimulus (coding) dimension (e.g., sound frequency or geometric forms), the coding requires the selection of the *specific* stimuli of the general class of stimuli that are to represent specific elements of information.

Sensory Channel and Coding Dimension

For many types of information, the stimulus, or coding, dimension (and, therefore, the general class of stimuli) is virtually dictated, or at least strongly suggested, by the nature of the information and the situation in which it is to be used. In other situations, however, there may be more flexibility in the selection of the particular sensory modality and coding dimension to be used. In any given situation, the selection of the coding dimension needs to be made on such considerations as the form in which the information is already available, the appropriateness of the various modalities in the operating situation, the transmission advantage of one over another, the relative loads on the various senses, and the like. Some of the earlier discussions of different senses may have some bearing on this, including the table on the amount of information in absolute judgments (Table 4-1).

Unidimensional and Multidimensional Coding

Probably most typically when codes are used, they are unidimensional, which means that a given code symbol has a given meaning such as the red, yellow, and green colors of traffic lights, the unique "you're out!" arm signal of the baseball umpire, or the $, ¢, ?, &, =, and ! signs of the printed page. On the other hand, in some contexts two or more coding dimensions are used in some combination. The notion of multidimensionality of codes is sometimes rather elusive, since you can mix them or match them in many ways. For our purpose, however, we will stick to fairly clear-cut cases. Multidimensional codes can vary in their degree of *redundancy*. A completely redundant code is one in which each thing to be identified has two (or more) unique attributes (such as a unique color *and* a unique shape). A partially redundant code is one in which the individual codes of a dimension are not uniquely associated with individual codes of another dimension; for example, in certain states rectangular road signs are used to show speed limits, but numerals give actual limits in specific areas. There are varying ways in which two or more dimensions can be partially redundant, and such codes are frequently used.

It has been suggested by some individuals [Miller, 33; Pollack and Ficks, 44] that the use of multidimensional codes generally increases information transmission, although this is not universally the case, and in any event there are limits beyond which additional dimensions are not useful.

Discussion

For what will have to be a broad-brush discussion of coding, we shall first give some examples of research on coding systems, particularly visual and auditory, to illustrate how such research can contribute to the selection or development of reasonably optimum coding systems for particular purposes. Other examples of coding systems will be included in later chapters, especially Chapter 5. Second, we shall discuss some general considerations and guidelines in the selection of codes for certain purposes.

Visual Codes

Much of the information that is presented visually (probably too much) is linguistic in nature (using alphabetical characters such as used in printed material, instructions, and labels) or quantitative in nature (such as temperature, pressure, velocity, and bills to be paid). In such cases, the type of coding systems to be used—alphabetical and/or numerical (i.e., alphanumeric)—is virtually predetermined, and there is, therefore, no problem in the selection of the appropriate code, although there usually is some option in the selection of the specific design features of the letters, numerals, and displays used. The possible use of other types of visual codes (and of the optional use of alphanumeric codes where their use is not predetermined) typically occurs when the requirement is that of symbolizing each of various independent items or concepts. The range of such codes includes colors, geometric shapes, configurations, symbolic representations, and alphanumeric symbols. The possible uses of such codes include symbols on graphic displays (maps, charts, etc.); CRT (cathode-ray-tube) displays; signs; warnings and the identification of hazards; the identification of objects, materials, and ideas; and trademarks.

Unidimensional visual codes. A good many types of individual visual codes (quaintly called *alphabets*) have been tried out in experimental studies for comparing their use by human beings. A few such studies will be cited for illustration, although the reader should be warned to expect some apparent contradictions (but .we shall come back to this point later). One such study was made by Hitt [27]. Five different codes were used, as shown in Figure 4-4: numerals, letters, geometric shapes,

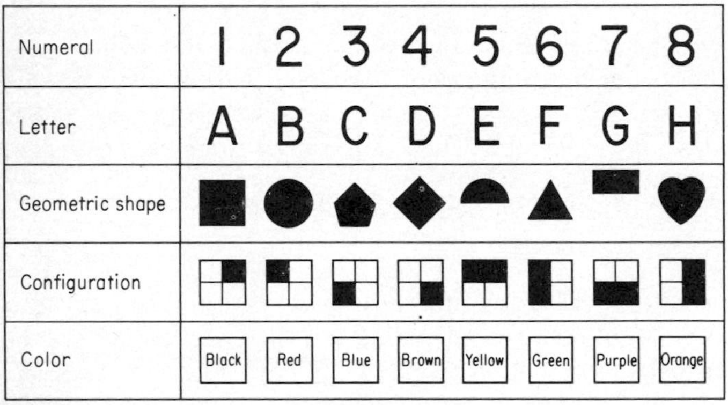

Numeral	1	2	3	4	5	6	7	8
Letter	A	B	C	D	E	F	G	H
Geometric shape	■	●	⬠	◆	◗	▲	▬	♥
Configuration	⊞	⊞	⊞	⊞	⊞	⊞	⊞	⊞
Color	Black	Red	Blue	Brown	Yellow	Green	Purple	Orange

Figure 4-4 Illustration of code symbols used in comparing coding of targets. [From Hitt, 27.]

configuration, and color, each having eight categories. These were used in displays on 30- by 22-in. cardboard posters with 8 columns and 5 rows, making 40 cells in each display. These displays might be thought of as maps, and the eight different categories of one class were used to symbolize, under different experiment conditions, eight types of buildings, radar units, aircraft, defense units, and industries (such as an aluminum plant, a petroleum refinery, and a lead refinery). Each display included only one type of code, but under different conditions the number of different symbols of the type was varied (2, 4, and 8) and the density (the number of different symbols in a cell) was varied (1, 2, and 3).

The subjects performed five different tasks with these displays: *identification* (e.g., identify type of industry in one cell); *location* (e.g., locate cell that includes only one steel mill); *counting* (e.g., count number of aluminum plants in row C); *comparison* (e.g., compare number of petroleum refineries in one cell with that in another); and *verification* (e.g., the plant in a given cell is a steel mill—true or false). Their chore was to go down a list of questions of these types and record their answers. A summary of the major results is given in Figure 4-5, which shows, for the different codes and tasks, the number of correct responses per minute.

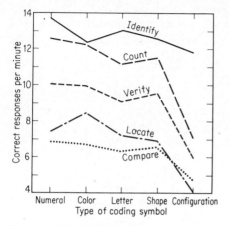

Figure 4-5 Relationship between coding method and performance in five tasks comparing targets on a display. [Adapted from Hitt, 27.]

In general terms it can be seen that the numerical and color codes were best for most of the tasks and that the configuration code was consistently the last, with the letter and shape codes falling in between, in some instances comparing reasonably well with the numerical and color codes.

Another study [Alluisi and Muller, 2] had a comparison of 10 symbols of each of 7 types of visual codes (a couple of styles of numerals; three variations of *inclination* codes that represented angular direction, such as hands on a clock; an *elipse-axis ratio* code of variations of elipses ranging from thin to fat; and colors). At any given session the subjects were presented with codes of only one kind and were asked either to press a key corresponding to the symbol or to answer verbally (read out). The subjects participated in the experiment during 32 sessions involving different codes and methods of response (motor versus verbal) and also conditions of forced pacing versus self-pacing. The performance of the subjects was measured in bits per units of time. Very briefly, they found that the numerical code was best, this being in line with the previous study by Hitt [27], but that the color code was worst, in contrast with Hitt's study. We shall come back to this point shortly, but might add now, incidentally, that, aside from coding considerations, verbal responses under forced-pacing conditions transmitted almost twice as many bits as did motor responses under forced pacing. (Verbal and motor responses did not differ under self-pacing conditions.)

But returning to the matter of coding, let us look at one more study of unidimensional visual codes, this one by Smith and Thomas [49]. They used four codes, namely, color, military symbols, geometric forms, and aircraft shapes as shown in Figure 4-6, each code having five symbols.

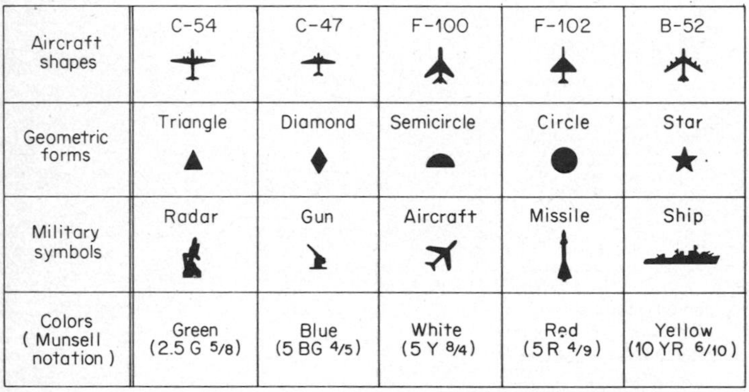

	C-54	C-47	F-100	F-102	B-52
Aircraft shapes	✈	✈	✈	✈	✈
Geometric forms	Triangle ▲	Diamond ◆	Semicircle ◗	Circle ●	Star ★
Military symbols	Radar	Gun	Aircraft ✈	Missile	Ship
Colors (Munsell notation)	Green (2.5 G 5/8)	Blue (5 BG 4/5)	White (5 Y 8/4)	Red (5 R 4/9)	Yellow (10 YR 6/10)

Figure 4-6 Four sets of codes used in a study by Smith and Thomas [49] (copyright 1964 by the American Psychological Association and reproduced by permission). The notations under the color labels are the Munsell color matches [57] of the colors used.

For each of the three shape codes, displays were prepared with 20, 60, or 100 symbols of the type in question, these being randomly allocated to any of 400 positions in a 20-by-20 imaginary matrix. In various parts of the study three sets of displays were used (each set having separate displays for each of the three shape codes), as follows:

1. Sets with shape symbols colored randomly
2. Sets with shape symbols all the same color (but with different displays for each color)
3. Sets in which each of the five symbols of a shape class was coded a unique color, with different displays of each symbol-color combination

The task of the subjects was to count the number of items of a predesignated *target* class, such as *red*, *gun*, *circle*, or *B*-52, depending upon the set of displays used in the particular phase of the study. Both time and errors were recorded. For our present purpose, two results are particularly pertinent, these being shown in Figure 4-7. This figure shows (a) a comparison of the mean time to count the four classes of symbols and (b) the percent of errors. In both instances it can be seen that colors fared best, and aircraft shapes worst.

As would be expected, time and errors increased with density (number of items in the display). A few passing observations might be added. In the first place, green targets took significantly more time than the other colors to count, followed by blue, white, red, and yellow (although these four did not differ significantly). In the second place, there were differences in the time and errors in counting certain of the individual shape symbols of a given class. And in the third place, in the case of color and shape codes that were held constant (when each of the five symbols of a class had a distinct, completely redundant color code), it was found that the color of the shape codes tended to reduce the time and errors in counting the shape codes, as indicated by the ✕'s in Figure 4-7.

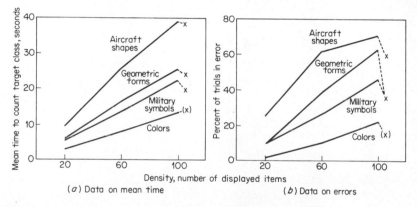

Figure 4-7 Mean time *a* and errors *b* in counting items of four classes of codes as a function of display density [Smith and Thomas, 49] (copyright 1964 by the American Psychological Association and reproduced by permission). The ✕'s indicate comparison data for displays of 100 items with color (or shape) held constant.

Multidimensional visual codes. For our examples of multidimensional visual codes we shall describe two investigations that used color and numeric or alphanumeric codes, and actually provided a comparison of multidimensional with unidimensional codes. In the first of these [Anderson and Fitts, 4] the following codes were used:

Single coding dimensions:
(a) Color patches: 9 different colors were used.
(b) Black numerals: 0, 2, 3, 4, 5, 6, 7, 8, 9 (excluding 1).

· Combination coding dimension:

    (*c*)   Color-numeric: black numerals on color patches.

By varying the number of items of each of these three classes, it was possible to vary the amount of stimulus information to be presented in different "messages" from 9.51 to 25.36 bits. Without going into details, each message (i.e., specific colors, numerals, or combinations) was presented to the subjects for 0.1 sec, and the subjects recorded what they thought the message was. The results of the study, summarized in Figure 4-8, show the total information transmitted to the subjects (actually

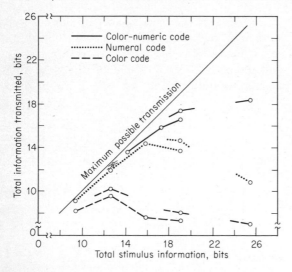

Figure 4-8 Amount of total information transmitted by three codes (color, numeric, and combination color-numeric) in relation to amount of stimulus information. [Adapted from Anderson and Fitts, 4.]

received by them) when varying amounts of stimulus information were presented. For any given amount of stimulus information, the amount transmitted generally was higher for the redundant color-numeric code than for either the color or the numeric code.

    Substantial confirmation of this comes from a study by Smith [48] in which specific alphanumeric items in a field of 20, 60, or 100 such items were to be searched or counted. In some displays the items to be searched or counted were color coded, and in others they were not. The major results are summarized briefly below; in particular this shows the percent of *reduction* in mean time when the alphanumeric items were color coded as opposed to being black and white. The reduction was generally greater in the case of the counting task.

| Display density | Reduction in time, % | |
	Search task	Counting task
20 items	45	63
60 items	64	70
100 items	70	69

Color Coding

Certain of the studies discussed above dealt with the use of colors, in most cases only a few colors, such as five. But one might be curious about how many colors one *could* differentiate on an absolute basis. Jones [29] in an excellent survey of color coding suggests that the normal observer can identify about 9 surface colors, these varying primarily in hue. Along these lines, Conover and Kraft [11] proposed, on the basis of their study, four sets of color codes for coding purposes when absolute recognition of the symbols is required (barring the possibilities of participation of extremely color-blind people). These sets include, respectively, 8, 7, 6, and 5 colors, with the possibilities of using every other color of the 8-color set if only 4 color codes are to be used as a set. In forming these sets the colors in any one set were selected so that there would be virtually equal discriminability between adjacent colors. Since this book is not in color, the Munsell notations are given in Table 4-2.

Table 4-2 Munsell Designations of Colors Recommended for Four Sets of Color Codes

| 8-color code | | 7-color code | | 6-color code | | 5-color code | |
n^*	p^*	n	p	n	p	n	p
1R	999	5R	1008	1R	999	1R	999
9R	892	3YR	890	3YR	890	7YR	884
1Y	946	5Y	1128	9Y	1131	7GY	960
7GY	960	1G	1103	5G	1101	1B	1093
9G	1099	7BG	1095	5B	1087	5P	1007
5B	1087	7PB	1133	9P	1005		
1P	1135	3RP	1003				
3RP	1003						

* n = book notation of Munsell color system; p = Munsell production number.
SOURCE: Adapted from Conover and Kraft [11].

However, this is not the whole story in the use of color codes. The typical color cone (Figure 3-2, Chapter 3) is, of course, three dimensional in form, which suggests the use of hue, saturation, and brightness as dimensional variations. Feallock et al. [20] pursued this line starting with a sample of 36 colors from a total of 381 of the Federal Standard on Colors [58], as authorized for use by the Federal government. These varied in hue, saturation, and brightness. In this investigation the subjects (26 with normal color vision and 8 who were color-blind) first were put through a rigorous training program to learn the identity of each color. Then they were shown the individual colors and asked to identify each. It was then possible to determine how many subjects confused each color with every other color. This was done under four light conditions.

In the responses of the 26 normal-color-vision subjects under all four conditions, 10 colors were found to be identified perfectly, and 11 were confused in fewer than 5 percent of the judgments. Furthermore, by careful selection, a set of 24 colors was so chosen that only two of those within the set were confused with others in the set. Thus, it appears that with trained observers, the number of possible identifiable colors is substantially greater than indicated by previous studies.[4]

Although color seems to be a very useful coding dimension in some contexts, it is apparent that it is not a universally preferred scheme, since in certain studies above (and others), other coding dimensions were found to be superior or at least equal. These differences must lead us to conclude that the context in which color codes are used has a bearing on their potential utility. An important condition is the task that the observer is to perform. In particular, the bits of evidence point to its being the attention-getting characteristic that makes color particularly useful for searching and counting tasks. A comparison of color and alphanumeric codes for these tasks is shown in Figure 4-9 [Smith, 48]. Although searching takes less time than counting, in both cases color coding resulted in lower mean times than did alphanumeric coding. A word of possible caution should be added, however. Presently available data are based on studies with relatively few colors being used, such as five or ten. It may be that the use of larger numbers of color might reduce the advantage of color in searching and counting tasks. By way of contrast, it appears that color generally loses its advantage in such tasks as the identification of specific symbols that represent individual things or concepts.

While on the subject of color, it is interesting to note that the use of various nonrelevant colors used in a display may tend to complicate a visual searching task.

[4] The Federal Standard identifications of these colors, with an identification ($\times$) after those that were identifiable under all lighting conditions, are 32648 ($\times$), 31433, 30206 ($\times$), 30219, 30257, 30111, 31136 ($\times$), 31158, 32169, 32246, 32356, 33538 ($\times$), 33434, 33695, 34552 ($\times$), 34558, 34325 ($\times$), 34258 ($\times$), 34127, 34108 ($\times$), 35189, 35231 ($\times$), 35109, and 37144 ($\times$).

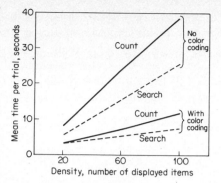

Figure 4-9 Mean counting and search times for alphanumeric items as a function of display density, with and without color coding [Smith, 48]. Note the marked reduction in mean time with color coding.

Auditory Codes

In a way, every sound has more than a single attribute or dimension, these including, according to the situation, combinations of frequencies, intensities of various frequencies, total intensity, any modulation of intensity, duration, and time intervals between signals. For example, the Morse code consists of a given spectrum, but differentiates by a *time* dimension, namely, the intervals between signals. Individual piano notes consist of relatively pure tones (plus their overtones), but the "touch" of the pianist produces variations in duration and intensity. In the use of auditory codes one can basically hold all dimensions constant except one, and so produce a variation in only one dimension, as in the Morse code. Where the amount of information in individual signals should be high, it is possible to use combinations of dimensions. This was demonstrated by Pollack and Ficks [44] in their study in which individual signals were identified by a combination of six dimensions (frequency, intensity, rate of interruption, on-time fraction, total duration, and spatial location). They found that it was possible for subjects to identify 150 combinations (7.2 bits) with high accuracy.

A different twist in the use of multidimensional auditory codes was tried by Mudd [37] when he used auditory cues as an aid in directing visual attention to certain areas of a visual display. The display consisted of 32 dials, each of which had a null position. At different times a pointer on one dial was caused to deviate from this null position, and the operator was to identify the deviant dial and throw a corresponding switch, which automatically recorded the time required to identify the deviant dial. Each of the 32 dials was the deviant one twice during an experimental run. Different auditory cues were used individually and in combination as codes to direct visual attention to the sector of the panel that contained the deviant dial. A sketch of the panel and of one of the combinations of auditory codes is shown in Figure 4-10. The auditory

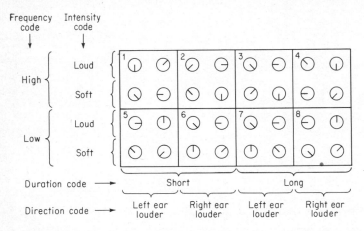

Figure 4-10   Illustration of some of the auditory coding cues used as aids in a visual monitoring task of identifying deviant dials on a display panel. Under different conditions, the auditory cues were presented individually and in combinations. For example, the intensity codes and direction codes, when used by themselves, differentiated the top and bottom halves and the left and right halves, respectively. [From Mudd, 37.]

coding dimensions were frequency, intensity, duration, and direction (difference in intensity between signals to left and right ear). The results are shown in Figure 4-11, this giving mean response times in identifying the deviant dials when various auditory cues were used as aids in the visual searching task. Search time was reduced from 8.84 sec with no cues, to 6.10 with one, to 4.10 with two. But the two cues represented a point of diminishing returns; the addition of a third and fourth coding dimension generally did not reduce search time further. With respect to the individual sound dimensions, it can be seen that frequency was most effective (5.70 sec) and direction least (6.94 sec).

General Guidelines in the Use of Coding Systems

Generalizations in an area such as the use of codes are pretty treacherous, especially since this ball park is a very complicated one. (We have seen above a few instances of apparent inconsistencies in research findings.)

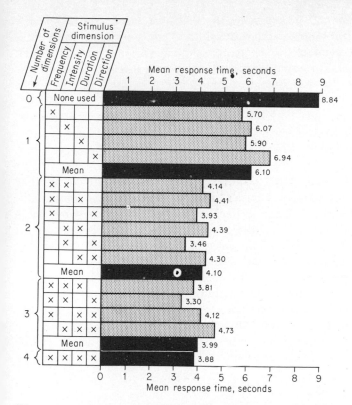

Figure 4-11   Mean response times in identifying deviant dials on visual display when various auditory cues were used singly and in combinations as aids in directing visual attention to the section of the display that contained the deviant dial. The × or ×'s at the left of each bar show the dimension or dimensions used in the experimental condition. The dark bars distinguish the means of all values where 1, 2, or 3 dimensions were used, as well as the values when no cueing dimensions were used and when all four were used. For an indication of the way in which they were used, refer back to Figure 4-10. [Adapted from Mudd, 37.]

Further, in any given situation it may be necessary to effect a trade-off of one advantage for another. There may be, however, a few guidelines that clearly stick out, and others that might be teased out, of the available evidence, that may have some generality.

Detectability of codes.   To begin with, any stimulus used in coding information needs to be detectable; specifically, it has to be of such a nature that it can be sensed by the sensing mechanism in the situation at hand. In the lingo of the psychologists this is an "absolute threshold." But "absolute" thresholds are not absolute. They vary from individual to individual, and even for the same individual at various times, and vary with the situation. For example, noise, competing stimuli, time limitations, physiological or psychological stress, etc., can raise the threshold of detectability.

Discriminability of codes. Further, every code symbol, even though detectable by the sensory mechanism, needs to be discriminable from other code symbols. This may be a problem where differences along some stimulus dimension are to be used, such as sound frequency and brightness. As discussed above, stimulus dimensions vary in the number of levels that can be identified on an absolute basis. In this connection, the *degree* of difference between adjacent stimuli may have some influence on the effectiveness of a coding system, even if all differences used are reasonably discriminable. One phase of the study cited earlier will illustrate this [Mudd, 37]. In that study three different levels of interstimulus difference were used with each of the four auditory cueing dimensions. Thus, with the intensity dimension, three pairs of intensities were used at different times, one pair having a wide difference, another a moderate difference, and the third a still smaller difference (but one which still resulted in fairly adequate discrimination). The degree of difference was found generally to result in differences in response time, as indicated below:

Interstimulus difference in loudness	*Mean response time, sec*
Largest	4.57
Average	5.68
Smallest	6.81

Especially under conditions of stress or time pressure, the greater the distinction between two stimuli, the more likely is the distinction to be recognized quickly, at least in many circumstances. A couple of qualifications must be added. First, any stimulus must be well within the range of human sensitivity, for example, intensity. Second, this apparently does not apply necessarily to all dimensions. In this study, for example, the frequency dimension, when used by itself, did not bring about this type of differential effect; even the smallest difference used was about as effective as the largest.

Compatibility of codes. *Compatibility* is a very generalized concept that has substantial applicability to human factors engineering. For our general purposes it might be defined thus: Compatibility refers to the spatial, movement, or conceptual relationships of stimuli and of responses, individually or in combination, which are consistent with human expectations. This concept is most straightforward in the case of stimulus-response (S-R) compatibility, and will be discussed further in Chapter 9. The concept of compatibility is also applicable where there is simply information transfer (as in the use of certain coding systems) in the absence of any corresponding physical response. In this connection, the degree of compatibility of coding systems generally tends to be maximum when the task at hand requires a minimum amount of information transformation—encoding or decoding. Thus, we should take advantage of associations that are already built into the repertoire of people; these may be either natural or learned associations. To refer again to traffic lights, if someone were foolish enough to change the system from red, yellow, and green to, say, violet, chartreuse, and azure, we should really be loused up—*because* of the need to recode from one system to another, at least until we learned the new one.

As an interesting twist, there presumably are some cross-modality compatibility relationships, as suggested by a study by Mudd [38]. He transmitted pairs of sound signals to subjects by the use of earphones, these two differing only in one dimension at a time. (Four dimensions were used at different times.) The subject faced a pegboard and was asked to consider a plug in the center as a reference point in space to represent the first (standard) stimulus of the pair; when the second sound was presented, he was asked to transfer the plug from the center (reference) hole to any other hole on the pegboard that he considered to represent the second sound. An analysis was then made of the directions and distances from the center that the plugs were moved by the subjects for the various pairs of sounds, and, lo and behold, there was at least moderate consistency across subjects. The primary *stereotypes* (i.e., conceptual association of auditory dimensions with spatial dimensions) are given below:

Stimulus dimension	Specific stimulus	Stereotype response
Frequency	High frequency	Up (or right)
	Low frequency	Down (or left)
	(of the two associations, the up-down was more dominant than the right-left)	
Intensity	High intensity	Up
	Low intensity	Down
Duration		No systematic pattern
Direction	High intensity, right ear	Right
	High intensity, left ear	Left

In connection with these particular stereotypes (associations of auditory and spatial dimensions), the associations were used as the basis for developing the auditory coding pattern used by Mudd [37] as an aid in a visual searching task (Figure 4-10). A quick test for you, the reader, may illustrate the implications of compatibility in coding:

HCIHW NAC UOY DAER RETSAF?
WHICH CAN YOU READ FASTER?

Symbolic associations of codes.   Closely related to compatibility as a coding principle—in fact, perhaps simply a special case of compatibility—is the principle of using symbolic associations in the selection of codes, especially in the use of codes that are physical representations of physical objects. For example, maps sometimes have airplanes to represent airports, some road signs represent the road ahead (rather than saying "right turn"), and the knobs of control devices sometimes are shaped like the devices they control, such as a miniature wheel being used as the handle of a landing-gear lever.

Standardization of codes.   When coding systems are to be used by different people in different situations (as in the case of traffic signs) their standardization will facilitate their use when people shift from one situation to another.

Use of multidimensional codes.   In very general terms, the use of two or more coding dimensions in combination tends to facilitate the transfer of information to human beings, probably especially so where there is complete redundancy, and perhaps less so with partial redundancy.

Discussion

One thing that keeps cropping up in various investigations of coding systems is the fact that the context in which a coding system is used, particularly task and environmental variables, sometimes have a bearing on the use of the coding system. It would be nice if one could set forth a clear-cut, unambiguous set of specifications on when to use what code system. Life is not that simple, and if one factor stands out above all others in the entire human factors engineering domain, it is that invariant rules are few and the answers to many questions may well be qualified by "it depends." The selection of codes is no exception to this rather universal qualification. But, with this skeptical point of view, Table 4-3 may provide some leads to some of the pros and cons of a few types of coding dimensions.

ORGANIZATION OF INFORMATION

As pointed out earlier, a display does not transmit information as such, but rather presents stimuli which may be meaningful to the receiver. Thus, when we discuss the *organization* of information, this is really

Table 4-3 Summary of Certain Visual and Auditory Coding Methods
(Numbers refer to number of levels which can be discriminated on an absolute basis under optimum conditions.)

Alphanumeric	Single numerals, 10; single letters, 26; combinations, unlimited. Good; especially useful for identification; uses little space if there is good contrast. Certain items easily confused with each other.
Color	Hues, 9; hue, saturation, and brightness combinations, 15–24. Particularly good for searching and counting tasks; poorer for identification tasks; trained observers can use many codes (up to 24). Affected by some lights; problem with color-defective individuals.†‡
Geometric shapes	15 or more. Generally useful coding system, particularly in symbolic representation; good for CRTs. Shapes used together need to be discriminable; some sets of shapes more difficult to discriminate than others.
Visual angle	24. Generally satisfactory for special purposes such as indicating direction, angle, or position on round instruments like clocks, CRTs, etc.§
Size of forms (such as squares)	5. Takes considerable space. Use only when specifically appropriate; preferably use less than 5.
Visual number	6. Use only when specifically appropriate, such as to represent numbers of items. Takes considerable space; may be confused with other symbols.
Brightness of lights	3–4. Use only when specifically appropriate. Preferably limit to two levels; weaker signals may be masked.¶
Flash rate of lights	4. Limited applicability. Preferably limit to two levels; combination of individual flashes and controlled time intervals may have special application, such as lighthouse signals and naval communications.
Sound frequency	5. For untrained listeners, use less than 5 levels; space widely apart, but avoid multiples and low and high frequencies; intensity should be 30 dB above threshold. Frequency changes easier to detect than single frequencies; combinations usually require training except for clearly distinguishable sounds such as bells, buzzers, and sirens.¶
Sound intensity	4. Preferably use less than 4. Intensity changes easier to detect than single intensities; for pure tones restrict to 1000–4000 Hz, but preferably use wide band.
Sound duration	Use clear-cut differences, preferably 2 or 3.
Sound direction	Difference in intensity to two ears should be distinct; particularly useful for directional information (i.e., right versus left).

† Feallock et al. [20].
‡ Jones [29].
§ Muller et al. [39].
¶ Morgan et al. [34].

a euphemistic way of referring to the organization of stimuli—such as their temporal characteristics and their number and type. It should be noted here, again, that human abilities to deal with a barrage of stimulus inputs depend very much on perceptual and mediation processes rather than exclusively on sensory processes.

Speed and Load of Stimuli

Load refers to the variety of stimuli (in type and number) to which the receiver must attend. Thus, if an individual has several different types of visual instruments, or several of the same type, to which he must give his attention, the load on the visual system will be greater than if there were fewer. On the other hand, *speed*, when used in this context, relates to the number of stimuli per unit of time or, conversely, to the time available per stimulus. Incidentally, one could contemplate speed and load for a single sensory modality (such as vision or hearing) or for combinations of the sensory modalities.

The effects of load and speed have been investigated in a number of experiments. In one study, for example, Mackworth and Mackworth [32] had a visual search task which utilized a display panel with 3 clocks at the top and 50 columns with numbers in certain positions of each. The clocks presented a three-digit number that increased its value one number at a time, this change being identified by a loud click. The chore of the subject was a rather complicated one that need not be described in detail, but basically consisted of a continuous comparison of the changing number of the clocks with the numbers in the columns; he then reported the column (i.e., *A*, *B*, *C*, etc.) in which the most nearly corresponding number appeared. Load was varied by controlling the number of columns to be used (5, 10, 15, 20, 30, 40, or 50). Speed was controlled by the rate at which the clocks kept changing. Subjects were scored on the number of errors made.

Some of the results are shown in Figure 4-12. In particular, this showed for fast speeds (6 decisions per minute) and slow speeds (3 decisions), the relationship between load and errors. The dotted lines (which reflect the general relationships) indicate that errors increase with load (even though the number of decisions per minute remains constant) and that speed also is related to errors.

Interestingly enough, in studies of load and speed it has been found that the arithmetic product of these two variables typically results in a linear relationship with performance. This was illustrated, for example, by one of a series of studies by Conrad [12], with a *clock-watching* task in which the subjects were to press a key as a pointer approached 12 or 6 o'clock on any of the clock dials used. In the experiment in question, under different circumstances two, three, or four dials were used, and

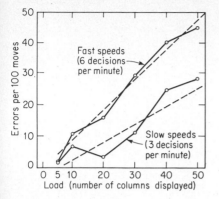

Figure 4-12 Effects of increasing load (number of channels used) upon errors. These data come from the study using a panel of various numbers of columns of comparison numbers. The fast and slow speeds were, respectively, 6 and 3 decisions per minute. [From Mackworth and Mackworth, 32.]

speed was varied. The relationship between the product of speed and load on the one hand, and omissions on the other, is shown in Figure 4-13. It can be seen that this relationship is essentially linear over the primary range of speed times load.

Speed stress and load stress. The patterns of results illustrated above have been confirmed in a number of studies. Such results have led Conrad to postulate the existence of *speed stress* and of *load stress* in situations such as these [15]. He suggests that *speed stress* is essentially a reaction on the part of a person working on a task, that has the effect of worsening his performance beyond what might be expected from the physical characteristics of the display. *Load stress*, on the other hand, changes the character of the task. As the number of signal sources (visual displays) is increased, more time is needed to make judgments simply because of the greater scanning coverage required.

Time-phasing of Signals

When stimuli are presented briefly and in close temporal sequence, perceptual failure may occur because of speed or load stress. Some of the aspects of timing of auditory signals have been investigated by Conrad in a series of studies [13 to 17] using the multidial displays mentioned earlier.

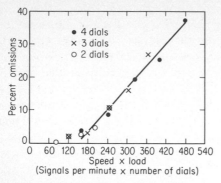

Figure 4-13 Relationship between speed × load and "omissions" in a clock-watching study. The load is the number of sources (in this case two, three, or four clock dials). The relationship is essentially linear over the primary range of speed × load values. [From data from Conrad, 12, as adapted by Mackworth and Mackworth, 32.]

In one of these studies [Conrad, 16], for example, four of his dials were used, signal speed being varied from 40 to 160 signals per minute. A detailed analysis was made of the number of occasions that a response was late (pressing a key corresponding to the dial *after* the pointer had passed the 12 or 6 o'clock positions). Further, this analysis was made in relation to the timing of adjacent stimuli and response. Considering two stimuli and their responses, one could envision this usual sequence of events (S = stimulus; R = response):

S_1, R_1, S_2, R_2

An analysis of the timing of the intervals between these events indicated that if stimuli are close together, or if they come in bunches, the responses to them frequently are missed, delayed, or otherwise affected. While absolute limits of minimum desirable interstimulus intervals are a bit hard to come by, it has been pointed out that where such an interval is shorter than, say, 0.5 sec, the stimuli are likely to be confused; in fact, an individual may then respond to the two as though they were one.

Self-pacing of signals. With frequently occurring signals, it is of course sometimes (perhaps usually) the case that their spacing is completely outside the control of the individual in question. In some circumstances, however, as with certain types of industrial machinery, the individual can control the rate at which materials (or other stimuli) are presented. In a study of self-pacing with the same type of multidial display [Conrad, 16], it was possible for the subjects to change the intersignal intervals, to delay signals or to speed them up. Without elaborating on the details, it can be said that the subjects typically tended to *slow down* the signals in the case of short intervals and to *speed up* the signals in the case of longer intervals. The data made it possible to compare the extent to which individuals did this. Those who did this most would produce a distribution of self-paced intervals which would be rather narrow; slowing down the signals which otherwise would come fast would increase those intersignal intervals, and speeding up the signals which otherwise would come slowly would reduce those intervals, which tended to cause the intervals to be relatively comparable. Those individuals who did this generally performed better than those whose self-paced intervals were more variable.

Discussion

The implication for equipment design of some of the research relating to speed, load, and timing of frequently occurring visual signals is relatively clear. Where visual signals would likely occur fairly rapidly, it would seem desirable (where possible) to provide that the information be presented over a limited number of sources (channels) rather than over many. In addition, the rate of presentation should be within acceptable bounds as far as human response performance is concerned. Further, where the spacing of signals to be presented can be controlled, one should avoid short intersignal intervals, the bunching of signals, and short intervals between signals following previous responses. Where feasible, it seems desirable to permit the individual to control the rate of signal input.

MULTIPLE–SENSORY INPUTS

The above discussion of load, speed, and related variables generally was restricted to frequently occurring visual signals of some general class. Sometimes, however, workers are subject to several inputs through the same sensory channel and to inputs through the various senses, especially vision and hearing. The relationships of the multiple inputs can be of several kinds, including the following: (1) *time sharing* (potentially meaningful and relevant information may be coming from different sources through the same sensory channel or through two or even more sensory channels); (2) *multiple-sensory channels for reinforcing* (two or more senses may be used to reinforce each other by transmitting identical or supporting information, usually simultaneously, or at least in very close temporal sequence); and (3) *noise* (noise generally refers to some irrelevant, usually undesirable, stimuli; although we usually think of noise as auditory in nature, the term is also applied to other irrelevant stimuli such as *visual* noise, or clutter).

There have been many studies of the effects of different inputs into the same sensory channel and of inputs from one sensory modality upon inputs from another. It is not feasible here to discuss these effects extensively, although a few points may be made.[5]

Time Sharing

Time sharing refers to situations in which the human being has two or more chores to which he has to alternate his attention. In a strict sense an individual cannot give simultaneous attention to two or more aspects of a situation. In performing various functions simultaneously, such as steering a car, controlling the accelerator, and keeping our eyes peeled on the traffic, we actually keep shifting our conscious attention from one to another, sometimes very rapidly. Time sharing can take many forms, such as receiving two or more sensory inputs, taking various physical actions, or combinations of these. Where these demands press the limits of individuals, load or speed stress, or both, may occur. In effect, then, time sharing can induce a form of load and speed stress.

Time sharing of visual inputs. The investigations of speed and load discussed above dealt with relatively homogeneous tasks and therefore did not involve time sharing. Many activities of the real world, however, do not conform to the pristine patterns of many laboratory tasks. This disparity led Weisz and McElroy [54] to study the effects of task-induced stress (really speed stress) on actually time-shared tasks that would be more representative of human operator activities in man-machine systems, including tasks that required storage and integration of information over time (which most laboratory tasks in this domain have not required). Since the details of their experimental procedures might serve as an overload to the reader, we shall try to boil the study down to its major features. The experiment involved five different response tasks to be carried out in response to different stimuli presented simultaneously by frames with a CRT. The range of stimuli consisted of variations of each of four geometric forms (i.e., rectangle, trapezoid, triangle, parallelogram); the stimuli were generated on a CRT by the use of a computer, and appeared in the guadrants of the CRT. Each form had seven variations, such as very tall, thin rectangles ranging down to low, wide rectangles. The five types of tasks (events) to be responded to were:

[5] For a more extensive discussion of this topic the reader is referred to Mowbray and Gebhard [36] and Howell and Briggs [28].

- *Mean:* Estimate of the *population mean* of rectangles in upper right quadrant, with the seven variations in rectangles considered as forming a scale. Every frame.
- *Form:* Identification of *form value* (i.e., the specific variation) of trapezoid in lower right quadrant. Every frame.
- *Pair:* Detection of pairs (and triplets) of identical forms in frame.
- *Run:* Remembering of location of pairs of stimuli and checking for appearance of a similar pair on the next frame in the same quadrant location (short-term memory required).
- *Line:* Searching the lower two quadrants for *extreme* variations of trapezoid or triangle in dotted-line form (rather than solid line).

These tasks were time shared, and after a breaking-in period the subjects performed the composite tasks for a series of 18 trials (9 on each of two days), these being varied at 3 rates, namely, 10 sec per frame, 7 sec per frame, and 4 or 5 sec per frame. The subjects recorded their responses by the use of special push buttons. Performance was measured by errors, including a composite (total) of errors in all tasks. Let us now see what gleanings there are from this study. Figure 4-14 shows the

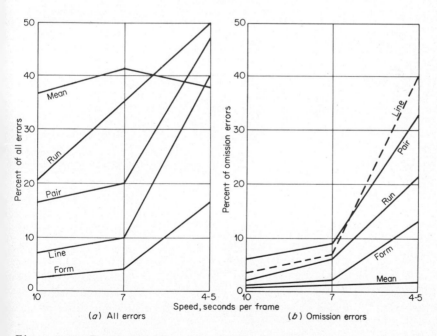

Figure 4-14 Percent of all errors *a* and omission errors *b* in tasks as a function of speed (in seconds per frame as presented to subjects). See text for description of tasks. [From Weisz and McElroy, 54.]

percent of all errors (*a*) and the percent of omissions (*b*) for each of the tasks. Certain points stand out from these figures. Obviously, the highest level of speed stress was accompanied by an increase in total error. (Although not shown in the figure, the total composite error index, as computed, increased from 18.4 to 22.4 and 32.2 percent.) But the interesting thing is that percent of all errors (*a*) on the *mean* task, despite the difficulty of the task, did not increase like the others, and the *form* task did not increase as much as the others. These two tasks also had the lowest percent of omission indexes (*b*), especially the *mean* task, its omission rate being virtually unaffected by speed stress.

Some speculations about the results by the investigators may cast some light on the effects of speed stress on time-shared visual-input tasks. To begin with, it should be noted that the *mean* and *form* tasks required a response for each frame, but the others did not; the others involved some temporal or spatial uncertainty which required search of the display to identify significant events, and were more "probabalistic" in the sense that an event calling for a response did *not* occur in every frame. Further, the *run* task (the one that suffered most from speed stress) also depended on short-term memory from one frame to the next. Thus, it appears that in time sharing of tasks such as these, speed stress does not affect all tasks equally, but rather affects particularly the tasks with greater uncertainty and those which depend on short-term memory.

Another relevant study is that conducted by Olson [40]. The experimental task was sharing visual attention between two tasks. One of these was a tracking task, consisting of a moving "road" that was curved and a wheel that controlled a pointer on the road; the object was to keep the pointer on the road. The other task was the identification of dial pointers (on anywhere from 6 to 18 dials) which deviated randomly from a neutral position. While the primary concern was more with other variables (dial arrangement, load, speed, etc.), one aspect of the study was related to the present topic of time sharing. Performance on the two tasks was correlated only to the extent of .20, which suggests the hypothesis that individuals tended to adopt their own priority strategies of giving primary attention to one task or the other. Whether the priority strategies are fairly common to most subjects because of the intrinsic nature of the task (as in the case of the study by Weisz and McElroy) or self-selected (as presumably in the study by Olson), it nonetheless appears evident that when the pressures of speed stress tax the capacities of people, something has to give, specifically performance on some of the time-shared tasks.

Time sharing of auditory inputs.   Essentially the same adverse effect of time sharing (i.e., of something having to give) is apparent in the case of auditory inputs, such as when two or more inputs occur simultaneously, overlap each other to some degree, or occur very close together in time. If an individual is listening, say, for verbal messages, and two messages occur at the same time, only one of them usually will get through. If, however, there is a slight lag in one, the first typically is identified more accurately than the second [Webster and Thompson, 53]. But if there is a distinct intensity difference, the second being the more intense, it will tend to have priority on the receiver's attention, even though it may lag after the first by as much as 2 sec.

The adverse effect of simultaneous messages occurs even when only one is relevant and needs to be attended to [Broadbent, 8]. There is also some evidence that when a competing (and irrelevant) message is relatively similar in content and words to a relevant message, the interference effects on the relevant message are greater than if the irrelevant message is rather different [Peters, 41].

Time sharing of auditory and visual sensory channels. Since vision and audition are the most important senses for receiving information from displays, it would be particularly in order to compare these two senses in the degree to which they are influenced by interference from each other and from other possible work activities. Present inklings suggest that when both visual and auditory inputs are being time shared, the auditory channel is more resistant to interference effects than the visual channel [Mowbray, 35].

Discussion. It is fairly evident that there are bounds beyond which the time sharing of sensory inputs typically results in some degradation of performance. When circumstances permit (and they sometimes do not), efforts should, of course, be made to so manipulate the situation that degradation will at least be minimized if not eliminated. As in other contexts, we need to accept generalizations with caution. From this point of view, research evidence—some cited above and some not—seems to suggest a few general guidelines, although these may not be applicable across the board:

1. Where possible, the number of potentially competing sources should be minimized.
2. Where time sharing is likely to impose speed or load stress, the receiver should be provided with some inklings about priorities, so his strategy in giving attention to first things first can take these into account.
3. Where possible, the requirements for use of short-term memory and for dealing with low-probability events should be minimized.
4. Where possible, input stimuli which require individual responses should be separated temporally and presented at such a rate that they may be responded to individually. Extremely short intervals

(say, less than 0.5 or 0.25 sec) should be avoided if at all possible. Where possible, the receiver should be permitted to control the input rate.

5. Where a choice of sensory modalities is feasible in a situation where a sensory input has competition, the auditory sense is generally more durable and is less influenced by other inputs.

6. Some means of directing attention to relevant and more important sources will increase the likelihood of their priority in the receiver's attention; for example, in some situations visual stimuli (such as lights) might be used as advance cues to the location of relevant auditory sources, or vice versa.

7. Where two or more auditory inputs might have to be time shared, it would be desirable to schedule relevant messages or signals so they do not occur simultaneously, to separate physically the sources (such as speakers) of relevant versus irrelevant messages [Poulton, 45], to filter out (if possible) any irrelevant messages, and where they cannot be filtered out, to make them as different as possible from those that are relevant such as by making the relevant stimuli more intense or using clearly distinct spectral characteristics.

8. Especially when repetitive manual tasks are time shared with non-related sensory inputs, the greater the learning of the manual task, the less will be its possible effect on the reception of the sensory input.

Multiple Sensory Channels for Reinforcing

Two or more sensory channels, such as vision and audition, can be used in combination, the one reinforcing the other in transmission of information.

Redundant coding with different senses. While the evidence about some aspects of human behavior is a bit inconclusive, there is virtually no question but that the use of redundant visual and auditory coding (the simultaneous presentation of identical information to both senses) increases the odds of reception of the information. Different specific studies could be used to illustrate this point. We shall use a vigilance study to illustrate this [Buckner and McGrath, 10].

Three of the conditions in the study were as follows:

· Visual task: Subjects detected an increment in the brightness of an intermittent light that was on for 1 sec and off for 2.

· Auditory task: Subjects detected an increment in the loudness of an intermittent 750 Hz tone that was on for 1 sec and off for 2.

· Combined visual and auditory task: The visual and auditory tasks were combined, and both visual and auditory signals occurred simultaneously.

In this task the subjects were supposed to detect 24 signals during 60-min vigilance watches; during any watch, only visual or auditory or combined signals were used. Figure 4-15 shows the percent of signals

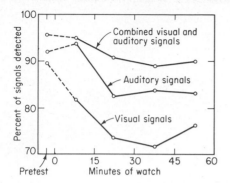

Figure 4-15 Performance on a visual, auditory, and combined visual-auditory vigilance task. The vertical scale shows the percentage of signals that were detected during certain periods of a 1-hr watch. [From Buckner and McGrath, 10.]

of each type that were detected. The consistent advantage of the combined signal is evident throughout the hour watch; performance in this task did not deteriorate nearly as much as in the individual auditory and visual tasks.

In another study that reinforces this point, subjects were to press one of three keys (left, center, or right) in response to a visual signal (red, orange, or green, respectively), an auditory signal (100, 700, or 5000 Hz), or a combined signal [Klemmer, 30]. The percentages of correct responses were as follows: visual signal, 89 percent; auditory signal, 91 percent; and combined visual and auditory signal, 95 percent.

In connection with the use of two or more sense modalities in combination, Symons [52] presents some fairly persuasive evidence that the sensitivity of (at least) the visual sense is enhanced by the simultaneous stimulation of various other sense organs. This increased sensitivity may then be one of the factors that tends to enhance the reception of such redundant stimuli. Although there seems to be general support for the principle that the human reception of information is enhanced with redundant coding for the visual and auditory senses, such a principle should not be applied indiscriminantly, since there are exceptions to most principles. In this case, for example, there might be situations where one of the modalities might itself be overloaded, in which case it might be wise not to overburden it further with a redundant coding system.

Use of one channel in cueing another. A second way in which two or more sensory channels can be used in combination is by the use of signals from one channel to serve as cues in facilitating the use of the other sensory channel. In one experiment, for example, verbal messages could be transmitted over any of three auditory channels, referred to as A, B, and C [Speith, Curtis, and Webster, 50]. A small jewel lamp was used to indicate over which auditory channel a message addressed to the operator was about to appear; this cue virtually eliminated the misidentification of channels. A reverse twist of this is represented by the study by Mudd [37], previously cited, in which auditory cues were found to facilitate a visual searching task.

Noise

In most situations noise—auditory, visual, or otherwise—tends to cause degradation in the reception of meaningful stimuli. The effects of auditory noise will be dealt with in Chapter 16.

SUMMARY

1. Information serves either of two roles in systems: (*a*) it is a necessary part of the process of bringing about the output, or (*b*) in certain systems (such as a communications system) it serves as both input and output.
2. Information theory deals with the measurement of information. The unit of measure of information is the *bit;* in the simple case the number of bits is equal to the logarithm to the base 2 of the number of possible alternatives.
3. Information theory (and information measurement) has been applied in such contexts as stimulus dimensions (the information in different visual and other sensory dimensions), language, motor activities, and human memory.
4. The original sources of information about objects and events in our environment are called *distal stimuli.* In turn, *proximal stimuli* consist of some form of energy (light, sound waves, etc.) coming from the distal stimuli; the energy of the proximal stimuli stimulate the sensory mechanisms.
5. *Channel capacity* has been characterized as the limit of information that can be transmitted in a given context (with a certain stimulus dimension) and also as the upper limit of information that can be received in any and all dimensions and sensory modalities. Whether there really is an upper limit in the latter sense may not be known,

but there are, in any event, certain factors that influence the amount of information actually received in given situations.

6. Sensory inputs from most displays involve the coding of stimuli of some stimulus dimension.

7. For any given sensory modality, such as vision, the amount of information that can be transmitted varies with the method of coding. The use of multidimension codes typically results in greater information transmission than single-dimension codes.

8. For some coding dimensions, information transmission is increased with greater degrees of difference between adjacent stimuli (such as loudness of sounds).

9. Following are some guidelines for selecting code systems: The stimuli used need to be *detectable* by the sense organ and *discriminable* one from another, and preferably should be *compatible*, have *symbolic associations*, and be *standardized*.

10. *Load* in information transmission refers to the variety of stimuli to which an individual should be sensitive. *Speed* refers to the number of stimuli per unit of time. The information-processing abilities of people are affected adversely by both load and speed above certain levels.

11. Information reception usually is improved if the information is presented through a limited number of information sources (minimum load).

12. Where there must be time sharing of sensory inputs, human performance usually is better when the signals are separated temporally (preferably by 0.5 sec or more), when the receiver can control the input rate (self-pacing), and when the receiver has some method of identifying the more important input (if he has to make a choice); auditory signals generally are more durable than visual signals.

13. The simultaneous presentation of identical information by two sensory modalities usually increases the probability of reception.

14. Signals presented on one channel (such as audition) can serve as cues to facilitate the use of another channel.

15. The time sharing of visual tasks adversely affects primarily those tasks with greater uncertainty and those which depend on short-term memory; thus speed stress does not affect all tasks equally.

REFERENCES

1. Abramson, N.: *Information theory and coding*, McGraw-Hill Book Company, New York, 1963.
2. Alluisi, E. A., and P. F. Muller, Jr.: *Rate of information transfer with seven symbolic visual codes: motor and verbal responses*, USAF, WADC, TR 56–226, May, 1956.

3. Alluisi, E. A., P. F. Muller, Jr., and P. M. Fitts: An information analysis of verbal and motor responses in a forced-paced serial task, *Journal of Experimental Psychology*, 1957, vol. 53, pp. 153–158.

4. Anderson, N. S., and P. M. Fitts: Amount of information gained during brief exposures of numerals and colors, *Journal of Experimental Psychology*, 1958, vol. 56, pp. 362–369.

5. Annett, J., C. W. Golby, and H. Kay: The measurement of elements in an assembly task—the information output of the human motor system, *Quarterly Journal of Experimental Psychology*, 1958, vol. 10, pp. 1–11.

6. Attneave, F.: *Applications of information theory to psychology*, Holt, Rinehart and Winston, Inc., New York, 1959.

7. Beebe-Center, J. G., M. S. Rogers, and D. N. O'Connell: Transmission of information about sucrose and saline solutions through the sense of taste, *Journal of Psychology*, 1955, vol. 39, pp. 157–160.

8. Broadbent, D. E.: Listening to one of two synchronous messages, *Journal of Experimental Psychology*, 1952, vol. 44, pp. 51–55.

9. Broadbent, D. E.: *Perception and communication*, Pergamon Press, New York, 1958.

10. Buckner, D. N., and J. J. McGrath: *A comparison of performances on single and dual sensory mode vigilance tasks*, Human Factors Research, Inc., Los Angeles, Calif., TR 8, ONR Contract Nonr 2649(00), NR 153–199, February, 1961.

11. Conover, D. W., and C. L. Kraft: *The use of color in coding displays*, USAF, WADC, TR 55–471, October, 1958.

12. Conrad, R.: Speed and load stress in a sensori-motor skill, *British Journal of Industrial Medicine*, 1951, vol. 8, pp. 1–7.

13. Conrad, R.: Missed signals in a sensori-motor skill, *Journal of Experimental Psychology*, 1954, vol. 48, pp. 1–9.

14. Conrad, R.: Adaptation to time in a sensori-motor skill, *Journal of Experimental Psychology*, 1955, vol. 49, pp. 115–121.

15. Conrad, R.: Some effects on performance of changes in perceptual load, *Journal of Experimental Psychology*, 1955, vol. 49, pp. 313–332.

16. Conrad, R.: The timing of signals in skill, *Journal of Experimental Psychology*, 1956, vol. 51, pp. 365–370.

17. Conrad, R., and B. A. Hille: Self-pacing performance as a function of perceptual load, *Journal of Experimental Psychology*, 1957, vol. 53, pp. 52–54.

18. Engen, T., and C. Pfaffmann: Absolute judgments of odor quality, *Journal of Experimental Psychology*, 1960, vol. 59, pp. 214–219.

19. Eriksen, C. W.: Location of objects in a visual display as a function of the number of dimensions on which the objects differ, *Journal of Experimental Psychology*, 1952, vol. 44, pp. 56–60.

20. Feallock, J. B., J. F. Southard, M. Kobayashi, and W. C. Howell.: Absolute judgments of colors in the Federal Standards System, *Journal of Applied Psychology*, 1966, vol. 50, pp. 266–272.

21. Fitts, P. M.: The information capacity of the human motor system in controlling the amplitude of movement, *Journal of Experimental Psychology*, 1954, vol. 47, pp. 381–391.

22. Garner, W. R.: An informational analysis of absolute judgments of loudness, *Journal of Experimental Psychology*, 1953, vol. 46, pp. 373–380.

23. Geyer, B. H., and C. W. Johnson: Memory in man and machines, *General Electric Review*, March, 1957, vol. 60, no. 2, pp. 29–33.

24. Hake, H. W.: "A note on the concept of 'channel capacity' in psychology," in H. Quastler (ed.), *Information theory in psychology*, The Free Press of Glencoe, Ill., Chicago, 1954, pp. 248–253.

25. Hake, H. W., and W. R. Garner: The effect of presenting various numbers of discrete steps on scale reading accuracy, *Journal of Experimental Psychology*, 1951, vol. 42, pp. 358–366.

26. Halsey, R. M., and A. Chapanis: On the number of absolutely identifiable spectral hues, *Journal of the Optical Society of America*, 1951, vol. 41, pp. 1057–1058.

27. Hitt, W. D.: An evaluation of five different abstract coding methods, *Human Factors*, July, 1961, vol. 3, no. 2, pp. 120–130.

28. Howell, W. C., and G. E. Briggs: *Information input and processing variables in man-machine systems: A review of the literature*, NAVTRADEVCEN, Port Washington, N.Y., TR 508–1, October, 1959.

29. Jones, M. R.: Color coding, *Human Factors*, 1962, vol. 4, pp. 355–365.

30. Klemmer, E. T.: Time sharing between frequency-coded auditory and visual channels, *Journal of Experimental Psychology*, 1958, vol. 55, pp. 229–235.

31. Kristofferson, A. B.: *A time constant involved in attention and neural information processing*, Bolt, Beranek, and Newman, Inc., Cambridge, Mass., 1966.

32. Mackworth, N. H., and J. F. Mackworth: Visual search for successive decisions, *British Journal of Psychology*, 1958, vol. 49, pp. 210–221.

33. Miller, G. A.: The magical number seven, plus or minus two: Some limits on our capacity for processing information, *Psychological Review*, 1956, vol. 63, pp. 81–97.

34. Morgan, C. T., J. S. Cook, III, A. Chapanis, and M. W. Lund (eds.): *Human engineering guide to equipment design*, McGraw-Hill Book Company, New York, 1963.

35. Mowbray, G. H.: Simultaneous vision and audition: The detection of elements missing from overlearned sequences, *Journal of Experimental Psychology*, 1952, vol. 44, pp. 292–300.

36. Mowbray, G. H., and J. W. Gebhard: *Man's senses as information channels*, Report CM–936, Johns Hopkins University, Applied Physics Laboratory, Silver Spring, Md., May, 1958.

37. Mudd, S. A.: *The scaling and experimental investigation of four dimensions of pure tone and their use in an audio-visual monitoring problem*, unpublished Ph.D. thesis, Purdue University, Lafayette, Ind., June, 1961.

38. Mudd, S. A.: Spatial stereotypes of four dimensions of pure tone, *Journal of Experimental Psychology*, 1963, vol. 66, no. 4, pp. 347–352.

39. Muller, P. F., Jr., R. C. Sidorsky, A. J. Slivinske, E. A. Alluisi, and P. M. Fitts: *The symbolic coding of information on cathode ray tubes and similar displays*, USAF, WADC, TR 55–375, October, 1955.

40. Olson, P. L.: *Display arrangement, number of channels and information speed as related to operator performance*, unpublished Ph.D. thesis, Purdue University, Lafayette, Ind., July, 1959.
41. Peters, R. W.: *Competing messages: the effect of interfering messages upon the reception of primary messages*, USN School of Aviation Medicine, Report NM 001 064.01.27, 1954.
42. Pierce, J. R., and J. E. Karlin: Reading rates and the information rate of a human channel, *Bell Telephone Technical Journal*, 1957, vol. 36, pp. 497–516.
43. Pollack, I.: The information of elementary auditory displays, *Journal of the Acoustical Society of America*, 1952, vol. 24, pp. 745–749.
44. Pollack, I., and L. Ficks: Information of multidimensional auditory displays, *Journal of the Acoustical Society of America*, 1954, vol. 26, pp. 155–158.
45. Poulton, E. C.: Listening to overlapping calls, *Journal of Experimental Psychology*, 1956, vol. 52, pp. 334–339.
46. Raisbeck, G.: *Information theory: an introduction for scientists and engineers*, The M.I.T. Press, Cambridge, Mass., 1963.
47. Shannon, C. E., and W. Weaver: *The mathematical theory of communication*, The University of Illinois Press, Urbana, 1949.
48. Smith, S. L.: *Display color coding for visual separability*, MITRE Corp., Bedford, Mass., MITRE Report MTS-10, August, 1963.
49. Smith, S. L., and D. W. Thomas: Color versus shape coding in information displays, *Journal of Applied Psychology*, 1964, vol. 48, pp. 137–146.
50. Speith, W., J. F. Curtis, and J. C. Webster: Responding to one of two simultaneous messages, *Journal of the Acoustical Society of America*, 1954, vol. 26, pp. 391–396.
51. Steinbuch, K.: *Information processing in man*, paper presented at IRE International Congress on Human Factors in Electronics, Long Beach, Calif., May, 1962.
52. Symons, J. R.: The effect of various heteromodal stimuli on visual sensitivity, *Quarterly Journal of Experimental Psychology*, November, 1963, vol. 15, pt. 4, pp. 243–251.
53. Webster, J. C., and P. O. Thompson: Responding to both of two overlapping messages, *Journal of the Acoustical Society of America*, 1954, vol. 26, pp. 396–402.
54. Weisz, A. Z., and L. S. McElroy: *Information processing in a complex task under speed stress*, Decision Sciences Laboratory, Electronics System Division, AFSC, USAF, Report ESD-TDR 64–391, May, 1964.
55. Welford, A. T.: The measurement of sensori-motor performance: Survey and reappraisal of twelve years' progress, *Ergonomics*, 1960, vol. 3, pp. 189–230.
56. Wiener, N.: *Cybernetics*, John Wiley & Sons, Inc., New York, 1948.
57. *Munsell book of color*, Munsell Color Co., Baltimore, 1959.
58. New Federal Standard on Colors, *Journal of the Optical Society of America*, 1957, vol. 47, pp. 330–334.

There are of course many sources of information, i.e., distal stimuli, that people should receive which cannot adequately be sensed *directly* by human sensory and perceptual processes and for which displays of some type should be used. In this chapter we shall begin by a rather general discussion of displays and shall then discuss and illustrate particular types of visual displays. In the next chapter we shall deal with auditory and tactual displays.

WHEN DISPLAYS SHOULD BE USED

Some of the circumstances in which information should be presented indirectly, by displays, are the following:

1. When distal stimuli are of the type that humans generally can sense, but cannot sense adequately under the circumstances because of such factors as:

    (a) Stimuli at or below threshold values (e.g., too far, too small, or not sufficiently intense) that need to be amplified by electronic, optical, or other means

    (b) Stimuli that require reduction for adequate sensing (e.g., large land areas converted to maps)

    (c) Stimuli embedded in excessive noise that generally need to be filtered or amplified

    (d) Stimuli far beyond human sensing limits that have to be converted to another form of energy for transmission (e.g., by radio and TV) and then reconverted to the original form or converted to another form

    (e) Stimuli that need to be sensed with greater precision than people can discriminate (e.g., temperature, weights and measures, and sound)

    (f) Stimuli that need to be stored for future reference (e.g., by photograph and tape recorder)

    (g) Stimuli of one type that probably can be sensed better or more conveniently if converted to another type in either the same sensory modality (e.g., graphs to represent quantitative data) or a different modality (e.g., auditory warning devices)

(*h*)   Information about events or circumstances that by their nature virtually require some display presentations (such as emergencies, road signs, and hazardous conditions)

2.   When distal stimuli are of the type that humans generally cannot sense or that are beyond the spectrum to which humans are sensitive, and so have to be sensed by sensing devices and converted into coded form for human reception (e.g., certain forms of electromagnetic energy and ultrasonic vibrations)

In these and other types of circumstances, it may be appropriate to transmit relevant information (stimuli) *indirectly* by some type of display. For our purposes we shall consider a display to be any method of presenting information indirectly, either in *reproduced* or *coded* (symbolic) form. If a decision is made to use a display, there may be some option regarding the sensory modality and the specific type of display to use, since the method of presenting information can influence, for better or worse, the accuracy and speed with which information can be received.

ERRORS IN USE OF DISPLAYS

The errors that can occur in the use of displays were first dramatized by the results of a survey by Fitts [24] of errors made by United States Air Force pilots in responding to instruments and signals. Of the types of errors made, the following were the most frequent: misinterpreting multirevolution instruments, 18 percent; misinterpreting direction of indicator movement, 17 percent; failing to respond to warning lights or sounds, to radio-range signals, etc., 14 percent; and errors from poor legibility, 14 percent. This survey and other evidence of errors in the use of displays were the instigations for a substantial amount of research in the intervening years that has been directed toward the improvement of display design to minimize such errors.

Although substantial progress has been made in this direction in the last quarter century, it should be recognized that even the best designed displays do not eliminate all human errors in their use. Human fatigue, shifting attention, lack of motivation, limited sensory and perceptual skills, and other factors will probably always lead to some human error. The use of displays that are optimally designed in terms of human considerations, however, can at least help to minimize errors.

MAJOR TYPES AND USES OF DISPLAYS

Displays can be generally described as either *dynamic* or *static*. Dynamic displays are those that continually change or are subject to change through time, and include the following types: displays that depict the status or condition of some variable, such as temperature and pressure gauges, speedometers, and altimeters; certain CRT displays such as radar, sonar, TV, and radio-range signal transmitters; displays that present intentionally transmitted information, such as record players, TV, and movies; and those that are intended to aid the user in the control or setting of some variable, such as the temperature control of an oven. (It might be observed, incidentally, that there are some devices that do double duty as both displays and controls; this is especially the case with devices used for making settings, such as an oven control.)

Static displays, in turn, are those that remain fixed over time. Most such displays are visual, such as signs, charts, graphs, labels, and various forms of printed or written material.

Uses of Displays

Some of the primary uses of displays are given below, expressed in terms of the types of information obtained from them.

- *Quantitative information:* Using a display to obtain the actual quantitative value of some variable. In most instances the variable is a dynamic, changeable one, such as temperature; various types of measuring devices and nomographs, however, provide quantitative information, but of a static nature.

- *Qualitative information:* Using a display to obtain information about the approximate value of some continuous changeable variable, or about its trend, rate of change, or direction of deviation from a given value. Examples are observing the fluctuations of a speedometer around some desired speed and listening to sonar returns on a ship to determine if some other object is approaching or going away.

- *Check information:* Using a display to determine if the value of a continuous changeable variable is normal, or within an acceptably normal range, as in the case of pressure gauges.

- *Status and warning information:* Using a display to identify the specific status, or condition, of some system, component, or situation. Typically each status or condition is independent, or discrete, as in the use of on and off (dichotomous) indications, emergency and warning devices, stop-caution-go lights, and indications of independent conditions of some class such as a TV channel. In certain instances, however, discrete display indications may actually reflect distinct ranges of some underlying continuous variable, such as a red light indicating the discharge of a battery. Thus, some such indications tend to approximate qualitative or check information.

- *Tracking information:* Using a display to obtain information about the movement of a "target" in a tracking task. The target represents the desired output of the system. It may be an actual target or a representation of a changing value that needs to be matched by a response with the use of some control mechanism.

· *Time-phased information:* Using a display to receive meaningful pulsed or time-phased signals, e.g., signals that are controlled in terms of duration of the signals and of intersignal intervals, and of their combinations, such as the Morse code and blinker lights.

· *Pictorial and graphic information:* Use of pictorial or graphic representations of objects, areas, or other configurations. Certain displays may present dynamic images, such as TV or movies, or may present symbolic representations, such as heart beats shown on an oscilloscope. Others may present static information, such as photographs, maps, charts, diagrams, blueprints, and graphic representations such as bar graphs and line graphs.

· *Identification information:* Using a display to identify some condition, situation, or object, such as the identification of hazards, traffic lanes, and color-coded pipes. The identification usually is in coded form.

· *Alphanumeric and symbolic information:* Use of verbal, numerical, and other related coded information in many display forms such as signs, labels, placards, instructions, music notes, printed and typed material, and computer print-outs. Although in most instances such information is static, in certain circumstances it may be dynamic, such as news bulletins displayed by moving lights on a building.

· *Speech and related auditory information:* The use of radios, telephones, record players, or other communication systems to transmit speech, music, and related auditory signals. (By a slight stretch of the term we can consider the output components of such systems as "displays.")

We probably could set forth some other types, but the above will cover most of the types of information that are presented by displays. The kinds of displays that would be preferable for presenting certain types of information are virtually specified by the nature of the information in question, but for presenting most types of information there are options regarding the kinds of displays and certainly about the specific features of the displays. In this connection, considerable research has been carried out relating to the effects of display designs on human performance in their use. It is not feasible here to summarize this research or to illustrate the many types of displays that have been developed. For at least certain classes of displays, however, the results of some pertinent research will be summarized briefly to illustrate the effects of differences in design upon human performance, and, to a degree, to illustrate the emergence of principles of design that are conducive to adequate human performance. This chapter is largely restricted to a consideration of visual display design as it relates to receiving the information presented. Considerations of displays for which the display-control relationship is important will be deferred to Chapter 11.

QUANTITATIVE VISUAL DISPLAYS

The most extensive research in the domain of information displays has been carried out with dynamic quantitative visual displays, such research generally dealing with the basic shape or configuration and with the specific features such as scale size and markings. Some of the types of displays used for presenting dynamic quantitative information are shown in Figure 5-1. There are three basic types of quantitative scales, namely, (1) fixed scales with moving pointers, (2) moving scales with fixed pointers (or in some cases lubber lines), and (3) counters (in which the numbers of mechanical counters click into position, as mileage readings on many speedometers).

Readability of Quantitative Scales

Studies have been carried out regarding the readability of different types of quantitative scales. Although there have been some apparent differences in the results of these studies, such differences generally can be attributed to the particular experimental conditions.

Comparison of scales. Of the studies on the readability of different types of scales, one involved a comparison of five types, these being (1) vertical as shown in d of Figure 5-1, (2) horizontal as shown in e, (3) semicircular as shown in c, (4) circular as shown in a, and (5) openwindow as shown in the lower-right example of g [Sleight, 55]. The first four were fixed-scale moving-pointer designs and the fifth was a moving scale, fixed-pointer design. All these scales were comparable in size of graduations, size and form of numerals, and dimensions of the pointers. Each of 60 subjects read 17 scale settings of each of the 5 types, with each setting being shown for 0.12 sec by use of a tachistoscope (an instrument that controls viewing time). The subjects were scored by errors in their readings, the percentage of errors being given in Table 5-1. While this table also gives results from three other studies (to be mentioned later), it can be seen that in this study the open-window design had the fewest errors (0.5 percent), followed by the circular (10.9 percent), the semicircular (16.6 percent), the horizontal (27.5 percent), and the vertical (35.5 percent). The clear advantage of the open-window design (under the restricted time conditions) probably can be attributed to the fact that this design did not require any visual search time to locate the pointer; the subjects were already fixated at the pointer position before the tachistoscopic presentation. With the other (moving-pointer) designs it was

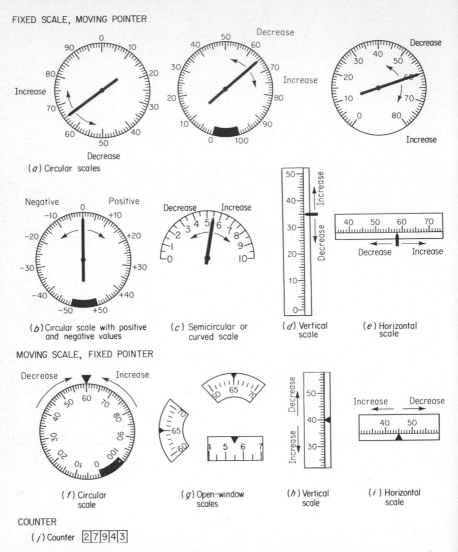

FIXED SCALE, MOVING POINTER

(a) Circular scales

(b) Circular scale with positive and negative values

(c) Semicircular or curved scale

(d) Vertical scale

(e) Horizontal scale

MOVING SCALE, FIXED POINTER

(f) Circular scale

(g) Open-window scales

(h) Vertical scale

(i) Horizontal scale

COUNTER

(j) Counter   2 7 9 4 3

Figure 5-1   Examples of certain types of displays used in presenting quantitative information. (Reference will be made later to certain features of these scales.)

necessary first to change the visual fixation to the pointer setting and then read the setting—all in 0.12 sec. This study, however, has been thoroughly pulled apart, for example, by Murrell [44, p. 160], particularly on the grounds that in the normal use of quantitative scales there is no time limit for viewing the scale; rather, the observer takes whatever time he needs.

A few other studies have been directed toward the analysis of readability under conditions that circumvented either the visual-search aspect or the time constraint aspect of Sleight's study. In a study by Thomas [58], for example, the same five types of scales were all reduced in size so that the eye could fixate at the (now) smaller area for any design. He showed the scale settings for 0.50, 0.10, 0.04, and 0.02 sec and presented rankings of the instruments by accuracy (this information is also given in Table 5-1). In this situation, the open window was second

Table 5-1   Summary of Different Studies of Instrument Designs Used in Quantitative Reading

Experimenter and variable given	Type of instrument				
	Open-window	Semi-circular	Circular	Hori-zontal	Vertical
Sleight (time, 0.12 sec):					
Errors, %	0.05	16.6	10.9	27.5	35.5
Rank order	1	3	2	4	5
Thomas: rank order					
Exposure time, sec					
0.50	2	5	3	1	4
0.10	2	5	3	1	4
0.04	5	4	2	1	3
0.02	5	4	2	1	3
Graham (time, 0.50 sec):					
Errors > ±0.1, %			8.9	6.4	14.4
Rank order			2	1	3
Elkin: errors					
Overall, %	8.1		14.3		21.0
Subject-terminated,* %	0.85		2.50		3.35
0.12 sec,* %	10.00		19.15		43.35
Rank order	1		2		3

* These data from Elkin are limited to one display-response complexity, in which subjects used instruments graduated by 5s and read to the nearest 5.

SOURCES: See references at end of chapter: Sleight [55], Thomas [58], Graham [26], and Elkin [23].

under conditions of 0.50 and 0.10 sec, but last under the shorter time exposures. The other four maintained their *relative* rankings (relative to each other) under all exposure conditions.

In another study of the horizontal, vertical, and circular shapes, with a 0.50-sec exposure time [Graham, 26], the vertical was less accurate than the others, but there was not much difference in accuracy between the horizontal design and the circular design. In still another study

[Elkin, 23], a comparison was made of the vertical, circular, and open-window designs, also under varying time conditions (0.12, 0.36, and 1.08 sec and subject-terminated). This study resulted in an overall ranking by accuracy of (1) open-window, (2) circular, and (3) vertical. The results of these studies also are summarized in Table 5-1 for comparative purposes. And in still another study [Jones et al., 35], the circular scale fared quite well.

Discussion. The unraveling of these apparently twisted skeins may not be as difficult as it would appear, although there still are a few loose ends unaccounted for. To begin with, there is fairly persuasive evidence, such as that from Elkin [23], to support the principle that for most accurate scale reading, the display information should be present for whatever time is required to complete the reading when this is feasible. Under the subject-terminated time condition (when the image remained in view until the subject responded), *time* to respond was actually *less* from the time the instrument was shown than it was with *controlled* times, and *errors* were *fewer*.

The differences among the various types of scales probably can be accounted for in part by differences in visual-search time as suggested by Murrell [44, p. 162], in part by the specific features of the scales used (for example, the poor showing of the semicircular scale used by Thomas might be attributed to its cramped features), and in part by exposure times. In any event, under one condition or another each of the scales—except the vertical one—turned out to be relatively readable. This leads one to concur with Murrell in his conclusion that the specific type of scale may be of less consequence than the specific design features of the scale.

There are, however, some strong hints both from studies and from experience that a scale that may be very readable in one situation is not necessarily equally readable in others. In considering other types of quantitative displays, for example, fixed scales with moving pointers generally seem to be preferable to moving scales with fixed pointers [Christensen, 16], probably because the position of the pointer (in a fixed scale) adds a perceptual cue that is missing in a moving-scale design. In this connection, many basically quantitative scales are also used in a qualitative manner, such as in noting approximate deviation from a desired value or in noting the rate and direction of change in observing altitude. The use of fixed-scale moving-pointer designs certainly facilitates this purpose. Fixed scales, however, have their limitations, especially when the range of values is so great that it cannot be shown on the face of a relatively small scale. In such a case certain moving-scale fixed-pointer designs, such as rectangular open-window and horizontal and vertical scales, have the practical advantage of occupying a small panel space, since the scale can be wound around spools behind the panel face, with only the relevant portion of the scale exposed.

Considering further the specific use of certain designs, one could envision situations in which, say, the vertical scale (which generally has not fared well in experiments) might be the most appropriate, as when two or more scales might be used in a comparative manner to show corresponding values for similar mechanisms such as the rpm's of two or more engines; in such a case a single scanning could indicate if all mechanisms have approximately the same readings or not (this would almost be like a check-reading task). (Further mention will be made of vertical scales a bit later.)

As an additional comment, it should be pointed out that the counter generally is superior for obtaining individual quantitative values when the values are not subject to continual change. The superiority of counters over conventional dials and related scales for obtaining quantitative values was clearly indicated by the results of a study by Zeff [64], in which the mean response times to presentations of values by these two methods were as follows:

Counter (digital display)	0.94 sec
Circular scale	3.54 sec

Further, in 800 trials there were only 4 reading errors with a counter, but 50 with a circular scale. Needless to say, a counter cannot be used if the numerals are continually changing, as with a roulette wheel; the visual processes simply cannot fixate on the numerals under such circumstances.

Readability of Altimeters

The design of altimeters has posed a special problem for aviation for far too long a time. There have been numerous instances in which aircraft accidents have been attributed to misreading of the commonly used model. That model consists of three pointers representing, respectively, 10,000, 1,000, and 100 ft, like the second, minute, and hour hands of a watch. In an early experimental study of altimeters Grether [28] found that that model was read with more errors and took more time than most other (experimental) designs, presumably because the reader had to *combine* three pieces of information. In turn, a combination counter (for thousands of feet) and pointer (for hundreds) proved to be used most accurately and with fewest errors (excluding a counter used as a standard of comparison and a couple of moving-scale varieties that would not lend themselves to practical use in altimeters). In more recent times essentially the same conclusions have been reached by an investigation sponsored by the Department of Defence [Hill and Chernikoff, 31]. The four altimeter models used in this study are shown in Figure 5-2. Their investigation included both laboratory tests and standardized flight tests; in addition, pilots completed questionnaires in which they indicated their preferences for the types of models. On the basis of all the data collected, the investigators ranked the four models in the order shown in Figure 5-2. Actually the first two did not differ by much in the tests

Counter–drum–pointer
model
(CDP)

Counter–pointer
model
(CP)

Drum–pointer
model
(DP)

Three–pointer
model
(3-P)

Figure 5-2   Illustration of four models of altimeters used in a comparative study of the designs. The rank order of results of laboratory and field tests and of expressions of pilots' preferences is: upper left; upper right; lower left; and lower right. [From Hill and Chernikoff, 31].

themselves, but because of strong pilot preference the counter-drum-pointer (CDP) model was recommended for adoption. It is interesting to note that the model corresponds fairly well to the best one reported by Grether [28] and that the three-pointer (3-P) model (the one like most altimeters in actual use) came out at the bottom in both studies.

Some further inklings about altimeter designs come from a study by Simon and Roscoe [54] as reported by Roscoe [49], in which four types of instruments were compared for use in displaying present altitude, predicted altitude (in 1 min), and command altitude (the altitude at which the plane is supposed to be flying); these are shown in very simplified form in Figure 5-3. The displays were intended to provide comparisons of three design variables, namely, (1) vertical versus circular scales, (2) integrated presentations (of the three altitude values mentioned above) versus separate presentations, and (3) *spatial-analogue* presentations ver-

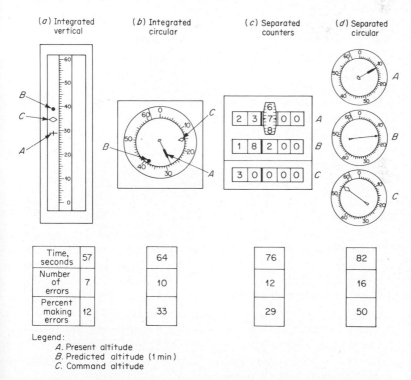

	(*a*) Integrated vertical	(*b*) Integrated circular	(*c*) Separated counters	(*d*) Separated circular
Time, seconds	57	64	76	82
Number of errors	7	10	12	16
Percent making errors	12	33	29	50

Legend:
    A. Present altitude
    B. Predicted altitude (1 min)
    C. Command altitude

Figure 5-3 Four display designs for presenting (*A*) present altitude, (*B*) predicted altitude (in 1 min), and (*C*) command altitude and three criteria (mean time for 10 trials, number of errors, and percent of 24 subjects making errors). The displays are shown in overly simplified form. [Adapted from Simon and Roscoe, 54, as presented by Roscoe 49.]

sus digital counters. Time and error performance scores of pilots in solving altitude-control decision problems and the percent of pilots making errors in using the four designs are also given in Figure 5-3. The clear and consistent superiority of design *a* (the integrated vertical-scale display) is apparent. The explanation for this given by Roscoe [49] is primarily its pictorial realism in representing relative positions in vertical space by a display in which *up* means *up* and *down* means *down*. (This is again an example of compatibility.) Design *b*, which represents vertical space in a distorted manner (around a circle), did not fare as well as *a*, but it was generally superior to *c* and *d*, both of which consisted of *separate* displays of the three altitude values rather than an *integrated* display. Thus, we can derive a strong hint that integrated displays (where they are indeed appropriate) generally are preferable to displays that have distinctly separate indications for the various values.

As an aside, this example reinforces the point that hard and fast generalizations are fairly treacherous. We have seen, for example, that *in this case*, a vertical scale was clearly best and the use of separate counters was manifestly inappropriate, presumably because of the requirement to envision relative positions in vertical space.

Specific Features of Quantitative Scales

The ability of people to make visual discriminations (such as those that are required in the use of quantitative scales) is influenced in part by the specific features that are to be discriminated. Some of the relevant features of quantitative scales are length of scale unit, scale markers (how many and size), numerical progressions of scales, and the design of pointers.

Length of scale unit.   The length of the scale unit is the length, on the scale, that represents the numerical value that is the smallest unit to which the scale is to be read. For example, if a pressure gauge is to be read to the nearest 10 lb, then 10 lb would be the smallest unit of measurement; the scale would be so constructed that a given length (in inches, millimeters, etc.) would represent 10 lb of pressure. (Whether there is, or is not, a marker for each such unit is another matter.)

A number of studies have been carried out relative to the accuracy of instrument readings when the length of the scale unit has been varied. In these, there have been some curious differences between the results of certain United States studies [Grether and Williams, 29; Kappauf et al., 36] versus certain British studies [Jones et al., 35, Murrell et al., 45]. In the study by Grether and Williams [29], it was found that at a normal reading distance of 28 in., errors were reasonably optimum when the scale units ranged from about 0.04 up to about 0.10 in. These results are shown in Figure 5-4. These values correspond reasonably well with the

LABORATORY BULLETIN #2-82

SUBJECT: Blood Bank Policy

1. The temperatures of all water baths and refrigerators (with the exception of the OR Blood Bank refrigerator) listed on the Blood Bank Daily Quality Control sheet must be checked and recorded on Saturdays, Sundays, and holidays.

2. The quality control testing on the Blood Bank reagents must be performed on Saturdays, Sundays and holidays.

3. There is a written procedure in the Blood Bank Quality Control Manual for the preparation and use of the "weak" antibody used in the daily quality control program.

4. There are now written instructions in the Blood Bank Quality Control Manual for the testing of the Blood Bank refrigerators and freezer alarm system and for the calibration of the hemolators.

5. In the typing/screening of donor and patient samples and in the cross-matching of donor units, we have established the use of sequential numbers for each 24-hour period starting at 8:00 a.m. of each day. Each tube used in typing and screening or cross-matching will have a number on it that corresponds to its position on the log sheet.

6. The labels and test results of each in-house donor unit is to be rechecked

February 25, 1982

7. we are putting the title and address on the master copy of ea[ch] Blood Bank form and stamping the name and address on all standard gove[rn]ment forms that do not have it.

8. You are to store <u>thawed</u> fresh frozen plasma at 2°-4°C until used.

WENDELL W. CISCO
Supervisor, Blood Bank/Hematology

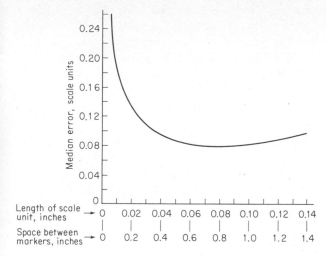

Figure 5-4   Median error in scale units in quantitative reading in relation to length of scale unit and space between graduation markers. [After Grether and Williams, 29.]

results of the study by Kappauf et al., which suggested a range of optimum values from about 0.044 to 0.088 in. Such studies have served as the basis for the acceptance of a fairly standard minimum value of 0.05 in. for scales that are used under optimum viewing conditions of good contrast and adequate illumination (1 fc or more on the dial face) and a reading distance up to about 28 in. [Morgan et al., 43, p. 102]. In turn, the generally accepted value for less than optimum conditions (such as low illumination) is about 0.07.

By contrast, the British investigators [Jones et al., 35; Murrell et al., 45] have reported substantially lower scale-unit values, as shown in Figure 5-5. (These data are based on interpolations by subjects to the nearest value of the scale setting where there were five scale units or *called divisions* between scale markers.) This figure shows a distinct flattening of the curve at about 2 minutes of arc, the equivalent of 0.016 in. at a reading distance of 28 in., less than half the value reported in the United States investigations. Data such as these have served in part as the basis for the establishment of standards by the British Standards Institution. Actually, their recommendations if applied to a reading distance of 28 in. would prescribe a minimum scale unit (minimum called spacing) of about 0.02 in.

Such disparities press one for possible explanations. Murrell [44, pp. 169 and 170] postulates three, namely: (1) practice (it has been found in

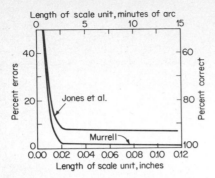

Figure 5-5 Percent of error in reading quantitative scales in relation to length of scale unit. The scale unit values in inches are converted from minutes of arc for a reading distance of 28 in., and the figure has been changed from a log scale to a linear scale for better comparison with Figure 5-4 by Grether and Williams. [Adapted from Murrell et al., 45, as given in Murrell, 44, and from Jones et al., 35, table on p. 32.]

various studies that practice does increase accuracy in the reading of scales); (2) number of types of scales used (in certain of the United States studies subjects have shifted from one type of scale to another; this switching may increase errors in the use of any one type); and (3) vision of subjects (the British subjects all had reasonably normal vision; the visual acuity of the United States subjects is not reported).

Even allowing for these factors, there probably is still an unresolved question relating to the matter of the minimum scale unit. If one were forced to make a choice as far as a *general* practice is concerned, the writer would cast his lot with the prevailing practice (i.e., larger scale units) in the United States on the following grounds: some users may have below-normal vision or may not be fully experienced in the use of a particular scale; scales may sometimes be used under adverse viewing conditions such as pressure of time or inadequate illumination; and frequently a user will have a number of different visual displays to use rather than a single type. By implication, there could indeed be favorable circumstances in which smaller scale units could be fully justified, as with individuals with good vision, when the users would be fully experienced, when the viewing conditions typically would be satisfactory, and perhaps where accuracy requirements would not be stringent.

Scale markers. Another consideration in instrument design is whether or not a scale should have a scale marker for each scale unit (the smallest unit to which the scale is to be read). If the scale is to be read to some degree of precision *beyond* that identified by markers, it of course is necessary for the reader to interpolate. While there has been some conflicting evidence on the question of interpolation, data from Elkin's study [23] shed some light on this matter. In that study the subjects used scales under the following four conditions of *display-response complexity:*

Scale-graduation unit	Scale to be read to nearest	Legend
a. 5 (0, 5, 10, 15, . . . , 100)	5	G-5/R-5
b. 1 (1, 2, 3, 4, 5, . . . , 100)	5	G-1/R-5
c. 1 (1, 2, 3, 4, 5, . . . , 100)	1	G-1/R-1
d. 5 (0, 5, 10, 15, . . . , 100)	1	G-5/R-1

To consider only the results from the subject-terminated condition, the errors for the above four combinations of display-response complexity produce a pattern such as that shown in Figure 5-6. It is very evident that having *more* markers than are required (line *G-1/R-5*) is not nearly as error producing as having *fewer* markers than the values to which the scale is to be read (line *G-5/R-1*); however, there is no advantage in hav-

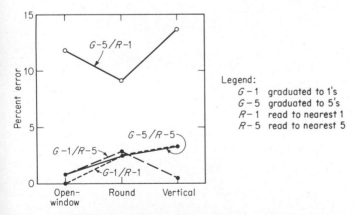

Legend:
G – 1   graduated to 1's
G – 5   graduated to 5's
R – 1   read to nearest 1
R – 5   read to nearest 5

Figure 5-6 Errors in a quantitative reading task for four conditions of display-response complexity, for three types of instruments. The four lines represent different combinations of scale graduations (graduated to 1s and to 5s) and precision of the response to be made (to be read to 1s and to 5s). [Adapted from Elkin, 23.]

ing more markers than required for the reading task. If the scale units have to be shortened much below the recommended length, however, it is generally desirable to use a scale with fewer graduation markers, which therefore requires some interpolation, since a compressed scale with a marker for each scale unit is too crowded for people to read accurately. In such a case, however, it is desirable to have a marker for every other scale unit (such as 0, 2, 4, 6, 8, and 10) rather than for every fifth or tenth unit. A further note should be added about the use of markers. Where high accuracy is required (as with certain test instruments and fine measuring devices), a marker should be placed at every scale unit, even though this requires a larger scale or a closer viewing distance.

Numerical progressions of scales. Every quantitative scale has some intrinsic numerical-progression system that is characterized by the numerical difference between adjacent graduation markers on the scale and by the numbering of the major scale markers. In general, the garden variety of progression by 1s, of 0, 1, 2, 3, etc., is the easiest to use. This lends itself readily to a scale with major markers at 0, 10, 20, etc., with intermediate markers at 5, 15, 25, etc., and with minor markers at individual numbers. Progression by 5s is also satisfactory, and by 2s is moderately so. Some examples of scales with these progressions are shown in Figure 5-7. The off-beat progression system by 4s and others by 2.5s,

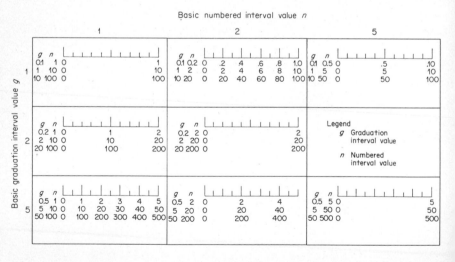

Figure 5-7 Examples of certain generally acceptable quantitative scales with different numerical-progression systems (1s, 2s, and 5s). The values to the left in each case are, respectively, the graduation scale interval g (the difference between the minor markers), and the numbered scale interval n (the difference between numbered markers). For each scale there are variations of the basic values of the system, these being decimal multiples or multiples of 1, 2, or 5.

3s, 6s, etc., usually give trouble and should be avoided except under special circumstances that distinctly justify them. The general superiority of progressions by 1s and 5s, and to some extent 2s, is undoubtedly due to our numerical habits. As in other aspects of human factors engineering, it is desirable to take advantage of compatibility with our learned responses. Where large numerical values are used in the scale, the relative readabilities of the scales are the same if they are all multiplied by 10, 100, 1000, etc. Decimals, however, make scales more difficult to use, although for scales with decimals the same relative advantages and disadvantages hold for the various numerical progressions. The zero in front of the decimal point should be omitted when such scales are used [Vernon, 62].

The design of pointers. The few studies that have dealt with pointer design leave some unanswered questions, but, in general, suggest that discriminations are facilitated when the pointer is about the width of a scale unit (instead of narrower or wider) [Churchill, 18; Kappauf et al., 36]; when the pointer meets, but does not overlap, the smallest marker; when the tail of the pointer is the same color as the dial face; and when the pointer is close to the dial face (to avoid parallax).

Combining scale features. Several of the features of quantitative scales discussed above have been integrated into relatively standard formats for designing scales and their markers, as shown in Figure 5-8. These are based on scale unit lengths of (a) 0.05 in. for use under normal viewing conditions (with adequate illumination) and (b) 0.07 in. for low

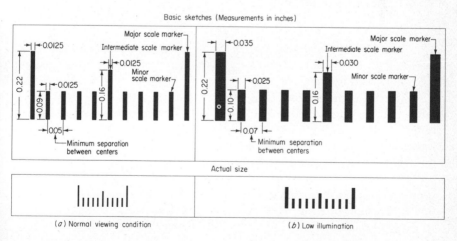

Figure 5-8 Recommended format of quantitative scales, considering length of scale unit and graduation markers. Format *a* is proposed for normal illumination conditions under normal viewing conditions, and *b* for low illumination. [Adapted from Morgan et al., 43.]

illumination, and for viewing at about 28 in. Such formats should, of course, be considered basic guides rather than rigid requirements. (And we should keep in mind the still unexplained disparity between the United States studies and those of our British cousins.) Although the formats in Figure 5-8 are shown in a straight horizontal scale, they can, of course, be adapted to other styles, such as circular and semicircular.

Scale size and viewing distance. In the discussion of visual acuity in Chapter 3, reference was made to the visual angle, that is, the angle that is subtended at the eye by some specific feature being viewed, such as the component elements of the big E on a vision text chart. The concept of visual angle has relevance to our discussion above about quantitative scales. The specific features of such scales have at times been specified in terms of visual angle as in Figure 3-6. If *not* so specified, it is necessary to specify *size* of the specific feature and *viewing distance*. Our discussion above generally was predicated on a viewing distance of 28 in., which approximates that used with many instrument panels and is an average preferred viewing distance. However, when quantitative displays are used at farther distances, the display features need to be enlarged in order to maintain, at the eye, essentially their same visual angles. At distances less than 28 in., the features can be reduced accordingly. When using the dimensions discussed above (such as those shown in Figure 5-8), the following formula can be used to derive display features that would produce the same visual angles at some other viewing distance:

$$\frac{\text{Viewing distance in inches}}{28}$$

The assumption that equal visual angles—regardless of distance—result in equal discriminability, however, is subject to at least a bit of suspicion [Chapanis and Scarpa, 15].

The total dimensions of a scale, however, must take into account not only the length of the scale unit necessary for the viewing distance, but also the total range of values (i.e., the number of scale units) that must be provided for.[1] Figure 5-9 shows, for a circular scale, the diameters that would be required to provide a constant visual angle of scale units at various viewing distances for certain total numbers of scale units (for example, 50, 100, 200, and 300). In particular this shows the diameters that would provide, at various viewing distances, visual angles equal to those of 0.05- and 0.07-in. scale units when viewed at 28 in.

[1] For a thorough discussion of scale design (based on British practice) taking into account range of values, viewing distance, numerical progression, etc., see Murrell [44, chap. 9].

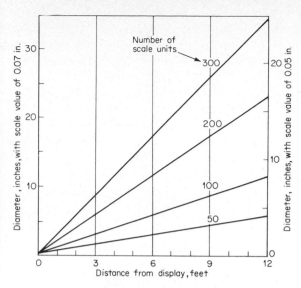

Figure 5-9 Diameters of circular quantitative scales required to provide constant visual angle of scale units at various viewing distances for certain total numbers of scale units. The visual angles on which this figure is based correspond to a scale value of 0.05 in. and of 0.07 in. when viewed at 28 in. Each line represents the diameter values for a scale with a specified number of scale units (for example, 50, 100, 200, and 300). Interpolations can be made between these values. These values can be extrapolated on a straight-line to longer distances or can be applied to derive the length of straight scales by multiplying diameter values by 3.1416. All diameter values allow 0.01 in. for height of graduation markers.

Concentric Quantitative Displays

Quantitative readings are sometimes made from indications on different types of concentric displays, such as military plotting tables and radar scopes with polar coordinates and certain continuous recording charts. In such displays the distance from the center of an indication (e.g., a tracing or radar blip) represents a quantitative value such as temperature and range. The angle from the center, in turn, represents such parameters as hour of the day and azimuth of a target. Certain principles that are relevant to the design of other types of quantitative displays also are relevant to concentric types. For example, numerical progressions by 1s are most compatible with human perceptual and mediation processes, and by

5s and by 2s are also reasonably acceptable. Further, it has been found that displays with a fair amount of detail in the concentric markers generally can be read more accurately than those that require considerable interpolation [Green and Anderson, 27].

However, there is a possible trap in providing such detailed features. Although the specific errors of estimating values close to the indications are reduced by the use of more ring markers rather than fewer, the *gross* errors tend to increase with the number of concentric markers; the gross errors are those of mistaking one marker for another. The solution to this problem is the use of heavy markers for major divisions, such as every fifth ring, to make them perceptually distinct from the lighter individual rings.

QUALITATIVE VISUAL DISPLAYS

Displays that are read in a qualitative manner (i.e., to derive an approximate value or to determine a rate of change or trend) are basically quantitative, in that the variable depicted by the display is a quantitative variable. In a very rough sense we can regard qualitative reading operations as falling in one of two classes (although there undoubtedly are exceptions).

Qualitative Reading of Quantitative Displays

In the first place some displays that are used mainly for obtaining quantitative values are also used for qualitative purposes, such as determining a trend or rate of change. We may use a speedometer in this way, for example, in glancing at it to see if the speed is somewhere within a general range we wish to maintain (such as between 55 and 60 mph) or to determine if we are below a given speed limit (such as 60 mph). At other times we may want to ascertain what the exact speed is. (A traffic officer would use his speedometer to ascertain the exact speed of a car he is following.)

There is some evidence, however, to suggest that a display that may be best for a quantitative reading task is not necessarily best for a qualitative reading test. The clearest evidence for this comes from a study previously cited, in which open-window, circular, and vertical designs are compared [Elkin, 23]. In one phase of this study, subjects made qualitative readings, as follows:

Pointer setting	Response to be made by subject
Above 60	High
40–60	OK
Below 40	Low

The accuracy of the readings was very high (only 3 errors were made in 1440 readings). A comparison of time taken in making the readings, however, is interesting, especially when compared with the *lowest* average reading times for the quantitative reading task (which was with scales graduated to 5s and read to the nearest 5).

Type of scale	Average reading time, sec (scales graduated to 5s)	
	Qualitative	Quantitative
Open-window	115	102
Circular	107	113
Vertical	101	118

Thus, while the open-window design took the least time (of the three types) for quantitative reading, it took the longest time for qualitative. Another study, however, adds some additional light to this matter [Hamer and Critchley, 30]. They used three types of simulated speedometers, namely, circular, semicircular, and horizontal, in a time-sharing tracking task. One task was keeping a "car" on a simulated road that kept moving, and the other was keeping the "speed" within 2 miles, plus or minus, of a designated speed, when the speed was subject to external change by a cam. They found no appreciable difference in the speed control when the three types of speedometer displays were used. Now, to refer back to Elkin's study, it will be recalled that the open-window design, a moving-scale display, took the longest time for a qualitative reading, 115 sec. The circular and vertical models, both fixed-scale models, took less time, 107 and 101 sec. Since all three scales used by Hamer and Critchley were fixed-scale designs, it might well be suggested that, for qualitative reading, one should use a fixed-scale type, avoiding a moving-scale type, such as an open-window. Two possible explanations are offered for the apparent advantage of the fixed-scale design. In the first place, with a moving-scale display the observer presumably has to read the scale quantitatively and *then* qualitatively (to ascertain if the value is within acceptable bounds). In the second place, the position of the moving pointer on a fixed-scale display may itself provide a perceptual clue as to whether the value is within bounds or not.

Qualitative Reading of Precoded Displays

The second type of qualitative reading is that in which the distinct ranges of values with different meanings can be predetermined and accordingly precoded on the display itself. Where this is possible, color coding is frequently used, as illustrated in Figure 5-10. Some illumination conditions,

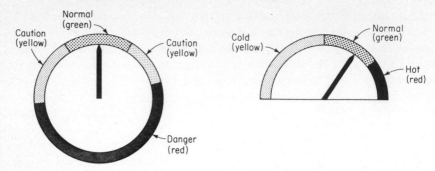

Figure 5-10   Illustration of color coding of sections of instruments that are to be read qualitatively.

however, may preclude the use of color, such as in aircraft when the red illumination is used to maintain peripheral dark adaptation. In such a circumstance, zones on an instrument can be shape coded. In this connection, it is desirable (if feasible) to take advantage of any natural associations people may have with designs or shapes. One study was directed toward determining what, if any, such associations people had with each of seven different coding designs [Sabeh, Jorve, and Vanderplas, 50]. After having solicited a large number of designs initially, the investigators selected the seven shown in Figure 5-11. These were presented to 140 subjects, along with a list of seven "meanings" as follows: caution,

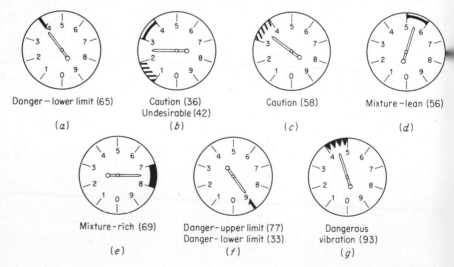

Figure 5-11   Association of coded zone markings with subjective "meaning," showing the number of individuals (out of 140) who reported significant associations. [Adapted from Sabeh, Jorve, and Vanderplas, 50.]

undesirable, mixture—lean, mixture—rich, danger—upper limit, danger —lower limit, and dangerous vibration. Figure 5-11 shows the number of subjects out of 140 who selected the indicated meaning to a statistically significant level. This is another illustration of the concept of compatibility (in this case the compatibility of association with symbol "meanings") as applied to a design problem.

CHECK–READING DISPLAYS

In the design of instruments that are to be used under check-reading circumstances, two considerations probably are dominant. In the first place, individual instruments should have clearly distinguishable characteristics to identify the null (neutral, normal, satisfactory) condition or the undesirable condition (whatever that may be). In dial design this frequently is done by having, around the circumference, a marking that characterizes such a condition.

An interesting variation of design for check reading a qualitative instrument has been investigated by Kurke [41]. He used simulations of three variations of a quantitative instrument in which a given range of readings indicated a "danger" condition which required attention. These three variations, no indication, a red line, and a red wedge, are shown in Figure 5-12, along with mean-time scores of a group of subjects. Such

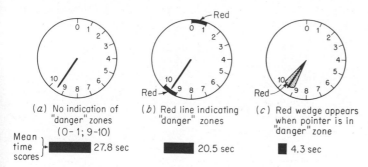

Figure 5-12 Three designs of quantitative instruments used in a check-reading situation where the "danger" condition was to be identified. Design *a* has no indication of the danger zone, *b* has a red line at the circumference, and *c* shows a red wedge when the pointer is in the danger zone. [From Kurke, 41.]

data tend to support the hypothesis that design features such as the red wedge apparently facilitate the perception of specified instrument settings.

In the second place, where several check-reading instruments are to be used in the same instrument panel, the arrangement of the instruments (especially of their null positions) can facilitate the task. In one study, for example, a comparison was made of the time taken to check-read two panels of dials. In one panel the dial pointers, when at normal, were aligned at the 9 o'clock position; in the other, the dial pointers, when at normal, were pointing in different directions (though in each case toward a 45° arc of red line on the circumference) [Senders, 53]. Briefly, it was found that subjects could check-read 32 dials aligned with pointers in the 9 o'clock position in the same time it took to check-read only four dials when the null positions were not so aligned.

This is, of course, essentially a perceptual process, and in a few subsequent studies comparisons have been made of the time (and errors) in check-reading dials with various pointer arrangements. For example, it was found in another study that the 12 o'clock position of alignment did not differ significantly from the 9 o'clock position in terms of errors [Dashevsky, 21]. And in still another investigation it was reported that patterns of pointer "symmetry" lent themselves to check-reading as rapidly as alignment in the 9 o'clock position [Johnsgard, 34]. The symmetry consisted of horizontal or vertical double rows of dials with the null positions pointed toward an imaginary line between the rows.

Some other patterns were also investigated by Dashevsky [21], these being shown in Figure 5-13. Some of these (*d*, *e*, and *f*) incorporated lines extending from one dial to another to form continuous lines when the pointers were in their null positions. The errors resulting from this comparison are given below:

	Arrangement		
	12 o'clock	*Subgroups*	*Subgroups rotated*
Open	*a.* 53	*b.* 193	*c.* 201
Extended line	*d.* 8	*e.* 15	*f.* 41

This study demonstrated that extending the lines made by the pointers can result in noticeable reduction of errors, since single deviant pointers then show up as breaks in an otherwise continuous line. However, this is not the whole story. In a subsequent study Oatman [46] worked in another variation that offers considerable promise of facilitating the perception of deviant pointers in a check-reading task. In particular he used extended-pointer displays in which the pointer was the length of the diameter of the dial as shown in Figure 5-14*c* and *d*. He found modest, but significantly higher, detection rates in an extended-pointer design *c* as contrasted with short-pointer designs *a* and *b*; one of these, *a*, was an open design, and the other an extended-line design. In a

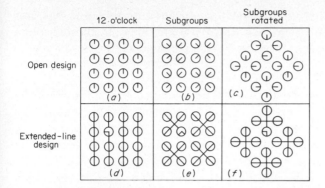

Figure 5-13 Patterns of panels of check-reading dials used in study by Dashevsky [21]. (Copyright 1964 by the American Psychological Association and reproduced by permission.) In this study the 12 o'clock extended-line pattern *d* resulted in the lowest number of errors.

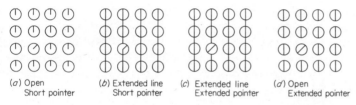

Figure 5-14 Patterns of panels of check-reading dials used in a study by Oatman [46]. In this study extended-pointer designs (such as *c* and *d*) resulted in fewer errors than those without extended pointers.

subsequent study he found no significant difference between extended-pointer patterns with and without the extended-line feature. He concluded, then, that factors making the deviant *dial* more conspicuous (e.g., length of pointer) are apparently more significant in reducing check-reading errors than factors which make the display pattern simpler (e.g., extended line between the dials).

Though we grant some unreported inconsistencies in the results of these and other studies, there nonetheless seems to be evidence that the configuration presented by panels of check-reading displays can facilitate the identification of deviant instruments. Certainly, extended pointers contribute to this perceptual detection, and the extension of lines between pointers does not hinder it and probably helps by providing a visual field that is broken up by a deviant pointer. A postscript should be added to this discussion. If, in fact, the exclusive use of a check-reading display is *simply* detecting deviant conditions, it may be that one could use a warning light instead of a dial. But some check-reading dials may *also* be used in a qualitative manner; in such a case the dial is of course preferable.

VISUAL INDICATORS AND WARNING SIGNALS

Visual displays used to present dynamic indication information and warnings are most typically light signals, although other devices such as signal flags and semaphores are used. There has apparently been little research relating to such signals, but we can infer some general principles from our knowledge of human sensory and perceptual processes that might be helpful. In the first place, if such signals are to *attract* attention, they need to be within the convenient visual span of the individual who is to attend to them; if in actual practice they do not need to attract attention, but are *referred* to by the individual as necessary, location is unimportant. Location, however, is especially important in the case of warning indicators, which usually should be within about 30° of the normal line of sight. In the second place, if attention getting is important, certain perceptual principles might be useful, such as using high intensity lights, large lights, or flashing lights. (Incidentally, a flash rate of about 10 Hz is reasonably optimum; higher rates tend to result in lowered detectability.) Third, the signals should be individually identifiable. If colored lights are used, for example, only a limited number of colors should be used, and those that might be confused (such as red and orange) should not be used together; if color-blind people might use the signals, some supplementary coding system might be useful, such as location coding. Fourth, their meaning should be clear to the user; if there is any possible doubt about the meaning of such signals, they should have short, unambiguous labels. And, fifth, signal lights should be clearly distinguishable against their backgrounds.

Aside from features of the signals that may affect the perception thereof, the background in which they are imbedded can also affect their perception. For example, lights generally cannot be discriminated well against a bright background, like the stars that we cannot see during the daylight hours; signal lights then should be in areas with sharply contrasting backgrounds, such as under hoods. Likewise, lights cannot be discriminated well when other background lights are somewhat similar. (Traffic lights in areas with neon signs and Christmas tree lights represent very serious deviations from this principle.) And still another background characteristic relates to the steady versus flashing state of any background lights. In an interesting investigation of these, Crawford [20] used both steady and flashing signal lights against backgrounds of *irrelevant* lights (what we might call *noise*), these being all steady, all flashing, or some admixture of steady and flashing lights. Very briefly, his results indicated that the average time to identify the signal lights was minimal when the background-noise lights were all steady (this was especially so when the signal light was itself flashing); that the advantage of a flashing signal light (contrasted with a steady light) was completely lost if even one background-noise light was flashing; and that steady signals were more effective (could be identified more quickly) than flashing signals if the proportion of the noise lights that were flashing was any greater than 1 out of 10. In other words, flashing lights against other flashing lights really make life difficult for the viewer.

PICTORIAL AND GRAPHIC DISPLAYS

The range of pictorial and graphic displays—both of a dynamic and a static nature—is fairly wide. Whether of a rather strictly pictorial nature (intended to *reproduce* an object or scene, as on a TV scope or in an aerial photograph) or more graphic form (i.e., illustrative or symbolic), the intent is one of conveying a visual impression that requires little or no interpretation. One important type of pictorial and graphic display is a cathode-ray tube (CRT). There are many design problems associated with CRTs, such as the size and brightness of the scope, the adjustment of brightness, the size of targets and their contrast with the scope, the viewing distance and size, and the use of various aids such as transparent overlays and pantographs in estimating such variables as range and azimuth. It is not feasible here to discuss these in detail, so the reader is referred to other sources, such as Morgan et al. [43, pp. 109–120]. A few other types of pictorial and graphic displays will be discussed for illustrative purposes, along with some supporting research relating to them.

Aircraft-position Displays

The problem of representing the position and movement of aircraft has haunted designers for some years, and the advent of jet aircraft (and soon in numbers, supersonic aircraft) serves to accentuate this problem. Most current approaches, however, are toward some type of physical representation that could immediately convey to the pilot a reliable impression of the position of the aircraft in space. Actually, there are a number of components of information that may need to be represented, such as altitude, attitude (e.g., heading, roll, and pitch), and speed. Quite a few schemes have been developed to present such information. In one simulated situation, four variations of instruments were used to depict the position of the aircraft in relationship to the horizon, and these were compared by performance criteria [Bauerschmidt and Roscoe, 6]. The displays used are shown graphically in Figure 5-15. Without going into the details, it can be said that the subjects (28 pilots) performed best on type IV. This model (a pursuit-type moving-aircraft display with space-stabilized error dot) depicts the horizon, the airplane, and its "error" from a desired direction in the manner that is most compatible with our common frame of reference of an aircraft in relation to a fixed horizon;

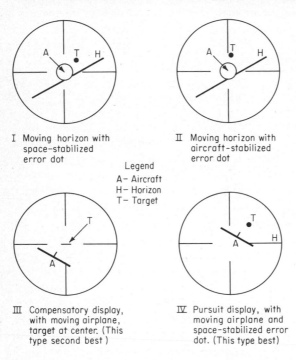

I Moving horizon with
  space–stabilized
  error dot

II Moving horizon with
  aircraft–stabilized
  error dot

Legend
A– Aircraft
H– Horizon
T– Target

III Compensatory display,
  with moving airplane,
  target at center. (This
  type second best)

IV Pursuit display, with
  moving airplane and
  space–stabilized error
  dot. (This type best)

Figure 5-15 Sketches of experimental pursuit moving-airplane steering displays. See text for description. [Adapted from Bauerschmidt and Roscoe, 6.]

this is sometimes called an *outside-in* display. The moving-horizon frame of reference (I and II) is clearly less compatible; this is sometimes called an *inside-out* display.

A more sophisticated type of aircraft-position display is what is called a *contact analog*. It provides, on a TV tube, an electronically generated representation of the terrain with the aircraft superimposed in proper position [Balding and Süskind, 4]. Figure 5-16 is an illustration of this type of display, showing an aircraft in three different relations to its predetermined *altitude-hold* flight path. This type of display can be interpreted as another example of the principal of compatibility, since it

  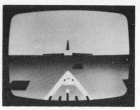

(a) On selected altitude   (b) Below selected altitude   (c) Above selected altitude

Figure 5-16 Illustration of a contact analog type of aircraft-position display, the Kaiser Flight-Path System. The display can present an electronically produced representation of the aircraft with its altitude and direction related to the landscape and its flight path. This particular illustration of the display shows the aircraft in an intended altitude-hold flight path (its "highway in the sky"). [Photographs courtesy of Kaiser Aerospace and Electronics, Palo Alto, Calif.]

is consistent with our common experiences in viewing objects in three-dimensional space.

Principles of aircraft-position displays. In connection with aircraft flight and navigation display systems, Roscoe [49] has teased out of the relevant research a few principles of display that he proposes have substantial validity, despite sometimes conflicting research results. Most of these principles have been touched on before, in one context or another, but they will nevertheless be reiterated here.

1. *The principle of display integration:* The notion of display integration requires that *related* information be presented in a common display system which allows the relationships to be perceived directly (as illustrated in Figure 5-3, design *a*). This principle does *not* apply to the haphazard combining of unrelated information in a common display.
2. *The principle of pictorial realism:* This principle relates to the presentation of graphic relationships in such a manner that the encoded symbols can be readily identified with what they represent; in effect, the symbols are an analog of that which they represent.
3. *The principle of the moving part:* In the use of aircraft displays this principle is in conformity with the outside-in (i.e., moving-aircraft) display rather than with the inside-out (i.e., moving-horizon) display. In more general terms, it seems preferable for the image of the *moving part* (i.e., an aircraft or symbol representing any other moving object) to be displayed against a fixed scale or coordinate system.
4. *The principle of pursuit tracking:* In pursuit tracking the index of *desired* performance (sometimes referred to as the target) and the index of *actual* performance move over the display against a common scale or coordinate system. Generally this scheme results in better

performance than does compensatory tracking, in which the index of either the desired or actual performance is fixed, with the moving index showing only the *error*, or difference.

5. *The principle of frequency separation:* This principle relates to the *dynamics* of the display, especially to the high-frequency components of the display's indications. (The indications of a display typically are the composite of various movement components, each having its own "frequency" of movement.) In simple operational terms this principle provides that the elements of a display that respond immediately to a control input should move in the "expected" direction.

6. *The principle of optimum scaling:* This principle deals with the physical relationship (really the ratio) of the physical dimensions of that to be represented (i.e., features of the surface of the earth) to the dimensions on the display that represent such features, such as the number of millimeters or inches on the display that represent a mile on the earth. This problem gets all intertwined with precision requirements, but taking these into account, some optimum relationship is possible.

Although the above principles were crystallized because of their relevance to the problems of aircraft flight and navigation displays, they probably are equally valid for numerous other reasonably corresponding display problems.

Configurations

There has been quite a spate of research over the years on the problems of extracting information from complex configurations such as aerial photographs, CRT displays, maps, and charts, in which there are many shadings of specific portions of the configuration or variations in the amount of detail that is presented. Within limits it is possible to control certain features of these displays, such as the *grain* of photographic reproduction, the number of *lines* of a TV picture, and the contrast and brightness of the detailed features, in order to incorporate those features that are most consistent with human sensory and perceptual abilities. Although it is not yet feasible to specify all of the features of such displays that are preferable for human use, some such variables have been investigated. A few of these will be discussed.

Use of simulated configurations in research.   Because of the complexity of many literal reproductions of pictorial material (i.e., photographs, radar, etc.), in some studies of such displays simulations have been used rather than reproductions such as photographs. (Although the variables are easier to control with simulations, this procedure does remove one a step further from the real world.) Such an approach was used, for example, by Baker et al. [3] in studying certain aspects of searching for visual targets against a complex background. They used a matrix of 300 × 300 cells (90,000) as the background in searching for 20 specified targets. The background consisted of a random pattern of white cells on a black ground. Each target consisted of a different group of adjacent white cells. These displays were then modified through photographic processes to produce four different degrees of visual resolution (blur), as shown in Figure 5-17. The task of the subject was to identify, from displays such

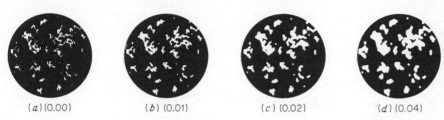

   (a)(0.00)   (b)(0.01)   (c)(0.02)   (d)(0.04)

Figure 5-17 Illustration of displays used in study of target recognition. These four displays represent samples of differences in resolution (blur). The blur indexes, given in parentheses, are derived from the specific method of creating the blur conditions. The subjects were asked to identify in the display a certain target which had been shown to them as a "briefing target." [From Baker, Morris, and Steedman, 3.]

as those shown in that figure, 1 of the 20 specific targets. Portions of the total matrix were used, these being 6, 12, 16, and 24 in.^{2}, respectively.

Some of the results of this study are summarized in Figure 5-18, which indicates that the time needed to identify the specified targets increases with the search area (this, in turn, is roughly proportional to the number of irrelevant forms among which the targets to be identified were imbedded). These data suggest the possible desirability in complex-area displays of filtering out (such as in radar) or otherwise minimizing irrelevant display features. Further, in this study it was found that time and errors were lowest when the resolutions of the reference (briefing) targets and the targets to be searched for in the display were the same. The implication of this, in simpler terms, is that if a person is trying to search for some specific target in a complex visual field, he is more apt to find it if he is provided with an image (in this case, a "briefing" target) that closely resembles the way in which the target actually will appear in the field, even though it is badly blurred.

As an aside, it was also found that time and errors in identifying targets were highly correlated with target size (specifically, the ratio of target area to the area of the smallest circle which could contain the target). In a subsequent study it was found that a target (such as those used) must have a minimum size of 0.042 in. as displayed in order to expect relatively accurate and rapid recognition [Steedman and Baker, 57].

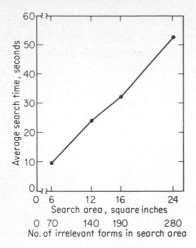

Figure 5-18 Relationship between the number of irrelevant forms in a complex display and the time required to identify a specified target. [Adapted from Baker, Morris, and Steedman, 3.]

Another example of simulation in the use of configurations was a map-reading study in which the intent was to examine the effects on the use of maps of various backgrounds and target codes [Christner and Ray, 17]. The three target codes were color, numerals (enclosed in squares), and geometric shapes (enclosed in squares), there being eight targets of each type. The different types of targets were placed on "maps" of five types of backgrounds, namely, colored, white, patterned, solid gray, and mixed gray. Subjects performed the same tasks as in the study by Hitt [32, also mentioned in Chapter 4], namely, to identify, locate, count, compare, and verify targets. In general, the various background codings were not systematically related to the rate of responding, but certain target codes were, at least for certain tasks. For the identification task, number coding was superior to color coding (in agreement with Hitt's study). For the locating and counting tasks, however, color coding was superior to numbers. Thus, in general, it appears that the selection of a background for such displays could be based on such factors as ease of construction and economy, whereas the choice of target codes perhaps should depend in part upon the specific task to be performed.

Photo-interpretation. The photo-interpretation of aerial photographs involves the extraction of information from such displays for a good many purposes, including the identification of different types of targets for military intelligence. Although the perfect identification of all relevant targets is well beyond human perceptual abilities, there are various factors that can influence the performance of photo-interpreters. Such variables (sometimes called *enabling variables* in contrast to *operator-ability variables*) cover a wide assortment, but three are of particular concern, namely: resolution (this is a function of lens quality, type of film, exposure, film processing, etc.); scale factor or representative fraction RF (this is a function of the focal length of the lens and the altitude of the aircraft and can be thought of as a measure of distance on the ground represented by a specified distance on the photograph); and image contrast (this is the contrast in brightness between features of a photograph and its background). Within at least moderate limits some such variables can be controlled.

Reasonable contrast is, of course, necessary for making any visual discriminations. In studying the relative effects of resolution and the scale factor on the ability of photo-interpreters to identify targets, Williams et al. [63] asked observers to identify certain features of aerial photographs that varied in the combination of these two factors, as follows:

· Resolution: 55, 26, and 13 ft. (The resolution of the original prints was 13 ft, which means that two objects on the ground, separated by 13 ft or more, could be seen as two images on the prints; the other resolution values were produced by optical degradation of the original prints.)
· Scale factor: 1:40,000; 1:27,000; 1:12,000.

In this comparison it was found that improving the resolution rather systematically resulted in improvement in the identification of images, but that reducing the scale factor did not consistently result in better performance. Thus, it appears that contrast and resolution of photographs probably have more effect in photo-interpretation than the scale factor.

In photo-interpretation circles there is something of a controversy about the value of stereoscopic viewing of photographic imagery. Stereoscopic images are produced by the use of two cameras simultaneously taking aerial pictures in such a manner that the ground covered by one camera overlaps part of that covered by the other. These two images are then viewed with stereoscopic equipment. The basis for the controversy can be seen in the apparently inconsistent results of investigations such as those of Zeidner et al. [65] and Davidson [22]. Zeidner et al., for example, using military photographs, found that a group of nonstereo photo-interpreters identified more objects (e.g., more of the objects that presumably were in the photographs) but that the stereo group made fewer errors (e.g., they reported fewer "objects" that presumably were *not* present in the photographs); thus, nonstereo viewers tended to "see" images that were *not* there, but also to detect more images that *were* there. On the other hand, Davidson [22], using, as subjects, graduate students in a forestry aerial photogrammetry course, found a significant difference in performance between those using stereo and those using nonstereo procedures, in favor of the stereo group. The stereo group had significantly *fewer* wrong responses and a higher percent of total accuracy. All the bits of evidence on stereo photo-interpretation have not yet been obtained through research to date. But the evidence that is available probably can be interpreted, in a somewhat negative way, as implying that the claimed advantages of stereo processes have not been as yet firmly established.

Graphic Representations

The format of some of the graphic representations (bar charts, pie charts, line charts, etc.) that find their way into newspapers and other publications lead one to hope that there *must* be better ways of presenting the information that the graphs presumably are intended to convey. Although there has been relatively little research to date regarding the design of graphic representations, two investigations will be summarized to demonstrate that research in this area may actually have some practical application. One of these dealt with a comparison of three formats for depicting trend data, as illustrated in Figure 5-19a, b, and c, a line format, a vertical-bar format, and a horizontal-bar format [Schutz, 51]. For each format, there were variations in numbers of points depicted (6, 12, or 18) and in the number of missing values. The subjects were required to estimate the trend of the data and were scored on the time required to make such estimates and on the accuracy of their estimates. On both scores the line graph proved to be preferable, as indicated by the following mean scores:

Format	Mean relative time	Mean accuracy score
Line	6.81	1.72
Vertical bar	7.36	1.64
Horizontal bar	8.91	1.40

In a subsequent study [Schutz, 52], multiple-line graphs were used exclusively in two tasks, namely, *point reading* (reading vertical-axis values from specified horizontal-axis values) and *comparing* (determining the line with the highest value on the vertical axis for specified values on the horizontal axis). Two types of displays were used, namely, multiple-graph (2, 3, or 4 separate graphs, each with a single line, as illustrated in

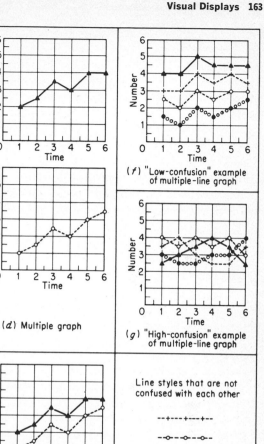

Figure 5-19 Illustrations of formats of multiple-trend charts investigated by Schutz. Illustrations *a*, *b*, and *c* show formats that were compared on time and accuracy of reading; some examples had 12 or 18 points instead of the 6 shown (line graphs generally were superior). Illustrations *d* and *e* were used in comparing multiple-graph figures with multiple-line graphs. Illustrations *f* and *g* show low-confusion and high-confusion examples of multiple-line graphs. See text for discussion. [Adapted from Schutz, 51, 52.]

Figure 5-19*d*), and multiple-line graphs (a single graph with 2, 3, or 4 lines, as shown in Figure 5-19*e*); in the multiple-line graphs two degrees of confusion were introduced by varying the extent to which the lines crossed each other, as shown in Figure 5-19*f* and *g*. The 4 styles of lines used were selected from 25 that were tried out in a previous study in which the 4 were found to be the lines not confused with each other. For the point-reading task the two formats were about equal in time needed to complete the task (relative time scores were 18.95 for the multiple-line format and 18.97 for the multiple-graph format). The comparing task, however, could be performed in less time with the multiple-line format, 19.40 versus 28.49, probably because the task tends to depend more on perception (simple comparison of the lines) than on a mediation process of keeping the value of one graph in mind momentarily while looking at the other and then comparing the *values* rather than the *configurations*. The addition of mental operations to a task typically adds a piece of time of the task.

ALPHANUMERIC CHARACTERS

Alphanumeric characters are used in various contexts, such as in printed copy (books, newspapers, etc.), in typewritten and handwritten form, in tabular form (telephone directories, tables of numerical data, computer print-outs, etc.), and in identification labels (on control panels, traffic signs, license plates, etc.). Over the years there has been a great deal of research relating to various facets of the business of communicating by written material, including content, writing style, and typography. Since certain aspects of typography have particular pertinence to human factors engineering, we shall touch especially on this topic, but this does not imply that the other aspects are outside the ken of human factors engineering.[2]

Ironically, even in discussions of alphanumeric information there is a fair quota of confusion in the use of words. For our purposes we will adopt the following definitions:

· *Visibility:* The quality of a character or symbol that makes it separately visible from its surroundings. (This is essentially the same as the term *discriminability* as used in Chapter 4.)

· *Legibility:* The attribute of alphanumeric characters that makes it possible for each one to be identifiable from others. (This depends on such features as strokewidth, form of characters, contrast, and illumination.)

· *Readability:* A quality that makes possible the recognition of the information content of material when represented by alphanumeric charac-

---

[2] For an excellent survey of legibility and related aspects of alphanumeric characters and related symbols, the reader is referred to Cornog and Rose [19].

ters in meaningful groupings, such as words, sentences, or continuous text. (This depends more on the spacing of characters and groups of characters, on their combination into sentences or other forms, on the spacing between lines, and on margins, than on the specific features of the individual characters.)

Strokewidth of Letters and Numerals

The strokewidth of letters and numerals is usually expressed as the ratio of the thickness of the stroke to the height of the letters or numerals. Some examples of strokewidth-to-height ratios are shown in Figure 5-20.

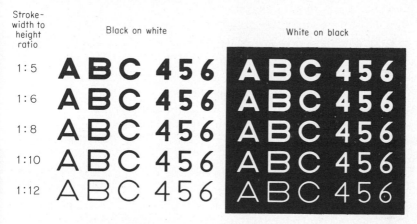

Figure 5-20 Illustrations of strokewidth-to-height ratios of letters and numerals.

Investigations of the legibility of strokewidth still leave some loose ends, although there are certain reasonably stable implications that have emerged from a few studies. A pair of rather basic studies are those by Berger [7, 8]. In these studies, white numbers on a black background and black numbers on a white background were used under daylight conditions outdoors, with the strokewidth-to-height ratios ranging from a very thin 1:40 to a heavier 1:5. The criterion used was the average distance at which the subjects could read the numerals; these averages are shown in Table 5-2. In the case of the white numerals, the optimum legibility (i.e., the greatest reading distance) occurred with the 1:13.3 ratio numerals, but there was a fairly wide range of ratios that gave reasonably comparable results. With black letters on a white background, however, the optimum ratio was distinctly lower, around 1:8, with a range from about 1:5.8 to 1:10 giving reasonably comparable results; higher ratios were poorer. This difference in optimum strokewidth of white and of black characters has been confirmed in other investigations, and is

Table 5-2   Average Distances in Meters at Which Numerals of Different Strokewidth-to-Height Ratios Can Be Read

Color of numerals	Strokewidth-to-height ratio*							
	1:40	*1:20*	*1:13.3*	*1:10*	*1:8*	*1:6.6*	*1:5.8*	*1:5*
White	33.9	35.8	36.5	35.5	34.7	33.4	31.4	29.4
Black	25.2	28.0	31.1	32.7	33.5	33.1	32.1	29.9

* The numerals were 42 by 80 mm in size.
SOURCE: Adapted from C. Berger, I. Stroke-width, form and horizontal spacing of numerals as determinants of the threshold of recognition, *Journal of Applied Psychology*, 1944, vol. 28, pp. 208–231, Table 1.

attributable to a phenomenon of irradiation, in which white features appear to spread into adjacent black areas, but not the converse. The phenomenon is especially accentuated with highly illuminated display features (in the studies by Berger, for example, under floodlighting conditions, the optimum strokewidth of white luminous numerals on black was 1:40!). Dark adaptation of individuals also tends to accentuate the effect. Because of this effect, white characters on black should have thinner strokewidths than black on white.

There have been numerous studies relating to the legibility of alphanumeric characters of varying strokewidth-to-height ratios, especially capital letters and numerals [Brown et al., 12; Brown and Lowery, 11; Kuntz and Sleight, 40; Soar, 56]. There have indeed been some inconsistencies in the results of these studies, but the consistencies tend to outweigh the inconsistencies, and for fairly conventional uses of alphanumeric characters under normal illumination conditions, it is possible to set forth some generalizations, as follows:

Color of character	Strokewidth-to-height ratio
Black on white	1:6–1:8
White on black	1:8–1:10

Under high illumination, higher ratios of white on black would tend to combat the irradiation effect. Table 5-3 presents one set of recommended strokewidth ratios for various styles of type and viewing conditions. These jibe fairly well with the ones given above.

Width-Height Ratio

The relationship between the width and height of alphanumeric characters is described either as width-height ratio (expressed as a ratio, such as 4:5, or as a percent, such as 80 percent) or conversely as height-width

Table 5-3   One Set of Recommended Styles of Print and Strokewidth of Capital Letters and Numerals for Various Specified Conditions

Condition	Variety of style	Strokewidth
Low level of illumination	Bold	1:5
Low contrast with background	Bold	1:5
Contrast value of 1:12 and up		
Black letters on white	Medium bold–medium	1:6–1:8
White letters on black	Medium–light	1:8–1:10
Dark letters on illuminated background	Bold	1:5
Illuminated letters on dark background	Medium–light	1:8–1:10
Highly luminous letters	Very light	1:12–1:20
Characters to be read at great distances or of		
below optimum size	Bold–medium bold	1:5–1:6

SOURCE: From *Design for legibility of visual displays,* prepared by Human Factors Group, Bendix Radio Division, Bendix Aviation Corp., Feb. 15, 1959.

ratio (such as 1.25 percent). Certain examples are shown in Figure 5-21. On logical grounds it has been proposed that, in general, the width-height ratio should be about 3:5 (width 60 percent of height) by reason of the geometric structure of many letters [66]. Many letters have what might

		Legibility of letters	
	Description	Daylight	Transillumination
FOX	Garamond Bold	Poor	Acceptable
FOX	W^* 55% of H^*	Poor	Poor
FOX	W 70% of H	Optimum	Acceptable
FOX	W 85% of H	Optimum	Acceptable
FOX	W 100% of H	Acceptable	Optimum

$^*W$ = width; H = height

Figure 5-21   Legibility of capital block letters of various width-height ratios and of Garamond Bold letters. [Adapted from data from Brown, 10.]

be considered three width elements and five height elements such as a block B. Since normal visual acuity under adequate viewing conditions makes it possible for people to see features that subtend a visual angle of 1 minute of arc, it follows that a letter with three width and five height elements would have to subtend at *least* 3 by 5 minutes of visual angle to be visible. (In practice, the sizes should be larger.) Also, in practice, this 3:5 ratio has been found to be a bit on the low side for capital letters, and ratios from about 2:3 up to 1:1 have been found to be more legible. This is indicated by the results of a study by Brown [10] in which Garamond Bold font and block-letter fonts of four ratios were used experimentally (with width 55, 70, 85, and 100 percent of the height). An evaluation of this comparison, in overly simplified form, is shown in Figure 5-21, for both daylight and transillumination conditions (in which the letters appeared red against a dark background). The upshot of various such studies has led to the adoption by the military services of a set of capital letters of a width-height ratio of 1:1, as shown in Figure 5-22.

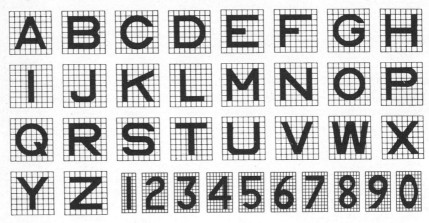

Figure 5-22   Letter and numeral font of United States Military Specification No. MIL-M-18012B (July 20, 1964); also referred to as NAMEL (Navy Aeronautical Medical Equipment Laboratory). The letters as shown have a width-height ratio of 1:1 (except for I, J, L, and W). These ratios can be reduced to about 2:3 without any appreciable reduction in legibility. The numerals have a width-height ratio of 3:5 (except 1 and 4).

For the accompanying numerals the ratio is 3:5. The narrower acceptable width of numerals probably can be attributed to the fact that the configurations of their elements are somewhat less intricate than those of letters, and are thus more legible in their narrower form than are letters.

Font of Alphanumeric Characters

Actually, most conventional fonts of alphanumeric characters (and many of the offbeat styles) can be read with reasonable adequacy under normal conditions where size, contrast, illumination, and time permit. There are, however, significant differences in the legibility and readability of different type fonts when viewing conditions are adverse, where time is important, or where accuracy is important. In this connection, the font of capital letters and numerals shown in Figure 5-22 (United States Military Specification No. MIL-M-18012B) has been rather widely tested and found to be generally satisfactory. Although specifically designated for aircrew station displays, the characters, of course, have a wide range of applicability. Another set of characters that are rather widely used by the military services is that shown in Figure 5-23 [MIL Standard MS

A B C D E F G H I J K L M
N O P Q R S T U V W X Y Z
O I 2 3 4 5 6 7 8 9

Figure 5-23 United States Military Standard letters and numerals MIL Standard MS 33558 (ASG), Dec. 17, 1957. The basic strokewidth-to-height ratio is 1:8, and the width is about 70 percent of the height. These are sometimes referred to as AND (Air Force–Navy Drawing 10400).

33558 (ASG)]. These are sometimes referred to as AND (Air Force–Navy Drawing 10400). The numerals of these two sets (Figures 5-22 and 5-23) were included with a third font, that of Berger [7, 8], in a comparative study of legibility [Atkinson et al., 2] with the following average numbers of errors:

	Average errors (*for two conditions*)	
Font	*Daylight*	*Transillumination*
NAMEL (Figure 5-22)	5.5	11.4
Berger	8.1	13.3
AND (Figure 5-23)	9.7	14.7

The fairly well-established legibility of such capital letters and numerals as the NAMEL set does not necessarily imply that they are the most legible forms possible. Curiosity keeps prodding people out of the leave-well-enough-alone attitude. In letter and numeral design, for example, some experimental forms have been developed. One set of numerals, developed by Lansdell [42] and later revised by him, was tested by Foley [25]. These numerals were compared experimentally for legibility with those of Mackworth [Bartlett and Mackworth, 5]; the two sets are shown in Figure 5-24.

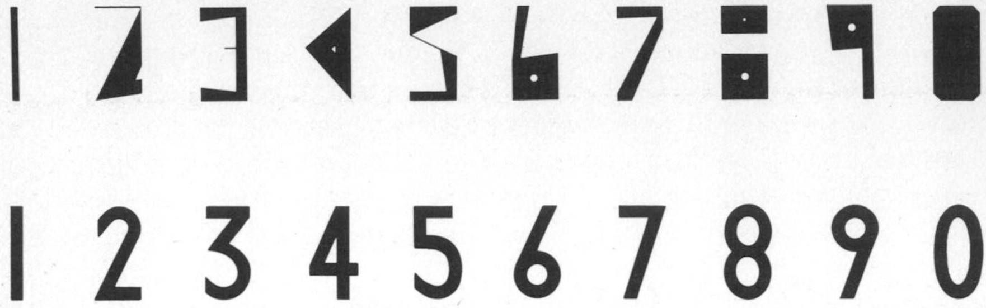

Figure 5-24 Illustration of Lansdell numerals (top row) and Mackworth numerals (bottom row). It was found that people could learn to identify the Lansdell numerals quite readily. The Lansdell design was found to be more legible under various conditions of illumination and viewing angle. [From Foley, 25. Numerals by Lansdell, 42, as later revised.]

While the Lansdell numerals cause one to look twice, it was found that people could learn to identify them readily. In comparison with the Mackworth numerals, it was found that the legibility of the Lansdell numerals was significantly better, both under straight-ahead viewing conditions and at different angles of view. It might be added that under low levels of illumination, the Lansdell numerals were most legible when presented as white on black, the reverse combination being best at high levels of illumination.

Size of Alphanumeric Characters

In the earlier discussions of visual acuity in Chapter 3, it was pointed out that the ability of people to make visual discriminations depends on such factors as size, contrast, illumination, and time. The size of alphanumeric characters that could be discriminated would, of course, depend on many conditions. A systematic method has been proposed by Peters and Adams [47] for taking certain such conditions into account in determining heights of letters and numerals for use as labels and markings on panels. In particular, their method is predicated on a formula that brings in adjustments for illumination, reading conditions, viewing distance, and the importance of reading accuracy, as follows:

H (height of letter, in.) $= 0.0022D + K_1 + K_2$

where D = viewing distance

K_1 = correction factor for illumination and viewing condition

K_2 = correction for importance (for important items such as emergency labels, $K_2 = .075$; for all other conditions, $K_2 = .0$)

This formula has been applied to various viewing distances, in combination with the other variables, to derive the heights of letters and numerals for those conditions as given in Table 5-4. The lower bounds of these values for a reading distance of 28 in. (for K_1 values of .06) are 0.12 in. for nonimportant and 0.20 in. for important markings. These correspond well with the minimum values proposed by Morgan et al. [43] of 0.10 in. for noncritical and 0.20 in. for critical or adverse reading situations; these values are also within the range proposed by Brown [10] for two sizes of letters, namely, $\frac{9}{64}$ in. (0.14 in.) for the bulk of letters and $\frac{11}{64}$ in. (0.17 in.) for emphasis. The application of the Peters and Adams formula for higher K_1 values (of 0.16 and 0.26), however, gives

Table 5-4 Table of Heights H of Letters and Numerals Recommended for Labels and Markings on Panels, for Varying Distance and Conditions, Derived from Formula H (in.) $= 0.0022D + K_1{}^* + K_2$

Viewing distance, in.	*0.0022D value*	*Nonimportant markings, $K_2 = .0$*			*Important markings, $K_2 = .075$*		
		$K_1 = .06$	$K_1 = .16$	$K_1 = .26$	$K_1 = .06$	$K_1 = .16$	$K_1 = .26$
14	0.0308	0.09	0.19	0.29	0.17	0.27	0.37
28	0.0616	0.12	0.22	0.32	0.20	0.30	0.40
42	0.0926	0.15	0.25	0.35	0.23	0.33	0.43
56	0.1232	0.18	0.28	0.38	0.25	0.35	0.45

* Applicability of K_1 values:

$K_1 = .06$ (above 1.0 fc, favorable reading conditions)
$K_1 = .16$ (above 1.0 fc, unfavorable reading conditions)
$K_1 = .16$ (below 1.0 fc, favorable reading conditions)
$K_1 = .26$ (below 1.0 fc, unfavorable reading conditions)

SOURCE: Based on formula of Peters and Adams [47]; see text.

noticeably higher recommended letter heights, ranging up to 0.40 in. Such sizes, however, would be derived from the formula only for important markings, with less than 1 fc of illumination, under unfavorable reading conditions. For greater viewing distances, of course, letter and numeral sizes need to be increased.

Directions of Contrast

The conventional white on black direction of contrast has, of course, demonstrated its general utility in a wide range of uses of alphanumeric characters. It is in some of the unconventional circumstances that questions might arise about the pros and cons of white on black versus black on white. Let us take a quick look at a couple of such circumstances. The desirable direction of contrast of alphanumeric characters presented on TV scopes, such as might be used in high-performance aircraft and spacecraft, is tied in with the level of ambient illumination [Kelly, 38]. Under very low levels of illumination (such as on night flights, or in low-illuminated cockpits), white characters on black are the more legible, whereas under high illumination (such as daylight flights, or in high-illuminated cockpits), black on white is more legible. As a result of this investigation it was proposed that where TV presentations are to be made under various ambient illumination conditions, there should be provision (either mechanical or electronic) for changing the direction of contrast. Somewhat corresponding results were reported in a study of reflectorized highway signs [Allen and Straub, 1]. These investigators reported increasing legibility of black letters on white under conditions of high illumination (100-fL), whereas the legibility of white letters on black actually was reduced under the 100-fL conditions, as contrasted with 10-fL conditions. The results of these two studies are undoubtedly the consequence of irradiation, as discussed before, and argue for avoiding white on black contrast under high levels of illumination.

Readability

As indicated earlier, readability refers generally to the recognition of the information content of material such as in words, sentences, continuous text, or other meaningful groupings of words or alphanumeric characters. The readability of printed or typed text, and its comprehension, is a function of a wide assortment of factors such as type style (font), type form (capital, lower case, bold face, italics, etc.), size, contrast, leading (spacing) between lines, length of lines, and margins. This is obviously a very broad spectrum of variables, and it is not in our province here to pull together and synthesize the research in this area.[3] However, to illustrate such research, let us summarize the results of one particular study [Poulton, 48]. In the experiment, a comparison was made of the comprehension of material printed in the four types and formats shown in Table 5-5. The subjects (275 scientists) were tested on their comprehension of the subject matter, and their average percentage scores are shown in that table in the last line. Condition A was considerably easier to comprehend than the others, probably owing largely to style of type (compared with C which was comparable otherwise in point size, number of columns, etc.), and perhaps in part to point size and number of columns (compared with B, the same style but in smaller point size and with two columns instead of one). Such a study is, of course, far from conclusive,

---

[3] For treatment of readability (as well as other aspects of alphanumeric characters) the reader is referred to such sources as Burt [13], Cornog and Rose [19], Klare [39], and Tinker [61].

Table 5-5   Effects of Printing Types and Formats on Reading Comprehension

	Style of type			
Condition	A 7 Modern Extended No. 1	B 7 Modern Extended No. 1	C 101 Imprint	D 327 Times New Roman
Point size	11	9	11	9
Letter height, in., total	0.15	0.13	0.15	0.13
Letter height, lower case x	0.07	0.05	0.07	0.06
Leading between lines, in.	0.03	0.01	0.02	0.01
Line length	5.2	2.8	5.0	2.8
Columns per page	1	2	1	2
Comprehension scores, %	63	56	58	58

SOURCE: From Poulton [48].

but at least it suggests some of the variables that may influence readability and comprehension.

As another example of factors that influence readability, Tinker [60] had some subjects read regular type such as this (actually roman type), *had other subjects read italicized type such as this,* AND HAD OTHER SUBJECTS READ CAPITALIZED TYPE SUCH AS THIS. The regular-type group read significantly faster than those who read material in all capital letters. These results undoubtedly are due in large part to familiarity of people with the conventional upper and lower case type in continuous text.

In the use of words as labels (such as on instrument panels, for identification), however, the shoe is on the other foot; words in all capital letters in this type of situation generally are more readable than those in lower case or of mixed type [Hodge, 33]. (As another indication of the effects of topography on readability, this study also indicated that readability of words used as labels was optimum with a spacing between letters of about 75 percent of the letter width.)

Readability under conditions of limited space. In some circumstances there is limited space for letters or numerals, as in the case of instrument panel windows. This raises the question as to how the available space should be used, considering the size of characters and the possible use, or nonuse, of a border. The results of an experiment by Bridgman and Wade [9] indicate that where there is restricted space for letters or numerals, readability is greater if the figures nearly use up the available space, perhaps with only a strokewidth or less between the figures and their surrounding contrasting border (figures which actually touch the border may be misread). An illustration of this is given in Figure 5-25, showing numerals of two sizes within the same limited space. It can be added that even for letters or numerals of a *given* size where space is not at a premium, a border surrounding the figures tends to improve readability, but a large surround (leaving more space between it and the figures) is better than a small one. This may seem inconsistent with the point made above regarding filling up most of the space to the border. The explanation, however, is fairly clear. Within a *given* limited space, the figures should be as large as possible (leaving some clearance); for a given *size* of figure, a larger surrounding border contributes to readability. Some of these relationships are shown in Figure 5-25.

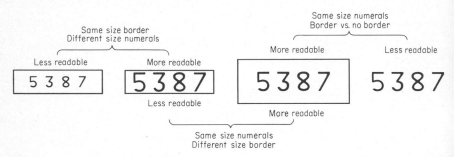

Figure 5-25   Illustration of numerals varying in degrees of clearance with border and of numerals without border.

Readability of groupings of letters and numerals.   In some circumstances combinations of letters and numerals are used for coding or identification. The ability of people to read (and remember) various types of combinations with various numbers of characters, as they might be used on automobile license plates, was investigated by Karmeier et al. [37]. Using groups of college students and police observers, he tested 16 different combinations, shown in Table 5-6. The results for these two groups are shown in that table, both in rank order of correctness of identification and in percent correct. As indicated in this table, the groupings of 5 characters generally contributed to greater accuracy than did groupings of 6 or 7. This is undoubtedly the consequence of what is sometimes referred to as *digit span* (or in this case what we might call *alphanumeric span*), since identification of characters is distinctly related to the number presented. Otherwise, however, there seems to be a tendency for people to do better when letters precede numbers (perhaps because of our familiarity with this form, as in telephone numbers) and when the numerals have their common groupings by 3s (such as 123 456), or—

Table 5-6   Rank Order and Percent of Correct Identification of 16 Different Groupings of Letters and Numerals

Form	College students		Police officers	
	Rank	% correct	Rank	% correct
12 345	1	92.0	1	98.7
A 1234	2	90.8	3	89.4
AB 123	3	89.8	2	92.7
123 AB	4	86.8	4	82.7
123 A4	5	83.0	8	79.4
1A 234	6	80.4	4	82.7
123 456	7	75.0	4	82.7
AB 1234	8	70.0	7	82.0
1234 AB	9	58.8	13	38.7
1A 2345	10	58.6	9	58.0
1234 A5	11	55.6	14	36.0
ABC 123	12	53.4	12	40.7
A1 2345	13	51.0	10	47.3
1234 5A	14	32.8	16	21.3
1234 567	15	28.9	11	42.2
1234567	16	26.0	15	30.7

SOURCE: From Karmeier et al. [37, Table 1, p. 423, and Table 5, p. 426].

when 7 digits are used—to have a space between the first 4 and the last 3 (as 1234 567 versus 1234567).

There is some support for the notion that people tend to use "natural" groupings of numerals, as suggested by the results of a study by Thorpe and Rowland [59] in which they asked subjects to repeat orally sets of 7, 8, and 9 numerals that they had previously learned. The patterns of oral response that were most commonly used were:

Number of digits	Most common patterns used		
7	3, 3, 1	3, 4	3, 2, 2
8	3, 3, 2	2, 2, 2, 2	
9	3, 3, 3		

(In the case of 7 digit numbers, the 3, 4 pattern was more commonly used than the 4, 3 pattern that Karmeier et al. [37] reported, but Karmeier did not include the 3, 4 pattern in his set of possible patterns.)

Saying What Is Meant

Your experiences in everyday reading and writing will confirm the fact that recorded verbal material does not always convey reliably the meaning that is intended. The unambiguous, understandable use of language is very pertinent to various tangents of human factors engineering, including the preparation of training materials, job aids, instructions, directions, and labels. Although we shall not presume to provide instant insight into the art of clear writing, we shall give a couple of horrible examples of unclear writing that are reported by Chapanis [14], along with some suggested modifications that obviously express more clearly what was intended. These examples, shown in Figure 5-26, are of a notice beside an

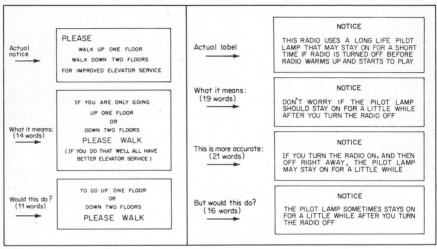

(a) Notice beside elevator (b) Label on AM–FM radio

Figure 5-26   Examples of an elevator sign and of a label on a radio, and of suggested revisions. [Adapted from Chapanis, 14.]

elevator and a label on a radio. As another example he cites the instructions included in an instruction manual, as follows:

> WARNING: The batteries in the AN/MSQ-55 could be a lethal source of electrical power under certain conditions.

On the equipment itself, however, was a much more expressive translation that someone had printed in large red letters:

> LOOK OUT! THIS CAN KILL YOU!

The moral of these examples needs no elaboration.

SYMBOLIC CODES

In Chapter 4, and earlier in this chapter, we discussed types of visual codes, including brief mention of symbolic codes. Such codes are used in graphic displays, CRTs, displays to present status information such as signs and identifications, and printed materials, etc.

Road Signs

Road signs represent a natural situation for use of symbolic codes, such as the shape of signs and the shape of configurations on them. Such signs, of course, have to be visible at appropriate distances and under many conditions and discriminable from other signs, and any alphanumeric characters on them must be readable. However, there are two other characteristics that it would be desirable for road signs to have. First, insofar as possible they should be visually suggestive of that which they are intended to symbolize, in order to minimize the "recoding" of symbols. And second, they should generally be standardized across those geographical boundaries that are commonly traversed. In both of these considerations, the International Road Signs used in most European countries and in many others would generally receive higher marks than those in the United States, especially since there is still considerable diversity in the design of road signs from one state to another. Some examples of the International Road Signs are shown in Figure 5-27.

$Figure\ 5\text{-}27$ Examples of a few International Road Signs. These signs conform to two useful design principles, namely, they are directly symbolic of their meaning, and they are standardized across countries.

DISCUSSION

In this chapter we have discussed and illustrated a variety of visual displays. The sample of displays covered, however, is only suggestive of the wide range of visual displays that are in actual use. There are also many specialized types of displays that it has not been possible to include. In the selection or design of visual displays for certain specific purposes, the basic type of display to use is sometimes virtually dictated by the nature of the information to be presented and the use to which it is to be put. In other circumstances, however, options may be available. And in some situations the designer must use considerable ingenuity to work out an optimum design that is consistent with human sensory, perceptual, and mediation characteristics.

Although it is not possible to provide specific guidelines to follow in resolving every design problem, there are certain general guidelines and principles that can be followed in many situations. A few are given below, along with some words of caution that circumstances sometimes will justify deviations from these. These guidelines deal largely with some of the more conventionally used visual displays.

General Guidelines in Selecting Visual Displays for Certain Purposes

Quantitative reading

· Counter or open window preferable if values remain long enough to read.

· Fixed scale and moving scale acceptable.

· For long scales, moving scale with tape on spools behind panel or a counter plus circular scale has practical advantage over fixed scale.

· If possible, display information for time determined by reader (usually 0.50 sec is adequate).

· For values subject to continuous change, display all (or most) of range used (as with circular or horizontal scale).

· If two or more items of *related* information are to be presented, consider integrated display.

· Smallest scale unit to be read should be represented on scale by about 0.05 in. or more.

· Preferably use marker for each scale unit, unless scale has to be very small.

· Use conventional progression system of 1, 2, 3, 4, etc., unless there is reason to do otherwise.

Qualitative reading

· Fixed scale preferable (moving pointer of scale, especially circular scale, shows trend and rate of change).

· Avoid moving scale and open-window designs.

· If certain ranges of quantitative scale have specific meanings, precode each such range.

Check reading

· Use indicator light if feasible and if used *strictly* for check reading.

· Otherwise, fixed scale satisfactory; identify null position clearly.

· For groups, use circular scales, and arrange null positions systematically for ease of visual scanning, as at 9 o'clock or 12 o'clock positions.

· Preferably use extended pointers, and possibly extended lines between scales.

Setting quantitative values

· Fixed scale preferable (provides better visual cues of rate and relative position).

· Moving scale can be used in some circumstances.

Tracking

· Fixed scale preferable to moving scale.
· Counter is unsatisfactory.

Warning

· Flashing lights, shutters, or other clearly visible signals (auditory signals usually preferable).

Indicating current condition

· Depending on situation, use signal lights, semiphore, shutters, etc., so each signal is clearly distinguishable.
· If underlying variable is quantitative, fixed scale can be used with clearly identified ranges.

Status information (usually static status)

· Signs, labels, color codes, lights, etc., depending on situation.

Codes (used in many situations)

· Need to be visually detectable.
· Need to be discriminable each from the others.
· Preferably should be compatible.
· Preferably should symbolize what they represent.
· Preferably should be standardized.

SUMMARY

1. Displays, which present information indirectly to people, should be used when relevant distal stimuli (e.g., original stimuli) cannot be sensed at all by people or cannot be sensed adequately.
2. *Dynamic* displays transmit information that is subject to change; *static* displays present information that is not subject to change.
3. Displays are used to obtain information of various types, such as (*a*) quantitative information; (*b*) qualitative information; (*c*) check information; (*d*) status and warning information; (*e*) tracking information; (*f*) time-phased information; (*g*) pictorial and graphic information; (*h*) identification information; (*i*) alphanumeric and symbolic information; and (*j*) speech and related auditory information.
4. The speed and accuracy of reading quantitative instruments are related to the basic design of the instrument, its detailed features, and the context in which the instrument is used.
5. In the design of aircraft-position displays the following principles are proposed: (*a*) display integration (the combining of *related* infor-

mation in a common display); (*b*) pictorial realism; (*c*) moving part (the aircraft should be depicted as the moving part rather than the horizon); (*d*) pursuit tracking (as contrasted with compensatory tracking); (*e*) frequency separation (see text); and (*f*) optimum scaling (scale dimensions should be reasonably optimum for adequate reading). Although these principles are specifically for aircraft-position displays, they probably also apply to other display problems.

6. Graphic displays that depict trends are read most accurately and rapidly if the trends are depicted by lines, rather than by series of bars (either vertical or horizontal); for presenting comparative information, multiple-line graphs (one graph with two or more lines) are generally preferable to the use of two or more separate graphs.

7. Capital letters and numerals used in visual displays are read most accurately (*a*) when the ratio of strokewidth to height is about 1:6 to 1:8 for black on white and somewhat higher (up to 1:10) for white on black, and (*b*) when the width is at least two-thirds the height.

8. The desirable size of letters and numerals depends upon the viewing distance, the illumination level, reading conditions, and other variables.

9. *Irradiation* is a phenomenon in which white lines on black tend to appear wider than they are and black lines on white tend to appear narrower. It occurs primarily when the eyes are adapted to darkness.

REFERENCES

1. Allen, J. M., and A. L. Straub: *Sign brightness and legibility*, Virginia Council of Highway Investigation and Research, Reprint 16, October, 1956.

2. Atkinson, W. H., L. M. Crumley, and M. P. Willis: *A study of the requirements for letters, numbers, and markings to be used on trans-illuminated aircraft control panels. Part 5. The comparative legibility of three fonts for numerals*, Naval Air Material Center, Aeronautical Medical Equipment Laboratory, Report TED NAMEL–609, part 5, June 13, 1952.

3. Baker, C. A., D. F. Morris, and W. C. Steedman: Target recognition of complex displays, *Human Factors*, May, 1960, vol. 2, no. 2, pp. 51–61.

4. Balding, G. H., and C. Süskind: Generation of artificial electronic displays, with application to integrated flight instrumentation, *IRE Transactions on Aeronautical and Navigational Electronics*, September, 1960, vol. ANE–7, no. 3, pp. 92–98.

5. Bartlett, F. C., and N. H. Mackworth: *Planned seeing*, H. M. Stationery Office, London, 1950.

6. Bauerschmidt, D. K., and S. N. Roscoe: A comparative evaluation of a pursuit moving-airplane steering display, *IRE Transactions on Human Factors in Electronics*, 1960, vol. HFE–1, pp. 62–66.

7. Berger, C.: I, Stroke-width, form and horizontal spacing of numerals as determinants of the threshold of recognition, *Journal of Applied Psychology*, 1944, vol. 28, pp. 208–231.

8. Berger, C.: II, Stroke-width, form and horizontal spacing of numerals as determinants of the threshold of recognition, *Journal of Applied Psychology*, 1944, vol. 28, pp. 336–346.

9. Bridgman, C. S., and E. A. Wade: Optimum letter size for a given display area, *Journal of Applied Psychology*, 1956, vol. 40, pp. 378–381.

10. Brown, F. R.: *A study of the requirements for letters, numbers, and markings to be used on trans-illuminated aircraft control panels. Part 4. Legibility of uniform stroke capital letters as determined by size and height to width ratio and as compared to Garamond Bold*, Naval Air Material Center, Aeronautical Medical Equipment Laboratory, Report TED NAMEL–609, part 4, Mar. 10, 1953.

11. Brown, F. R., and E. A. Lowery: *A study of the requirements for letters, numbers, and markings to be used on trans-illuminated aircraft control panels. Part 1. The effect of stroke width upon the legibility of capital letters*, Naval Air Material Center, Aeronautical Medical Equipment Laboratory, Report TED NAMEL–609, part 1, Sept. 26, 1949.

12. Brown, F. R., E. A. Lowery, and M. P. Willis: *A study of the requirements for letters, numbers, and markings to be used on trans-illuminated aircraft control panels. Part 3. The effect of stroke-width and form upon the legibility of numerals*, Naval Air Material Center, Aeronautical Medical Equipment Laboratory, Report TED NAMEL–609, part 3, May 24, 1951.

13. Burt, C. L.: *A psychological study of typography*, Cambridge University Press, London, 1959.

14. Chapanis, A.: Words, words, words, *Human Factors*, February, 1967, vol. 7, no. 1, pp. 1–17.

15. Chapanis, A., and L. C. Scarpa: Readability of dials at different distances with constant visual angle, *Human Factors*, 1967, vol. 9, no. 5, pp. 419–425.

16. Christensen, J. M.: *Quantitative instrument reading as a function of dial design, exposure time, preparatory fixation and practice*, USAF AML Report AF 52/116, 1952.

17. Christner, Charlotte A., and H. W. Ray: An evaluation of the effect of selected combinations of target and background coding of map-reading performance—Experiment V, *Human Factors*, 1961, vol. 3, pp. 131–146.

18. Churchill, A. V.: The effect of pointer width and mark width on the accuracy of visual interpolation, *Journal of Applied Psychology*, 1960, vol. 44, pp. 315–318.

19. Cornog, D. Y., and F. C. Rose: *Legibility of alphanumeric characters and other symbols: II. A reference handbook*, National Bureau of Standards, Miscellaneous 262–2, Superintendent of Documents, Washington, D.C., February, 1967.

20. Crawford, A.: The perception of light signals: the effect of mixing flashing and steady irrelevant lights, *Ergonomics*, 1963, vol. 6, pp. 287–294.

21. Dashevsky, S. G.: Check-reading accuracy as a function of pointer alignment, patterning, and viewing angle, *Journal of Applied Psychology*, 1964, vol. 48, pp. 344–347.

22. Davidson, C. E.: *Photographic interpretation: social influence on stereoscopic vs. nonstereoscopic viewing*, M.S. thesis, Purdue University, Lafayette, Ind., 1963.

23. Elkin, E. H.: *Effect of scale shape, exposure time and display complexity on scale reading efficiency*, USAF, WADC, TR 58–472, February, 1959.

24. Fitts, P. M.: Engineering psychology and equipment design, in S. S. Stevens (ed.), *Handbook of experimental psychology*, John Wiley & Sons, Inc., New York, 1951.

25. Foley, P. J.: Evaluation of angular digits and comparison with a conventional set, *Journal of Applied Psychology*, 1956, vol. 40, pp. 178–180.

26. Graham, N. E.: The speed and accuracy of reading horizontal, vertical, and circular scales, *Journal of Applied Psychology*, 1956, vol. 40, pp. 228–232.

27. Green, B. F., and Lois K. Anderson: Speed and accuracy of reading coordinates on a horizontal plotting table, *Journal of Applied Psychology*, 1955, vol. 39, pp. 227–236.

28. Grether, W. F.: Instrument reading: I. The design of long-scale indicators for speed and accuracy of quantitative readings, *Journal of Applied Psychology*, 1949, vol. 33, pp. 363–372.

29. Grether, W. F., and A. C. Williams, Jr.: "Speed and accuracy of dial reading as a function of dial diameter and angular separation of scale divisions," in P. M. Fitts (ed.), *Psychological research in equipment design*, Army Air Force, Aviation Psychology Program, Research Report 19, 1947.

30. Hamer, H., and H. Critchley: *Dial study of simulated speedometers*, unpublished paper, Department of Psychology, Purdue University, Lafayette, Ind., August, 1967.

31. Hill, J. H., and R. Chernikoff: *Altimeter display evaluation: final report*, USN, NEL Report 6242, Jan. 26, 1965.

32. Hitt, W. D.: An evaluation of five different coding methods, *Human Factors*, July, 1961, vol. 3, no. 2, pp. 120–130.

33. Hodge, D. C.: Legibility of uniform-strokewidth alphabet: I. Relative legibility of upper and lower case letters, *Journal of Engineering Psychology*, 1962, vol. 1, pp. 34–46.

34. Johnsgard, K. W.: Check-reading as a function of pointer symmetry and uniform alignment, *Journal of Applied Psychology*, 1953, vol. 37, pp. 407–411.

35. Jones, J. C., A. J. Ward, and P. W. Haywood: *Reading dials at short distances*, AEI Engineering (Associated Electrical Industries Ltd., London), Jan.–Feb., 1965, vol. 5, no. 1, pp. 28–32.

36. Kappauf, W. E., W. M. Smith, and C. W. Bray: *A methodological study of dial readings*, Princeton University, Department of Psychology, Report 3, August, 1947.

37. Karmeier, D. F., C. G. Herrington, and J. E. Baerwald: "A comprehensive analysis of motor vehicle license plates," in H. O. Orland (ed.), *Highway Research Board, Proceedings of the thirty-ninth annual meeting*, NRC Publication 773, 1960, vol. 39, pp. 416–440, National Academy of Sciences, Washington, D.C.

38. Kelly, R. B.: *The effect of direction of contrast of TV legibility under varying ambient illumination*, Dunlap and Associates, Inc., Darien, Conn., Contract Nonr 1076(00), June, 1960.

39. Klare, G. R.: *The measurement of readability*, Iowa State University Press, Ames, 1963.

40. Kuntz, J. E., and R. B. Sleight: Legibility of numerals: the optimal ratio of height to width of stroke, *American Journal of Psychology*, 1950, vol. 63, pp. 567–575.

41. Kurke, M. I.: Evaluation of a display incorporating quantitative and check-reading characteristics, *Journal of Applied Psychology*, 1956, vol. 40, pp. 233–236.

42. Lansdell, H.: The effect of form on the legibility of numbers, *Canadian Journal of Psychology*, 1954, vol. 8, pp. 77–79.

43. Morgan, C. T., J. S. Cook, III, A. Chapanis, and M. W. Lund (eds).: *Human engineering guide to equipment design*, McGraw-Hill Book Company, New York, 1963.

44. Murrell, K. F. H.: *Human performance in industry*, Reinhold Publishing Corporation, New York, 1965.

45. Murrell, K. F. H., W. D. Laurie, and C. McCarthy: The relationship between dial size, reading distance and reading accuracy, *Ergonomics*, 1958, pp. 182–190.

46. Oatman, L. C.: Check-reading accuracy using an extended-pointer dial display, *Journal of Engineering Psychology*, 1964, vol. 3, pp. 123–131.

47. Peters, G. A., and B. B. Adams: These three criteria for readable panel markings, *Product Engineering*, May 25, 1959, vol. 30, no. 21, pp. 55–57.

48. Poulton, E. C.: *Effects of printing types and formats on the comprehension of scientific journals*, Applied Psychology Research Unit, Cambridge, England, Report APU 346, 1959.

49. Roscoe, S. N.: Airborne displays for flight and navigation, *Human Factors*, 1968, vol. 10, no. 4, pp. 321–332.

50. Sabeh, R., W. R. Jorve, and J. M. Vanderplas: *Shape coding of aircraft instrument zone markings*, USAF, WADC, Technical Note 57–260, March, 1958.

51. Schutz, H. G.: An evaluation of formats for graphic trend displays—Experiment II, *Human Factors*, 1961, vol. 3, pp. 99–107.

52. Schutz, H. G.: An evaluation of methods for presentation of graphic multiple trends—Experiment III, *Human Factors*, 1961, vol. 3, pp. 108–119.

53. Senders, V. L.: *The effect of number of dials on qualitative reading of a multiple dial panel*, USAF, WADC, TR 52–182, 1952.

54. Simon, C. W., and S. N. Roscoe: *Altimetry studies: II. A comparison of integrated versus separated, linear versus circulatar, and spatial versus numerical displays*, Hughes Aircraft Company, Culver City, Calif., Technical Memorandum 435, May, 1956.

55. Sleight, R. B.: The effect of instrument dial shape on legibility, *Journal of Applied Psychology*, 1948, vol. 32, pp. 170–188.

56. Soar, R. S.: Stroke width, illumination level, and figure-ground contrast in numeral visibility, *Journal of Applied Psychology*, 1955, vol. 39, pp. 429–432.

57. Steedman, W. C., and C. A. Baker: Target size and visual recognition, *Human Factors*, August, 1960, vol. 2, no. 3, pp. 120–127.

58. Thomas, D. R.: Exposure time as a variable in dial reading experiments, *Journal of Applied Psychology*, 1957, vol. 41, pp. 150–152.

59.   Thorpe, C. E., and G. E. Rowland: The effect of "natural" grouping of numerals on short term memory, *Human Factors*, 1965, vol. 7, pp. 38–44.

60.   Tinker, M. A.: Prolonged reading tasks in visual research, *Journal of Applied Psychology*, 1955, vol. 39, pp. 444–446.

61.   Tinker, M. A.: *Legibility of print*, Iowa State University Press, Ames, 1963.

62.   Vernon, M. D.: *Scale and dial reading*, Cambridge University, England, Medical Research Council, Unit in Applied Psychology, Report APU 49, June, 1946.

63.   Williams, A. C., C. W. Simon, R. Haugen, and S. N. Roscoe: *Operator performance in strike reconnaissance*, USAF, WADD TR 60–521, 1960.

64.   Zeff, C.: Comparison of conventional and digital time displays, *Ergonomics*, 1965, vol. 8, no. 3, pp. 339–345.

65.   Zeidner, J., R. Sadacca, and A. I. Schwartz: *Human factors studies in image interpretation: the value of stereoscopic viewing*, USA, HFRB Technical Research Note 114, June, 1961.

66.   *Design for legibility of visual displays*, Bendix Aviation Corp., Bendix Radio Division, Human Factors Group, Baltimore, Feb. 15, 1959.

AUDITORY

AND

TACTUAL

DISPLAYS

In the days gone by, man received most of his information directly, via the original (natural) distal stimuli; thus, the wailing of wolves at night was sensed via sound waves generated by the wolves, and the presence of approaching ships was sensed by direct vision. But now it is possible to change one form of energy into another and thus to alter completely the nature of stimuli from some source. We use radar to detect ships and aircraft; we can show "pictures" of different speech sounds; we can sense the presence of people by an olfactometer that shows a reading on an instrument; we can detect fish by electronic devices; and we can cause a buzzer to warn a blind person of some physical obstacle. The possibilities of converting stimuli that intrinsically are associated with one sensory modality into a form that stimulates another sensory mechanism open wide the doors to some interesting new uses for certain senses, in particular the auditory and tactual senses.

Actually, we depend heavily upon these senses for many mundane aspects of daily life. But technology is presenting these senses with new kinds of opportunities. In this chapter we shall discuss the use of auditory displays and then touch briefly on the embryonic but developing field of tactual displays. (We shall hold in abeyance a discussion of speech communication until the next chapter.)

AUDITORY DISPLAYS

Auditory signals are particularly appropriate for certain uses such as the following: as warning devices or to attract attention; to transmit information when the receiver moves from place to place or frequently shifts visual attention (and, therefore, cannot always be watching for visual signals); when vision cannot be used (such as at night); to relieve the visual modality when it is overburdened; in many instances to transmit signals that originate in auditory form; and to transmit directional information (as to aircraft pilots). Although the auditory modality is a fairly versatile sense, in normal circumstances it is most useful for transmitting relatively short, simple messages.

The gamut of auditory displays includes conventional sound-generating devices such as bells, buzzers, and sirens; various kinds of electronic systems such as sonar and aircraft radio range systems; and communication equipment used either for voice communications (which will be discussed in the next chapter) or for other types of signals such as radiotelegraph communications. The more conventional devices are heard directly by the receivers, whereas most electronic and communication equipment usually requires earphones or loudspeakers.

Considerations in Selection of Auditory Displays

In the design or selection of auditory displays for any given situation, at least three types of considerations are relevant. In the first place are those relating to the intended function of the display. In considering the intended function, Harris and Levine [5] suggest that the following aspects (here presented in the form of a series of questions) may be pertinent:

1. Is the signal to be a warning, a call, or an instruction?
2. Will the signal indicate an emergency or routine situation?
3. How much time will be available to take action?
4. Will the signal sound at regular intervals or infrequently?
5. Must it function near other signals with which it might be confused?
6. Will the signal be protecting life or valuable property?

The second type of consideration relates to the situation, with particular focus on other meaningful auditory inputs and the ambient noise (its intensity, spectral characteristics, etc.). And the third (and the most important) type of consideration relates to the human receivers of the signals; the signals obviously need to be within the repertoire of the sensory, perceptual, and information-handling abilities of the receivers. Since a recitation of data relevant to these considerations probably could fill the rest of this book, we shall discuss briefly a few examples of auditory displays and then pull together a few general principles on the use of auditory displays as predicated on human abilities.

Nonelectronic Sound-generating Devices

Some of the conventional nonelectronic sound generators are listed below, along with brief comments about their characteristics.

1. Buzzers. Buzzers have about the lowest decibel ratings of any warning devices and usually operate in the lower frequency range, from about 150 to 400 Hz. There are, however, certain types that have important frequency components up to about 1500 Hz. They are

particularly suitable for fairly quiet situations where a distinctive sound commands attention yet does not cause alarm. Coding can be accomplished by combinations of signals and time intervals.

2. Bells. Bells normally have a higher intensity level and frequency range than buzzers and are therefore usually more suitable for areas with somewhat higher-intensity ambient noise, especially low-frequency noise. They can be used as warning devices or for transmitting other types of information.

3. Horns. Horns generally are described as "growlers" (with high intensity, 90 to 100 dB, and low frequency) and "screamers" (also with high intensity, but with higher frequencies, say, up to 3000 to 5000 Hz). They typically are designed so as to direct the energy in one direction, but by rotation they can cover a wide range. They penetrate noise well and attract attention.

4. Sirens. Sirens with wailing frequencies are particularly useful as warning devices because they can attract attention and penetrate ambient noise. However, they do not lend themselves to very elaborate coding of information, and high-intensity sirens can approach the upper limit of tolerance, especially for nearby receivers.

5. Diophones. The most common use is as foghorns. They are quite satisfactory for attracting attention. They can penetrate high-frequency ambient noise better than low-frequency noise.

6. Chimes. Chimes are not as alerting as certain other devices, and so generally should not be used as emergency signals. The frequencies selected should be those that are minimal in the ambient noise. By using different frequencies, chimes can be used for perhaps four or five different coding purposes.

7. Whistles. With their high intensities and potential range of frequencies whistles lend themselves to satisfactory use as warning devices and for other information purposes.

Table 6-1 shows the intensity ranges and predominant frequencies of a few devices, separated into those that are most suitable for large and for small areas.

Signaling devices for various situations. Some very general suggestions have been made concerning the appropriateness of certain types of signaling devices for specific circumstances [adapted in part from Harris and Levine, 5]:

- Fairly quiet locations (50 to 60 dB): heavy-duty buzzer or chime; for signaling one person, such as a secretary, a light-duty buzzer or 1-in. bell
- Open factory, such as light assembly area (70 to 80 dB): 6-in. heavy-duty bell usually sufficient; to signal one or two persons, a light-duty 3-in. bell or heavy-duty buzzer

Table 6-1 Intensity Ranges and Predominant Frequencies of Certain Auditory Signal Devices

Type of device	Average intensity level, dB		Predominant audible frequency
	At 10 ft	At 3 ft	
For large areas, high-intensity coverage			
4-in. bell	65–77	75–83	1000
6-in. bell	74–83	84–94	600
10-in. bell	85–90	95–100	300
Horn	90–100	100–110	5000
Siren	100–110	110–121	7000
For small areas, low-intensity coverage			
Heavy-duty buzzer	50–60	70	200
Light-duty buzzer	60–70	70–80	400–1000
1-in. bell	60	70	1100
2-in. bell	62	72	1000
3-in. bell	63	73	650
Chime	69	78	500–1000

SOURCE: Harris and Levine [5].

- Noisy factory area, such as machine shop or punch-press room (90 to 100 dB): horn or 10-in. bell required
- Outdoors: siren

Audio Warning Signals for Air Force Weapon Systems

Certain work situations require particular consideration in the design or selection of auditory signals, for example, the design of warning signals for Air Force weapon systems. Because of the importance of this problem in aircraft and because of some of the problems associated with such warning systems, an analysis of research and experience in the use of auditory signals for such systems was made by Licklider [10]. A summary of his analysis will be given below. While certain aspects mentioned will have unique applicability to Air Force systems, some of the aspects have much broader implications and probably can then be extrapolated (where pertinent) to other situations.

Functions of warning signals. Licklider points out that, in the Air Force context with which he was concerned, warning signals should embody three components, as follows:

- A (for attention or alert). To attract and, if necessary, hold the operator's attention.

· G (for general category). To designate the general category of exigency (or type of emergency).
· S (for specific condition or suggestion). To identify the condition specifically and/or to suggest appropriate action.

Further, he suggests that warning signals that symbolize these three components could be combined or arranged in the following possible patterns:

1. AGS. One signal to serve all three purposes (many signals in use are of this type)
2. A, G, S. Three signals in sequence, one for each function
3. A, GS. Two signals, one to attract attention, A, the other to identify both the general category and the specific condition or suggestion, GS.
4. AG, S. Two signals, one to attract attention and to identify the general category, AG, the other to indicate the specific condition or suggestion, S.

On the basis of a rational analysis of knowledge of hearing and attention, plus experience, Licklider comes to the conclusion that the most promising arrangement for at least certain Air Force warning systems is type 4, a nonspeech signal fulfilling the AG function and a recorded speech signal fulfilling the S function. Such a system would, of course, require special equipment so programmed that appropriate speech material would be transmitted for any given exigency. Although this type of system is actually in operation, such elaborations are, of course, not practical in most circumstances in which warning signals are used, and, in fact, may not be warranted in many situations. However, this type of dissection of the functions of warning signals has implications for certain other types of circumstances where less elaborate devices might be used. One can readily conceive of circumstances in which several types of exigencies might arise, which would suggest that the warning signals used should attract attention, A, and differentiate between different general types of exigencies, G, and in some instances indicate the specific situation, S.

Audio Information Displays

Most auditory signals are used to transmit *warnings*. Some such signals, however, are used to transmit other types of information; a couple of examples of *information* signals will be given.

Sonar.   There are two types of sonar used in naval operations. Passive sonar consists of equipment that is designed to sense underwater sounds that are generated by ships, torpedoes, and other objects with mechanical sound sources. The sensed vibrations are converted electronically into auditory signals that are transmitted to operators through earphones. On the other hand, in the use of active sonar (also called *echo-ranging sonar*), a high-frequency sound is transmitted underwater; if it strikes an object, the sound will be reflected back from the object to the ship, where it will be picked up and reproduced. If the object that the soundwaves strike is moving away, the returning sound will have a lower frequency than the one transmitted; if moving toward the original source, the returning sound will have a higher frequency. This is predicated on the Doppler effect. The sonar operator then has to judge whether the outgoing or returning signal is higher in pitch in order to determine whether the object is going away or coming closer. In both types of sonar the auditory discriminations to be made require careful selection of operators for their auditory-discrimination skills, and extensive training of operators to enable them to interpret the signals received. To aid the operators in making the discriminations required, a number of schemes are used to modify or clarify the signals, such as filters to reduce noise or the translation of high frequencies to lower frequencies that usually can be discriminated more accurately.

Radio range signals.   Another example of a situation that requires discrimination between sounds is the reception of radio range signals in aircraft, in which the A (dot-dash) or the N (dash-dot) is heard if the pilot is to the left or to the right, respectively, of the center beam. The signal in one beam is on when that in the other is off. When the beams are of equal strength, the pilot hears a continuous signal meaning that he is on the center beam. Under adverse noise conditions the difference between the A and N signals may not be properly identified, and the pilot may think he is to the right of the beam when actually he is to the left, or vice versa.

Some of the aspects of this particular problem have been investigated under simulated signal and noise conditions that were somewhat like those encountered in an airplane [Flynn et al., 2]. Without discussing the details of this investigation, the results probably have important implications for various auditory displays. In particular the signal-to-noise ratio was found to be a much more critical factor in effective auditory discriminations than the intensity of the signal itself is.

Principles of Auditory Display

As in other areas of human factors engineering, most guidelines and principles have to be accepted with a few grains of salt, since specific circumstances, trade-off values, etc., may argue for their violation. With such reservations in mind, a few guidelines for the use of auditory displays are given below. These generally stem from research findings and experience. Some of them are drawn in part from Mudd [11] and Licklider [10].

1. General principles

(A) *Situationality:* The efficiency of any given auditory display is dependent upon the total situation in which the display is used. The design of such displays should take into consideration other relevant characteristics of the system in which the display is found (e.g., noise levels, and type of response controlled by the auditory signal).

(B) *Compatibility:* Where feasible, the selection of signal dimensions and their encoding should exploit learned or natural relationships of the users, such as high frequencies being associated with up or high, and wailing signals, with emergency. Where appropriate, the stimulus signals should "explain" the responses to be executed so that compatibility is optimized. For example, a pure tone signal to "lift up" would be a tone of *increasing* pitch rather than a tone of *decreasing* pitch.

(C) *Approximation:* Two-stage signals should be considered when complex information is to be displayed and a verbal signal is not feasible. The two stages should consist of:

(1) Attention-demanding signal: to attract attention and identify a general category of information.

(2) Designation signal: to follow the attention-demanding signal and designate the precise information within the general class indicated above.

(D) *Dissociability:* Auditory signals should be easily discernible from any ongoing audio input (be it other meaningful input or noise). For example, a warning bell would not stand out if the operator happened to be a Swiss bell ringer; in such a case, some other type of signal probably would be better, such as a siren.

(E) *Parsimony:* Input signal to the operator should not provide more information than is necessary to carry out the proper response. For instance, instructions signaled to one position in a system need not be signaled to other positions if the positions are autonomous.

(F) *Forced entry:* When more than one kind of information is to be presented to the receiver, the signals should be designed to prevent the receiver from listening to only part of the total signal.

(G) *Invariance:* The same signal should designate the same information at all times.

2. Principles of presentation

(H) *Avoid extremes of auditory dimensions:* This will generally keep the signals more within the range of discriminability of people.

(*I*) *Establish intensity relative to ambient noise level.*

(*J*) *Use interrupted or variable signals:* Where feasible, avoid steady-state signals and, rather, use interrupted or variable signals. This will tend to minimize perceptual adaptation.

3. Special principles concerning quantitative information

(*K*) *Use standard reference tone for comparative purposes:* If, say, frequency is to indicate a quantitative value, a reference tone should be used to indicate a "base line."

(*L*) *Use cueing signal to warn of "on-course" position:* Where auditory signals are used in some continuous tracking task, use a continuous scale indication, or one composed of small steps, to warn of approaching on-course information.

4. Special principles concerning warning signals

(*M*) *Alerting:* Preferably use high-intensity sudden-onset signals. Where earphones are used, consider *dichotic* presentation (alternating signal from one ear to the other). Variable-frequency sounds (such as sirens) are more alerting than steady-state sounds.

(*N*) *Nondistracting:* Although warning signals should be alerting, they should not distract or frighten receivers. This is possible in part by avoiding signals with unnecessarily high intensities or that have strong affective associations.

(*O*) *Quick acting:* The signal should convey critical information as quickly as possible. Alerting should be accomplished in less than 1 sec, and if subsequent signal components are to convey other information (such as the G or S components above), these should be presented within the next 2 sec.

(*P*) *Discriminability of different signals:* If different warning signals or signal components are used, each should be clearly discriminable from the others.

(*Q*) *Nonmasking:* Warning signals should be of such a nature that they will not mask other important signals or be masked by other signals. Dichotic signals may be useful when earphones are used. Avoid frequencies that are used for other purposes.

(*R*) *Separate communications system:* Where feasible, use a separate communication system for warnings, such as loudspeakers or other devices not used for other purposes.

(*S*) *Nonpainful and nondamaging:* Avoid intensities that approach painful or damaging levels. When extremely high intensities must be used, avoid concentration in single or restricted frequencies; use low (rather than high) frequencies and short durations.

5. Principles of installation of auditory displays

(T)   *Test signals to be used:* Such tests should be made with a representative sample of the potential user population to be sure the signals can be detected by them.

(U)   *Avoid conflict with previously used signals:* Any newly installed signals should not be contradictory in meaning to any somewhat similar signals used in existing or earlier systems.

(V)   *Facilitate changeover from previous display:* Where auditory signals are to replace some other mode of presentation (such as visual), preferably continue both modes for a while, to help people become accustomed to new auditory signals.

(W)   *Emphasize reliability:* An auditory signaling system should be as error-free as possible, in order to obtain and maintain user confidence.

Comparison of Visual and Auditory Modalities

In the development of systems, there sometimes is an option in the sensory modality to use in presenting information by displays. Most commonly this choice is between the visual and auditory modalities. Thus, a comparison of the advantages and disadvantages of these two modalities may aid in deciding which to use or in deciding whether to use them in combination.

Relative characteristics of vision and hearing. In comparing vision and hearing in their suitability for displaying information, Henneman [7] characterizes these modalities as follows:

1. Auditory stimuli are essentially temporal; the information is extended through time. Visual stimuli, however, are characteristically spatial, having a position or location in space.

2. Auditory stimuli typically arrive sequentially in time, whereas visual stimuli may be presented either sequentially or simultaneously.

3. By reason of the sequential presentation of auditory stimuli, they have poor "referability," meaning that they usually cannot be kept continuously before the observer, although they can be repeated periodically. Visual stimuli offer good referability, because the information usually can be stored in the display.

4. Auditory stimuli offer fewer dimensions for the coding of information than visual stimuli do.

5. Speech (as one form of auditory stimuli) offers greater flexibility, such as off-the-cuff variations in connotations, nuances, and inflections. Visual stimuli, on the other hand, require advance coding.

6. The selectivity of messages in speech offers a time advantage, since the pertinent information is already selected for the receiver. With

visual stimuli, however, searching for information may be necessary, such as looking for information from tables, charts, and maps.

7. The rate of transmission of speech is limited to the speaking rate, whereas visual presentations can be faster.

8. Auditory stimuli are more attention-demanding; they break in on the attention of the operator. Visual stimuli, however, do not necessarily have this captive audience; the operator has to be looking toward the display in order to receive the stimulus.

9. Hearing is somewhat more resistant to fatigue than vision is.

Attention-demanding qualities of visual and auditory stimuli. Mention was made above of the attention-demanding advantage of auditory as opposed to visual stimuli. This advantage occurs primarily when there is some form of distraction. The results of an experiment by Henneman [7] will serve to illustrate this point. Subjects were presented with standard reading-test material by vision (slides) and by audition (tapes). Two forms of distraction were used, namely, visual (a motion picture that required occasional responses) and manual (a knob-turning task). The average scores of subjects on a reading test (including those of a control group) are given below:

Distraction	Auditory presentation	Visual presentation
Visual	2.44	2.10
Manual	2.65	2.05
None (control)	2.85	3.24

The results of this study suggest that a visual method of presentation may be more susceptible to adverse effects than an auditory method of presentation, although this may not apply across the board to all types of information and situations.

TACTUAL DISPLAYS

The sense of touch has of course been used as a means of communication for the blind through the use of material printed in Braille, and in a few other situations such as in the shape coding of control knobs that are to be identified by touch. In recent years, however, there has been a flurry of interest in the possible use of the sense of touch for other communication purposes, especially in situations where the visual and auditory channels are already burdened. This interest, in part, has been ignited by a few exploratory studies that have suggested considerable promise for this method of communication, both in static and dynamic contexts.

Static Tactual Displays

Static tactual displays might be used for identification of control knobs (as mentioned above), for labels on displays, and perhaps for other identification purposes. In such instances the problem is one of creating configurations that can be discriminated tactually. As a case in point, Austin and Sleight [1] tested the ability of people to identify letters and numerals made of ¼-in. masonite. They found that about 60 percent of the symbols were identified correctly at or above a 90 percent accuracy level. The 11 symbols identified most accurately, in order of accuracy of identification, were C I V O 7 L U J D E T.

A study in the use of raised letters in words was carried out by Seminara [13]. He used letters of ¼-in. brass sheet metal, formed into 90 test words ranging from two to seven letters per word. The subjects "read" these words by touching the letters. Their time and errors were recorded and are shown below:

No. of letters in word	Mean time per word, sec	Errors in 90 trials
2	5.5	4
3	8.4	3
4	11.0	0
5	15.9	2
6	17.9	3
7	24.5	3

Studies such as the above indicate the feasibility of communication by the use of static tactual displays. The efficiency of such a static system, however, is admittedly low (as shown by the fact that a seven-letter word takes an average of 24.5 sec to read).

Dynamic Tactual Displays

The more exciting possibilities in the use of tactual displays are in transmission of dynamic information. Gilmer [in Hawkes, 6, pp. 76–84] suggests the possible use of tactual communications for transmitting information of the following classes: quantitative information; coordinates in space; direction; rates; language; attention demanding; and vigilance.

In discussing the types of energy to which the nerve endings are capable of responding, Geldard [3] mentions the following: (1) mechanical; (2) electrical; (3) thermal; and (4) chemical. Considering the effects on the skin, the adaptation of the skin to stimuli, and other factors, it is probable that mechanical and electrical stimulation offer the greatest promise.

Mechanical stimuli.   There are two general approaches to the use of mechanical stimuli in tactual communications. The first of these consists in the use of some type of vibrator affixed to the surface of the skin and used to transmit coded vibrations. The coding of vibrations can be based on such physical parameters as location(s) of vibrator(s), frequency of vibrations, intensity, and duration. The selection of specific parameters should, of course, be predicated on the sensory and perceptual skills of people in discriminating the various stimuli that might be used in a coding system. In this connection, the exploratory research of Geldard and his associates has been particularly significant [Geldard, 3; Geldard and Sherrick, 4; and Sherrick, in Hawkes, 6, pp. 147–158]. His experiments in tactual communications have been concerned with vibrations on the chest. After some preliminary studies he developed a tactual communication system that used five locations on the chest, three levels of intensity of the vibrations at these locations, and three durations. Combinations of these (5 × 3 × 3) provided the basis for a vibratory language of 45 unique patterns, which in turn were used for coding the 26 letters, 10 numerals, and 4 frequently used words (of, the, in, and). These codes are depicted in Figure 6-1.

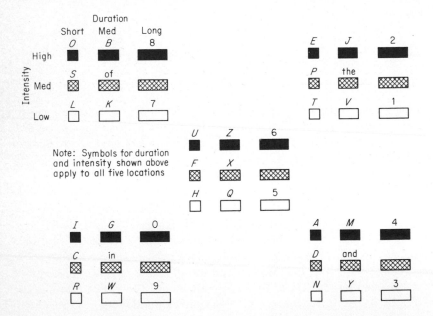

Figure 6-1 Coding of the vibratory language used by Geldard [3]. Each group of nine·symbols is to be thought of as belonging to a single vibrator which varies in intensity (three steps) and duration (three steps). [From Geldard, 3.]

In experiments with this system, subjects were able to learn the code system fairly well within about a 12-hr period. From rates of reception, it was estimated that a well-trained subject might receive as many as 67 words per minute by this system, well over the most demanding military requirement for Morse code reception of 24 words per minute.

Considering the various possible parameters for use in such communications, Alluisi [in Hawkes, 6, pp. 115–116] suggests that man's tactual discrimination is limited to two or three levels of intensity, two or three steps of *change* of intensity, three or four steps of duration, and six or seven positions (loci) on the chest. However, considering the possible areas of the body, a number of other locations might be identified that would permit accurate discrimination of vibrations, perhaps (as Alluisi suggests) as many as 26—one for each letter of our alphabet.

The second general approach to the use of mechanical stimuli is by the transmission to the skin of amplified speech sounds transmitted by a single vibrator [Myers, 12]. In the various phases of the experiment, 16 selected words and numbers and 17 basic speech sounds (phonemes such as e, g, th, s, z, i, and ng) were recorded on tape. The frequencies of these recordings were then reduced by rerecordings to one-half (for the words and numerals) or one-eighth (for the phonemes). These reduced-frequency recordings were then amplified and transmitted to the skin via a vibrator. To indicate the possible utility of their procedure, two subjects were able to receive the phonemes with a median accuracy of 91 percent after eight practice sessions.

Electrical stimuli. A basic problem in the possible use of electrical stimuli is that they are, as Gilmar [in Hawkes, 6, pp. 76–84] suggests, between the boundaries of "pain" and "painless pulses." These boundaries take in several parameters such as intensity, polarity, duration, and interval, and also electrode type, size, and spacing. A few of these parameters have been explored. For example, Gilmer suggests that it probably

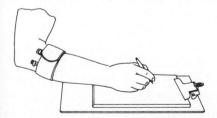

Figure 6-2 Cutaneous transducer assembly on forearm of receiver preparatory to receiving message by electrocutaneous communication system. [From Hennessy, 8.]

would be possible to use three levels of intensity, six loci (three on each palm), and two *duration-depth* dimensions in an electropulse communication system, which would allow for 36 discrete symbols such as 26 letters and 10 digits. Additional locations would, of course, increase the range of possibilities. Although considerable progress has been made in the development of hardware components for electrocutaneous communication systems [Hennessy, 8], such systems are not as yet at the stage of practical use—although that time is coming. In connection with such systems, an illustration of a cutaneous transducer assembly is shown in Figure 6-2.

Discussion

The most demanding use of tactual communications probably would be for the transmission of alphanumeric information; each character would require its own code, and this code would have to be learned. In contemplating the potential uses of tactual displays, one can easily be persuaded to agree with Howell [in Hawkes, 6, pp. 103–113] in taking a dim view of the prospects of extensive use of this mode for transmitting this type of information. He points out that the tactual system is not as well equipped to resolve small intensive and qualitative differences as are the visual and auditory systems, and that there would be little point in investing the necessary months in training that would be required to bring the tactual mode up to the level of performance of the visual and auditory systems.

On the other hand, the tactual system seems to be well suited for readily transmitting a limited number of discrete stimuli such as might be needed in warning or alerting situations. Further, the attention-demanding characteristic of tactual stimuli would seem to recommend it for possible use in vigilance tasks, if the infrequently occurring signals of such tasks could be converted to, say, electrocutaneous form. Howell [in Hawkes, 6] and Hirsch and Kadushin [9] have explored the use of tactual stimuli as directional cues in tracking and aircraft-control tasks, with rather encouraging results. And when the visual and auditory channels are overburdened, a tactual communication system would be especially handy for getting important messages through. If the interest and enthusiasm of the "tickle talk" investigators of the past few years are any indication of things to come, the prospects for the use of tactual communications in the lives of at least some of us are not far away.

SUMMARY

1. Since various types of sound-generating devices have their own characteristics, the selection of one for a particular purpose should be based on its unique characteristics as related to the specific purpose.

2. In the use of auditory displays, the signal-to-noise ratio is a more critical factor in detection of signals than signal intensity is.

3. Auditory signals usually are more *attention getting* than visual signals and thus lend themselves for use as warning signals.

4. Audio warning signals in some circumstances (such as for certain Air Force systems) need to perform three functions, namely, attract attention, indicate the general nature of the emergency, and indicate the specific conditions or suggest appropriate action. When the situation requires such a warning system, one signal component can perform the first two in combination, followed by a second signal to perform the third function.

5. In the design of auditory displays, various principles can serve as guidelines where they are appropriate. Some of the most important are the following: compatibility (using signals that already have meaning for the purpose at hand), dissociability (discernibility from any other ongoing audio input), forced entry (signal should be such that the receiver cannot ignore any aspect of it), and invariance (the same signal should designate the same information at all times).

6. Of the forms of energy that might be used for tactual displays, mechanical and electrical stimulation offer the greatest promise.

REFERENCES

1. Austin, T. R., and R. B. Sleight: Accuracy of tactual discrimination of letters, numerals, and geometric forms, *Journal of Experimental Psychology*, 1952, vol. 43, pp. 239–247.

2. Flynn, J. P., S. J. Goffard, I. P. Truscott, and T. W. Forbes: *Auditory factors in the discrimination of radio range signals: collected informal reports*, Harvard University, Psycho-acoustic Laboratory, OSRD Report 6292, Dec. 31, 1945.

3. Geldard, F. A.: Adventures in tactile literacy, *American Psychologist*, 1957, vol. 12, pp. 115–124.

4. Geldard, F. A., and C. E. Sherrick: Multiple cutaneous stimulation: The discrimination of vibratory patterns, *Journal of the Acoustical Society of America*, 1965, vol. 37, no. 5, pp. 787–801.

5. Harris, E. A., and W. E. Levine: How to specify audible signals, *Machine Design*, Nov. 9, 1961, vol. 33, no. 23, pp. 166–174.

6. Hawkes, G. R. (ed.): *Symposium on cutaneous sensitivity*, USA Medical Research Laboratory, Ft. Knox, Ky., Report 424, Dec. 22, 1960.

7. Henneman, R. H.: Vision and audition as sensory channels for communication, *Quarterly Journal of Speech*, 1952, vol. 38, pp. 161–166.

8. Hennessy, J. R.: Cutaneous sensitivity communications, *Human Factors*, 1966, vol. 8, no. 5, pp. 463–470.

9. Hirsch, J., and I. Kadushin: *Experiments in tactile control of flying vehicles*, The Sixth Annual Conference on Aviation and Astronautics, Tel Aviv and Haifa, Feb. 24–25, 1964.

10.   Licklider, J. C. R.: *Audio warning signals for Air Force weapon systems*, USAF, WADD, TR 60–814, March, 1961.
11.   Mudd, S. A.: *The scaling and experimental investigation of four dimensions of pure tone and their use in an audio-visual monitoring problem*, unpublished Ph.D. thesis, Purdue University, Lafayette, Ind., 1961.
12.   Myers, R. D.: *A study in the development of a tactual communication system*, Symposium on the Air Force Human Engineering, Personnel, and Training Research, 1960, NAS, NRC, Publication 783 pp. 238–243.
13.   Seminara, J. L.: Accuracy and speed of tactual reading, an exploratory study, *Ergonomics*, 1960, vol. 3, pp. 62–67.

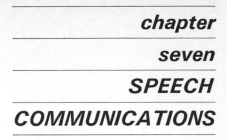
Language, of course, is a system of coding, each language representing a different system. Written and spoken forms are parallel systems for presenting meaning by the visual and auditory senses. In discussing speech communications Denes and Pinson [5] refer to the "speech chain" as comprising different forms, or levels, in which a spoken message exists in its progress from the mind of the speaker to the mind of the listener, as follows:

Speaker

1. *Linguistic level:* neural activity in speaker's brain
2. *Physiological level:* the neural and muscular activity in forming speech sounds

Transmission system

3. *Acoustic level:* transmission of sound waves from speaker to receiver (*Note:* If some communication equipment is used, the sound waves may be converted to some other form, such as electromagnetic energy, and then converted back to sound waves.)

Receiver

4. *Physiological level:* neural activity in hearing and perceptual mechanisms
5. *Linguistic level:* recognition of words and sentences

Although most speech communications involve such direct transmission, communication systems such as the telephone, intercommunication systems, radio, and TV serve as intervening mechanisms in this process, especially to extend the range of the human voice and to aid in communications under conditions of noise. Such systems, of course, do not always transmit signals with high reliability. In this connection, although the unreliable telephone transmission of some initially unreliable gossip should not be the cause for very much alarm, the criticalness of some communications, such as in airport control towers, imposes extremely high standards of reliability in the transmission of speech. The possibilities of enhancing the transmission of speech probably can be considered in terms of the major "components" of the speech communications system in question, namely, the message to be transmitted, the talker (the person transmitting the message), the transmission system (i.e., telephone, radio, intercom, etc.) including whatever noise is present, and the receiver (the person receiving the message). Before discussing these, however, we should mention briefly certain characteristics of speech and the measurement of speech intelligibility.

CHARACTERISTICS OF SPEECH

The speech of any given language consists of speech sounds called *phonemes*. These phonemes are generated by the *articulators* (the lips, tongue, teeth, and palate) as these interact to interrupt or constrict the breath stream. In the formation of possible phonemes by the processes of articulation, there are five types of articulation [Miller, 10, p. 17]:

· *Plosives*, or *stops*, produced by completely stopping the passage of air, as *p* in *pop*
· *Fricatives*, or *spirants*, produced by forming a narrow slit or groove for air passage, as *th* in *their*
· *Laterals*, formed by closing the middle line of the mouth, and leaving an air passage around one or both sides, as *l* in *let*
· *Trills*, caused by rapid vibration of an articulator, as the trilled *r* in certain European languages
· *Vowels*, produced by an unobstructed air passage

Each basic type of articulation has several possible variations that can be produced by changes in the positions of the articulators, which gives a wide range of possible phonemes. The English language includes about 16 different vowel sounds, the specific number depending on whether certain closely related sounds are counted as the same or different; and it includes about 22 consonant sounds, making a total of about 38 phonemes. Other languages employ some phonemes that we do not use, and vice versa.

Each phoneme, then, has its own unique phonetic characteristic in its combination of frequencies and intensity (or speech power), and these form the building blocks of the spoken English language. The physical characteristics of speech sounds give rise to corresponding sensations, as follows:

Speech characteristic	Corresponding sensation
Intensity, or speech power, usually dB	Loudness
Frequency	Pitch
Harmonic composition	Quality

Intensity of Speech

The average intensity, or speech power, of individual phonemes varies tremendously, with the vowels generally having much greater speech power than the consonants. For example, the *o* as pronounced in *talk* has roughly 680 times the speech power of *th* as pronounced in *then* [Fletcher, 6]. This is a difference of about 28 dB.

The overall intensity of speech, of course, varies from one person to another and also for the same individual. Fletcher [6, pp. 76–77], for example, shows that when one talks almost as softly as possible, speech has a decibel level of about 46, and when one talks almost as loudly as possible, it has a decibel level of about 86. The results of a survey by Fletcher [6, p. 77] of the telephone-speech levels of a good-sized sample of people showed a range from about 50 to about 75 dB, with a mean of about 66 dB. There was, however, a heavy concentration of cases from about 60 to 69 dB, with about 40 percent falling within 3 dB above or below the mean.

Frequency of Speech

As indicated above, each phoneme has its own unique spectrum of several frequencies, although there are, of course, differences in these among people, and each individual can (and does) shift his spectrum up and down the register depending in part on the circumstances, such as talking in a quiet conversation, when the pitch and loudness are usually relatively low, or talking to a group, when the pitch and loudness typically are higher, or screeching at the children to stay out of the wet varnish on the floor (when the pitch and loudness usually are near their peak). Across the board, men have a somewhat lower spectrum than women. Figure 7-1 shows as line *S* the average speech spectrum for a group of men [Beranek, 1]. This figure divides the range of frequencies from 200 to 6100 into 20 bands that contribute equally to what is called an *articulation index*. (Reference will be made later to line *N* of this figure.)

Quality of Speech

The quality of speech is really the mixture of pitch and loudness. In a sense every phoneme has its own distinguishing quality, and the speech of individuals also has its own quality that makes it possible to identify the voices of individuals.

SPEECH INTELLIGIBILITY

In an actual communication situation the criterion of voice communication is generally its intelligibility to a receiver. For evaluating speech communication under different conditions (e.g., under noise, with differ-

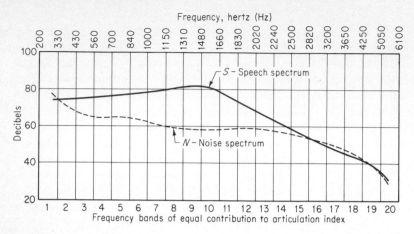

Figure 7-1 Average speech spectrum for men, with superimposed noise spectrum, to illustrate derivation of articulation index. [Adapted from Beranek, 1.]

ent communication systems, and at various distances), and for research, we need some measure of the intelligibility of speech.

Tests of Speech Intelligibility

The most straightforward method of measuring speech intelligibility is by the use of some test. Such tests usually are by the transmission of speech material to a receiver, who is then asked to repeat what he hears. The material may consist of complete sentences, of independent words, or of nonsense syllables. One fairly common type of test consists of *phonetically balanced* (PB) word lists (see Chapter 16 on noise for examples). These consist of 50 words of one syllable that have been carefully selected so that the complete list represents fairly well the phonemes used in conventional speech. The score of an individual receiver is based on the number of words in the list that he is able to recognize. Speech intelligibility tests actually can serve several purposes. Our present interests lie in the uses mentioned above, those of speech research and of evaluating speech communications under various conditions. In addition, however, the tests can be used to measure the speech effectiveness and the hearing abilities of individuals.

The Articulation Index

Although speech intelligibility tests can be used in the evaluation of communication systems, the use of such tests is time-consuming and usually requires numerous subjects with carefully controlled experimental conditions. Since the use of such tests is then not always practical, some of the experimenters of the Bell Telephone Laboratory have developed a method of estimating the intelligibility of speech [Fletcher, 6, French and Steinberg, 7]. These procedures call for the computation of an articulation index (AI). While the development of the AI need not be described here, it may be pertinent to see how this index is obtained and also to see how it can be used. To help illustrate the method, refer to Figure 7-1. In this figure, line S represents the average speech spectrum for a sample of men, and line N represents the overall noise spectrum of a particular situation. In deriving the AI for such a situation the following procedures are used:

1. For each of the 20 frequency bands, determine the difference in decibels between the speech and noise level for the band, using the midpoint of the band; this is the speech-to-noise ratio for that band.
2. Obtain weight for each band from Figure 7-2; read across the base scale to determine the point that represents the decibel difference for

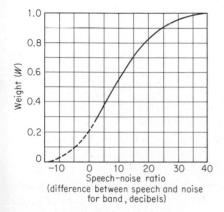

Figure 7-2 Weighting factor for various speech-to-noise ratios for deriving articulation index (AI). For each frequency band, the decibel difference between the speech and overall noise level in that band, obtained from Figure 7-1, is multiplied by the weight W obtained from this figure. [Adapted from French and Steinberg, 7.]

the band, and then trace this position vertically to the point where it intersects the curve; then read the weight W on the vertical scale that corresponds to this point. This is the weight for the band.
3. Multiply the weight for each band by 0.05.
4. Add these products for all 20 bands. This sum is the AI.

This method of deriving AIs is most applicable for conditions in which the noise is steady, rather than widely variable, and for conditions in which the noise has a wide spectrum, rather than a narrow spectrum.

Intelligibility Score

An intelligibility score usually is simply an estimate of the percentage of spoken material that can be understood. It is probably more meaningful than the AI, and in fact, the AI frequently is converted into an intelligibility score. One illustration of this conversion is shown later in this chapter in Figure 7-5; (that particular example deals particularly with the effects on intelligibility of the context of speech material). Incidentally, Figure 7-5 can be used to estimate the percentage of intelligibility (for sentences, isolated words, or syllables) for any given AI. Thus, upon deriving an AI for any given communication situation or system, it is possible to obtain an approximation of the percentage of intelligibility that might be expected under the circumstances. It has been suggested that any system for which the AI is less than 0.3 might be thought of as inadequate, although the importance of reasonably complete intelligibility of the communication would be pertinent to setting any minimum level of acceptability [Beranek, 1].

Speech Interference Level

The speech interference level (SIL) is another index that is used in estimating the effects of noise on speech intelligibility, and has been used by engineers as a gross basis for comparing the relative effectiveness of speech transmission under different environments of reception. For any given situation, it is actually the simple numerical average of the decibel level of noise in three octave bands, namely, those with centers at 500, 1000, and 2000 Hz. Earlier practice provided for the SIL to be the average of the following bands: 600 to 1200, 1200 to 2400, and 2400 to 4810 Hz and in some instances 300 to 600 Hz. The results of the SIL based on the earlier and current practices are similar, but some shift in the reference values is necessary [Webster, 14]. If, in either case, the decibel levels in the three octave bands are 70, 80, and 75 dB, respectively, the SIL would be their average, or 75 dB. Although the SIL is a fairly handy measure to use, and is useful as a rough index in estimating the effects of noise on intelligibility of speech in many circumstances, it can be misleading under other circumstances. It is most appropriate when the noise spectrum is reasonably flat across the three octave bands. If the spectrum is markedly different from a flat one, the SIL will not give a close estimate of intelligibility [Rosenblith and Stevens, 13, pp. 200–207]. In particular, it can be misleading if there are intense low-frequency components, if the spectrum is irregular, if the noise consists primarily of pure tones, if the noise has *square* waves or pulses or is irregular or interrupted. In addition, the adequacy of the SIL depends on the character of the voice of the speaker, his location relative to the receiver, and the nature of the material being spoken. In the use of the SIL, then, these limitations need to be kept in mind.

Noise Criteria Curves

A further elaboration of the SIL theme resulted in the development of a set of *noise-criteria* (NC) curves [Beranek, 2] as shown in Figure 7-3. The derivation of these will not be discussed, but in general they take into

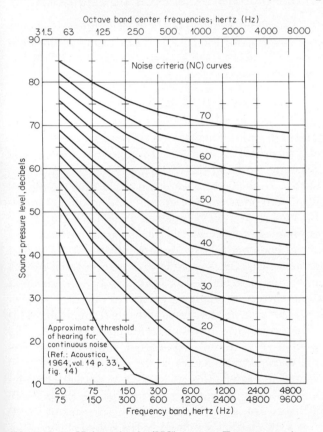

Figure 7-3 Noise criteria (NC) curves. For any specified standard, such as an NC of 30, no frequency band of the noise should exceed the NC curve in question. Each NC curve has a loudness level in phons that is 22 units greater than the SIL in decibels that is expressed by the NC number of the curve. [Adapted from Beranek, 2.]

account the SIL and loudness level in phons (as discussed in Chapter 3); each NC curve has a loudness level in phons that is 22 units greater than the SIL in dB, which is expressed by the NC number of the curve. In use, individual NC curves are recommended as permissible noise levels in the eight octave bands; no single octave band should exceed the NC curve used as the standard. Each recommended NC value (such as given later in Table 7-1) is intended to permit adequate speech communications required for the room or space in question and to minimize annoyance in that area.

COMPONENTS OF SPEECH COMMUNICATIONS

If we need to do something about improving the intelligibility of speech communication in some system, we need to do so in terms of the individual components involved, such as the message itself, the talker, the transmission system, and the receiver.

The Message

Under adverse communication conditions such as noise, some speech messages or message units are more susceptible to degradation than others. Under such conditions, then, it behooves one to construct messages in such a way as to increase the probabilities of their getting through.

The vocabulary used.   If we picked a word randomly from the dictionary, the probability of your guessing the correct one would be infinitesimally small. (P.S. Would you have guessed *pedantic?*) But if we arbitrarily restrict our language to two words (for example, *pedantic* and *jostle*), your probability of guessing the right one would be 50–50. Under extremely noisy conditions, when it is difficult to make out the speech sounds, the total number of possible words that *might* be used has a marked influence on the correct recognition of the words—the smaller the possible vocabulary, the greater the probability of recognition. This general principle has been confirmed in experiments such as the one by Miller, Heise, and Lichten [11]. In this experiment subjects were presented with words from vocabularies of various sizes (2, 4, 8, 16, 32, and 256 words, and unselected monosyllables). These were presented under noise conditions with signal-to-noise ratios ranging from −18 to 9 dB. The results, shown in Figure 7-4, show clearly that the percent of words correctly recognized was very distinctly related to the size of the vocabulary that was used.

In certain types of operational situations (such as control-tower operations, and voice radio) the vocabularies that are used are in fact very restricted; actual experience probably had demonstrated the principle that speech reception is better when there is a limited assortment of possible words or other components of the message.

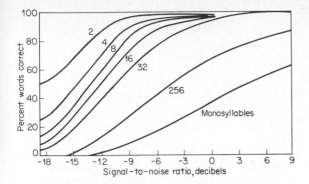

Figure 7-4 Intelligibility of words from vocabularies of different sizes under varying noise conditions. The numbers 2, 4, 8, etc., refer to the number of words in the vocabulary used. [From Miller, Heise, and Lichten, 11. Copyright 1951 by the American Psychological Association and reproduced by permission.]

The context of the message. Closely related to the size of the vocabulary used is the context of the message or message components. This is essentially a problem of expectancy. If one were to hear the expression "a rolling————gathers no moss," but failed to distinguish the third word, we could readily supply it in the context of the expression. But it would be hard for us to fill in this blank: "On Wednesday he ————." The effects of context on intelligibility are illustrated in Figure 7-5. In that figure it can be seen that, for any given AI (as derived by the method discussed earlier in this chapter), sentences are more intelligible than isolated words, and isolated words, in turn, are more intelligible than separate syllables. Thus, components in the context of a meaningful conceptual message stand a better chance of being picked up against a noisy background than do those that have no such contextual backdrop.

Phonetic aspects of the message. As indicated earlier, some speech sounds have more speech power than others, and thus can more likely get through adverse noise conditions or inadequate communication systems than sounds with lower speech power. Under adverse conditions, then, speech should include predominantly speech sounds that have high probabilities of being heard and recognized. In this connection, the military services of the United States and the United Kingdom have adopted a set of alphabetic equivalents (such as the now familiar *Roger*) in which a word is used to represent a letter. This list is referred to as the *US-UK phonetic alphabet.* The International Civil Aviation Organization has also adopted a phonetic alphabet, called the *ICAO list,* to be used in interna-

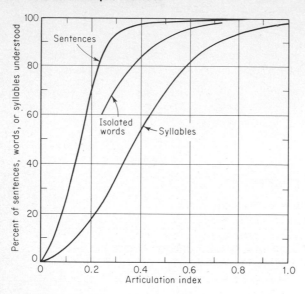

Figure 7-5 Approximate relationship between articulation index (AI) and percent of intelligibility for
sentences, isolated words, and syllables. [Adapted from
French and Steinberg, 7.]

tional aviation communications. The words selected to represent the
various letters have not been pulled out of a hat but rather have been
selected on the basis of thorough and extensive research on their phonetic
characteristics, in order to identify those words that are least likely to be
confused with others under adverse communication conditions and that
can penetrate adverse noise conditions. It might be added that these phonetic alphabets also have the advantages of comprising a limited "vocabulary" and of providing a familiar context to aid in their intelligibility.

The Talker

As would be expected, the intelligibility of speech depends to a considerable extent upon the character of the talker's speaking voice, such as its
intensity, enunciation, dialect, and the like. Included in the research
relating to talkers, one study dealt with the speech characteristics of
good talkers [Bilger, Hanley, and Steer, 4]. A group of highly intelligible
and a group of extremely unintelligible subjects were selected, and
recordings made of their speech. Characteristics of their speech were
then analyzed from these recordings, and comparisons made between
the two groups. The two groups were significantly different in certain
speech characteristics, but not in others. The characteristics in which
they differed significantly were the following:

1. Syllable duration (in seconds). The superior group had longer average syllable duration.
2. Syllable intensity (in decibels). The superior group spoke with greater intensity.
3. Percentage of speech time. The superior group utilized proportionately more of the total time with speech sounds and less with pauses.
4. Pitch variability. The individuals in the superior group varied more in the pattern of fundamental vocal frequencies.

There are probably two general factors that bring about differences in the intelligibility of the speech of individuals. In the first place, the structure of the articulators of individuals certainly influences the phonetic characteristics of speech and, therefore, its intelligibility to others. And in the second place, the speech habits people have acquired usually are fairly well ingrained. These two factors probably operate against the possibility of major improvement in the speech of many individuals, but at the same time it has been found that at least moderate improvements in the intelligibility of the speech of most people can be expected following appropriate speech training programs.

The Transmission System

For our purposes we can regard anything that intervenes between the speaker and receiver as part of the transmission system, including communicating mechanisms such as telephones, radios, etc., ambient noise in the environment, and in-line noise in any communication equipment. High fidelity in communication equipment is generally assumed to contribute to optimum intelligibility of speech. However, this is not necessarily so, as demonstrated by Licklider [8]. He found that certain types of distortion (actually amplitude distortion) left speech very intelligible, and in fact, under some circumstances, distorted speech was even more intelligible than undistorted speech. Since some forms of distorted speech can be intelligible, and since high fidelity systems are very expensive, it is useful to know just what effects various forms of speech distortion have on intelligibility to be able (it is hoped) to make better decisions about the design or selection of components of communication equipment. The components of such equipment can produce a variety of forms of distortion, such as frequency distortion, filtering, amplitude distortion, modifications of the time scale, and displacement of the speech spectrum along the frequency scale. For illustration we shall discuss the effects of filtering and amplitude distortion.

Effects of filtering on speech. The filtering of speech consists basically in blocking out certain frequencies and permitting only the remaining frequencies to be transmitted. Filtering may be the fortuitous, unintentional consequence, or in some instances the intentional consequence, of the design of a component. Most filters eliminate frequencies *above* some level (a *low-pass* filter) or frequencies *below* some level (a *high-pass* filter). Typically, however, the cutoff, even if intentional, usually is not precisely at a specific frequency, but rather tapers off over a range of adjacent frequencies. A given filtering affects the intelligibility of certain phonemes more than others. Fletcher [6, pp. 418–419], for example, points out that long *e* is recognized correctly about 98 percent of the time if the frequencies either above or below 1700 Hz are eliminated. But while *s* is affected only slightly by eliminating frequencies below 1500, its intelligibility is practically destroyed by eliminating frequencies above 4000. Most short vowels have important sound components below 1000 Hz, and 20 percent error in their recognition occurs when frequencies below that level are eliminated; on the other hand, the elimination of frequencies above 2000 has little effect on their intelligibility.

However, the intelligibility of normal speech does not depend entirely upon intelligibility of each and every speech sound. For example, filters that eliminate all frequencies above, or all frequencies below, 2000 Hz will *in the quiet* still transmit speech quite intelligibly, although the speech does not sound natural [Rosenblith and Stevens, 13, p. 112]. The distortion effects on speech of filtering out certain frequencies are summarized in Figure 7-6. This shows, for high-pass filters and for low-pass filters with various cutoffs, the percentage of intelligibility of the speech that

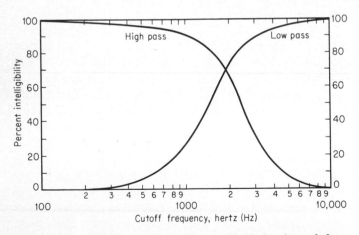

Figure 7-6 Effect on intelligibility of elimination of frequencies by the use of filters. A low-pass filter permits frequencies below a given cutoff to pass through the communication system, and eliminates frequencies above that cutoff. A high-pass. filter, in turn, permits frequencies above the cutoff to pass and eliminates those below. [Adapted from French and Steinberg, 7.]

is transmitted. It can be seen, for example, that the filtering out of frequencies above 4000 Hz or below about 600 Hz has relatively little effect on intelligibility. But look at the effect of filtering out frequencies above 1000 Hz or below 3000 Hz! Such data as those given in Figure 7-6 can provide the designer of communications equipment with some guidelines to follow when trying to decide how much filtering can be tolerated in the system.

Effects of amplitude distortion on speech. Amplitude distortion has been defined as the deformation which results when a signal passes through a nonlinear circuit [Licklider, 8]. One form of such distortion is *peak clipping*, in which the peaks of the sound waves are clipped off and only the center part of the waves left. Although peak clipping is produced by electronic circuits for experiments, some communication equipment has insufficient amplitude-handling capability to pass the peaks of the speech waves and at the same time provide adequate intensity for the lower-intensity speech components, thus reducing intelligibility. Since peak clipping impairs the quality of speech and music, it is not used in regular broadcasting, but it is sometimes used in military and commercial communication equipment. In such cases premodulation clippers are built into the transmitters, thus reducing the peaks, but the available power is then used for transmitting the remainder of the speech waves. Center clipping, on the other hand, eliminates the amplitudes below a given value and leaves the peaks of the waves. Peak clipping is more of an experimental procedure than one used in practice. In either peak or center clipping the amount of clipping can be controlled through the electronic circuits. Figure 7-7 illustrates the speech waves that would result from both forms of clipping.

The effect of these two forms of clipping on speech intelligibility are very different, as shown in Figure 7-8. It can be seen that peak clipping

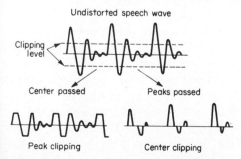

Figure 7-7 Undistorted speech wave and the speech waves that would result from peak clipping and center clipping. [From Miller, 10, p. 72.]

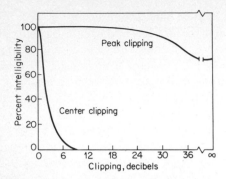

Figure 7-8 The effects on speech intelligibility of various amounts of peak clipping and of center clipping. [From Licklider and Miller, 9.]

does not cause major degradation of intelligibility even when the amount of clipping (in decibels) is reasonably high. On the other hand, even a small amount of center clipping results in rather thorough garbling of the message. The reason for this difference in the effects of peak and center clipping is the difference in phonetic characteristics of the vowels (which generally have more speech power) and of the consonants (which generally have lower speech power). Thus, when the peaks are lopped off, we reduce the power of the vowels, which are less critical in intelligibility, and leave the consonants essentially unscathed. But when we cut out the center amplitudes, the consonants fall by the wayside, thus leaving essentially the high peaks of the vowels. Since intelligibility is relatively insensitive to peak clipping, the communications engineers can shear off the peaks and repackage the wave forms, using the available power to amplify the weaker, but more important consonants.

Effects of noise on speech. Some noise is ever with us, but in large doses it is a potential bugaboo as far as speech communications are concerned, whether it is ambient noise (noise in the environment) or in-line noise in a communication system. The effects of various levels of ambient noise on speech are shown in Figure 7-9. That figure shows the SIL for voice levels at various distances. For any given distance and SIL, it is possible to see how loud one would have to talk or yell to be heard. For example, at a distance of 12 ft with a SIL of 66, it would be necessary to shout to be understood. If the noise level is much above this, it would probably be best to get out the boy scout wigwag flags or send up Indian smoke signals.

The subjective reactions of people to noise levels in private offices and in large offices (secretarial, drafting, business machine offices, etc.) were elicited by Beranek and Newman [3] by questionnaires. The results

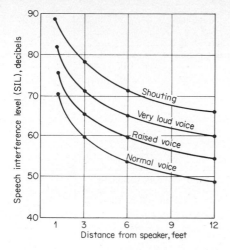

Figure 7-9 Speech interference levels for voices of different levels at various distances. In any given situation the SIL should be less than that given in order to have reliable communications at the distances and voice levels shown. The SIL used is based on the average decibel levels of octave bands with centers at 500, 1000, and 2000 Hz. [Adapted from Peterson and Gross, 12, Table 3-4.]

of this survey were used to develop a rating chart for office noises, as shown in Figure 7-10. The line for each group represents the SIL (base lines) that were judged to exist at certain subjective ratings (vertical scale). The dot on each curve represents the judged upper limit for intelligibility. The judged limit for private offices (normal voice at 9 ft) was slightly above 45 dB, and for larger offices (slightly raised voice at 3 ft) was 60 dB.

The judgments with regard to telephone use are given below:

SIL, dB	*Telephone use*
Less than 65	Satisfactory
65–80	Difficult
Above 80	Impossible

These judgments were made for long-distance or suburban calls. For calls within a single exchange, about 5 dB can be added to each of the above levels, since there is usually better transmission within a local exchange.

Criteria for control of background noise in various communication situations have been set forth by Peterson and Gross [12] based on earlier

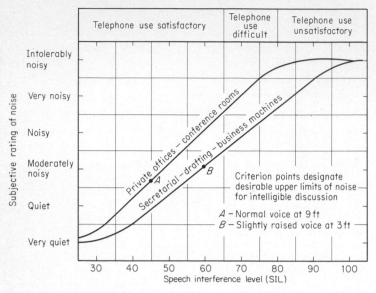

Figure 7-10 Rating chart for office noises. [Based on data from Beranek and Newman, 3, as modified by Peterson and Gross, 12, to reflect the current practice of using octave bands with centers at 500, 1000, and 2000 Hz.]

standards by Beranek and Newman [3]. These, expressed as SIL, are given in Table 7-1, along with the corresponding NC recommended subse-

Table 7-1 Speech Interference Levels and Noise Criteria Recommended for Certain Types of Rooms

Type of room	Maximum permissible level (measured in vacant rooms)	
	SIL	NC
Secretarial offices, typing	60	50–55
Coliseum for sports only (amplification)	55	50
Small private office	45	30–35
Conference room for 20	35	30
Movie theater	35	30
Conference room for 50	30	20–30
Theaters for drama, 500 seats (no amplification)	30	20–25
Homes, sleeping areas	30	25–35
Assembly halls (no amplification)	30	
Schoolrooms	30	25
Concert halls (no amplification)	25	15–20

SOURCE: SIL data from Beranek and Newman [3] as modified by Peterson and Gross [12, p. 65] to reflect current practice of using octave bands with centers at 500, 1000, and 2000 Hz; NC data from Beranek [2].

quently by Beranek [2]. (Aside from certain minor differences, the systematic difference of about 5 dB is due to the fact that the SIL data have been modified to reflect the current practice of using bands with centers at 500, 1000, and 2000 Hz.)

Effects of reverberation on speech. Reverberation is the effect of noise bouncing back and forth from the walls, ceiling, and floor of an enclosed room. As we know from experience in some rooms or auditoriums, this reverberation seems to obliterate speech or important segments of it. Figure 7-11 shows approximately the reduction in intelligibility that is

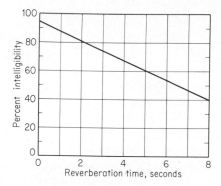

Figure 7-11 Intelligibility of speech in relation to reverberation time. The longer the reverberation of noise in a room, the lower the intelligibility of speech. [Adapted from Fletcher, 6.]

caused by varying degrees of reverberation (specifically, the time in seconds that it takes the noise to die down). This relationship essentially is a straight-line one.

Effects of earplugs on speech. In a sense, earplugs are part of the transmission system, since they intervene between the environment and the receiver. Although their purpose is to prevent or minimize hearing loss, one might expect that they would also reduce the intelligibility of speech. In fact, however, earplugs may actually increase the intelligibility of speech under high noise levels. Under low noise levels, however, the use of earplugs may impair speech intelligibility somewhat. It is, of course, in high-noise-level situations, however, that there would be a greater likelihood that earplugs would be worn, and in such conditions they can be effective. The explanation for this is that at high noise levels a point is reached where additional intensity cannot be discriminated; at such levels the difference between the intensity of the signal (in this case speech) and of its background noise cannot be discriminated. The value of earplugs under such circumstances is to bring the level of both the signal and the background noise down to the point where the difference between them can be discriminated.

The Receiver

The receiver is the last link in the communication chain. For receiving speech messages under noise conditions, the receiver should have normal hearing, should be trained in the types of communications to be received, should be reasonably durable in withstanding the stresses of the situation, and should be able to concentrate on one of several conflicting stimuli.

DISCUSSION

When feasible, of course, speech communications should be carried out under favorable conditions, uncluttered with noise. However, in many circumstances it is not possible to reduce noise at its source; (one cannot stop the rolling mills of a steel mill for people to communicate with others). Under these and other circumstances it is necessary to look to other elements of the total communication system, rather than to the noise source itself, for possibilities of improving the intelligibility of speech. On the engineering design side of the coin, the possibilities to consider are those of minimizing the transmission of noise if possible (through acoustical treatment and other means), improving the design of the communication equipment, and modifying the nature of the messages to be used; and on the personnel side of the coin, the possibilities are those of selection and training of talkers and receivers, where these are feasible.

SUMMARY

1. The *speech chain* comprises different forms or levels in which a spoken message is transmitted, including a linguistic and physiological level of the speaker, an acoustic level of the transmission system, and a physiological and linguistic level of the receiver.
2. The types of *phonemes* that can be generated by the human speech mechanisms are: (*a*) plosives, or stops; (*b*) fricatives, or spirants; (*c*) laterals; (*d*) trills; and (*e*) vowels.
3. The English language has about 38 phonemes out of many hundreds that people can generate. Phonemes vary from one another in their spectral characteristics, including the combination of *frequencies* and *intensities*.
4. *Speech intelligibility tests* can serve such purposes as (*a*) to evaluate the adequacy of a communication system, (*b*) to determine the

communication efficiency of individuals within a system, and (c) to determine hearing loss for speech.

5. In evaluating the adequacy of communication systems, the *AI* can be used to estimate the intelligibility of speech. An AI, in turn, can be converted into an *intelligibility score*, which is the estimated percentage of the spoken material that can be understood under a given set of conditions.

6. The *SIL* is an index of noise level (specifically the average of three octave bands); it is used as a measure of the *destructiveness* of noise in the reception of speech.

7. The *NC* curves are sometimes used as the basis for recommendations of permissible noise levels for specific communication situations.

8. The intelligibility of speech under adverse conditions depends in part on the nature of the *message*, including its vocabulary, its context, and its phonetics.

9. Different types of distortion of speech can be caused (intentionally or unintentionally) by the transmission system; however, the intelligibility of speech does not depend on high fidelity. The effects of filtering depend on the frequencies that are filtered out by either high-pass or low-pass filters. *Center* clipping affects intelligibility much more than *peak* clipping.

10. The use of *earplugs* can improve speech intelligibility under some circumstances by reducing the level of both the speech and the background noise to the point where the difference can be better discriminated.

REFERENCES

1. Beranek, L. L.: The design of speech communication systems, *Proceedings of the Institute of Radio Engineers*, New York, 1947, vol. 35, pp. 880–890.

2. Beranek, L. L.: Revised criteria for noise in buildings, *Noise Control*, 1957, vol. 3, no. 1, pp. 19–27.

3. Beranek, L. L., and R. B. Newman: Speech interference levels as criteria for rating background noise in offices, *Journal of the Acoustical Society of America*, 1950, vol. 22, p. 671.

4. Bilger, R. C., T. D. Hanley, and M. D. Steer: *A further investigation of the relationships between voice variables and speech intelligibility in high level noise*, TR for SDC, 104–2–26, Project 20–F–8, Contract N6ori–104, Purdue University, Lafayette, Ind. (mimeographed).

5. Denes, P. B., and E. N. Pinson: *The speech chain*, Bell Telephone Laboratories, 1963.

6. Fletcher, H.: *Speech and hearing in communication*, D. Van Nostrand Company, Inc., Princeton, N.J., 1953.

7. French, N. R., and J. C. Steinberg: Factors governing the intelligibility of speech sounds, *Journal of the Acoustical Society of America*, 1947, vol. 19, pp. 90–119.

8. Licklider, J. C. R.: Effects of amplitude distortion upon intelligibility of speech, *Journal of the Acoustical Society of America*, 1946, vol. 18, pp. 429–434.

9. Licklider, J. C. R., and G. A. Miller: The perception of speech, in S. S. Stevens (ed.), *Handbook of experimental psychology*, chap. 26, John Wiley & Sons, Inc., New York, 1951.

10. Miller, G. A.: *Language and communication*, Paperbacks ed., McGraw-Hill Book Company, New York, 1963.

11. Miller, G. A., G. A. Heise, and W. Lichten: The intelligibility of speech as a function of the context of the test materials, *Journal of Experimental Psychology*, 1951, vol. 41, pp. 329–335.

12. Peterson, A. P. G., and E. E. Gross, Jr.: *Handbook of noise measurement*, 6th ed., General Radio Co., New Concord, Mass., 1967.

13. Rosenblith, W. A., and K. N. Stevens: *Handbook of acoustic noise control. Vol. II. Noise and man*, USAF, WADC, TR 52–204, June, 1953; Report PB 111, 274, U.S. Department of Commerce, Office of Technical Services.

14. Webster, J. C.: Speech communications as limited by ambient noise, *Journal of the Acoustical Society of America*, 1965, vol. 37, pp. 692–699.

part
three
MEDIATION
ACTIVITIES

eight
MEDIATION
PROCESSES

A theme that was introduced before is the notion that the dominant function of human beings in systems is that of control, this being directed toward the achievement of certain system objectives or goals (i.e., the purposive modification of the future course of events or of conditions in the environment). Although the exercise of such system control depends upon the reception and interpretation of relevant stimuli and the responses to effect the intended control, the determination of the response(s) to be made, and the human control of their execution, are the consequence of some combination of mediation processes on the part of the controller. In their various and sundry forms, the system control functions that human beings perform require the exercise of a wide range of human mediation functions such as the following:

Information storage

Long term: The *learning* that is required for performing the system functions

Short term: Remembering for short periods of time information that is relevant to a specific operational situation, such as a message to be transmitted

Information retrieval

Recognition: Essentially a perceptual process involving the recognition or detection of relevant stimuli or signals

Recall: Including the recall both of previously learned factual information, procedures, processes, sequences, and other such classes, and of information in short-time storage, mentioned above

Information processing: Categorizing, calculating, coding, computing, interpolating, itemizing, tabulating, translating, etc.[1]

Problem solving and decision making: Analyzing, calculating, choosing, comparing, computing, estimating, planning, etc.[1]

Control of physical responses: The exercise of control over a wide range of physical responses including conditioned responses, selection of

---

[1] These enumerations of behaviors were developed by Berliner, Angell, and Shearer [3] and reflect the consensus of several people as representing reasonably unambiguous mediation processes.

responses appropriate to specific stimuli, sequences of responses, and continuous control responses

Practically every one of these (and perhaps other) mediation processes serves as the subject of a major field of knowledge and research in its own right, and it is not possible in this text to discuss the role of each in human performance in systems. Collectively, however, their manifestations in human performance are through the processes of the acquisition of skills and knowledge (learning) and of transfer of learning to actual work situations. In this chapter, then, we shall take something of an overview of the processes of learning and the transfer of learning. In addition, we shall discuss further the principle of compatibility as it relates to human performance, because of its significance in the processes of learning and in the execution of learned responses. As we discuss some aspects of the mediation processes, we should again remind ourselves that the information-receiving, mediation, and action functions of mankind are so intricately intertwined that their neat separation is a will-of-the-wisp. These inextricable interrelationships were evident, for example, in our earlier discussion of information input (Chapters 3 to 7) and will be further evident in this chapter and in the next chapter (although the major discussion of motor processes will follow in Chapter 10).

ACQUISITION OF SKILLS AND KNOWLEDGE

Learning consists of a relatively permanent change in an individual as manifested by his behavior. Whether an individual has learned something cannot be determined directly, but only by his subsequent performance. The spectrum of human performance abilities is of course rooted in the knowledge and skills that people have learned. Although staggeringly varied in their specific nature, the skills of people probably can be grouped into such classes as the ones suggested by Fitts and Posner [10, p. 4]: gross bodily skills (walking, maintaining equilibrium, etc.), manipulative skills (including those of continuous, sequential, and discrete types), perceptual skills, and language skills (including conventional communications, mathematics, metaphor, and other representations people use in thinking and problem solving, and in coding computer languages, for example).

Types of Learning

The nature of the performance or material to be learned (the task, operation, activity, content, etc.) virtually predetermines the type of learning involved. Gagné [14, 15] has postulated certain types of learning that form what he refers to as a *cumulative learning sequence*, with each level depending on lower levels in a building-block fashion, somewhat as shown in Figure 8-1. Although Gagné postulates this hierarchical scheme for human development, we can probably see some relevance of each of these levels to certain types of activities in the operation of systems. For example, the stimulus-response level applies to such straight-forward tasks as turning a control switch when a signal is heard; the chain relationship, to following certain specified motions; the multiple dis-

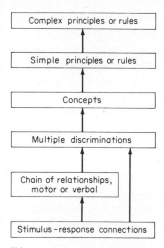

Figure 8-1 A generalized illustration of the cumulative learning sequence proposed by Gagné [15, p. 182]. In this model the learning at any given level depends on relevant learning at the lower levels.

criminations, to the selection of the appropriate response to each of several stimuli (such as traffic lights); and the application of principles, to certain maintenance problems.

Conditions that Contribute to Learning

The conditions and methods that contribute most beneficially to learning will not be discussed in detail here.[2] However, a few such conditions will be mentioned briefly: motivation on the part of the learner; knowledge of results (i.e., feedback to the individual regarding his performance); distribution of training periods (usually some spacing of learning periods is

---

[2] For extensive discussions of learning, the reader is referred to such sources as Gagné [14], Fitts and Posner [10], and Deese and Hulse [7].

desirable, although the optimum duration and spacing of learning periods is unique to the knowledge or skill being learned); and the types of incentives used (usually positive incentives are more effective than negative incentives and usually *intrinsic* incentives, i.e., those associated with the activity, are more effective than *external* incentives).

Transfer of Training

Much of that which we learn in one situation is *transferred* to another. This is of course the premise on which all education is founded—that what is learned in school will be transferred to relevant contexts in the real world. In our study of human factors we are of course interested in the transfer-of-learning process as it is used in transferring previously learned knowledge and skills to the human performances needed in systems. Although this transfer can be from previous life and educational experiences, of more specific interest is the transfer from training programs to the systems for which the training was intended.

In experimental situations and in certain actual training situations it is possible to measure the extent of transfer (with appropriate performance criteria, tests, etc.). The amount of such transfer can be presented in a number of ways, including the model by Ferguson [9] for which he proposes the concept of a simple mathematical transfer function such as

$$y = \phi(x)$$

in which y and x are measures of performance of the transfer task and training tasks, respectively, and ϕ is a mathematical function that shows the relationship for the specific situation. It might also be pointed out that transfer can be viewed in an information-theory context, as discussed in Chapter 4.

Theories of transfer of training.   Various theories of transfer have been set forth. The first major formulation was the concept of *identical elements* proposed by Thorndike [33], i.e., that transfer from one situation to another occurred to the extent that there were identical elements in the two situations. Another general theory is that of transfer through principles [Bass and Vaughn, 2, p. 40]. This theory postulates that positive transfer results when an individual applies to new situations the principles learned in previous specific situations which have sufficient generality to cover the class of stimuli that the previous and the new situations have in common. Since neither theory seems to fully explain transfer of learning in all types of situations, Deese and Hulse [7, p. 349] suggest other more general ways of describing the nature of transfer, in particular by stimulus and response analysis. As they point out, any task can be characterized by its stimulus components and the responses (overt and implicit) made to those stimuli. In turn, any two tasks (such as a learning task and the real situation task) can be described in terms of the resemblance of their stimuli and of their responses. This focus on the comparison of the degree of similarity of the stimuli and the responses in the transfer-of-training context has served as the basis for variations of this theme. Three such variations seem to be particularly relevant, as pointed out by Muckler et al. [26]. These include the theories of Wilie as presented by Woodworth [35, pp. 201–202], of Osgood [28], and of Gagné, Baker, and Foster [16]. Very briefly, Wilie proposed two general laws on transfer, as follows: (1) Transfer effect will be positive when an old response (acquired during original learning) is transferred to a new stimulus (the learned response must be associated with a new stimulus); (2) transfer effect will be negative when a new response is learned to an old stimulus (if the stimulus remains the same, but a new response is required, negative transfer will occur).

Osgood [28] has proposed the idea of a *transfer surface* to characterize his theoretical formulation. It is based on the conception of transfer effects being a continuous function of the degree of similarity of the two stimuli (that of the training task and that of the transfer task) and of the similarity of the two responses. This transfer surface is shown in Figure 8-2. It will be noted that the amount of positive transfer (vertical scale) is shown to be optimum when both the stimulus and the response of the transfer task are identical with those of the training task. The great-

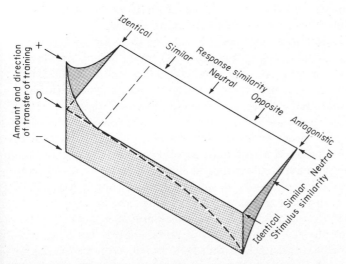

Figure 8-2 Illustration of transfer surface to characterize how the relationship between stimulus similarity and response similarity is related to degree of transfer from a training task to a transfer task. [Adapted from Osgood, 28.]

est amount of negative transfer occurs when the stimuli are identical but the response of the transfer task is antagonistic to that of the training task. Intermediate degrees of positive and negative transfer are depicted with varying degrees of stimulus and response similarity in combination with each other.

The theory proposed by Gagné, Baker, and Foster [16] is focused more on stimulus and response similarity *within* each of the two tasks than on the similarity between the training task and the transfer task (the details, however, will not be given here).

Simulation in Training

Some form of simulation is virtually part and parcel of most training programs; perhaps the only exception is that of on-the-job training. The forms of simulation used in training include situational simulation, computerized equipment, management games, and a whole assortment of trainers and related devices. Although terminology in this area is generally used rather promiscuously, the Air Force [36, part E, 3-2] differentiates among the following kinds of training equipment: *simulators*, relatively complex items of equipment used to reproduce, functionally, the conditions necessary for an individual to accomplish an operational "mission" synthetically; *training devices*, equipment used to train people in certain more specific operating functions; *training aids*, such as animated panels, training charts, and schematics;[3] and *training attachments*, items of equipment which must be used in conjunction with other equipment, such as visual attachments for night flying simulation.

Evaluation of simulation devices.   Whatever their nature and level of sophistication, however, the intent in using any kind of simulation device (simulators, trainers, training aids, etc.) is to facilitate the learning process so the learning can be transferred to the actual work situation.

Fidelity of simulation.   There has been an on going discourse about the degree of *fidelity* that should be achieved in training-simulation situations. Although this question is particularly important in the case of elaborate simulators (since the cost may depend on how accurately they simulate the real equipment), the question is also relevant to almost any type of simulation device, elaborate or simple. This, of course, harks back to the concept of transfer of training. In discussing fidelity, however, we should differentiate between fidelity of physical simulation, which refers to the similarity of physical features and physical operational characteristics, and fidelity of psychological simulation, which refers to the degree to which the human behaviors learned with the simulation are similar to those which would be involved in the true-life situation. Although there

---

[3] For discussions of training aids, see A. A. Lumsdaine in Folley [12, chap. 11] and W. C. Biel in Gagné [13, chap. 10].

still are some dangling theoretical questions relating to transfer of training, there are some observations that can be tossed into the discussion hopper. First, it is clear that the objectives of simulation should be toward the development of psychological simulation, since we are concerned with transfer of training, especially of the *critical* aspects of performance. Next, it can be said that fidelity of physical simulation does not always ensure fidelity of psychological simulation, although it usually would not detract from fidelity of psychological simulation. There are, however, inklings that in at least isolated circumstances high physical fidelity can result in less transfer than does less physical fidelity. In a tracking study, for example, it was found that training under a tracking rate that was less than the tracking rate of the transfer task produced more transfer than training under the same rate as that of the final task [Ammons, Ammons, and Morgan, 1]. Next, it is possible to have fidelity of psychological simulation with the physical simulations being very limited. But while this may be probable in some contexts, it probably is *not* true in other contexts; unfortunately, we cannot yet specify very well *when* one can achieve psychological fidelity without physical fidelity. It can also be said that many efforts to achieve a high degree of physical simulation probably have resulted in simulators that are much more elaborate (and costly) than would be necessary to achieve the desired objectives, no matter what they are.

Decisions regarding the design of simulation devices become particularly sticky in the case of rather costly simulators, since the degree of fidelity of physical simulation is frequently related to costs; the increasing costs with increasing fidelity usually are considered disproportionate to the possible benefits. While the relationships among cost, physical fidelity, and psychological fidelity certainly would vary from one kind of simulation to another, a generalized (and hypothetical) relationship has been proposed, as shown in Figure 8-3 [Miller, 25]. In this formulation, it is assumed that cost increases at an increasing rate with increasing degrees of physical simulation and that the amount of transfer of training tends to level off in the range of high physical fidelity; an optimum degree of simulation would be somewhere around the point where costs rise sharply and transfer effects are not subject to major additional increase.

Guidelines in simulation. Although there are many areas of ignorance concerning simulation, there nonetheless are at least a few straws in the wind that may have some utility in solving practical problems in the development of simulated training devices; however, these should be taken with several grains of salt.

1. It is probable that a point of diminishing returns in transfer (as traded off against, say, cost) frequently argues for something less than perfect physical fidelity.

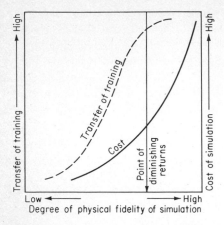

Figure 8-3 Generalized and hypothetical relationship among degree of fidelity of simulation, transfer of training, and cost. [Adapted from Miller, 25.]

2. Reversals of relationships between stimulus and response (as between the training and transfer situations) generally are conducive to negative transfer or to reduction in the amount of transfer.
3. In some simulation situations there are incidental physical or situational features of the transfer situation that probably have little or no relevance to transfer of learning.
4. Frequently it is possible to use a representation of a feature or considerable variation of a feature that presumably would have virtually no effect on the degree of psychological simulation. Work-space layouts, for example, might be of different materials. Photographs or diagrams might represent situational features.
5. In simulation of tasks for the training of people on important control tasks (such as flight simulators), control-feel fidelity is increasingly important in simulators which (*a*) require considerable time sharing among displays and controls, (*b*) are used for extensive training, or (*c*) are used to maintain high-level proficiency of experienced personnel such as pilots [Briggs and Wiener, 5].

Training

The conventional conception of training relates to organized training programs, especially of employees of an organization. It is proposed here, however, that we also regard as training other less organized and less direct schemes for helping people to learn to use the equipment, facilities, or other things that are designed for human use, such as lawn mowers, hair curlers, dishwashers, or farm tractors. Although these efforts usually do not involve direct contact of instructor and learner, the instruction manuals, operational manuals, illustrations, job aids, etc., that are used certainly are directed toward instruction, typically self-instruction.

Training content.   However the training is to be implemented (by formal training programs, self-instruction, or other), the objectives are the development of specific skills, knowledge, or attitudes that the learners are expected to acquire. It has been suggested that these can be grouped under the following headings:[4]

1.   Identifying (objects, locations, etc.)
2.   Knowing principles and relationships
3.   Following procedures
4.   Making decisions or choosing courses of action
5.   Performing skilled perceptual-motor acts
6.   Developing desirable motives and attitudes

More specifically, of course, the actual knowledge and skills that are required for performing the tasks must be identified and spelled out. For this, specific procedures and formats have been developed.[5] For equipment that is already in use, the tasks can be identified by task analysis of individuals presently using the equipment. For equipment that is being developed, it is necessary to predict what tasks will have to be carried out upon completion of the equipment. Such prediction is based on inferences from the specific design features of the equipment being developed, along with information about any procedures or operational policies that may also be evolving during the development phase.

An example of a part of a task analysis is given below for illustration. This particular example is from a task analysis of a flight engineer technician [36, fig. D, 3-6, part D, chap. 3]:

Duty: Operates aircraft power plant and systems controls.
Tasks
  (a)   Operates power-plant controls to provide desired economy of engine operations.
  (b)   Adjusts engine controls to control carburetor air, cylinder temperatures, engine rpm, and boost combinations as required by flight and load conditions.
  (c)   Controls jet-engine operation to maintain desired fuel and lube oil pressures and engine thrust and rpm.
  (d)   Regulates aircraft electrical system.

[4] Adapted from A. A. Lumsdaine [in Folley, 12, chap. 11] and the *Handbook of instructions for aerospace personnel subsystem design* [36].
[5] A Training Function Analysis procedure was developed by Purifoy and Fairman [29]. A comprehensive discussion of task analysis in deriving training and training equipment requirements has been prepared under USAF auspices [37].

Such tasks, in turn, can be further subdivided into subtasks and elements. In the case of complex systems the task analysis and the subsequent training programs sometimes become a part of what, in the Air Force, is referred to as the *personnel subsystem* (PSS).[6]

Self-instruction.   Some version of self-instruction is part of most organized training programs and of virtually all training that does not involve an instructor (i.e., the "training" of consumers in the use of products). Because of the current flurry of interest in programmed instruction (which is one form of self-instruction) we shall discuss this method briefly. With programmed instruction procedures, the individual is typically provided with a carefully prepared sequence of information with related questions or problems. This material is presented by some device (in some instances a machine), on a printed page, or by other means. The learner responds to each question or problem. The response required in some instances is one of several multiple-choice answers (as in a test); in others a *constructed* response is required, this being the answer a person gives in a completion type of question. The response may be made in pencil or by some mechanical means such as a stylus or a push button. But the type of device and the mechanics of giving responses are really incidental to the basic premises of programmed instruction procedures. Primarily they depend upon the learning principle of knowledge of results by providing immediate feedback regarding the correctness (or wrongness) of the response. Underlying the use of this principle, however, is the requirement that the material to be learned and the related questions or problems be properly programmed. Such material is presented in bite-size units arranged in a logical sequence, like building blocks, which makes it possible for the individual to *build up* the knowledge in question. The conventional use of programmed instruction is as an adjunct of organized training programs. Certain central features of it, however, can be worked into more operational circumstances (such as routine checkout procedures) or one-time functions (such as assembling a knocked-down doghouse) in which a sequence of activities is to be executed and in which step-by-step feedback is important.

Job Aids

A job aid has been described as something which guides an individual's performance to enable him to do something which he was not previously able to do, without requiring him to undergo complete training [J. J. Wulff and P. C. Berry, in Gagné, 13, chap. 8]. The most usual types of job aids are manuals, handbooks, checklists, wiring diagrams, motion

---

[6] For a discussion of the personnel subsystem of the United States Air Force and the formal procedures that are involved, the reader is referred to the *Handbook of instructions for aerospace personnel subsystem design* [36].

pictures, filmstrips, and recorded messages. In essence, job aids provide for long-term storage of information and for its retrieval at appropriate times. There obviously would be many jobs where they would be super-fluous. Where they are indicated, they reduce the amount of information that the trainee must learn. It is probable that even if a trainee learned the detailed information incorporated in some job aids, the recall of that information might be less reliable than its retrieval from a job aid.

Content of job aids.   Since job aids can be considered as being complementary to training, the question arises as to how much of the job information should be learned by the trainee as opposed to being stored in the job aid. As has been suggested by Folley [12, chap. 7], there is probably some optimum, such as that illustrated in Figure 8-4. If all the required

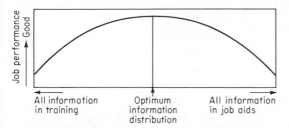

Figure 8-4   Generalized illustration of job per-formance as a function of distribution of job in-formation between training and job aids. [From Folley, 12, p. 113.]

information is incorporated in a job aid, performance might be impaired by requiring the worker to spend excessive time searching for desired information from an extensive assortment. If all required information must be learned, training time might be extended beyond reasonable bounds and there might be risks of unreliable recall.

Achieving an optimum (such as that illustrated) probably cannot be reduced to a formula, but certain guidelines have been suggested [Folley, 12, pp. 113–114]. These take into account:

1. The aptitude level of personnel, and therefore the level of training they can efficiently absorb.
2. Error tolerance of the system. Where errors are not permissible, a job aid can reduce the probability of error attributable to faulty memory.
3. The amount of information required for performance.
4. Opportunity, or lack of it, to refer to a job aid during performance.
5. Frequency of performance. Frequently performed activities usually are well learned and therefore typically do not require job aids.

The development of job aids.[7]   In developing job aids there is probably one paramount principle to follow. In line with the fact that human behavior follows relevant stimuli, job aids should be so designed that they provide, in convenient, clearly manifest form, some indication of the stimulus (the circumstances under which the pertinent information is applicable) and the response (the action or behavior that should be carried out under the circumstances in question). In this process, one should again base the decisions on the task requirements of the job, which spell out the actions that need to be performed to fulfill the system requirements.

Learning Ceilings

In the learning of many types of skills, especially motor skills and certain simple language skills, there is evidence that performance can be improved almost indefinitely through practice. This was demonstrated, for example, by Stevens and Savin [32] in their analysis of learning curves of many specific learning experiments. The rate of improvement, however, tapers off with continued practice (usually in a pattern that can be described mathematically as a power function) in such a manner as to produce, for practical purposes, a *ceiling* for future performance. It should be noted, however, that *actual* performance usually falls short of this (theoretical) ceiling of potential performance, this disparity probably being attributable to human foibles such as failure in information retrieval, lapse in attention, and—perhaps of particular importance— motivation. In this connection, Helson [21] suggests a hypothesis of par, or tolerance, predicated on the idea that individuals typically set some level of performance for themselves, and do not press themselves beyond this self-set standard. Incidentally, as we reflect about level of perform-ance, we should remind ourselves again of the ever-present fact of individual differences, such as in their aptitudes, learning abilities, theoretical ceilings, self-set standards, and of course actual performance.

COMPATIBILITY

In an earlier discussion (Chapter 5) the notion of compatibility was brought in, particularly in connection with certain features of displays. The concept of compatibility in human factors, however, has very wide implications. Compatibility refers to the spatial, movement, or conceptual features of stimuli and of responses, individually or in combination, which are most consistent with human expectations. Although there are many

---

[7] For further discussions of job aids, see J. J. Wulff and P. C. Barry in Gagné [13, chap. 4], and Folley [12, chap. 7].

different manifestations of compatibility, most instances probably can be considered to fall in one of three groups, namely, (1) *spatial* compatibility, i.e., the compatibility of the physical features, or arrangement in space, of certain items, especially displays and controls; (2) *movement* compatibility, the direction of movement of displays, controls, and system responses; and (3) *conceptual* compatibility, the conceptual associations that people have, such as green representing "go" in certain codes. In the context of perceptual-motor activities there is some presumption of compatibility of stimulus and response in combination. The term *stimulus-response compatibility* (*S-R* compatibility) was first used by Fitts and Seeger [11], following the earlier use of the term *compatibility* by Dr. A. M. Small. Fitts and Seeger characterized *S-R* compatibility as follows: "A task involves compatible *S-R* relations to the extent that the ensemble of stimulus and response combinations comprising the task results in a high rate of information transfer." In this information-theory context, the concept of compatibility implies a hypothetical process of information transformation, or recoding, in the activity and is predicated on the assumption that the degree of compatibility is at a maximum when the recoding processes are at a minimum.

Origins of Compatibility Relationships

Compatibility relationships stem from two possible origins. In the first place, certain compatible relationships are intrinsic in the situation, for example, turning a steering wheel to the right in order to turn to the right. In certain combinations of displays and controls, for example, the degree of compatibility is associated with the extent to which they are isomorphic or have similar spatial relationships. Other compatible relationships are culturally acquired, stemming from habits or associations that are characteristic of the culture in question. For example, in the United States a light switch is usually pushed up to turn it on, but in certain other countries it is pushed down. How such culturally acquired patterns develop is perhaps the consequence of fortuitous circumstances.

The Identification of Compatibility Relationships

If one wishes to take advantage of compatible relationships in designing or arranging equipment, it is of course necessary to know *what* relationships are *compatible*. There generally are two ways in which these can be ascertained or inferred. In the first place, certain such relationships are obvious or manifest; this is particularly true with many relationships that are intrinsic in the situation, such as the arrangement of corresponding displays and controls in juxtaposition to each other. In addition, certain culturally acquired relationships are so pervading that they, too, are obvious, such as the red, yellow, and green symbols of traffic lights. But when the most compatible relationships are not obvious, it is necessary to identify them on the basis of empirical experiments. Certain examples will be cited later, but in general, such experiments produce information on the proportion of subjects who choose each specific relationship of different possible relationships.

Spatial Compatibility

There are many variations of the theme of spatial compatibility, most of them spanning the gamut of physical similarities in displays and corresponding controls and their arrangement, and the arrangement of any given set of either displays or controls.

Physical similarity of displays and controls.   Sometimes there exists the opportunity to design related displays and controls so there is reasonable correspondence of their physical features, and perhaps also of their modes of operation. Such a case is well illustrated by Fitts and Seeger [11.] In this study three different displays and three different controls were used in all possible combinations. The displays consisted of lights in various arrangements. As a light would go on, the subject was to move a stylus along a channel (or channels) of the control that was being used to a location at the end of the appropriate channel; when that location was reached, an electric contact would turn the light off. The three displays and controls are illustrated in Figure 8-5. The experimental procedures will not be described, but in general, different groups of subjects used each combination of the three stimulus panels with the three response panels, their performance being measured in reaction-time errors and information lost. The results, also given in Figure 8-5, show that performance with any given stimulus or response panel was better when it was used in combination with its *corresponding* response or stimulus panel (S_a-R_a, S_b-R_b, and S_c-R_c) than when used in combination with a different configuration.

Physical arrangement of displays and controls.   Both experiments and rational considerations lead one to conclude that for optimum use, corresponding displays and controls should be arranged in corresponding patterns. This aspect of compatibility was well demonstrated by the results of a series of studies [Garvey and Knowles, 17; Garvey and Mitnick, 18; and Knowles, Garvey, and Newlin, 23] in which display-control combinations were used; most of the displays were small lights, and the controls in all displays were corresponding buttons to be pushed when a signal was given. The signals and corresponding push buttons represented letter-number combinations, the letters being from A to J, and the numbers from 1 to 10. Thus, a combination, such as B-3 was represented by a signal (or pair of signals) and also by a push button (or pair of push

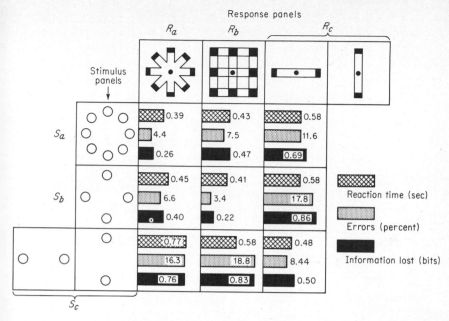

Figure 8-5 Illustrations of signal (stimulus) panels and response panels used in study by Fitts and Seeger [11]. The values in any one of the nine squares are the average performance measures for the combination of stimulus panel and response panel in question. The compatible combinations are S_a-R_a, S_b-R_b, and S_c-R_c, for which results are shown in the diagonal cells.

buttons). The signals were presented on different types of display panels (identified as D), and the push buttons were presented on control panels (identified as C), as shown in Figure 8-6. Upon receiving a signal from the display then being used, the operator responded by pushing the corresponding push button (or pair of push buttons) of whatever control arrangement he was then using. The subjects were scored on their rate of response.

Aside from the combined panel D-1 and C-1, the displays and controls were tried in various combinations. It was found that the combination panel D-1 and C-1 resulted in the most rapid response rate (about 0.7 responses per sec) as contrasted with the others (which ranged from 1.3 to 1.8 responses per sec). Thus, it appears that for optimum rate of response to a visual signal on a panel, the corresponding control should be adjacent to the display signal.

Another illustration—one of a more down-to-earth nature dealing with a gadget used morning, noon, and night (and sometimes in between) —is a study dealing with the arrangement of burner controls on a four-burner stove [Chapanis and Lindenbaum, 6]. The burners, in a sense,

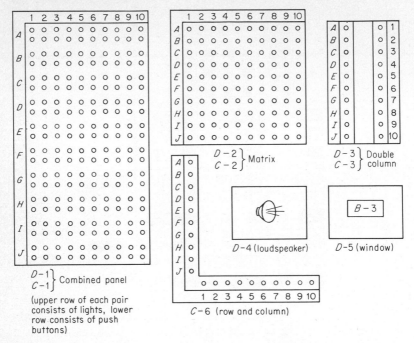

Figure 8-6 Illustrations of displays *D* and controls *C* used in studies by Garvey and associates [17, 18, 23]. Displays *D-1*, *D-2*, and *D-3* consisted of small lights as signals, *D-4* was a speaker, and *D-5* an open window with a reading of a letter-number combination. Controls *C-1*, *C-2*, *C-3*, and *C-6* were all push buttons. Since *D-2* and *C-2* were separate but of the same arrangement, the same illustration is used for both; likewise *D-3* and *C-3*.

can be thought of as displays. The four arrangements that were tried out are shown in Figure 8-7. Fifteen subjects each had 80 trials on each design. They were told what burner to turn on, and their reaction time was recorded along with any errors they made. The number of errors made on the four designs are given below:

Design	No. of errors, out of 1200 trials
I	0
II	76
III	116
IV	129

Design I was also best in reaction time (over the last 40 trials), with design II next best. Clearly, design I was the most compatible.

Arrangement of sets of similar devices. Sometimes an assortment of similar devices is used in combination, such as displays or controls that

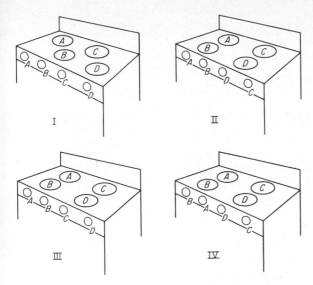

Figure 8-7 Control-burner arrangements of simulated stove used in experiment by Chapanis and Lindenbaum [6].

have some systematic relationship to each other. Here, again, we can capitalize on the concept of compatibility by arranging them in the most natural, or expected, pattern. One such example is the arrangement of push-button telephone key sets [Deininger, 8]. In this example, each push button serves both as a display and a control. Several different arrangements were experimented with initially, which resulted in the selection of five designs for further analysis. These are shown in Figure 8-8 along with information on the following criteria: keying time (seconds for punching a given sequence of numbers), errors (percentage of wrong punches), and preferences (votes for and against by the subjects).

The results illustrate the point that a particular design may not be best by all criteria and suggest that it is sometimes necessary to trade off certain advantages for others. In this particular instance the "telephone" design 4 was lowest in keying time, but was not lowest in errors (although only one other design was lower, namely, 3), was second in votes for, and had the fewest votes against. This design, incidentally, is similar to the present arrangement of numbers on the dial telephone; its familiarity may make it more compatible with the acquired response habits of dial-telephone users and so accounts for its general superiority in this experiment. However, this arrangement is not compatible with other conceptual associations of numbers increasing from left to right or in a clockwise direction. In any event, the design actually selected (and

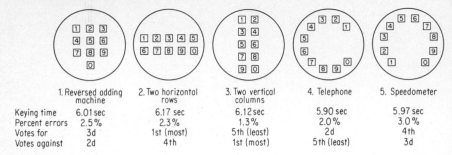

	1. Reversed adding machine	2. Two horizontal rows	3. Two vertical columns	4. Telephone	5. Speedometer
Keying time	6.01 sec	6.17 sec	6.12 sec	5.90 sec	5.97 sec
Percent errors	2.5%	2.3%	1.3%	2.0%	3.0%
Votes for	3d	1st (most)	5th (least)	2d	4th
Votes against	2d	4th	1st (most)	5th (least)	3d

Figure 8-8 Push-button arrangements for telephone key sets, with average keying time, errors, and "votes" for and against. [From Deininger, 8. Reprinted by permission of the copyright owner, American Telephone and Telegraph Company, and the author.]

now in use on some telephones) is 1, which by various criteria and trade-offs, seemed to be reasonably optimum.

Movement Relationships

Compatibility of movement relationships applies generally to perceptual-motor activities, but probably has some other ramifications as well, such as the responses of systems, for example, the movement of vehicles. Probably the most clear-cut model of such compatibility is one in which there is a clearly recognizable stimulus and a clearly recognizable response that is to be made following the stimulus. For example, if a blip on a plan-position indicator (PPI) radarscope veers to the right, the most compatible response with, say, a joy stick, would be to move it in a corresponding direction. There are, however, other perceptual-motor activities in which, there may be some question what the stimulus is, since there may be no single stimulus. And in some instances the response does not *follow* the stimulus in question, but may *control* the stimulus, as when a control device moves a pointer on a dial (this is somewhat suggestive of the chicken-and-the-egg conundrum). Further, in some circumstances the control device itself may very appropriately be considered a stimulus. On the other hand, referring back to the earlier discussion (Chapter 5) we shall recall that the concept of compatibility also applies to the direction movement of the indicators of visual display, even though no overt response is involved.

However, the compatibility of movement relationships of displays or controls (or both) in some circumstances become intertwined with their physical orientation to the user. In this connection, Gottsdanker and Senders [19] point out that there are two basic orientations of the *display plane* relative to the operator (the display plane may include some controls as well as displays.) One of these is a *front-wall* orientation (with the displays on a vertical plane facing the operator), and the other is a *table-top* orientation (with the displays on a horizontal surface). The other two planes are referred to as the *lateral cutting plane* (perpendicular to the display plane and in a lateral relation to it) and the *orthogonal cutting plane* (on a vertical, forward-aft axis). Such surfaces, in a work situation, could be either to the right or left of the operator.

Movement relationships of displays and rotary controls in same plane. In display-control systems in which a control movement is associated with a movement of some feature of the display (such as a pointer), a generally accepted principle is that with a rotary control in the same plane, a clockwise turn of a control device is associated with an increase in values, as in the left-hand examples of Figure 8-9, although

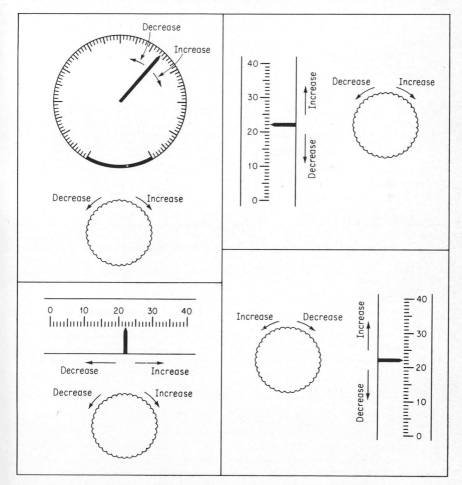

Figure 8-9 Illustration of desirable control-display movement relationships: fixed scales with moving pointer, with control in same plane as display.

there are certain exceptions. One exception to this is with a fixed horizontal or vertical scale with a moving pointer where a rotary control is beside the scale; in these scales the most compatible relationship has the indicator move in the same direction as the part (or side) of the control device nearest the indicator [Warrick, 34]; such examples are shown in the right-hand examples of Figure 8-9.

For moving scales with fixed pointers, it has been postulated that the following features would be desirable [Bradley, 4]:

1. That the scale rotate in the same direction as its control knob (i.e., that a direct drive exist between control and display)
2. That the scale numbers increase from left to right
3. That the control turn clockwise to increase settings

With the usual orientation of displays, however, it is not possible to incorporate all these three features in a conventional assembly. With the usual display orientation, only the two combinations of these features that are shown in Figure 8-10 as A, B, C, and D are possible. In a study relating to moving-display instruments and associated controls, these types and some variations of them were used experimentally, including

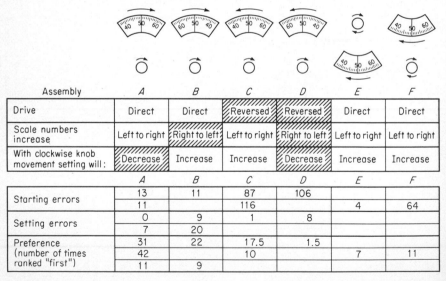

Assembly	A	B	C	D	E	F
Drive	Direct	Direct	Reversed	Reversed	Direct	Direct
Scale numbers increase	Left to right	Right to left	Left to right	Right to left	Left to right	Left to right
With clockwise knob movement setting will:	Decrease	Increase	Increase	Decrease	Increase	Increase

	A	B	C	D	E	F
Starting errors	13	11	87	106		
	11		116		4	64
Setting errors	0	9	1	8		
	7	20				
Preference (number of times ranked "first")	31	22	17.5	1.5		
	42		10		7	11
	11	9				

Figure 8-10 Some of the moving-display and control-assembly types used in study by Bradley [4]. The various features of these related to three desirable characteristics are given below the diagrams; crosshatching indicates an undesirable feature. With the usual display orientation, as in A, B, C, and D, all three desirable features are not possible. Some data on three criteria are given at the bottom of the figure, indicating the general preferability of A.

two *inverted* variations E and F, in which the fixed pointer (really the "lubber" line) is at the *bottom* (rather than top) of the dial. In these two cases (E and F) the three postulated desirable features are fulfilled. In various phases of this study each subject in the several different groups of subjects was told that the experimenter wanted to learn which of the control-display assemblies he (the subject) preferred. In order to get experience on which to base an evaluation, he was asked to make several "settings" with each type, and his settings served to identify three types of errors: (1) starting error (an initial movement in the wrong direction), (2) terminal overshoot (overshooting and then returning to the specified setting), and (3) setting error (incorrect setting). Some of the results are given at the bottom of Figure 8-10 (where two or three sets of data are given, they are from different groups used in different phases of the study, in which only certain assemblies were used in combination).

While assembly A incorporated the incompatible feature of a *counterclockwise* control turn bringing about an *increase* in scale value (rather than a decrease), nonetheless, this assembly was found throughout a series of subexperiments generally to be the best assembly, considering starting errors, setting errors, and preferences. While assembly E had low starting errors, it had the disadvantage of an arrangement in which the operator's hand covered the dial unless he assumed an awkward position, and it was not high in terms of preferences.

In evaluating the apparent superiority of assembly A over certain others (for example, C), it seemed clear that a *reverse drive* (such as in C and D) tended to give rise to more starting errors than a direct drive. And while E and F had all three of the postulated desirable features, they apparently had other undesirable features (for example, while the scale numbers increased from left to right, one might contemplate the fact that since they are at the bottom of the scale, the counterclockwise numerical progression *around* the scale might be thought of as another type of incompatibility).

While, in a sense, all this relates to a fairly incidental feature of display-control design, it probably does have broader implications. If within a situation, there might be different *kinds* of compatibility (as, indeed, there are in this experiment), it may be pertinent to know *what* type of compatibility is the more (or most) critical in case of some possible conflict (for example, *direct-drive* compatibility seems to be more critical than the *clockwise-increase* principle with moving scales).

Movement relationships of displays and rotary controls in different planes. In the relationship between movement of display indicators and of control devices, some control devices may be in a different plane from the displays with which they are associated, such as the lateral cutting plane or orthogonal cutting plane. In one such study, the control knobs that were used caused a pointer to move along a straight-line scale [Holding, 22]. The knob and pointer were in different planes, as illustrated in Figure 8-11. More than 700 subjects were used in the study and were asked to "twist the knob" in order to move the pointer, which was at one end of the scale in one series and at the other end in another series. The strongest relationships displayed by the subjects between direction of control movement and direction of pointer movement are shown in Figure 8-11. The results led to the conclusion regarding human

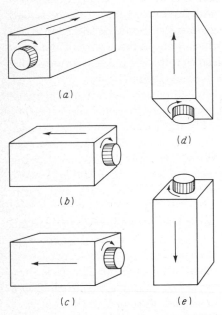

Figure 8-11 Illustration of some of the display-control relationships studied by Holding [22], showing the strongest relationships demonstrated by subjects.

performance in such situations that people's responses tend to be one of two types: (1) a generalized clockwise tendency; and (2) a helical, or screwlike, tendency in which clockwise rotation is associated with movement away from (as with screws, bolts, etc.), and counterclockwise is associated with movement toward, the individual.

A somewhat similar investigation was carried out with the rotary control knob in each of the following three locations (as characterized in terms of the three planes mentioned before) with respect to the display [Ross, Shepp, and Andrews, 30]: on the horizontal lateral cutting plane, on the vertical display plane, and on the right orthogonal plane (on the right side of the control box). In all locations the following relationships were confirmed: Clockwise rotation was associated with right display movement, and counterclockwise with left and with down. The clockwise-up relationship was significant for the right-hand knob, but for the horizontal and front vertical locations the relationship was ambiguous.

In some activities, such as in tracking in military operations, two rotary control devices are used simultaneously, usually one controlling the movement of a display in a lateral direction and the other in a vertical or forward-backward direction. Practical experience has suggested the desirability in such situations of having the right-hand control located vertically at the side and the left-hand control located vertically in front. But what are the most natural control-display relationships for such a setup? For pursuit tracking the following control-display relationship gave optimum time on target [Norris and Spragg, 27]:[8]

Crank	Clockwise rotation of crank moves display:
Left (facing operator)	To the right
Right (at side)	Away from subject

The same patterns of control-display relationships also held when both cranks were facing the operator, although accuracy was somewhat less than when the right-hand crank was at the side. These patterns are illustrated in Figure 8-12.

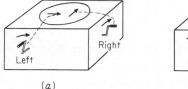 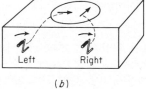

       (a)   (b)

Figure 8-12 Illustration of two patterns of control-display relationships for two-handed pursuit tracking with left and right cranks. The arrows on the circular target area indicate the directions of movement of the pointer or follower on the target face. Arrangement *a* (with right crank at side) is slightly more accurate than *b*. [Based on study by Norris and Spragg, 27.]

Movement relationship of stick-type controls. The compatibility of stick-type controls and movements of associated display indicators was investigated in connection with a tracking task [Spragg, Finck, and

---

[8] For compensatory tracking there was no clear-cut natural control-display relationship as there was for pursuit tracking [Green, Norris, and Spragg, 20].

Smith, 31]. Four combinations of control location and display movement were used, as illustrated in Figure 8-13. This figure also shows average tracking scores of subjects under these four conditions for one series of trials.

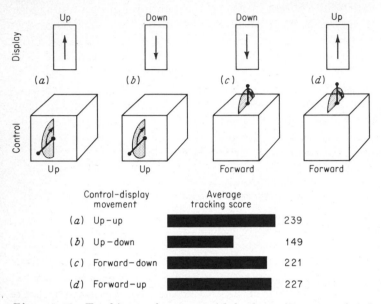

Control-display movement	Average tracking score	
(a) Up-up		239
(b) Up-down		149
(c) Forward-down		221
(d) Forward-up		227

Figure 8-13 Tracking performance with horizontally mounted and vertically mounted stick controls and varying control-display relationships. [Adapted from Spragg, Finck, and Smith, 31; data based on trials 9 to 16.]

For a horizontally mounted stick (on the vertical display plane), the superiority of the *up-up* relationship (control movement up associated with display movement up) over the *up-down* relationship is evident. For a vertically mounted stick (on the horizontal lateral cutting plane) there was less difference between the *forward-up* and *forward-down* relationships, although the forward-up relationship was slightly superior (but not significantly so). The investigators concluded that for the kind of tracking used it was about equally effective to mount the stick in a horizontal or vertical position, provided one *avoids* the up-down relationship with a horizontally mounted stick.

Corresponding results were reported in a study in which both horizontally mounted and vertically mounted control sticks were used in ascertaining the natural responses of subjects in controlling the direction of movement of a simulated missile in azimuth (bearing) and elevation [Lazar and Williams, 24]. While they found a strong up-up relationship for a horizontally mounted stick, there was considerable ambiguity in the movement relationships of the vertically mounted stick, since they found that a pull backward was associated with both an up and a down display-movement relationship.

Conceptual Compatibility

Probably the most common variety of conceptual compatibility relates to the associations in the use of coding systems, symbols, or other stimuli; these associations may be intrinsic (i.e., the use of visual symbols that represent things such as airplanes) or they may be culturally acquired. A number of illustrations were given in Chapter 4 and will not be discussed further here.

Discussion

Although different versions of compatibility involve the processes of sensation and perception and also response, the tie-in between these—the bridge between them—is a mediation process. We can see that the principle of compatibility is important in human factors engineering. Where compatible man-machine relationships can be utilized, the probability of improved system performance usually is increased. As with many aspects of human performance, however, there are certain constraints or limitations that need to be considered in connection with compatibility relationships. For example, some such relationships are not self-evident; they need to be ascertained empirically. When this is done, it sometimes turns out that a given relationship is not universally perceived by people; in such instances it may be necessary to "figure the odds," that is, to determine the proportion of people with each possible "association" or response tendency and make a design determination on this basis. In addition, there are some circumstances where trade-off considerations (of relative advantages and disadvantages of different possible solutions) may require that one forgo the use of a given compatible relationship for some other benefit.

SUMMARY

1. The various types of control functions that people perform in systems are predicated on human learning.
2. There are different types of learning that may be relevant to the activities people perform in systems, these being the learning of (a) stimulus-response relationships, (b) chains of relationships, (c) multiple discriminations, (d) concepts, (e) simple principles, and (f) complex principles.
3. The transfer of learning (from a *learning* task to the *real* situation) probably depends upon the relative similarities of the stimuli in the two situations and of their responses.

4. The transfer of training from *simulated* situations to real situations probably is predicated more on the fidelity of psychological simulation (i.e., the similarity of the behaviors) than on the fidelity of physical simulation.
5. The *content* of training in systems operations generally should be based on information about the actual knowledge and skills required; usually these can be identified by some method of task analysis.
6. Job aids provide for long-term *storage* of job information, and for its *retrieval* at appropriate times. Such aids should be designed to complement material that is to be learned through training.
7. Compatibility refers to the spatial, movement, or conceptual features of stimuli and of responses, individually or in combination, which are most consistent with human expectations. Some such features or relationships are intrinsic in the situation; others are culturally acquired (learned). In general, human performance is better (in speed, accuracy, etc.) if the stimuli and responses concerned are compatible with our expectations.

REFERENCES

1. Ammons, R. B., C. H. Ammons, and R. L. Morgan: *Transfer of training in a simple motor skill along the speed dimension*, USAF, WADC, TR 53–498, 1954.
2. Bass, B. M., and J. A. Vaughn: *Training in industry: the management of learning*, Wadsworth Publishing Company, Inc., Belmont, Calif., 1966.
3. Berliner, C., D. Angell, and J. W. Shearer: "Behaviors, measures and instruments for performance evaluation in simulated environments," in *Proceedings, Symposium on quantification of human performance, Aug.* 17–19, 1964, *Albuquerque, New Mexico*, M-5.7, Subcommittee on Human Factors, Electronic Industries Association.
4. Bradley, J. V.: *Desirable control-display relationships for moving-scale instrument*, USAF, WADC, TR 54–423, September, 1954.
5. Briggs, G. E., and E. L. Wiener: *Fidelity of simulation: I. Time sharing requirements and control loading as factors in transfer of training*, NAVTRADEVCEN, TR 508–4, Oct. 26, 1959.
6. Chapanis, A., and L. Lindenbaum: A reaction time study of four control-display linkages, *Human Factors*, November, 1959, vol. 1, no. 4, pp. 1–7.
7. Deese, J., and S. H. Hulse: *The psychology of learning*, 3d ed., McGraw-Hill Book Company, 1967.
8. Deininger, R. L.: Human factors engineering studies of the design and use of pushbutton telephone sets, *Bell System Technical Journal*, July, 1960, vol. 39, no. 4, pp. 995–1012.
9. Ferguson, G. A.: On transfer and the abilities of man, *Canadian Journal of Psychology*, 1956, vol. 10, pp. 121–131.
10. Fitts, P. M., and M. I. Posner: *Human performance*, Brooks/Cole Publishing Company, Belmont, Calif., 1967.

11. Fitts, P. M., and C. M. Seeger: *S-R* compatibility: spatial characteristics of stimulus and response codes, *Journal of Experimental Psychology*, 1953, vol. 46, pp. 199–210.

12. Folley, J. D., Jr. (ed.): *Human factors methods for system design*, The American Institute for Research, Pittsburgh, 1960.

13. Gagné, R. M. (ed.): *Psychological principles in system development*, Holt, Rinehart and Winston, Inc., New York, 1962.

14. Gagné, R. M.: *The conditions of learning*, Holt, Rinehart and Winston, Inc., 1965.

15. Gagné, R. M.: Contributions of learning to human development, *Psychological Review*, 1968, vol. 75, pp. 177–191.

16. Gagné, R. M., Katherine E. Baker, and Harriet Foster: *On the relation between similarity and transfer of training in the learning of discriminative motor tasks*, USN, SDC, TR 316–1–5, 1949.

17. Garvey, W. D., and W. B. Knowles: Response time patterns associated with various display-control relationships, *Journal of Experimental Psychology*, 1954, vol. 47, pp. 315–322.

18. Garvey, W. D., and L. L. Mitnick: Effect of additional spatial references on display-control efficiency, *Journal of Experimental Psychology*, 1955, vol. 50, pp. 276–282.

19. Gottsdanker, R., and J. W. Senders: *Compatibility of display and control*, Minneapolis-Honeywell Regulator Co., Minneapolis, Minn., MH Aero Document U-ED 6109, Feb. 24, 1959.

20. Green, R. F., E. B. Norris, and S. D. S. Spragg: Compensatory tracking performance as a function of the directions and planes of movement of the control cranks relative to movement of the target, *Journal of Psychology*, 1955, vol. 40, pp. 411–420.

21. Helson, H.: *Adaptation level theory*, Harper & Row, Publishers, Incorporated, New York, 1964.

22. Holding, D. H.: Direction of motion relationships between controls and displays in different planes, *Journal of Applied Psychology*, 1957, vol. 41, pp. 93–97.

23. Knowles, W. B., W. D. Garvey, and E. P. Newlin: The effect of speed and load on display-control relationships, *Journal of Experimental Psychology*, 1953, vol. 46, pp. 65–75.

24. Lazar, R. C., and J. R. Williams: *Investigation of natural movements in azimuth and elevation lever control adjustments for horizontal and vertical positions*, USA Ordnance Human Engineering Laboratory, Technical Memorandum 3–59, April, 1959.

25. Miller, R. B.: *Psychological considerations in the design of training equipment*, USAF, WADC, TR 54–563, 1954.

26. Muckler, F. A., J. E. Nygaard, L. I. O'Kelly, and A. C. Williams, Jr.: *Psychological variables in the design of flight simulators for training*, USAF, WADC, TR 56–369, January, 1959.

27. Norris, E. B., and S. D. S. Spragg: Performance on a following tracking task as a function of the relations between direction of rotation of controls and direction of movement of display, *Journal of Psychology*, 1953, vol. 35, pp. 119–129.

28.   Osgood, C. E.: The similarity paradox in human learning: a resolution, *Psychological Review*, 1949, vol. 56, pp. 132–143.

29.   Purifoy, G. R., Jr., and J. B. Fairman: *AN/AMQ-15 weather reconnaissance system training study planning report*, The American Institute for Research, Pittsburgh, 1959.

30.   Ross, S., B. E. Shepp, and T. G. Andrews: Response preferences in display-control relationships, *Journal of Applied Psychology*, 1955, vol. 39, pp. 425–428.

31.   Spragg, S. D. S., A. Finck, and S. Smith: Performance on a two-dimensional following tracking task with miniature stick control, as a function of control-display movement relationship, *Journal of Psychology*, 1959, vol. 48, pp. 247–254.

32.   Stevens, J. C., and H. B. Savin: On the form of learning curves, *Journal of the Experimental Analysis of Behavior*, 1962, vol. 5, pp. 15–18.

33.   Thorndike, E. L.: *The psychology of learning*, Teachers College Press, Columbia University, New York, 1913.

34.   Warrick, M. J.: "Direction of movement in the use of control knobs to position visual indicators," in P. M. Fitts (ed.), *Psychological research on equipment design*, Army Air Force, Aviation Psychology Program, Research Report 19, 1947.

35.   Woodworth, R. S.: *Experimental psychology*, Henry Holt and Company, Inc., New York, 1938.

36.   *Handbook of instructions for aerospace personnel subsystem design* (HIAPSD), USAF, AFSC Manual 80–3, 1967.

37.   *Use of task analysis in deriving training and training equipment requirements*, USAF, WADD, TR 60–593, December, 1960.

The many specific types of human performance that people exercise in most systems are directed toward the _control_ of the system in question to change the course of future events. Such human control involves, collectively, the following aspects of the conscious thought process of human beings [Kelley, 16, p. 41]: (1) _goal conception_, predicting possible future states of the controlled variable[1] that would occur with and without available control actions; (2) _goal selection_, planning the desired future state by choosing, from the range of possibilities, the goal or goals judged to be optimum according to appropriate criteria; (3) _programming_, programming the sequence of events and corresponding control actions required to bring about the desired state; and (4) _program execution_, carrying out the program sequence. As discussed earlier, many systems can be viewed as consisting of a hierarchy of systems within systems (i.e., of subsystems, sub-sub-systems, etc.). These same conscious processes apply, in one manifestation or another, to each level (what Kelley refers to as _loops_) of this hierarchy.

A THEORY OF CONTROL

A theory of control that is proposed [Kelley, 16, p. 234] is rooted in the belief that two properties of man distinguish him, in principle, from nonliving things, namely, (1) consciousness (men create internally in consciousness a dynamic model of themselves and their environment and create in consciousness possible future states of themselves and their environment, from which they make choices); and (2) controlled movement (the physical behavior is an expression of their conscious process, forming the means by which they are able to realize the choices

---

[1] The _controlled variable_ is the aspect of system performance that one wishes to control, such as the speed or location of a vehicle, the quality of a chemical product being produced, or—at a more specific level—the angle of turn of a steering wheel or the temperature of an oven.

they have made). This theory is at variance with the premises of cybernetics [Wiener, 27; Minski, 22] since it treats control systems as extensions of man's conscious control over the environment, rather than as autonomous mechanisms that independently control the environment. In line with this theory, even the most automatic of control systems can never be autonomous; rather, it serves simply as a mechanism to implement the conscious choices made by its creator.

CONSCIOUS PROCESSES IN CONTROL FUNCTIONS

If we accept the basic premise of Kelley's theory, we should of course view the human involvement in systems as including the goal conception, goal selection, programming, and program execution aspects. However, the nature and degree of the involvement of individuals in these various phases of the control process naturally hinges on the circumstance—the nature of the system, the level of the relationship of the individual to the system, the extent to which these phases of control process have been predetermined at higher levels, the designated role of the individual, the type of control mechanism(s) used, etc. For example, in some situations there is an intended one-to-one correspondence between some input information and the output response (such as in giving a telephone operator a number, or dialing a number, or in certain tracking tasks); in such instances the individual serves primarily as a transmission link and his primary control function is an internal one of endeavoring to match his output response (physical or communication) to the input information or signal. In other circumstances, certain input information is supposed to undergo some standard, or fairly standard, process before an output response is made (including invariant processes such as classification); such circumstances typically call for some form of information transformation or some form of filtering. In still other circumstances the input information, or stimuli, may be quite varied and unpredictable and require varied or innovative responses that are adapted to the specific situation in more of a problem-solving process. In still other circumstances (usually at higher levels in the hierarchy) individuals may be more in an overall planning (goal-setting) phase or in programming activities (such as computer programming) that result in specifying the goals or the procedures, or both, to be executed either by other people or by physical components of the system. In turn, the actual program-execution phase may be communicated to others for action, or it may be scheduled in time, or made contingent on and triggered by, a sensed event or state of the environment; if an individual exercises actual physical control, such control is effected by direct physical manipulation of objects or materials, by the use of tools, or by the use of control devices (automatic equipment may, of course, be employed at any point after the initiation).

Looking at the kinds and levels of human functions in this total process, the actual conscious physical control of system functions is one of primary concern because of some of the human problems that have cropped up as people carry out such control processes. This chapter will deal primarily with some aspects of human control processes, especially in connection with continuous control (i.e., closed-loop) processes such as tracking. Before getting into human performance in such processes, however, we should set the stage by discussing briefly the nature of continuous control operations. (We should also add the observation that although we shall become entangled with human motor processes, the major discussion of such processes will be deferred until the next chapter.)

THE NATURE OF CONTINUOUS CONTROL SYSTEMS

There are many types of systems with continuous control of some process or operation, these including the operation of all types of vehicles, tracking operations (as of military targets and aircraft), and certain production processes (as for chemicals and petroleum products).

Elements of Continuous Control Systems

In any system (or subsystem) some control system exists for exercising the desired control over the system or its subsystems. Such a control system is an integrated assemblage of human and mechanical elements of the following types [Kelley, 16, pp. 18, 19]: (1) *goal-selection (planning) system*, the human element(s) concerned with making choices about the desired modification of the environment; (2) *controller*, the element that produces a signal to the control junction to cause it to release or modulate the energies of control; (3) *power source*, the source of the energy of control; (4) *control junction*, the junction at which energy from the power source is released upon signal from the controller; (5) *control effector*, the element applying energy to modify the environment; and (6) *feedback sensor*. In general these elements also characterize the control systems at different levels in the hierarchy of a complex system.

Control Loops

In complex systems each level in the hierarchy, i.e., each *loop*, usually has two types of relationships with the outer (higher) loops and with the inner (lower) loops. On the one hand, the goal-selection and planning aspect of control typically permeates from the outer to the inner loops and thereby specifies, in a chain-reaction manner, the program-execution operations to be carried out. In the reverse direction, the controlled variable (i.e., the output) of any given inner-loop influences the next-higher loop.

These relationships are shown in Figure 9-1, this particular example showing the hierarchy of tasks in the admittedly difficult chore of ship steering [Kelley, 16, p. 29].

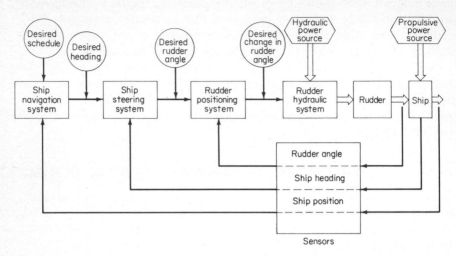

Figure 9-1   Hierarchy of tasks in steering a ship. Note that the desired goal at any given loop (e.g., the desired schedule) specifies the desired goal for the next-lower loop (e.g., desired heading), etc. Also, note the influence of any given loop (e.g., rudder hydraulic system) on the next-higher loop (e.g., rudder). [From Kelley, 16, p. 29.]

Inputs and Outputs in Closed-Loop Systems

For continuous, closed-loop systems, the input to the system specifies the system goal, or desired output; this may be constant (e.g., steering a ship at a specified heading or flying a plane at an assigned altitude), or variable (e.g., following a winding road or tracking a maneuvering aircraft). Such input typically is received directly from the environment and sensed by mechanical sensors or by people. If it is sensed mechanically, it may be presented to operators in the form of signals on some display. As input is received by *operators*, however, it may include not only the *system* input but also feedback to the operator of the system's output, which may have been affected by random disturbances and thus be more complex than the system input by itself. In a tracking task the input signal is sometimes referred to as a *target* (and in certain situations it actually is a target), and its movement is called a *course*. Whether or not the input signal represents a real target with a course, or some other changing variable such as desired changes in temperature in a production process, it usually can be described mathematically and shown graphically. While in most instances the mathematical or graphic representations do not depict the real geometry of the input (and the input-output relations) in spatial terms, such representations do have utility in characterizing the input. Inputs may change at a constant velocity (ramp), by steps, sinusoidally, or in more complex ways including randomly. Figure 9-2 illustrates some of these possibilities.

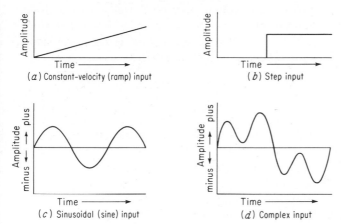

Figure 9-2 Types of inputs of tracking systems. Depending upon the circumstance, the amplitude can represent such variables as direction, distance, angle, voltage, velocity, and acceleration. Complex inputs frequently are the combination of two or more sinusoidal inputs.

In continuous control systems, the input to a system typically specifies the desired output of the system, such as curves in a road specifying the desired path to be followed by an automobile. The output is usually brought about by a physical response with a control mechanism (if by an individual) or by the transmission of some form of energy (if by a mechanical element). In some systems the output is reflected by some indication on a display, sometimes called a *follower* or a *cursor;* in other systems it can be observed by the outward behavior of the system, such as the movement of an automobile; in either case it is frequently called the *controlled element.* In fairly sophisticated systems with a hierarchy of loops, the input to any inner loop (which typically comes from the next-higher loop as its output) specifies the inner loop's desired output, which, in turn, becomes the input to the next-lower loop, etc.

Sinusoidal Functions

Since we shall be concerned with sinusoidal functions, it may be useful, as a reminder, to discuss sine waves briefly. A sine wave can be defined geometrically from the properties of a rotating vector (or a point on the circumference of a circle rotating at a constant angular velocity). This was illustrated in Chapter 3, Figure 3-8, in connection with the discussion of sound. While that particular figure represents changes in air pressure through time, the amplitude can depict many variables other than air pressure. As the point moves counterclockwise around the circle (starting from horizontal), its projection through time on a vertical axis forms a sine wave, and time is represented along the horizontal base. The sine of the angle, theta, Θ (sometimes called phase angle), is characterized as follows (see Figure 3-8 for meaning of NP and OP):

$$\sin \Theta = \frac{NP}{OP}$$

When $\Theta = 0$, the sine also is 0; as Θ increases up to 90°, the sine increases to unity. Subsequent changes in Θ bring about the characteristic temporal changes of the sine wave depicted, and plus and minus values are above and below the horizontal line.

The changes in amplitude depicted graphically occur in varying physical contexts, such as in vibrations, noise, and the motion of resonating bodies (at their resonating frequencies). These changes in amplitude also are characteristic of changes in velocity and of acceleration of objects in motion. Since changes in amplitude are associated with angular position, it is then possible to use Θ as a measure of amplitude of whatever variable quantity is involved in a system, such as an input or output quantity. It is also possible to give the amplitude of an error (the difference between actual output and desired output) in terms of Θ.

Relationships of angle, velocity, and acceleration. For sine functions, there is a specified relation among the variables of angle Θ, velocity, and acceleration, as depicted in Figure 9-3. (That figure also characterizes certain related tracking control "orders," but we shall mention these later.) While these relationships exist in any context in which a sine function is descriptive of continuous change of any type of variable, the relations probably can be most easily comprehended in the context of physical motion. As an example, let us envision a radarscope on which a target is moving up and down on a vertical axis, the motion being that of a sine function. The position (up or down) on this axis at any point in time is given by the vertical amplitude (up or down) of the angle Θ that designates that point in time. The *angle* curve in Figure 9-3, then, describes *position* through *time*. The *velocity v* curve, in turn, represents the velocity of the target at corresponding points in time; and, in turn, the *acceleration a* curve represents the corresponding rate of change in velocity (i.e., acceleration). Note that the velocity and acceleration curves are similar to the angle (i.e., position) curve, except that

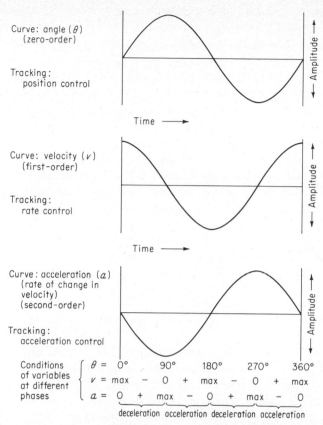

Conditions of variables at different phases

$\theta =$	0°		90°		180°		270°		360°
$v =$	max	−	0	+	max	−	0	+	max
$a =$	0	+	max	−	0	+	max	−	0

deceleration acceleration deceleration acceleration

Figure 9-3 Relationship among angle Θ, velocity, and acceleration curves and types of tracking tasks associated with these. The three curves are identical, but at different phases. The velocity curve is out of phase with the angle curve by 90°, and the acceleration curve is at opposite phase, being 180° different.

they have different phases; the velocity and the acceleration curves are out of phase with the angle curve by 90° and 180°, respectively.

In terms of derivatives, the phase-angle curve is a zero-order function, the velocity curve is a first-order function, and the acceleration curve is a second-order function. In other words, velocity v is the first derivative of angle Θ and thus requires one differentiation of Θ. In turn, acceleration a is the first derivative of velocity v and thus requires one differentiation of velocity, or the second derivative of angle and thus requires two differentiations of angle. Higher derivatives can be described as higher functions, i.e., third order, fourth order, etc.

Control Order

After this diversion into sine waves, let us get back to the matter of the input and output of continuous, closed-loop systems and to the ship we left adrift in Figure 9-1. The control of the system by the helmsman influences the rudder angle and, in turn, the ship heading and its position. But the nature of these influences is complex. Specifically, the *position* of the rudder control (which operates the rudder hydraulic system) produces a *rate of movement* of the rudder, and in a chain-reaction manner, the position of the rudder results in the angular acceleration of the ship and, in turn, the *rate of change of lateral position* with respect to the desired course. This hierarchy of control sequence represents various *control orders*. In continuous control processes that have a series of control loops (such as a ship), the sequence of chain-reaction effects can be described in terms of mathematical functions, such as a change in the *position* of one variables changing the *velocity* (rate) of the next, the *acceleration* of the next, etc. Almost any continuous control operation, then, can be characterized by its *control order* as predetermined by the mathematical derivative of the controlled variable, including zero order (position control), first order (rate or velocity control), second order (acceleration control), etc.

Control responses in relation to control order.   The control order, in effect, specifies the nature of the response that is to be made to various types of inputs. Some such responses are illustrated in Figure 9-4, for sine, step, and ramp inputs when position, rate, and acceleration control systems are used. In each case the dotted line represents the response over time (along the horizontal) that would be required for satisfactory tracking of the input in question. In general, the higher the order of control, the. greater is the number of controlled movements that need to be made by an operator in response to any single change in the input, as illustrated in Figure 9-4. This is illustrated further in Figure 9-5. If the input, changing over time, follows the pattern (line) shown, and if the control system (whatever it may be) is zero order (i.e., position control), then the movement of the control device by an operator should correspond exactly with that line. But if, instead, the control system is a first-order system (i.e., rate control), then the operator needs to anticipate and make the response movements shown. In turn, the other lines represent the changes that the operator would need to make with a control device if, in fact, the control system were a second-, third-, or fourth-order system. Some of the higher-order systems are, in fact, especially characteristic of certain vehicles, as of our ship. The rudder and flaps of aircraft serve to control the rate of change of heading and elevation and are second-order controls; and submarines typically have at least third-order control.

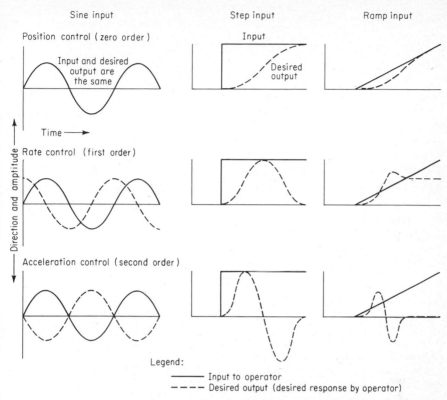

Sine input   Step input   Ramp input

Position control (zero order)

Figure 9-4   Tracking responses to sine, step, and ramp inputs which would be conducive to satisfactory tracking with position, rate, and acceleration control. The desired response, however, is not often achieved to perfection, and actual responses typically show variation from the ideal. In a positioning response to a step input, for example, a person usually overshoots and then hunts for the exact adjustment by overshooting in both directions, the magnitude diminishing until he arrives at the correct adjustment.

Mental functions related to control order.   The specific nature of the desired human response (output) in control systems is then specifically associated with the type of input (sine, step, ramp, etc.) and the control order that is used. With position control (zero order), the mental function required is essentially amplification (multiplication by a constant that represents the ratio of the input signal to the output response); this is shown in Figure 9-6a. In higher-order control operations other types of mental gymnastics are required, including those akin to mathematical differentiation, integration, and algebraic addition. A mental operation analogous to differentiation would be required, for example, in estimating velocity (rate), and to double differentiation in estimating

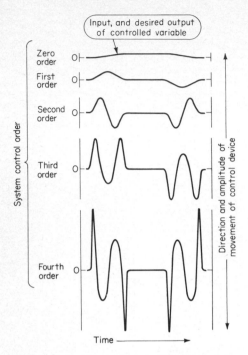

Figure 9-5 Control order illustrated by changes in a controlled variable and in its first four derivatives. Each line represents the changes over time that would have to be made with control systems of various control orders to make the controlled variable correspond to the input. [Adapted from Kelley 16, p. 31.]

acceleration. An operation analogous to integration would be required in estimating future position (at some specific time) from an estimate of a given velocity, and to double integration for estimating future position from an estimate of acceleration. Algebraic addition is of course the operation of adding two or more values (actually estimates of values), taking into account their sign (+ or −). Certain examples of these operations, as required of human operators in certain tracking tasks, are also diagrammed in Figure 9-6, along with the engineering symbols typically used to indicate these operations.

In general, people do not do well in these kinds of mental operations. In figuring velocities (a differential process), for example, it has been estimated that human error ranges from 10 to 50 percent [Birmingham and Taylor, 5]. There is some suspicion that people may do somewhat

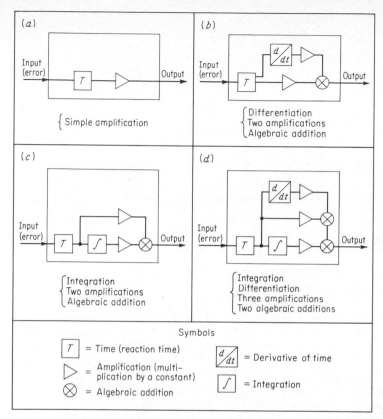

Figure 9-6 Functions of the human operator in various control operations, with engineering symbols used to represent such functions.

better in integration than in differentiation [Birmingham and Taylor, 4]. When one combines two or more operations, such as differentiation, integration, multiplication, and addition, the task becomes virtually impossible to perform. In some cases it is possible to transfer certain of these functions to a machine component and thereby relieve the operator of those functions. This is illustrated, for example, by Figure 9-7a. Other methods of facilitating human control of systems will be discussed in a later section of this chapter.

Transfer Functions

We are of course primarily concerned with the function of human beings in man-machine systems with a view to modifying the system to take advantage of the human performances which will contribute to adequate system performance. In this connection, some efforts have

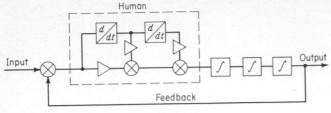

(a) Third-order control system

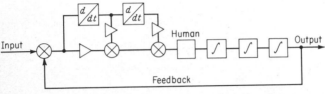

(b) Quickened version of third-order control system

Figure 9-7 (*a*) A symbolic representation of a third-order control system in which a human operator performs the following analog operations: (1) differentiating and double-differentiating (establishing first and second derivatives of error); (2) applying weighting factors to each of three terms (multiplying by a constant); and (3) adding the weighted factors algebraically in two cases. (*b*) A quickened version of that system which provides an analog processing of error information prior to feeding the signal to the human operator. (Quickening will be discussed later.) [Adapted from Birmingham and Taylor, 5.]

been made to describe human performance in essentially engineering terms, especially in continuous, closed-loop systems.

The physical components of systems cover, of course, a wide range of types, including electrical, electronic, and mechanical. These can include power sources, sensing mechanisms, control devices (such as servomechanisms and regulators), computers, power transmission and control components (such as gears, clutches, hydraulic and pneumatic components), structural components, and others. The operation of some engineering components can be described by a *transfer* function. A transfer function is a mathematical description of the ratio of the output of a component (such as a motor) to its input. It is derived from the differential equation relating the input and output signals of the component in question. The derivation will not be pursued here, but the concept itself is of concern in closed-loop systems because of the possibility of applying it to the human being, as well as to physical components.

Human transfer functions. As applied to human beings, the concept of a transfer function[2] would describe mathematically the relationship between sensory input (usually visual signals) and physical response (typically the operation of some control device). The intervening variables of human sensation, perception, mediation processes, and psychomotor control are all embedded in the framework of the *human transfer function*. If it were possible to represent the human operator's performance in terms of a mathematical transfer function, it would be conceivable that that performance could be reproduced by a mechanism. The possibility of borrowing the transfer function concept from engineering, however, hinges on the possibility of describing human beings in certain engineering (really mathematical) terms. The apparent vagaries and perversities of human behavior, however, have so far blocked major inroads in this direction. Probably a major stumbling block to this has been the fact that many aspects of human behavior (particularly psychomotor responses) are *nonlinear*, and nonlinear relationships are difficult to manipulate mathematically—especially the human varieties of nonlinearities. However, there may be certain classes of human responses which (at least within certain ranges) are sufficiently linear that, for practical purposes, it would be reasonable to treat them as such. In addition, in certain specific instances there may be some way around this problem, as by the use of *quasilinear* models [Licklider, 19] or by differentiating between the *linear* components of human responses and the nonlinear *remnant* [Cosgriff and Briggs, 8; McRuer and Krendel, 21; Senders, 25]. Besides the major barrier against the general use of human transfer functions imposed by the nonlinearity of human responses, Kelley [16, pp. 194–199] has posed what seem to be some other strong arguments against the general application of such mathematical models to design problems in manual control, as follows: (1) input narrowness (such models typically apply to single-input signals, whereas in most control processes human beings draw upon numerous sources of input information); (2) the models' lack of internal task representation (such models do not incorporate any explicit representation of the task; operators know a great deal about the nature of their tasks, such as a driver's "knowing" his automobile, and this knowledge is reflected in their performance); and (3) point-in-time limitation (such models imply that an operator's response at a point in time is a function of his input at that particular time, thus failing to represent human memory, planning, and prediction processes).

[2] Individuals interested in further readings on the human transfer function are referred to certain publications listed at the end of this chapter, especially the articles by Krendel and McRuer [18], Senders [25], Elkind [10], and Tustin [26], and the report by Diamantides and Cacioppo [9].

Discussion

There are, indeed, many operations and processes (particularly straight-forward ones) in which it is possible to predict in advance virtually every possible eventuality and to predetermine the control response that should be made in every such situation. To use a human being in such a situation converts the individual into a strictly data-transmission link (in which he is not very good, incidentally). Mechanisms can be designed to effect such operations. When the control process has essential goal-setting and planning functions, however, and the making of choices which are predicated on complex assortments of information and on accumulated experience, it appears that the operator's role is one that at least to date has not been duplicated by mathematical models or mechanical devices. People seem to have a monopoly in their mediation abilities to predict, to plan, to choose, to make decisions, to adapt, to innovate, and to make certain kinds of discriminations, evaluations, and judgments; efforts to develop control systems should take advantage of these unique human characteristics, both to increase the effectiveness of system performance and to create work activities that are personally satisfying and rewarding to the individuals involved.

DISPLAYS AND CONTROLS

Although displays were discussed earlier (Chapters 5 and 6) and controls will be discussed further later (Chapter 11), there are certain aspects of these that have unique pertinence to our present discussion of human control because of the way in which they predetermine the intermediate mediation functions and thus influence the quality of human performance.

Compensatory and Pursuit Displays

In tracking tasks, two types of information are usually displayed, namely, the input signal, or target, and the output signal, or the controlled element, follower, or cursor; the output signal shows the state of the system. These two types of signals frequently are represented on a visual display as blips on a CRT or as pointers on a dial or scale, or in some other way. In a *compensatory* display, one of the two indications (the target or the controlled element) is fixed and the other moves. When the two are superimposed, the controlled element is *on target;* any difference represents an error, and the function of the operator is that of manipulating the controls to eliminate or minimize that error. The source of any difference (error), however, cannot be diagnosed; whether the target has moved or changed course or whether the tracking has been inaccurate is not shown. An illustration is shown in Figure 9-8. In a pursuit display, both elements move, each showing its own *location* relative to the space represented by the display, as shown in Figure 9-8. With a pursuit display, the operator is presented with information about the actual location of both elements, whereas with a compensatory display he knows only the absolute error or difference. However, compensatory displays sometimes have a practical advantage in conserving space on an instrument panel since they do not need to represent the whole range of possible values or locations of the two elements.

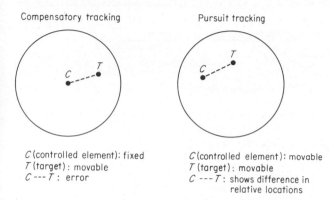

Compensatory tracking   Pursuit tracking

C (controlled element): fixed   C (controlled element): movable
T (target): movable   T (target): movable
C --- T : error   C --- T : shows difference in
                                              relative locations

Figure 9-8 Illustration of compensatory and pursuit tracking displays. A compensatory tracking display shows only the *difference* (error) between the target *T* and the controlled element *C*. A pursuit display shows the location (or other value represented) of both the target and the controlled element.

Comparison of compensatory and pursuit tracking and control order. The comparisons that have been made of tracking performance by using compensatory and pursuit displays have not demonstrated consistent superiority of one over the other. Rather, it seems that the human control performance with these two types of displays tends to vary with such variables as the control order and the nature of the course being tracked. In an investigation with certain of these relationships, the following variables were varied [Obermayer, Swartz, and Muckler, 23]: (1) compensatory and pursuit tracking and (2) control order, namely, position, rate, and acceleration. Three of the criteria of performance used were particularly pertinent, namely, average absolute error (AAE, average difference in inches between the input and output positions on the display), root mean square (rms, square root of the integrated square display error, in inches), and time on target (TOT, the time in seconds that the cursor was within $\frac{1}{10}$ in. of the target).

Some indication of the interactions of tracking mode and tracking control orders is indicated in Figure 9-9. The results for the three criteria

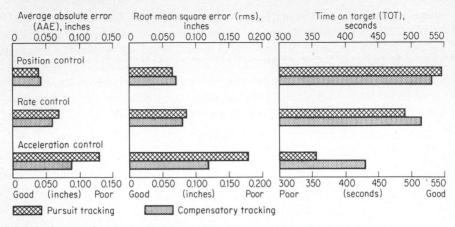

Figure 9-9 Tracking performance, by three criteria, on pursuit and compensatory tracking tasks with position, rate, and acceleration control orders. [Adapted from Obermayer, Swartz, and Muckler, 23.]

are relatively comparable—the *rank order* of performance of the six conditions was essentially the same for all three criteria, with performance generally being best with position control, next best with rate control, and poorest with acceleration control. But although there was no significant difference in performance for either position control or rate control between pursuit tracking and compensatory tracking, for acceleration control compensatory tracking was significantly better than pursuit tracking. The complexities of tracking performance are further illustrated by throwing a couple of other variables into the mix, in particular, course (input) *frequency* (e.g., the frequency of changes in the course, especially in the case of sine-wave inputs) and *amplitude* of input. For example, Chernikoff. and Taylor [7] report an advantage of rate control over position control at *lower* course frequencies (for both pursuit and compensatory tracking). Kelley and Prosin [17] also reported some interactions with one experienced subject, as follows: position control had an advantage with combinations of high-frequency and low-amplitude inputs, but with moderate to high amplitudes, rate control was better.

In pulling together the results of studies with pursuit and compensatory tracking, Briggs [6] concluded that under most experimental conditions, pursuit displays seem to be preferable, but he points out that the choice of a particular display must be made in the light of the kind of information required by the operator. For example, if the inputs are of very low frequencies, compensatory displays may have an advantage.

The interactions of variables such as the ones discussed above may illustrate some of the complexities that occasionally are encountered in the study of human behavior in circumstances that embody a number of variables (as tracking behavior does). Because of some of these intricacies, types of display and control systems have been developed that make life easier for the operator and that result in better system control.

Aiding

One such procedure is by the use of aiding. Aiding was initially developed for use in gunnery tracking systems and is most applicable to tracking situations of this general type, in which the operator is following a moving target with some device. Its effect is to modify the output of the control in order to help the tracker. In aided tracking, a single control adjustment usually affects two variables, specifically the position and the rate of the controlled element (and in some instances, also the acceleration). Let us suppose we are trying to keep a high-powered telescope directed exactly on a high-flying aircraft by using a rate-aided system. When we fall behind the target, our control movement to catch up again would automatically speed up the *rate* of motion of our telescope (and thus, of course, its position). Similarly, if our telescope gets ahead of the target, a corrective motion would automatically slow down its rate (and influence its position accordingly). Such rate-aiding would simplify the problem of quickly matching the rate of motion of the following device to that of the target and would thus improve tracking performance.

Aiding is illustrated in Figure 9-10 by before and after block diagrams of the tracking operation of responding to displayed error (a compensatory tracking task) by the movement of a damped joy stick [Birmingham and Taylor, 4], in which the fundamental human output is force. In the first illustration, *a*, the man is performing operations analogous to one differentiation, one amplification, one integration, and two algebraic additions. In effect, the joy-stick control (by its viscous damping) is performing essentially an integration operation, as shown in *a* and *b* of Figure 9-10. The change effected by aiding in this instance is that of shifting to the mechanism the differentiation, the integration, and the two addition operations, which leaves only an amplification operation for the man.

In the process of matching the rate of motion of the following device to that of the target, the ratio of these two changes is referred to as the *aided-tracking time constant* and (expressed in seconds) is as follows:

$$\text{Aided-tracking time constant} = \frac{\text{change in position per unit displacement of control}}{\text{change of rate per unit displacement of control}}$$

If viewed on a display, this constant is the ratio between the change in position of the follower (cursor) on the display and its rate of change.

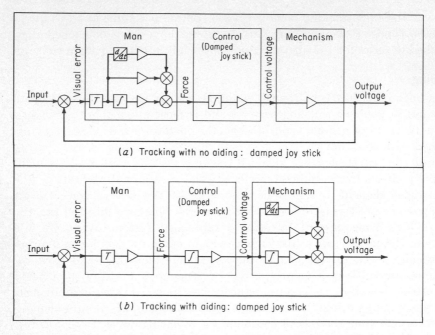

Figure 9-10 Block diagram of (*a*) an unaided and (*b*) an aided tracking system with a damped joy-stick control, in which the fundamental human output is force. In the aided system, the operations of differentiation, integration, and algebraic addition have been transferred to the mechanism, which leaves only the operation of amplification for the man. [From Birmingham and Taylor, 4.]

Since the movement of the follower is the direct consequence of the movement adjustment of the control device, this relationship can be given in terms of control movements. For example, a time constant of 0.5 sec means that a rotation of a handwheel control which produces a displacement of 10° in the position of the follower simultaneously generates a rate of follower movement of 20°/sec. The range of preferred aided-tracking time constants has generally been found to be within the values of 0.25 and 1.0 sec [Andreas and Weiss, 2] and the value of 0.5 is somewhere around the optimum for at least some tracking situations, but the optimum for any given system can be quite different from that of another system and it is preferable to determine the optimum experimentally. The effects of aiding on tracking performance in one experiment are illustrated in Figure 9-11. In this, the comparison is between unaided acceleration tracking and aided acceleration tracking. Comparable comparisons have been made of unaided and aided rate tracking. It should be added, however, that the effects of aiding depend upon a number of

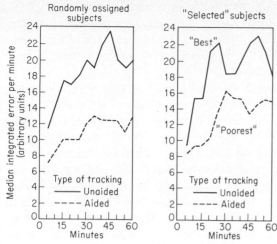

Figure 9-11 Comparison of performance of continuous tracking for 1 hr of unaided and of aided acceleration tracking. In the case of the selected subjects, the eight who were poorest on a pretest were assigned to the aided tracking task and the eight who were best were assigned to the unaided task; even the poorest subjects did better with aided tracking than the best ones with unaided. [Adapted from Garvey, 11.]

factors, such as the aided-tracking time constant used. Another factor is the nature of the input; aiding can be of greatest use where the input (whatever it may be) is low in frequency and high in amplitude rather than the reverse.

Augmented Displays and Controls

Some type of display or control augmentation is generally used for vehicular control systems that have higher-order control, such as when the input to the controlled element is a second, third, or fourth derivative of its output.

Quickening.   Quickening is one form of display augmentation and is used in what are called *command instruments* or *command systems*. In essence, quickening is a modification of a closed-loop system which reduces the need for the operator to perform analog differentiations or to sense and utilize derivative information separately [Birmingham and Taylor, 5]. It is particularly useful where the dynamics of the system are such that the apparent response of the system to control actions is delayed. The quickening is usually accomplished by appropriate modification of the information going into the display, so the displayed information is more easily and rapidly translatable in terms of the consequences of the operator's control actions. It indicates, in effect, what control action to take to achieve a desired system output.

By its nature, a quickened system is most appropriate where the consequences of the operator's actions are not immediately reflected in the behavior of the system, but rather have a delayed effect, frequently delayed by the dynamics of the system as in aircraft and submarines. Block diagrams of an unquickened and of a quickened system were shown above in Figure 9-7. If we refer back to Figure 9-5 and consider the responses that would be required for a second-order or third-order system (i.e., the lower curves of that figure), we see that in quickening the operator would still have to make those responses, but he would be shown what responses to make and would not have to go through the metal gymnastics of figuring what those complex movements should be (which, incidentally, he could not do).

A comparison of tracking performance with a quickened versus an unquickened system is reported by Birmingham, Kahn, and Taylor [3]. The six Navy enlisted men who served as subjects used, at different times, three nonquickened systems with one or two joy sticks to be moved in one or two dimensions, and a quickened system with a single joy stick. A summary of the results, shown in Figure 9-12, indicates a clear superiority of the quickened system over the others in time on

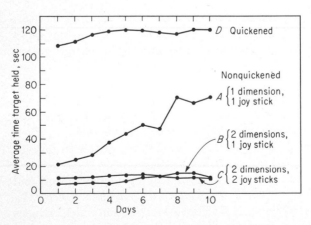

Figure 9-12 Comparison of tracking performance with a simulated quickened versus unquickened system. [Adapted from Birmingham, Kahn, and Taylor, 3.]

target (TOT). In fact, four of the six subjects made perfect scores with the quickened display.

Although quickening generally simplifies and improves some tracking tasks, it has certain possible disadvantages and limitations. For example, in a typical quickened system the operator is not provided with information regarding the current condition of the system, since his display shows primarily what control action he should take. It should also be kept in mind that quickening does not have any appreciable advantage in very simple systems, or in systems where there is no delay in the system effect from the control action and where there is already immediate feedback of such system response.

Predictor displays. A different type of display for use with manual control systems is a predictor display, developed by Kelley [14, 15, 16]. In effect, predictor displays use a fast-time model of the system to predict the future excursion of the system (or controlled variable) and display this excursion to the operator on a scope or other device. The predictive information is generated by a model of the system, operating repeatedly on an accelerated time scale. The model receives information from sensing instruments that are responsive to the existing conditions of the real system. Using this information, the model repetitively computes predictions of the real system's future, based on one or more assumptions about what the operator will do with his control, e.g., return it to a neutral position, hold it where it is, or move it to one or another extreme. The predictions so generated are displayed to the operator to enable him to reduce the difference between the predicted and desired output of the system.

While a command instrument (with quickening) in effect tells the operator what to do with his control, a predictor display tells him instead what he can expect to happen—what the future behavior of his system will be if he moves his control in a certain way.

A couple of examples of predictor displays are shown in Figure 9-13 [Kelley, 15]. These displays represent predicted depth errors of two high-speed submarines with different dynamic characteristics. Each shows the future predicted depth continuously up to 10 sec, assuming that the control is returned to its neutral (zero) position immediately. Note that the operator is not "programmed" as in a quickened system. But he is provided with very pertinent information about the predicted future response of the system, which makes it possible for him to take whatever corrective action is in order. In the first example the submarine will overshoot unless corrective action is taken well in advance; in the second example the submarine will undershoot a bit, and a moderate corrective adjustment will bring the submarine to the desired depth. With predictive information, in either such case, the operator has

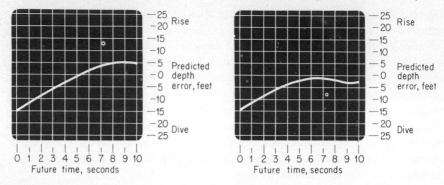

Figure 9-13 Examples of predictor displays for two submarines of different dynamic characteristics. These specific displays show predicted depth error, in feet, extrapolated to 10 sec, assuming that the control device would be immediately returned to a neutral position. [From Kelley, 15.]

the necessary predictive information available to take corrective action.

Predictive displays offer particular advantages for complex control systems in which the operator needs to anticipate several seconds in advance, such as with submarines, aircraft, and spacecraft. The advantages in such situations have been demonstrated by the results of experiments such as the one by McLane and Wolf [20] simulating the operation of a submarine in potential collision situations with four different displays. With a predictor display, collisions "occurred" in 4 percent of the trials, whereas with other displays, collisions occurred in 11, 13, and 18 percent of the trials.

Control augmentation. In display augmentation, such as with aided, command (e.g., quickened), and predictive displays, the information displayed to the operator is modified in some way in order to simplify his job of figuring out what control response to make. On the other hand, in control augmentation the operator makes a response that reflects the desired system output, and internal mechanisms of the system translate his signals into appropriate system control functions. If the operator is responding to changes in input (and thus specifying the desired output), all he needs to do is to track the input (such as making his responses correspond to the top curve of Figure 9-5); the control system then performs the mathematical operations required to convert the operator's simple input into more complex system maneuvers (such as those reflected by the lower curves of that figure).

Discussion. It does not take an Einstein to turn a switch when a light turns red, but learning the response required for some systems with complex high-order control tasks would be difficult, if not impossible, without some form of augmentation of the displays or controls or both.

Each variety of augmentation has its own advantages, as well as its own limitations and constraints. A command instrument (such as a quickened display), for example, simplifies life for the operator by virtually showing him what response to make. But, on the other hand, such an instrument simply does not tell the operator anything about the status of the system or the changes in that status, and therefore does not provide the information required to adapt to changing situations if, in fact, this should be desirable. (Incidentally, if, in fact, adaptation by the operator is not required and a quickened system could work, it would of course be possible to make the system completely automatic and thus throw out the operator.) When in the operation of a complex system, there is a premium placed on the adaptation by the operator to changing situations, however, the information presented by predictor displays is in a form that markedly simplifies the control process. In fact, it has been indicated that with a properly designed predictor display, an inexperienced individual can learn to operate a complex control system in as little as 10 min [Kelley, 15]. In turn, control augmentation tends to capitalize on man's capacity for planning and exercise of judgment, and to relieve him of the exercise of motor skill and strength, since his dominant role is that of deciding on the desired output and indicating this in a simple way. Thus in many circumstances control augmentation has an edge over display augmentation; but the complexity of control augmentation sometimes argues against its use even when it could be used, and the nature of some control tasks in turn argues for the use of some form of display augmentation.

Specificity of Displayed Error in Tracking

In certain compensatory tracking systems the error (the difference between input and output) can be presented in varying degrees of *specificity*. Some such variations are shown in Figure 9-14 as they were used in a tracking experiment by Hunt [13], these including: 3 categories of specificity (left, on target, and right); 7 categories; 13 categories; and continuous. The accuracy of tracking performance under these conditions (for two levels of task difficulty) indicates quite clearly that performance improved with the number of categories of information (greater specificity), this improvement taking a negatively accelerated form for tasks of both levels of difficulty. Although the results of other studies are not entirely consistent with these, the evidence suggests that in a tracking task, the mediating control functions are facilitated by the presentation of more specific, rather than less specific, display information.

Visual Noise in Tracking Displays

Noise in a visual display (such as a CRT) is a visual disturbance in the display, as in the background or in the target or response signals. Either variety of noise typically causes degradation in tracking performance.

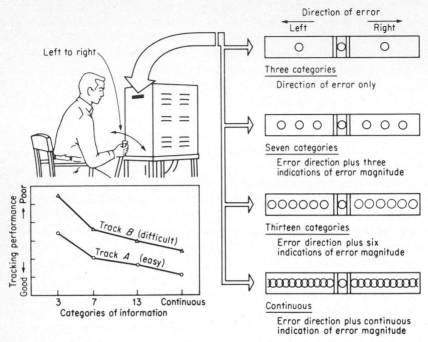

Figure 9-14 Compensatory displays used in study of the effects of specificity of feedback of error information, and tracking performance using such displays. The feedback error was presented by the use of lights (3, 7, 13, and continuous). [Adapted from Hunt, 13.]

The effects of the first type of noise (i.e., background noise) are shown in Figure 9-15. In this experiment the noise level was varied experimentally by varying the density of white blips (somewhat like the targets) on the black background. As shown in the figure, the effects were particularly noticeable for the short exposure times used experimentally; when the exposure time is long, the time for scanning is apparently sufficient to overcome some of the adverse effects of noise.

The effects of visual noise in the form of disturbances of the display signals are illustrated by the results of a study by Howell and Briggs [12] and are shown in Figure 9-16. This investigation was concerned with perturbations in pursuit tracking of the input signal I, the response signal R, and both I and R, and in compensatory tracking of the error E. The amplitude of the noise perturbations generated was such that, for three noise levels used, the average deviations of the signals produced by the noise were 0.03, 0.05, and 0.11 in., respectively. The most obvious thing about Figure 9-16 is that tracking performance was not noticeably affected when the visual-noise was of the response R display signal only;

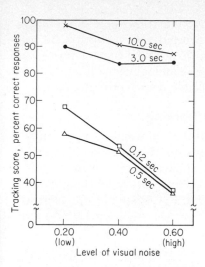

Figure 9-15 Relationship between visual noise level and tracking performance for four target exposure times. Exposure time is given in seconds of exposure of the film-strip that was used in presenting targets with visual-noise backgrounds. [Adapted from Wolf and Green, 28.]

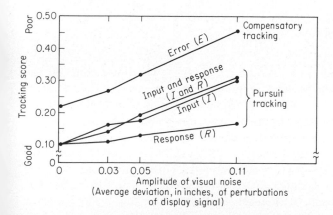

Figure 9-16 Effects on tracking performance of visual noise in the form of disturbances of error signals E in compensatory tracking and of input I and response R signals in pursuit tracking. Lines show tracking performance when the signal indicated (E, I, or R) was caused to deviate around its "true" position. [Adapted from Howell and Briggs, 12.]

apparently visually coded feedback information (the response signal) is less critical to operator performance then input information is. It has been suggested that experienced subjects in position control tasks can rely in part upon proprioceptive cues (pressure and movement) as a source of feedback and thereby maintain a reasonable degree of performance.

Compatibility of Displays and Controls in Tracking

In the mediation process of converting display information into control responses in tracking, certain manifestations of the principle of compatibility bob up. A couple of these are well illustrated by the results of a study by Regan [24] of both pursuit and compensatory tracking in two dimensions (vertical and horizontal) with both position and rate control. Three types of displays were used, as follows (see Figure 9-17): (1) two

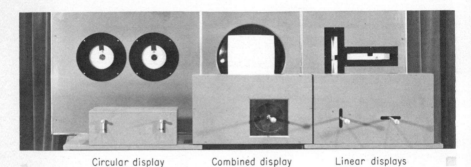

Circular display   Combined display   Linear displays

Figure 9-17   Three symbolic tracking displays used in tracking study. See text for description. [Adapted from Regan, 24.]

circular displays, each presenting one of the two dimensions of the input and each having a separate symbol for the target and the follower; (2) a combined circular display; and (3) two linear displays, representing, respectively, the vertical and horizontal dimensions. There were also three types of control mechanisms, namely, (1) two circular controls, each controlling one of two dimensions; (2) two linear controls; and (3) a single joy stick, which was free to move in both dimensions. Various combinations of displays and controls were used, as listed in Table 9-1. In considering the displays themselves, the results (analyzed by error scores) tended to show that the combined display (the center display) contributed to improved tracking performance compared with the displays which had separate indications, either circular or linear, for the vertical and horizontal dimensions. These results, then, support the hypothesis that compatibility of displayed information facilitates its assimilation and processing. (The combined display shows a target moving in two-dimensional space, which is the way we see things move in everyday life.)

In the various combinations of displays and controls, the superiority of the combined display with a joy stick was clearly evident, as shown in Table 9-1. But in addition (although not shown in that table) it was found

Table 9-1 Average Error Scores in Tracking Tasks Performed with Various Control-Display Relationships and with Other Tracking Variables

		Average error	
Display	*Control*	*Pursuit tracking*	*Compensatory tracking*
Circular, 2	Circular, 2 cranks	24.0	17.8
Linear, 2	Circular, 2 cranks	19.7	25.0
Circular, 2	Linear, 2 levers	25.2	25.5
Linear, 2	Linear, 2 levers	21.7	24.7
Combined, 1	Circular, 2 cranks	19.7	25.2
Combined, 1	Joy stick, 1	11.6	9.2
All displays and controls:			
Position control		10.3	14.5
Rate control		30.3	28.4

SOURCE: Data from Regan [24] based on last 9 of 27 trials. The data for the display-control combinations include both position and rate control. For the circular and linear models, two displays or controls were used—one for vertical movement, the other for horizontal movement.

that the compatible relationships (circular displays with circular controls and linear displays with linear controls) were significantly better than those in which the relationships were not thus compatible. As an aside, it was also evident that position control was, in this case, distinctly superior to rate (velocity) control.

DISCUSSION

A dominant function of human beings in systems is control—bringing about desired changes in the environment. Such control, of course, is essentially a conscious mediating process. The nature of these processes, however, frequently is influenced by—in fact, virtually predetermined by—the physical features of the system, especially the display and control system. Although the discussion in this chapter has illustrated some such effects, especially in tracking systems, it should be noted that there are numerous other variables that also can influence such processes, and that therefore should be taken into account in the design of systems.

SUMMARY

1. Human performance in most systems is directed toward the control of the system, to change the course of future events. Such control uses the following conscious thought processes: (*a*) goal conception, predicting future states of the system; (*b*) goal selection, choosing the desired future state; (*c*) programming; and (*d*) program execution.

2. A theory of control proposed by Kelley [16, p. 234] is rooted in the belief that man is distinguished from nonliving things by *consciousness* and the ability to exercise *controlled movement*. In line with this theory even the most automatic of control systems serves simply as a mechanism to implement the conscious choices of its creator.

3. In continuous control systems the input to the system specifies its desired output; in complex systems there may be a hierarchy of *loops*, with the output of one loop becoming the input of the next one.

4. Continuous control systems vary in their *control order*, the control order being determined in part by the sequential effects of a control movement and the dynamics of the system. Control orders include *position*, or *zero-order*, control (in which the control movement affects directly the position of the controlled variable); *rate*, or *first-order*, control (in which it controls a rate of change); and *acceleration*, or *second-order*, control (in which it controls the acceleration).

5. The control order in a control system predetermines the nature of the desired human control response to be made to effect the desired change in the controlled variables; with higher-order control systems the mental processes required may be analogous to mathematical differentiation, integration, and algebraic addition.

6. Compensatory displays used in continuous control (tracking) systems display only the error, or difference, between the input and the output signals; one signal is *fixed* and the other moves in relation to it. Pursuit displays show both the input and output signals in their own locations relative to the space represented by the display. Under most circumstances pursuit tracking is preferable to compensatory tracking.

7. Since human beings are not very adept at the operations analogous to mathematical operations that are required in some systems, various types of display and control modifications have been developed. These include: (*a*) aiding, in which the display signal of the controlled variable is modified in such a way that the operator simply serves as an "amplifier" of the displayed input signal; (*b*) quickening, in which the operator is in effect shown by the display what responses to make for the desired control; (*c*) predictor displays, in which the display indicates what the behavior of the system would be if the

control were returned to its neutral position; and (*d*) control augmentation, in which internal mechanisms translate the operator's response into appropriate system control functions.

8. Control in tracking tasks is facilitated if the indication of status of the input is more specific, rather than less specific.

9. Control in tracking tasks is facilitated when there is control-display compatibility.

REFERENCES

1. Adams, J. A.: Human tracking behavior, *Psychological Bulletin*, 1961, vol. 58, pp. 55–79.

2. Andreas, B. G., and B. W. Weiss: *Review of research on perceptual motor performance under varied display-control relationships*, University of Rochester, Rochester, N.Y., Science Report 2, Contract AF 30(602)–200, 1954.

3. Birmingham, H. P., A. Kahn, and F. V. Taylor: *A demonstration of the effects of quickening in multiple-coordinate control tasks*, USN, NRL, Report 4380, June 23, 1954.

4. Birmingham, H. P., and F. V. Taylor: *A human engineering approach to the design of man-operated continuous control systems*, USN, NRL, Report 4333, Apr. 7, 1954.

5. Birmingham, H. P., and F. V. Taylor: Why quickening works, *Automatic Control*, April, 1958, vol. 8, no. 4, pp. 16–18.

6. Briggs, G. E.: *Pursuit and compensatory modes of information display: a review*, USAF, AMRL, TDR 62–93, August, 1962.

7. Chernikoff, R., and F. V. Taylor: Effects of course frequency and aided time constant on pursuit and compensatory tracking, *Journal of Experimental Psychology*, 1957, vol. 53, pp. 285–292.

8. Cosgriff, R. L., and G. E. Briggs: *Accomplishments in human operator simulation*, Paper 60–AV–40, presented at Aviation Conference, ASME, June 5–9, 1960.

9. Diamantides, N. D., and A. J. Cacioppo: *Human response dynamics: GEDA computer application*, Goodyear Aircraft Corp., Report GER 8033, Jan. 8, 1957.

10. Elkind, J. I.: *Characteristics of simple manual control systems*, M.I.T., Lincoln Laboratory, Lexington, Mass., TR 111, Apr. 6, 1956.

11. Garvey, W. D.: *The effects of "task-induced stress" on man-machine system performance*, USN, NRL, Report 5015, Sept. 9, 1957.

12. Howell, W. C., and G. E. Briggs: The effects of visual noise and locus of perturbation on tracking performance, *Journal of Experimental Psychology*, 1959, vol. 58, pp. 166–173.

13. Hunt, D. P.: The effect of the precision of informational feedback on human tracking performance, *Human Factors*, 1961, vol. 3, pp. 77–85.

14. Kelley, C. R.: *A predictor instrument for manual control*, paper read before the Eighth Annual Office of Naval Research Human Engineering Conference, September, 1958, Ann Arbor, Mich., revised, January, 1962.

15. Kelley, C. R.: Predictor instruments look to the future, *Control Engineering*, March, 1962, pp. 86f.
16. Kelley, C. R.: *Manual and automatic control*, John Wiley & Sons, Inc., New York, 1968.
17. Kelley, C. R., and D. J. Prosin: "Frequency adaptive tracking systems," in C. R. Kelley (ed.), *Further research with adaptive tasks*, Dunlap and Associates, Inc., Santa Monica, Calif., TR to the USN ONR, July, 1967.
18. Krendel, E. S., and D. T. McRuer: A servomechanisms approach to skill development, *Journal of the Franklin Institute*, January, 1960, vol. 269, pp. 24–42.
19. Licklider, J. C. R.: "Quasi-linear operator models in the study of manual tracking," in R. D. Luce (ed.), *Developments in mathematical psychology*, The Free Press, New York, 1960.
20. McLane, R. C., and J. D. Wolf: *Symbolic and pictorial displays for submarine control*, paper presented at MIT-NASA Working Conference on Manual Control, Cambridge, Mass., Feb. 28–Mar. 2, 1966.
21. McRuer, D. T., and E. S. Krendel: *Dynamic response of the human operator*, WADC TR 56–524, October, 1957.
22. Minski, M. L.: Artificial intelligence, *Scientific American*, September, 1966, pp. 246–260.
23. Obermayer, R. W., W. F. Swartz, and F. A. Muckler: The interaction of information displays with control system dynamics in continuous tracking, *Journal of Applied Psychology*, 1961, vol. 45, pp. 369–375.
24. Regan, J. J.: *Tracking performance related to display control configurations*, USN, NAVTRADEVCEN, TR 322–1–2, Jan. 23, 1959.
25. Senders, J. W.: *Survey of human dynamics data and a sample application*, USAF, WADC, TR 59–712, November, 1959.
26. Tustin, A.: The nature of the operator's response in manual control and its implications for controller design, *Journal of the Institute of Electrical Engineers*, 1947, vol. 94, part IIA, pp. 190–202.
27. Wiener, N.: *The human use of human beings: cybernetics and society*, Doubleday & Company, Inc., Garden City, N.Y., 1950.
28. Wolf, Alice K., and B. F. Green, Jr.: *Tracking studies: II. Human performance on one-target displays*, M.I.T., Lincoln Laboratories, Lexington, Mass., Group Report 38–31, Nov. 20, 1957.

part
four
PHYSICAL
OUTPUT
ACTIVITIES

The direct *output* of people in many activities is some type of physical response, whether it is the operation of the control mechanism of a vehicle, the striking of typewriter keys, or the carrying out of the garbage. These and other outputs obviously depend upon the abilities and limitations of human beings in performing the motor activities that are required. In an overly simplified way, the nature of human motor abilities has implications relating to various aspects of human factors engineering, including (1) the design of control devices, (2) the design of hand tools and related devices, (3) the handling of materials, (4) the physical work layout, and (5) work methods and procedures.

BASES OF HUMAN MOTOR ACTIVITIES

Although, in human factors engineering, one is primarily concerned with the operational functioning of the body and body members, it is of course useful to have some understanding of the primary physical features and physiological processes that make people tick. Such activities depend essentially on the physical structure of the body (the skeleton), the skeletal muscles, the nervous system, and the metabolic processes.

The Skeletal Structure

The basic structure of the body, of course, consists of the skeleton, there being 206 bones that form the skeleton. Certain bone structures serve primarily the purpose of housing and protecting essential organs of the body, such as the skull (which protects the brain) and the rib cage (which protects the heart, lungs, and other internal organs). The other (skeletal) bones—those of the upper and lower extremities and the articulated bones of the spine—are concerned primarily with the execution of physical activities, and it is these that are particularly relevant to our subject. The skeletal bones are connected at body joints, there being two general types of joints that are principally used in physical activities, namely, synovial joints and cartilaginous joints. There are specific kinds of synovial joints, in particular (1) hinge joints (such as the fingers and knees), (2) pivot joints (such as the elbow, which is also a hinge joint), and (3) ball-and-socket joints (such as the shoulder and hip). The primary examples of cartilaginous joints are those of the vertebrae of the spine. These permit rotation and movement in virtually all directions. The amount of rotation and of movement of any given joint of the spine is relatively limited, but the combination of the many articulated vertebrae make possible, collectively, considerable rotation and considerable forward bending of the body.

The Skeletal Muscle System

The bones of the body are held together at their joints by ligaments. The *skeletal muscles* (also called *striated* or *voluntary* muscles) consist of bundles of muscle fibers that have the property of contractility; the muscle fibers serve to convert chemical energy into mechanical work. The two ends of each muscle blend into tendons which, in turn, are connected to different skeletal bones in such a manner that when the muscles are activated, they apply some form of mechanical leverage.

Neural Control of Muscular Activity

The activity of the muscles is under the control of the nervous system. The nerves entering a muscle are of two classes, namely, sensory nerves and motor nerves. Some of the sensory nerves are associated with the cutaneous senses (touch, heat and cold, pain, etc.). The other sensory nerves are proprioceptors, and it is these that provide kinesthetic feedback to aid in muscular control. Although the neural impulses of the proprioceptors typically do not reach the level of consciousness, they perform a dominant function in the control of muscular contraction. The motor nerves actually control the contraction (and relaxation) of the muscles. The signals to the motor nerves are routed via neutral centers. At these neural centers different routings of neural impulses are possible, but through frequent use certain connections become established, which increase the likelihood that subsequent neural impulses along the same primary pathway will follow the same routing at the neural centers along the way.

   Muscular control, via the motor nerves, is essentially voluntary. However, through practice we learn to carry out physical activities without conscious attention to them. The coordination in walking, eating, and routine job activities are examples. While many such activities are under conscious control in the sense that they can be consciously activated or terminated, once so activated (or terminated) we do not need to give conscious attention to the many muscular contractions and relaxations that are required in order to execute the "order" from the central nervous system.

Muscle Metabolism

The activation of muscles, of course, expends energy, the source of such energy being the foodstuffs we eat (carbohydrates, fats, and proteins). These substances are oxidized by the body and formed into products, especially glycogen, which can be used directly in muscle activity. Glycogen and other substances formed from food are stored in the body and are thus available for use, when needed, in performing physical work. The process of metabolism is the collective chemical process of the conversion of foodstuffs into two forms, namely, mechanical work and heat. Some of the mechanical work is of course used internally, in the basic body processes of respiration and digestion. Other mechanical work is used externally, as in the processes of walking and performing physical tasks. In either case, heat is generated, usually in amounts that are excessive to body needs; this surplus heat must be dissipated by the body.

As the body is caused to perform mechanical work, the glycogen and other materials which have been converted from foodstuffs are oxidized, and waste products (especially lactic acid) result from this process. This utilization of glycogen requires oxygen, and since the amount of oxygen available in the body is limited, the oxygen required for performing work generally must be supplied through the blood by the cardiovascular system. When an adequate supply of oxygen is thus provided, little or no lactic acid accumulates. If the level of physical activity requires more oxygen than is provided by the "normal" rate of blood flow through the cardiovascular system, the system adjusts itself to fulfill the increased demands. One adjustment is an increase in the breathing rate to bring additional oxygen into the lungs. In addition, the heart rate is increased to pump more blood through the "pipes" of the cardiovascular system; from the heart itself, the blood is pumped through the lungs, where it picks up a supply of oxygen, which is then carted by the blood to the muscles where the oxygen is needed. With at least moderate rates of work, the heart rate and breathing rate are normally increased to the level that provides enough oxygen to perform the physical activities over a continuing period of time. However, when the amount of oxygen delivered to the muscles fails to meet the requirements (as when the level of physical activity is high), lactic acid is formed and tends to accumulate in the blood. If the rate or duration of physical activity results in continued accumulation of lactic acid, the muscles will ultimately cease to respond.

At the initiation of physical activity, the muscles break down glucose into lactic acid which liberates the energy required for muscle activity; this phase of the process involves no oxygen; but the subsequent removal process of this lactic acid does require oxygen. If the rate of removal of lactic acid does not keep pace with its formation (owing to insufficient oxygen for the rate of activity), additional oxygen must be supplied after cessation of the activity to remove the remaining lactic acid. This is referred to as the *oxygen debt*. Since this debt has to be paid back, the heart rate and breathing rate do not immediately settle back to prework levels when work ceases, but rather slow down gradually until the borrowed oxygen is replaced.

Basal metabolism.   The basal metabolic rate is that which is required simply to maintain the body in an inactive state. Although it varies from individual to individual, the average for adults usually ranges from about 1500 to 1800 kcal/day [Tuttle and Schottelius, 72].[1] Considering the basal level plus the energy required for a relatively sedentary existence, Passmore [54] estimates that about 500 kcal are required for 8 hr in bed plus about 1400 kcal for nonworking time, adding up to a total of 1900 kcal per day; in turn, Lehmann [42] estimates the corresponding requirements (basal metabolism and leisure) at around 2300 kcal/day, and Tuttle and Schottelius [72] estimate that the typical adult who lives a fairly sedentary life utilizes about 2400 kcal/day. Thus, various estimates of the total nonworking calorie requirements range from about 1900 to about 2400 kcal/day. (The physiological costs of work will be mentioned later.)

Physiological Stress

In discussing human physical activities, Brouha [8] refers to an individual's "physiological capital" which provides him with a "physiological credit." As he carries out some physical activity, he uses his credit and may contract a "physiological debt," the amount of which varies with the nature of the activity and the environment. The greater the debt, the longer it will take to pay it back and to recover to a physiological resting level. This physiological debt is a reflection of the physiological stress that the individual has experienced. Such stress can be incurred as the consequence of the physiological work load of the physical activity, or of the environment (such as temperature), or a combination of both. Since a later chapter will deal with temperature variables, we shall focus

[1] The energy unit generally used in physiology is the kilocalorie (abbreviated kcal) or Calorie (with a capital C to distinguish it from the gram-calorie). The kilocalorie is the amount of heat required to raise the temperature of a kilogram of water from 15 to 16°C. The relation of the kilocalorie to certain other units of energy measurement is

1 kcal = 426.85 kg-m
1 kcal = 3087.4 ft-lb
1 kcal = 1000 cal = 1 C

here on physiological stress that occurs as the consequence of physical activity. As indicated above, if the rate of physical activity is low enough, a muscle or muscle group can function almost indefinitely, but at higher rates stress sets in, and if activity continues at the same rate, the muscle or muscle group will become completely fatigued and cease to function at all. This is shown in the example below from Tuttle and Schottelius [72, Table 6].

No. of contractions	Fatigued by	Work done
1 per 1 sec	14 contractions	0.912 kg-m
1 per 2 sec	18 contractions	1.080 kg-m
1 per 4 sec	31 contractions	1.842 kg-m
1 per 10 sec	No fatigue (no stress)	Almost indefinite

We can see that the rate of 1 contraction of the muscle per 10 sec did not produce complete fatigue and permitted almost indefinite continuation of work; faster rates (1 contraction per 4, 2, or 1 sec) produced such stress that the muscle ceased to function after 31, 18, or 14 contractions, respectively.

Efficiency of Work

The efficiency with which work is performed is essentially a ratio of output to input. In the context of efficiency of the work of the human body, the following equation is appropriate [Tuttle and Schottelius, 72]:

$$\text{Efficiency} = \frac{\text{work done (such as in foot-pounds)}}{\text{work done} + \text{heat produced}}$$

It has been estimated that the efficiency in bicycling is around 20 to 25 percent. The efficiency of human work generally would not exceed this range.

The efficiency with which an activity is performed, however, usually is related to the *rate* at which the activity is carried out. Typically, there is some *optimum* rate of performance for a given physical task, or at least a range of rates that is optimum. This is illustrated, for example, in Figure 10-1. This shows the relationship between rate of stair climbing and efficiency (expressed in arbitrary units). It can be seen that the efficiency was greatest when the task was completed in about 100 sec, the range from about 75 to 125 bringing about approximately comparable levels of efficiency. However, although there is something of an optimum rate of stair climbing (in terms of output-input efficiency), we should not fall

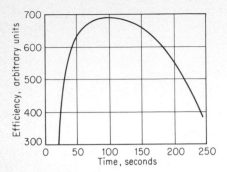

Figure 10-1 Efficiency of stair climbing at different speeds (speed is given by time in seconds to climb the stairs). [From Lupton, as presented by Tuttle and Schottelius, 72].

into the trap of assuming that stair climbing could be continued indefinitely at that rate.

Individual Differences and Age in Physical Functions

As would be expected, there are marked differences in the abilities of individuals to perform various physical functions. For example, Caldwell [13] reports the mean strength of a horizontal pulling action at a 150° elbow-angle as 162 lb with a range from 117 to 238 lb (more than twice the minimum value) and a standard deviation of 26 lb. In connection with age, we of more than two score years can testify that age takes its toll. Empirical support for this generally recognized down-grade trend is provided from many studies, such as the one by Malhotra et al. [44], who report that all the physical functions tested for a group of 879 soldiers started to show some deterioration after 30 years of age and continued to deteriorate after that age; the functions tested involved speed of running, abdominal muscle strength, agility, arm and shoulder strength, and capacity for short bursts of activity. As an additional example, Deupree and Simon [20] compared the reaction time and movement time of a group of elderly people (median age, 75) with a group of energetic sophomores (median age, 20); the reaction time of the older group was 11 percent slower, and the movement time 38 percent slower. The "aged," however can be consoled by the encouraging comments of Müller [50], who points out that whereas maximum (physical) work capacity may drop with age, *occupational* work capacity (in many occupations) is not affected materially in modern industry. In part (as Müller points out), this is due to the fact that modern industrial activities generally are less strenuous than in bygone days; in part, this physical deterioration is compensated for by increased experience and skill. However, engineers who design equipment that might be used by older persons need to take into account fully the fact that all people are not as spry as the engineers themselves were—or are—during their own tender years.

THE MEASUREMENT OF PHYSICAL ACTIVITY

If we wish to concern ourselves about human physical activity, in a human factors engineering context or otherwise, we need to deal with the ubiquitous criterion problem, that is, with the measurement of whatever facet is of concern. In an overly simplified way, such criterion measures are of two general classes, namely, those of an essentially physiological nature, and those of an operational nature (particularly those that reflect physical performance).

Physiological Criteria of Physical Activity

We have already alluded to certain physiological criteria of physical activities, such as heart rate and breathing rate. But the change from a resting state to a state of physical activity involves other physiological functions as well, such as blood pressure, pulmonary ventilation, oxygen consumption, carbon dioxide production, body temperature, perspiration rate, and chemical composition of the urine and blood. As Brouha states [8, p. 3], by measuring one or more of these physiological functions, it is possible to determine in what degree the *working level* differs from the *resting level*. (But any such variables need to be measured during the recovery period, as well as during the work period itself.) Brouha himself has used primarily the heart-rate recovery curve as a measure of physiological stress, this being the curve of the heart rate measured at certain intervals after work (such as 1, 2, and 3 min). Calorie requirements and oxygen consumption are also fairly common measures of physical activity. All these variables are influenced by physical activity and are therefore somewhat correlated with each other; some indications of the relationships of physiological measures (specifically of cardiac and respiratory functions) with work performed, and with each other, are reported by Brouha and Krobath [9]. However, the correlations of physiological measures are not consistently high; therefore, in measuring the physiological costs of work one needs to specify the particular variable that he is using as a criterion.

Most of the physiological variables mentioned above, although not perfectly correlated with each other, are indicative, in their own way, of the general level of physiological stress of the activity in question. There are, however, certain other essentially physiological criteria that can be related to the activity of the particular muscle or muscle group involved. One such scheme consists of recordings of electrical impulses from muscles during muscle activity. Such recordings, called *electromyographic* (EMG) recordings, are inked tracings of the electrical impulses; they provide estimates of the force of muscle contractions. An example of this technique is illustrated in Figure 10-2, based on a study by Lundervold [43]. In par-

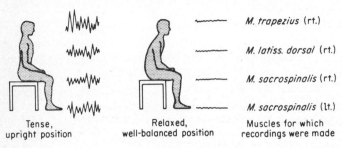

Tense, upright position	Relaxed, well-balanced position	*M. trapezius* (rt.) *M. latiss. dorsal* (rt.) *M. sacrospinalis* (rt.) *M. sacrospinalis* (lt.) Muscles for which recordings were made

Figure 10-2 Electromyographic recordings of electrical impulses from four muscles of the trunk of an individual typing in a tense, upright position and in a relaxed, well-balanced position. [From Lundervold, 43.]

ticular, this shows the EMG recordings that were made of four dorsal muscles of the trunk of a subject typing when in two different postures. The differences in the recordings under these two conditions are quite apparent for all four muscles in question.

Operational Criteria of Physical Activity

What we will label *operational* criteria include techniques for measuring or otherwise depicting the performance of the body or body members. Perhaps the most obvious, and most common, operational criteria relate to the performance of body members in making specific types of movements, such performance generally falling into the following groups: *range* of movement, *force* applied during the activity (i.e., strength), *endurance*, *speed*, and *accuracy*. For measuring these, various kinds of gadgetry are used, such as timing devices, motion pictures, strain gauges, and dynamometers. Aside from such criteria of body-member movements, there are certain techniques for depicting, or recording information about, the physical movements involved in activity. Such techniques include motion pictures, the use of interrupted-light photography (chronophotography), gliding cyclograms, and electronic techniques. Certain of these recording techniques will be discussed briefly.

　　Gliding cyclograms.　Gliding cyclograms are used in the biomechanical analysis of motion when the motion in question is repetitive.

It is photographic technique in which the movement of the body member (or some device being used) is photographed in such a way that the motion is depicted as a function of time [Drillis, 21]. The photographic image that is reproduced is either the light from a very small electric bulb that is affixed to the body member or to the device being used, or the reflection of light from a piece of Scotch light that is affixed. Figure 10-3 shows the movements being made in a nailing task, specifically the amplitude of hammer strokes, indicating their increase in the initial phase (the first five strokes), then their decline.

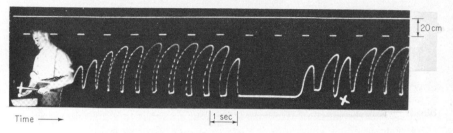

Time ⟶ |1 sec|

Figure 10-3 Recordings (gliding cyclograms) of movements in nailing. This shows the amplitude of hammer strokes along a time axis. In this particular illustration, the cadence of strokes was 1.72 per second or 103 strokes per minute. [From Drillis, 21.]

UNOPAR. An interesting device that has been developed for recording motions of body members is the Universal Operator Performance Analyzer and Recorder (UNOPAR) [Nadler, 52, pp. 417–428; Nadler and Goldman, 53]. The system consists of a physically small transducer attached to the subject's body member and radiating ultrasonic acoustic energy. The radiated energy, in turn, is received by three microphones along mutually perpendicular axes around the work space, as shown in Figure 10-4. The pickups of three-dimensional motion are converted, by appropriate circuitry, into tracings of velocity versus time for each of the three dimensions. These, in turn, can be subjected to various analyses, especially those relating to velocity, acceleration, displacement, and time variables.

Force platforms. Still another, quite different, operational criterion consists of recordings of total body forces during physical activity by the use of a force platform. A force platform is a small platform on which a subject stands when carrying out some physical activity. By the use of some sensing elements below the platform (such as piezoelectric crystals) it is possible to sense and then automatically record the forces generated by the subject in each of three planes, namely, vertical, frontal, and transverse. The original force platform was developed by Lauru [40];

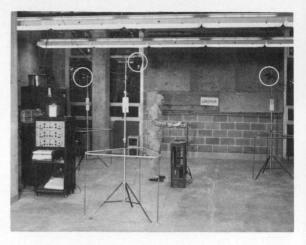

Figure 10-4 Universal Operator Performance Analyzer and Recorder, or UNOPAR. The left picture shows the transducer attached to the wrist. The right picture shows the location of the three microphones that pick up radiated energy. The system results in recordings of motions in each of the three dimensions. [From Goldman and Ross, 27.]

other platforms have been used experimentally by Barany [1], Barany and Greene [2], and Greene, Morris, and Wiebers [31]. Such devices are sensitive to slight differences in physical movements and can thus lend themselves to use in comparing the three-dimensional forces in different activities. Recordings of the forces in the operation of manual and of electric typewriters are shown in Figure 10-5 for comparison.

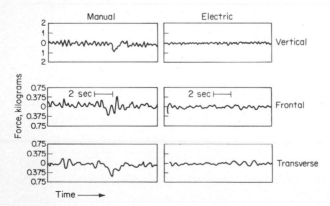

Figure 10-5 Forces in three dimensions (vertical, frontal, and transverse) in the operation of a manual and an electric typewriter, recorded with a force platform. [From Brouha, 8, p. 106.]

It has been proposed that such force-time recordings, as possible indices of energy expenditure, are nearly as accurate as metabolic measurements and can thus be used as a measure of physiological cost of a given motion [Brouha, 8, p. 103]. In fact, Brouha presents data for oxygen cost of certain work activities that show high correlations (ranging from .83 to .96) with data from the force platform (force-time *areas*, that is, the area under the curve produced by the platform). Although further data probably would be required to determine the generality of the relationship between force-platform measures of work and strictly physiological measures, the force platform does seem to offer considerable promise as a technique for studying work activities.

Universal motion analyzer. A device that has been rigged up by Smith and his associates [18, 32, 59, 63, 64, 74, 77] is the Universal Motion Analyzer, an example of which is shown in Figure 10-6. The device

Figure 10-6 Universal Motion Analyzer used by Smith and associates [59].

includes a panel of positions (the one shown has 34) for turn switches, knobs, or other devices arranged in rows and columns. By electrical methods it is possible to determine accurately the times required for elements or sequences of movements performed with the setup. Thus, it is possible to measure separately travel time and the manipulation time of any sequence. (Reference will be made to this later.)

ENERGY EXPENDITURE IN PHYSICAL ACTIVITIES

As indicated earlier, human beings are not now used as sources of energy nearly as much as in days gone by, since other sources of power have been developed for this purpose. However, we cannot disregard the physiological costs of at least some human activities. For example, some occupations still require substantial physical effort, at least at certain times or as accumulated over the work day, and in some countries the use of human beings as major sources of energy is almost dictated by economic considerations. When human physical activity in work is potentially dangerous to health and safety, some modification of the work is in order, whether by appropriate redesign of the equipment and work space, by modification of methods, or by other means, such as reduction of work periods or work pace.

Energy Expenditures of Gross Body Activities

In order to give some "feel" for the numerical values of energy expenditures for different kinds of physical activities, it may be useful to present, as some sort of a base, the physiological costs of certain everyday activities; the following examples are given in kilocalories per minute [Spector, 67, pp. 347, 348]: supine, 1.17; sitting, 1.8; standing at ease, 1.98; walking 2.4 mph, 4.3; and walking 4.8 mph, 10.7. In connection with rate of body movement (as in walking and running), the price per unit of work (in physiological costs) goes up with increasing rate. This is shown quite clearly in Figure 10-7, which shows the heart rate during and after marching a 1600-meter (about 1 mile) course, when the 18 subjects marched at speeds ranging from 4.9 to 15.2 km/hr (about 3.0 to 9.4 mph).

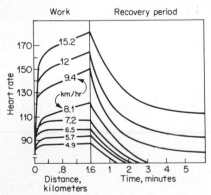

Figure 10-7 Heart rate during and after a march of 1600 km (1 mile) at various speeds, in kilometers per hour. [From Le Blanc, 41, as presented by Monod, 49.]

The increasing energy cost scoots up rather sharply at speeds of 8.1 km/hr and above. In addition, recovery time also increases markedly. The energy costs of body exercises of different intensities are shown further in Figure 10-8. The work loads were the consequence of running on a treadmill at

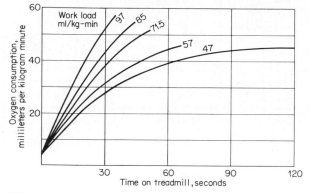

Figure 10-8 Oxygen consumption from the onset to the end of running on a treadmill with certain work loads that were the consequence of specified speeds and inclines; the energy requirements of the work loads are themselves expressed by an independently derived oxygen-consumption index. [From Margaria et al., 46.]

certain combinations of speed and inclines. The oxygen-consumption curves are obviously steeper the higher the work load; the lightest work led to exhaustion in about 3 min, the heaviest in about 30 sec. These and other examples clearly indicate the trade-offs in human work, in particular that the physiological price of work—per unit of work—is greater at higher rates of work than at more moderate rates.

Energy Expenditures in Carrying Loads

The discussion above dealt simply with the energy expenditures of lugging the body around, as, for example, at different paces, inclines, etc. Let us now, in our discussion of human work, shift to the common chore of carrying loads, in particular to illustrate the influence on energy costs of the work method, rate of work (speed), and work load. The effects of method are illustrated by a schoolboy in the commonplace task of carrying a school bag weighing, with its store of knowledge, 6 lb, while he is walking at 2.5 mph [Malhotra and Sengupta, 45]. The four methods of carrying the bag are shown in Figure 10-9, along with their relative oxygen requirements, with one least-demanding method (the rucksack) used as an arbitrary 100 percent base. The fact that the two methods that maintained a center of gravity (the rucksack and low back) required considerably less energy than the one-sided methods (across the shoulder, and one hand) jibes with a general principle proposed by Teeple [70] to

Rucksack	Lowback	Across shoulder	Hand
100%	137%	182%	241%

Relative oxygen required →

Figure 10-9 Oxygen consumption with four methods of carrying a 6-lb school bag. The relative oxygen requirements given are those in excess of the O_2 required for walking without the load, the rucksack method being considered 100 percent. [Adapted from Malhotra and Sengupta, 45.]

the effect that, in carrying loads, the best positions are those that affect the center of gravity the least. Teeple, however, had also investigated a balanced two-hand carrying method, which was quite satisfactory. As additional support for the center-of-gravity notion, an ancient but classic study by Bedale [6] resulted in lowest kilocalorie requirements for a yoke-carrying method (i.e., a fitted wooden yoke over the shoulders, with equal loads on the two ends), as contrasted with a one-hand and a hip-carrying method. The combined effects of the load being carried and speed are shown in Figure 10-10 [Passmore and Durnin, 55, based on other studies]. Here we see an accelerating cost in energy expenditure with increasing speed, this being true of all loads. In connection with loads and speeds, Teeple [70] suggests that the optimum load for an individual is about 35 percent of body weight and the optimum rate of carry is in the range of about 85 to 95 yd/min; these rates correspond to about 3 to $3\frac{1}{2}$ mph and are at points on the kilocalorie per minute curves of Figure 10-10 before those curves start their sharp upward accelerations.

Energy Expenditures of Specific Work Activities

Keeping in mind the energy expenditures of certain everyday activities of gross body movement and carrying loads, as given above, let us now cite some specific examples of energy costs of work activities; these are given in Figure 10-11. The energy costs of these examples range from 1.6 to 16.2 kcal/min. As we shall see shortly, some of these would, if continued for extended periods of time, do in even the hardiest among us.

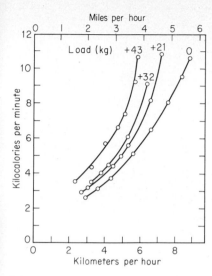

Figure 10-10 Energy expenditure when carrying various loads (43, 32, 21, and 0 kg) at different speeds. [Passmore and Durnin, 55.]

Keeping Energy Expenditures Within Bounds

It is usually within the scope of those who are concerned with the nature of human work activities (design engineers, industrial engineers, supervisors, administrators, etc.) to influence the level of physiological wear and tear in specific job activities. But if one is to try to keep energy costs within reasonable bounds, it is necessary to know both what those bounds are and what the costs are (or would be) for specific activities (such as shown in Figure 10-11).

Energy costs of grades of work. As a starter toward this, the following definitions of different grades of work may be helpful [Christensen, 14]:

	Energy expenditure		Approximate oxygen consumption, liters/min
Grade of work	kcal/min	kcal/8 hr	
Unduly heavy	over 12.5	over 6000	over 2.5
Very heavy	10.5–12.5	4800–6000	2.0–2.5
Heavy	7.5–10.0	3600–4800	1.5–2.0
Moderate	5.0–7.5	2400–3600	1.0–1.5
Light	2.5–5.0	1200–2400	0.5–1.0
Very light	under 2.5	under 1200	under 0.5

Figure 10-11 Examples of energy costs of various types of human activity. Energy costs are given in kilocalories per minute. [Data from Passmore and Durnin, 55, as adapted and presented by Gordon, 29.]

In considering such categories, it is important to recognize the considerable individual differences. A well-trained football player might much better endure, say, "very heavy" work of 10 kcal/min than, say, a middle-aged woman might endure "moderate" work of 5.0 kcal/min.

Still keeping in mind the matter of individual differences, however, we need to figure out a reasonable ceiling on energy expenditures over the period of the conventional working day. Lehmann [42], in synthesizing various types of relevant data, states that the maximum energetic output a normal man can afford in the long run is about 4800 kcal/day; subtracting his estimate of basal and leisure requirements of 2300 kcal/day leaves a maximum of about 2500 kcal/day available for the working day. This figures out to be about 5 kcal/min. But although he proposes this as a maximum, he suggests about 2000 kcal/day as a more normal load, this averaging out to be about 4 kcal/min. Edholm [23, p. 91] proposes a somewhat more conservative value, suggesting that the 2000 kcal/day expenditure should be considered as a maximum and that work levels preferably should be kept somewhat below this. Granting some modest differences between these and other physiological standards, we nonetheless get an impression of the general level of physiological costs that should not be exceeded.

Work and rest. If we accept some ceiling (such as 4 or 5 kcal/min) as a desirable upper limit of the average energy cost of work (and perhaps try to keep a bit below that limit), it is manifest that if a particular activity per se exceeds that limit, there must be rest to compensate for the excess. In this connection Murrell [51, p. 376] presents a formula for estimating the total amount of rest (scheduled or not scheduled) required for any given work activity, depending on its average energy cost. This formula (with different notations) is

$$R = \frac{T(K - S)}{K - 1.5}$$

in which R is rest required in minutes; T is total working time; K is average kilocalories per minute of work; and S is the kilocalories per minute adopted as standard. The value of 1.5 in the denominator is an approximation of the resting level in kilocalories per minute. If we adopt as S a value of 4 kcal/min and want to figure R for a 1-hr period ($T = 60$ min), our formula becomes

$$R = \frac{60(K - 4)}{K - 1.5}$$

Applying this to a series of values of K, we can obtain R values shown in the next to the top curve of Figure 10-12 ($S = 4$). The other curves (for values of $S = 3$, 5, and 6) are given for comparison when lower ($S = 3$ kcal/min) or higher ($S = 5$ or 6 kcal/min) standards of energy expenditure might seem to be appropriate. The lowest curve (for a value of $S = 6$), however, undoubtedly represents a level of activity that probably could not be maintained very long, except possibly by the hardiest among us. This general formulation needs to be accepted with a fair sprinkling of salt, in part because of individual differences in energy requirements of resting levels and of the level of energy that could be considered an acceptable standard S. Although there are other constraints that might be proposed, this scheme can provide at least a horseback approximation of the total rest requirements for various types of work.

The distribution of that rest (i.e., the duration and frequency of formal or informal rest periods) is another facet of the problem. In this connection, Murrell [51, p. 378] reminds us of Müller's research [50] which indicated that the readily available energy reserve of the average man is about 25 kcal, and that this reserve typically is not drawn on if the work remains below about 5 kcal/min. Murrell suggests as a rough guideline that rest pauses should be introduced before this reserve is exhausted; but he hastens to point out that although this guideline probably departs from precise physiological measures, it nevertheless would seem to provide sufficiently realistic results for use as the basis for work organization in industry. Although the curves in Figure 10-12 swing down to the zero

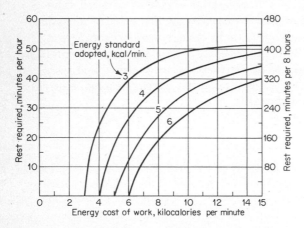

Figure 10-12 Total rest requirements for work activities of varying energy costs, for energy-expenditure standards (ceilings) of 3, 4, 5, and 6 kcal/min; a generally accepted standard is 4 kcal/min. The rest requirements for maintenance of the adopted standard are given per hour (left) and per 8-hr day (right). [Based on formulation of Murrell, 51, p. 376.]

rest-required line, we should keep in mind that this formulation deals only with the physiological costs of work. Because of *other* considerations, such as the more psychological factors of boredom and shift in attention, some rest must be provided for virtually any kind of continuous work, even though its physiological costs are nominal.

BIOMECHANICS OF MOTION

What we have referred to as operational criteria of physical activity are measures of the performance of body members in carrying out certain physical activities. This gets us into what is called *biomechanics*, which has to do with various aspects of the physical movements of the body and of the body members, including the range, strength, endurance, speed, and accuracy of movements.[2] The operation of the body members is basically a physical process and can be characterized in terms of kinematics (the science of motion). The bones connected at their joints, with their associated muscles, actually are levers; their functioning can be described in terms of force, space, mass, and time [Williams and Lissner, 78].

Data relating to the biomechanics of motion, of course, have potential application in the design of control mechanisms and tools, in the arrangement of the work space, and in the methods of work. For the moment we shall take an overview of certain aspects of biomechanics by presenting a few illustrative sets of data relating to body movements. Other examples will be brought out directly or indirectly in later chapters. In the application of biomechanical data to any given design problem, however, the designer should preferably use data from a sample of people relatively similar to the people by whom the equipment or facilities would be used (e.g., hale and hearty young males, elderly females, etc.).

Types of Movements of Body Members

There are, to begin with, certain classes of movements of the body members (arms, legs, etc.) that might be considered as being basic. Some of these, with their associated jargon in biomechanics, are given below [Damon, Stoudt, and McFarland, 17]:

· *Flexion:* bending, or decreasing the angle between the parts of the body
· *Extension:* straightening, or increasing the angle between the parts of the body
· *Adduction:* moving toward the midline of the body
· *Abduction:* moving away from the midline of the body
· *Medial Rotation:* turning toward the midline of the body
· *Lateral Rotation:* turning away from the midline of the body
· *Pronation:* rotating the forearm so that the palm faces downward
· *Supination:* rotating the forearm so that the palm faces upward

Essentially, the above describe the movements of body members in terms of the functioning of the muscles (e.g., flexion and extension) and of the direction of the movements relative to the body (e.g., adduction and abduction). However, in performing work or other physical activities, the movements of the body members can be described in more

---

[2] For a more extensive treatment of biomechanics, especially as related to equipment design, the reader is referred to Damon, Stoudt, and McFarland [17, especially pp. 187–252].

operational terms. There actually are different ways in which movements can be classified. One such scheme is given below:

- *Positioning* movements are those in which a body member such as the hand or foot moves from one specific position to another, as in reaching for a control knob.
- *Continuous* movements are those in which continuous control adjustments are made on the basis of changing stimuli associated with the task. Examples include such activities as operating the steering wheel of a car and guiding a piece of wood through a band saw.
- *Manipulation* movements involve the. use of or handling of parts, tools, certain control mechanisms, etc.; these are, of course, usually limited to finger or hand actions (although the writer has seen, in Pakistan, some amazingly facile meat cutters who ply their trade with the knives held between the toes!).
- *Repetitive* movements are those in which the same movement is repeated successively. Hammering, operating a screwdriver, and turning a hand-wheel are examples.
- *Sequential* movements are several relatively separate, independent movements in a sequence. The separate movements may be of essentially the same nature, such as those in typing, playing a piano, or operating a calculating machine, or they may be rather different, such as those in starting a car on a rainy night, which involves a series of manipulative movements such as turning on the ignition switch, pushing the starter button, turning on the lights, and starting the windshield wiper.
- A *static* adjustment is more the absence of a movement than it is a movement itself, since it consists in maintaining a specific position of a body member for a period of time. It uses muscular exertion, however, in maintaining proper muscular equilibrium.

As indicated above, various types of movements may be combined in sequence so that they blend one into another. For example, placing the foot on a brake pedal is a positioning movement, but this may be followed by a continuous movement of adjusting the amount of brake pressure to the conditions of the situation. Similarly, a continuous movement may include holding a position (a static adjustment) for a short time.

These operational categories are, of course, rather gross (or to be more semantically fashionable, we might refer to them as *macromotions*). For certain purposes, categories of a more *micro*motion nature are used. This is especially the case in industrial engineering practices in methods analysis, in which certain elemental motions are identified in work activities. Even for this purpose, there are different systems, but most of these stem from the original concept of therbligs developed by Gilbreth [28] many years ago. A more recent version of these systems is the Methods Time Measurement system [Maynard, 47].

RANGE OF MOVEMENTS

Movements of several members of the body are illustrated in Figure 10-13, along with the following values for each (as based on a sample of 39 men selected to represent the major physical types in the military services): mean angle (in degrees) and 5th and 95th percentile angles (computed from the standard deviations for the sample). In this, as in other aspects of biomechanics, there are the ever-present individual differences, including the effects of physical condition and the ravages of age.

STRENGTH AND ENDURANCE OF MOVEMENTS

Strength is defined as the peak force, or maximum possible exertion, achieved during an instant of time, and endurance as the ability to maintain a submaximal force over a period of time, whether seconds, minutes, or hours [Damon, Stoudt, and McFarland, 17]. As pointed out by Kroemer and Howard [39], however, such isometric force, as measured, does not depend only on the intrinsic muscle strength; it can be influenced by quite an assortment of factors such as the subject's motivation, the way the body is stabilized, how the reaction force is provided, the manner in which the force is exerted, the experimenter's instructions, and even the specific force values used (whether a single peak value; an average of, say, the three largest values; etc.). Kroemer and Howard [39] caution us further on the score that the lack of standardization across many sets of biomechanical data hampers comparisons and interpretations of such data.

But Kroemer [38] and Kroemer and Howard [39] have still other words of caution for us regarding the application of strength information. In particular, they raise questions about the validity of two common assumptions regarding the conditions that enable men to exert maximum forces (e.g., pulling up on a handle by the hip): (1) that such conditions are optimum for submaximal force exertion; and (2) that such conditions are optimum for dynamic work (as opposed to static work). Thus we should keep in mind that measurements of maximum static strength generally are not relevant in considering dynamic work.

Arm Strength

With the above cautions in mind, let us illustrate studies of maximum strength with some data from Hunsicker [36], who tested the arm strength of 55 subjects who made movements in each of several directions, with the

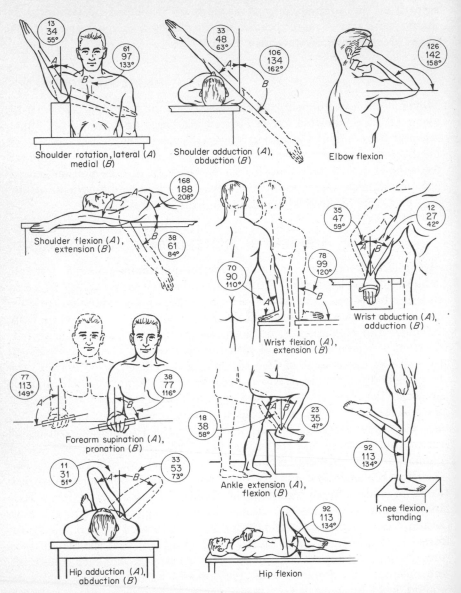

Figure 10-13   Range of certain movements of the upper and lower extremities, based on a sample of 39 men selected to represent the major physical types in the military services. The three values (in degrees) given for each angle are the 5th percentile, the mean, and the 95th percentile, respectively, of voluntary (not forced) movements. [Based largely on data from Dempster, 19, as reanalyzed by Barter et al., 5.]

upper part of the arm in each of five positions. The arm positions and directions of movements are illustrated in Figure 10-14. The strength of each subject for each movement was recorded automatically through an electrical system.

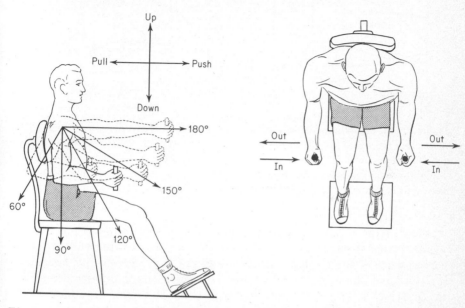

Figure 10-14 Side and top views of subjects being tested for their strength in executing six different movements, namely, push, pull, up, down, in (adduction), and out (abduction). All six movements were made at each of the five arm positions (180, 150, 120, 90, and 60°). See Figure 10-15 for results. [Adapted from Hunsicker, 36.]

Some of the results are shown in Figure 10-15. This figure shows, for the six movements, the maximum strength of the 5th percentile and the mean maximum strength. It is frequently the practice, in dealing with data relating to strength, to use the 5th percentile value as the maximum force to be overcome by users of equipment being designed, since this would in general ensure that 95 percent of the individuals in question would have that strength level or more.

When looking at the results, we can see that pull and push movements are clearly strongest, but that these are noticeably influenced by the position of the hand, with the strongest positions being at angles of 150 and 180°. The differences among the other movements are not great, but we can see certain patterns emerging, such as a tendency for up and down movements (especially the means) to be a bit higher at angles of about 120 and 90°, and for out movements to be generally the

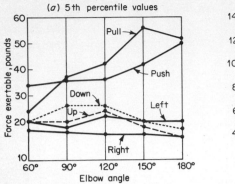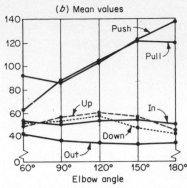

Figure 10-15 Maximum arm strength, in pounds of movements in various directions for different elbow angles of upper right arm. (a) shows 5th percentile values, and (b) the mean values (with compressed vertical scale of pounds) for 55 male subjects. See Figure 10-14 for illustrations of arm positions and of movements. [Based on data from Hunsicker, 36.]

weakest of the lot. These patterns, and that of the push and pull movements being stronger at the 150 and 180° angles, are undoubtedly the consequence of the mechanical advantages of such movements, considering the levers involved and the effectiveness of the muscle contractions in applying leverage to the body members. It might be added that although left-hand data are not shown, the strength of left-hand movements is roughly 10 percent below that of the right hand.

Endurance

As we all know from our own—sometimes discouraging—experience, the ability to maintain a given muscular force is related to the magnitude of the force. This is shown dramatically in Figure 10-16, which shows, for a sample of college students, the mean duration in seconds that the subjects could maintain a given force on a hand dynamometer, for varying percentages of maximum exertible strength. Mean duration of maximum (or near maximum) strength is obviously very nominal, which implies, of course, that people should not be expected to maintain near-maximum strength more than a moment.

If you were to test the endurance of people of varying strength, you would find that stronger individuals typically can maintain a *given* force for longer periods of time than weaker individuals. But when individual differences in strength are eliminated by testing each individual's endurance at a given percentage of his *own* maximum strength (such as 50 percent), the correlation virtually disappears [Caldwell, 13]. In other words,

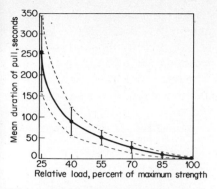

Figure 10-16 Duration of pull in seconds on an isometric dynamometer handle in front of the shoulder, at varying percentages of maximum exertible strength, for 18 male and 18 female college students. The solid line shows mean values, and the two dotted lines show the variability in individual duration times, specifically ±1 standard deviation. [Adapted from Caldwell, 13.]

a weak person can maintain, say, 50 percent of his *own* strength level about as long as a strong person can maintain 50 percent of *his* strength.

Acceptable Levels of Weights to be Lifted

Recognizing that the exertion of maximum strength should be avoided if at all possible, a number of organizations and individuals have concerned themselves with maximum weights that individuals should be permitted to lift. For example, the International Labour Office (ILO) [81] has established maximums of 80 lb and from 33 to 44 lb for adult males and females, respectively (with lower levels for adolescents). And certain states have passed laws governing such maximums, especially as applicable to females and minors. But as pointed out by Snook and Irvine [66] and the Bureau of Labor Standards [82], most such standards have failed to take into account relevant human variables (e.g., sex, age, training of worker, and physical fitness) or task variables (such as size of object, height from which and to which it is to be lifted, and frequency) and, therefore, have constrained applicability. As a step in the direction of bringing some of these variables into consideration, Snook and Irvine [66], on the basis of a well-controlled study sponsored by the Liberty Mutual

Insurance Company, developed a set of maximum acceptable weights of lift for males for three heights of lift, namely, (1) floor level to knuckle height, (2) knuckle height to shoulder height, and (3) shoulder height to arm reach (above the head). These maximums were based on the subjective judgments of male subjects (between 25 and 37 years of age in good physical condition) about the maximum weights that they could lift "comfortably . . . without straining yourself . . . and that you can lift once every 15 minutes." Figure 10-17 summarizes these maximums as

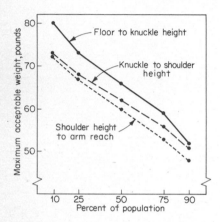

Figure 10-17 Maximum acceptable weights of lifts for young males in good physical condition. These maximums are based on the subjective judgments of subjects of the weights that could be lifted comfortably, without straining, once every 15 min. [Adapted from Snook and Irvine, 66, table 8.]

they would be applicable to percents of a corresponding population; unless one is in a position to be selective in personnel, the most relevant data are those applicable to 90 percent of the population. These particular data, of course, take account of only certain of the human and task variables mentioned above; a related study by Snook and Irvine [65], however, takes into account certain other task variables, in particular, frequency of lifts of different weights as these influence certain physiological criteria.

The data given in Figure 10-17 (with the lowest values hovering around 50 lb) jibe reasonably well with weight limits recommended by the Swiss Accident Insurance Institute and presented by the International Occupational Safety and Health Information Center (CIS) [80] for males of corresponding ages (20 to 35); their recommended reasonable weight limits for occasional lifting are as follows:

	Weight, lb, for Specified Age Groups					
	14–16	16–18	18–20	20–35	35–50	Over 50
Male	33	42	51	55	46	35
Female	22	26	31	33	29	22

Other Aspects of Strength and Endurance

This is not the place to reproduce the many sets of biomechanical data that are available. However, for illustrative purposes here, a few bits and pieces of data and a few unadorned generalizations from certain studies will be recapped below, without the details of the methods or results.

· *Grip strength:* The grip strength of a sample of 552 male industrial workers ranged from 75 to 170 lb, with a mean of 125 lb [Fisher and Birren, 25]. But university students do not do as well, as indicated by the following results [Tuttle et al., 71]: right hand, mean of 108 and standard deviation of 21; left hand, mean of 95 and standard deviation of 18. These differences point to the moral that in considering biomechanical data, one should be aware of differences for different samples of subjects and use data from subjects that are reasonably representative of the individuals for whom the equipment is being designed (e.g., housewives versus lumberjacks).

· *Strength of hand turn:* In turning movements with the hand (such as with handles on heavy refrigerator doors or stirrup-type devices on some garden tank-spraying equipment) the following generalizations can be made [Salter and Darcus, 60]: The forces that people can exert increase in both a pronation movement (turning the hand inward from a palm-up position) and a supination movement (turning the hand outward); in both of these movements the relation between hand position (as it is turned in or out) and force is greater when the elbow is flexed at about 90 or 150°, and least at about 30°.

· *Elbow flexion versus extension:* An elbow flexion action (bending) is about half again as strong as an extension action [Provins and Salter, 58].

· *Elbow versus shoulder forces:* Rotation action of the shoulder is about half again as strong as that of the elbow and has nearly three times as much staying power [Provins, 57].

- *Lifting action:* In lifting heavy objects to various levels, markedly heavier weights can be lifted to a low level of 18 in. (i.e., 124, 138, and 146 lb) by short, medium, and tall males from 17 to 32 years old than can be lifted to an intermediate level of 42 in. (73, 92, and 96 lb) or a high level (53, 65, and 67 lb) [Switzer, 69].

As indicated earlier, there are certain human variables that are related to muscle strength and endurance. Observations about a few of these, based on the discussion by Damon, Stoudt, and McFarland [17] are given below:

- *Age:* Strength reaches a maximum by the middle to late 20s and declines slowly but continuously from then on, until at about age 65 strength is about 75 percent of that exerted in youth.
- *Sex:* Women's strength is about two-thirds that of men.
- *Body build:* Although body build is related to strength and endurance, the relationships are complicated; for example, athletic-looking individuals generally are stronger than others, but less powerfully built persons may be more efficient; and for rapidly fatiguing, severe exercise, slender subjects are best, with obese subjects worst; and for moderate exercise, those with normal build are best.
- *Exercise:* Exercise can increase strength and endurance within limits, frequently these increases being in the range of 30 to 50 percent above beginning levels.

SPEED AND ACCURACY OF MOVEMENTS

Speed generally is the primary requirement in executing movements that are otherwise not difficult or demanding, such as in applying the brake pedal of an automobile or reaching for parts to be assembled. In turn, accuracy is the primary requirement in executing such movements as those in tracking (in which continuous control is required), in certain positioning actions that require precision and control, and in certain manipulative activities. However, in some circumstances both speed and accuracy may be required.

Reaction Time

Many movements are triggered by some external stimulus such as a changing traffic light or auditory warning signal. The time to initiate a movement following such a stimulus is referred to as *reaction time* or *lag time*. However, reaction time actually is a combination of delays; the nature of these delays and the range of typical times (in milliseconds) required for them have been summarized by Wargo [75] as follows: receptor delays, 1 to 38; neural transmission to the cortex, 2 to 100; central-process delays, 70 to 300; neural transmission to muscle, 10 to 20;

and muscle latency and activation time, 30 to 70. These add up to total delay, or reaction times, ranging from 113 to 528 msec. In considering physical responses we should differentiate between reaction time itself and *movement* time (the time from the *activation* of the muscles of a body member, such as a hand or foot, until the completion of the movement). Movement time would of course vary with the type and distance of movement, but it has been estimated [Wargo, 75] that a minimum of about 300 msec (0.30 sec) can be expected for most control activities. Adding this value to an estimated reaction time of 200 msec would result in a total response time of about 500 msec. However, the nature and distance and location of the response mechanism can influence the total time. For example, one part of a larger project dealt with the possible nature and location of emergency cutoff switches of agricultural tractors [Pattie, 56]. With response time (actually reaction time plus movement time) as a criterion, several types and locations of cutoff switches were investigated, including toggle switches (of three lengths, positioned on the steering wheel or outside and above the steering wheel, and with different planes of throw) and push buttons (inside the steering wheel near center and outside and above the steering wheel). The mean response times varied from 425 msec (a long toggle switch above the steering wheel with right-left plane of throw) to 619 msec (a small toggle switch to the right of the wheel post with a forward and back plane of throw), a difference of about 200 msec. (The range of mean response times of individual subjects on given devices was from about 320 to 800 msec.)

The range of the sources of delay in reaction times as such and in total reaction times is influenced by a number of variables, such as the nature of the stimulus, the number of choices, and the degree of expectancy.

Simple and complex reaction time. When a single stimulus and a single response are involved in a circumstance in which the stimulus is anticipated (as in conventional laboratory studies), reaction times usually range from about 150 to 200 msec (0.15 to 0.20 sec), with 200 msec being a fairly representative value; the value may be higher or lower depending on the stimulus modality and the nature of the stimulus (including its intensity and duration), as well as on the subject's age and other individual differences. In connection with stimulus differences, for example, Swink [68] reports the following mean reaction times (in seconds) for a light, a buzzer, an electropulse, and combinations: light, 0.241; buzzer, 0.225; electropulse, 0.207; light and buzzer, 0.202; light and electropulse, 0.196; buzzer and electropulse, 0.183; and light and buzzer and electropulse, 0.178. If there are several possible stimuli, each with its own response, the time goes up largely because of the additional central process time required to make a decision on what the correct response should be. Such a complex reaction time (also called *disjunctive* reation time) is pretty much a function of the number of choices available, as indicated below [summarized from various sources by Damon et al., 17, p. 239]:

Number of choices	1	2	3	4	5	6	7	8	9	10
Approximate reaction time, sec	0.20	0.35	0.40	0.45	0.50	0.55	0.60	0.60	0.65	0.65

There is some evidence to suggest that complex reaction time is linearly related to the number of bits of information involved (that is, the logarithm to the base 2 of the number of alternatives available) [Hick, 34; Hilgendorf, 35], even up to 10 bits (about 1000 alternatives) in the case of Hilgendorf's study.

Expectancy.   Most data on simple and complex reaction times come from laboratories in which the subject is anticipating a stimulus. (And in some industrial circumstances people actually are waiting for a stimulus.) However, when stimuli occur infrequently or when they are not expected, the ante is raised. This was illustrated, for example, in a study by Warrick et al. [76] in which typists at their regular jobs were asked to press a button whenever a buzzer sounded, the buzzer going off only once or twice a week over a period of 6 months. Data were also obtained from the same task when the subjects were given a 2 to 5 sec warning. Comparative data (in milliseconds) for certain percentiles are given below, along with the median (50th percentile) of the simple reaction time:

	Percentile		
	10th	50th	90th
(a) Unexpected signal	510	610	820
(b) Signal following warning	410	510	680
(c) Simple reaction time	. . .	200	

The simple reaction time c was determined by having the subjects respond to the buzzer when their fingers were actually resting on the button. The difference of 310 between that value and the median of 510 for condition b is essentially movement time. The unexpected condition in this study, a, added about 100 msec more to the total time.

Discussion.   In the case of responses to be made to unexpected stimuli (such as responding to sudden emergency conditions in driving a car), the total time to make a response (reaction time plus movement time) can easily be 0.75 or 1.0 sec or more. In fact, it has been estimated by Wargo [75] that in the case of pilots of 1800-mph supersonic aircraft, the total time to initiate a control response in suddenly sighting another supersonic aircraft on a collision course can be as long as 1.7 sec, this being the simple addition of 0.3 sec for visual acquisition of the other aircraft, 0.6 sec for recognition of the impending danger, 0.5 sec for selection of a course of action, and 0.3 sec for initiation of the desired control response. Add the response time to the aircraft itself, and it would be futile to take any action if the planes were closer than about 4 miles.

When time is, as they say, of the essence, one should not throw up his hands in complete despair of the time lag in human response. There are, indeed, ways of aiding and abetting people in responding rapidly to stimuli. Speed requirements, for example, can be reduced by taking actions such as using sensory modalities with shortest reaction time, using two or more senses to reduce reaction time, presenting stimuli in a clear and unambiguous manner, minimizing the number of alternatives from which to choose, giving advance warning of stimuli if possible, using body members that are close to the cortex to reduce neural transmission time, using control mechanisms that minimize response time, and training the individuals. In more exotic circumstances, one can even by-pass the human physical response by the direct use of electrical muscle-action potentials for effecting control responses [Wargo, 75].

Predetermined Time Systems

With particular reference to time involved in making different types of movements, mention should be made of some predetermined time systems that are used in industrial engineering for developing standard times for work operations.[3] Such systems (as the Methods Time Measurement system, MTM) set forth specific time values for specific elemental motions made during an operation; the standard time for the complete operation is based on the sum of the times for the individual elements, usually with certain "allowances" included. Most such systems are predicated on the accumulation of empirical data obtained from extensive time-study observations. More will be said of this later, but it should be noted here that there is evidence that some element times are not strictly additive because of possible interactions among element times. Some predetermined time systems provide for at least a few such interactions by the use of tables of data for a given type of motion for different conditions under which the motion is executed.

Positioning Movements

Positioning movements are made when an individual reaches for something or moves something to another location, usually by hand; in effect, then, they are travel movements. The criterion against which their per-

[3] Some of these systems are described briefly by Karger and Bayha [37, chap. 3], and one system, the MTM system, is described in detail. See also Maynard [47].

formance is measured is usually in terms of time, but in some cases may be in terms of accuracy. These criterion measures can be influenced by different features of the movement, most of which are predetermined by work layout and design factors. Some of these features are the nature of the stimulus that triggers the movement, distance and direction of the movement, single possible terminal versus alternative terminals, fixed terminal (with automatic stop) versus precise terminal position under control of an individual, and visual versus nonvisual (i.e., "blind") control. Depending on the nature of the movement, positioning movements can be dissected into two or three relatively distinct components, namely, reaction time (the time to initiate a response following the stimulus that triggers it), primary or gross travel time (to bring the body member near the terminal), and a secondary or corrective type of motion to bring the body member to the precise position desired. Where there is an automatic fixed terminal (such as on a typewriter carriage), the secondary, or corrective, component virtually drops out of the picture.

Time and distance of movements.   In the execution of positioning movements, reaction time is almost a constant value, unrelated to the distance of movement. This is shown, for example, in Figure 10-18, which is based on a pair of related studies in which the subjects moved a sliding device to a marked position when a buzzer was sounded [Brown and Slater-Hammel, 11, and Brown, Wieben, and Norris, 12]. Three different

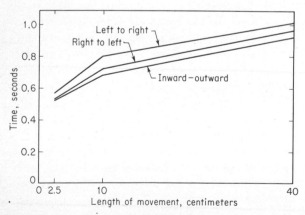

Figure 10-18   Times required for horizontal positioning movements of different lengths. Times given are from the sounding of a buzzer to completion of movement. Reaction time, which was essentially the same for all movements (about 0.25 sec), is included in time values. [Adapted from Brown and Slater-Hammel, 11, and Brown, Wieben, and Norris, 12.]

distances were used, namely, 2.5, 10, and 40 cm, the movements being left to right, and inward and outward. This figure also shows that movement time is related to distance, but is not proportionate to distance. This is demonstrated also in the results of a study by Barnes [3] in which the subjects moved a carriage on a slide back and forth (in and out) between mechanical stops positioned at 5, 10, and 15 in. The average times required for these movements are given below, the time for the 5-in. movement being arbitrarily set at 100:

Distance, in.	Time
5	100
10	115
15	125

This lack of linear relationship between distance and time probably can be attributed to the relatively constant reaction time, the time required for acceleration to the maximum speed, and (except where there is a mechanical terminal) the secondary, or corrective, movement in bringing the body member to the precise terminal.

Termination of movements. The effect on movement time of a mechanical stop at the terminal of the movement was mentioned briefly above; Barnes [3], for example, reported that the use of visual control required about 17 percent more time than did the use of a fixed mechanical terminal. Other variations in the termination of the movement can also have some effect on movement time. For example, in the MTM system mentioned earlier [Karger and Bayha, 37; Maynard, 47], different time units are allowed for *reach* movements of three types, namely, those in which the object to be reached for (1) is in a fixed location, (2) is in a location that may vary from cycle to cycle, and (3) may be jumbled with other objects. These times, based on empirical data, are shown in graphic form in Figure 10-19. We thus see that the movement (reach) time is influenced by the nature of the *termination* of the movement, the variation presumably being essentially a function of the time required to search for or select specific objects in their specific locations.

Direction of positioning movements. Differences in time and accuracy of positioning movements made in various directions is a function of the biomechanical functioning of the particular combination of body members and muscles that are brought into the act. The results of certain investigations will illustrate this effect, but we shall skip most of the experimental procedures and, rather, present certain of the results that illustrate the point. As a first example, Briggs [7] carried out a series of experiments with the movement of a stylus back and forth in a horizontal plane between a large (3-in.) buzzer and a target. The target could be *in*

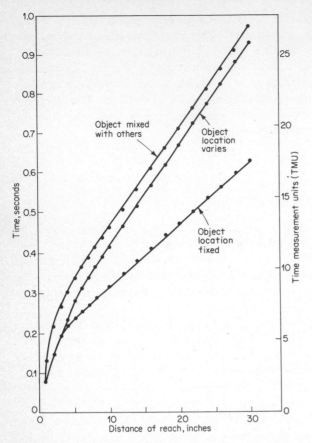

Figure 10-19 "Reach" times allowed under three conditions of terminating movements, for various distances of movement, as based on the MTM (Methods Time Measurement) system of predetermined times. [Reproduced with permission from *Basic Motions of MTM* by William Antis, John M. Honeycutt, Jr., and Edward N. Koch, published by the Maynard Foundation, copyright 1963, 1968, p. 2-2. Reprint of this material is prohibited without express permission of the MTM Association for Standards and Research, Fair Lawn, N.J.]

(in front of the body) or *out* (at different angles and distances from the body), with the buzzer being in the opposite location (out or in). The target circle was drawn on a piece of stretched paper through which the stylus was punched. Accuracy was determined by counting the stylus punches within the target, the targets being ¼, ½, ¾, or 1 in. in diameter.

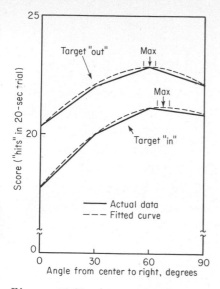

Figure 10-20 Accuracy scores of repeated positioning movements at various angles when the target is in (in front of body) and out (14 in. away from central point). [Adapted from Briggs, 7.]

Scores were based on the number of hits at the target within 20-sec trials and hence were a combination of accuracy and speed.

Figure 10-20 shows the accuracy scores of repeated 14-in. positioning movements at various angles to the right from the front of the body, when the target was in and out. This shows that accuracy is optimum with angles of movement of about 60° right from center, and also that accuracy is generally greater when the target is away from the body (out rather than in). Further elaborations on this theme, with the target at the in position and with the buzzer out at various angles and distances, produced results shown in Figure 10-21. In that three-dimensional figure, the left-to-right axis represents the angle of movement from the subject's center, shown as though we were facing the subject, so his right is to our left and vice versa; the distance of the movements is represented from front to back. The vertical dimension represents the average accuracy scores of the subjects for movements of the different angles and distances. An eyeball scanning of this figure shows up the following primary features: Accuracy is greatest for the short distance (7 in.); at far distances (35 in.) accuracy drops off consistently for angles from 120° right to 120° left; but for short distances (7 in.) the relationship has some undulations with a maximum around 60° right.

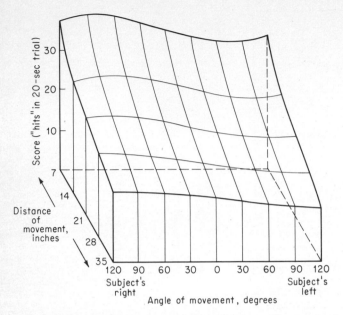

Figure 10-21 Accuracy scores of positioning movements of various lengths at different angles from center position. Height of the form at any vertical location indicates accuracy score for the angle and distance represented by the position. [From Briggs, 7.]

Keeping this study in mind, let us shift to another investigation of the time required to make 40-cm positioning movements in eight different directions from a center starting point [Schmidtke and Stier, 62]. The results, shown in Figure 10-22, indicate that time was generally shorter for movements in the area of a 55° angle, which is within the general range of Briggs' 60° optimum. Thus, although these two studies vary in part in their criteria (time versus accuracy) and in the location of the terminal of the movement (in toward the body versus out away from the body), nonetheless they reveal the same general pattern. This pattern suggests that in biomechanical terms, controlled arm movements that are primarily a pivoting of the elbow, with fairly nominal upper arm and shoulder action, tend to be more accurate and to take less time than those with a greater degree of upper arm and shoulder action.

As we look at two-handed simultaneous positioning movements (such as in reaching for parts in bins), however, a different pattern emerges, as reflected by a study by Barnes and Mundell [4]. The speed of such movements (with mechanical terminals) when made at angles of 0, 30, 60, and

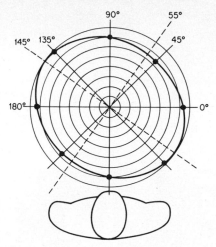

Concentric circles represent
equal time intervals

Figure 10-22 Average times of hand movements made in various directions. Data were available for the points indicated by black dots; the oval was drawn from these points and represents assumed, rather than actual, values between the recorded points. The concentric circles represent equal increments of time to provide a reference for the average movement times depicted by the oval. [Adapted from Schmidtke and Stier, 62.]

90° from a dead-ahead direction did not vary much with the angle, but errors in moving the hand to the terminal positions that needed visual control, increased very markedly from the 0 to 90° angles. The probable explanation for this lies in the requirement for visual control in terminating the movements; the closer the two terminals to the dead-ahead position, the more accurate the visual control of the termination of the movements.

Blind positioning movements. When visual control of movements is not feasible, the individual needs to depend on his kinesthetic sense for feedback. Probably the most usual type of blind positioning movement is one in which the individual moves his hand (or foot) in free space from one location to another, as in reaching for a control device when the eyes are otherwise occupied. The very well-known study by Fitts [26] probably provides the best available data relating to the accuracy of the *direction* of such movements in free space. He used the arrangement shown in Figure 10-23, with targets positioned around the subject at 0, 45, 90, and 135° angles left and right, in three tiers, namely, a center (reference) tier, and tiers 45° above and below the center tier. The blindfolded subjects were given a marker with a sharp point, which they pressed against each target when they tried to reach to it. A bull's eye was scored zero, and marks in subsequent circles were scored from 1 to 5, marks outside the circles being scored 6.

Figure 10-23 Head-on view of subject in blind-positioning study by Fitts [26]. Targets used in the study included three tiers. [Courtesy of USAF, AMRL, Behavioral Sciences Laboratory.]

Figure 10-24 shows the results. Each circle in this figure represents the subjects' accuracy in hitting the target in the corresponding position. The size of the circle is proportional to the average accuracy score for that target; the smaller the size, the better the accuracy. The circles within circles (those in the four quadrants) indicate, relatively, the proportionate number of marks that were made in each quadrant. From this figure it can be seen that blind positioning movements can be made with greatest accuracy in the dead-ahead positions and with least accuracy in the side positions. With regard to the level of the targets, the accuracy is greatest for the lowest tier, average for the middle tier, and

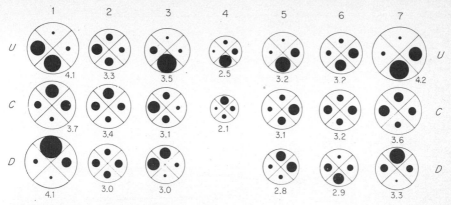

Figure 10-24 Relative accuracy scores for different areas in blind-positioning study by Fitts [26]. The position of the circles represents the location of targets, which ranged from 135° left (number 1) to 135° right (number 7), number 4 being straight ahead. The three tiers represent those up, center, and down. The size of each circle represents the relative number of errors, so small circles indicate greater accuracy. The relative size of the four dark circles within each light one is proportional to the errors in each quadrant of the target. [Courtesy of USAF, AMRL, Behavioral Sciences Laboratory.]

poorest for the upper tier. Further, right-hand targets can be reached with a bit more accuracy than left-hand ones.

Thus, in general, in the positioning of control devices or other gadgets that are to be reached for blindly, positions closer to the center and below shoulder height usually can be expected to be reached more accurately than those farther off to the sides or higher up.

The accuracy of the *distance* of blind positioning movements (as opposed to their direction) is shown in Figure 10-25 [from Brown, Knauft, and Rosenbaum, 10]. In this study, direction was controlled by having the subjects move a slide along a groove in a panel; although the movements were made in darkness, the apparatus was lighted momentarily so that the subjects could see the point along the groove to which the slide was to be moved. The figure shows, for each of six directions and for each of four distances, the average error in distance made by the subjects. The most obvious general trend was to overshoot short distances and to undershoot longer distances. This is called the *range effect*. (The only exception to this was condition 4, a movement from top to bottom, this exception probably being due to lack of compensation for the weight of the arm as it was moved down.) As an aside, in another analysis of the results, it was found that for longer distances (10 to 40 cm), horizontal movements (lateral and straight ahead) toward the body were more accurate than those away from the body.

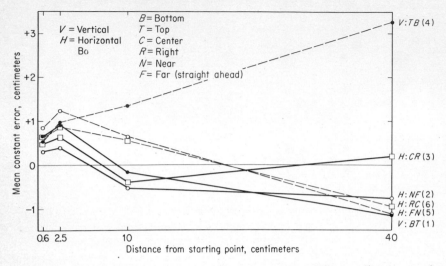

Figure 10-25 Mean constant errors for movements of different distances and directions. Movements with plus errors were too long and with minus errors, too short. [From Brown, Knauft, and Rosenbaum, 10.]

Continuous Movements

As indicated earlier, continuous movements are those in which the individual makes continuous adjustments in response to a changing situation, typically receiving feedback in some form that can serve in modifying the activity as it is carried out. There are, however, perhaps two types of continuous movements that we should distinguish: (1) single body movements in themselves (usually the hand) and (2) continuous control movements (usually with a control mechanism, as in a tracking task).

Single hand movements. These movements require continuous muscular control throughout their entire span, but have a distinct termination. A laboratory example of this type of movement was investigated by Corrigan and Brogden [15] by having subjects move a stylus along a 35-cm narrow track formed between two brass plates on a piece of glass; they were to control the stylus continuously, keeping up with a target of constant velocity that moved in view under the glass along the path. The direction of the path was positioned at 24 different angles around a center at 15° intervals. Errors were recorded automatically when the stylus touched either brass plate. The results are shown in Figure 10-26. This shows the error index at different angles and indicates that errors are least (and accuracy, therefore, greatest) at around 135 and 315° (1:30 and 7:30 o'clock positions), maximum errors being at around 45 and

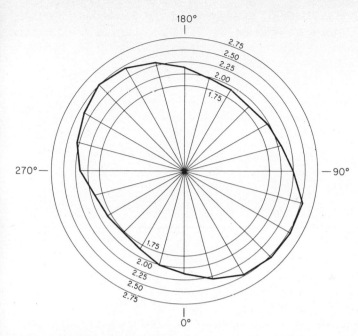

Figure 10-26   Errors in continuous movements (35 cm) made
at various angles from point of origin. Length of line from the
center indicates relative errors at the angle in question. [Adapted
from Corrigan and Brogden, 15, table 2.]

225°.[4] Therefore, continuous hand movements from lower left to upper
right (or vice versa) are more accurate than those from lower right to
upper left (or vice versa). A comparison of Figure 10-26 with Figure 10-22
(which shows average movement times in various directions) indicates
a remarkable similarity, which further supports the notion of the biomechanical advantage of an elbow pivoting movement in contrast with
one that requires greater upper arm and shoulder involvement.

Continuous control movements.   Certain aspects of continuous
control tasks (such as tracking) were discussed in Chapter 9, especially
as such control is influenced by the nature of the task. However, one
additional study will be cited here since it is germane to one or two points
made above. In this study the subjects tracked a moving dot with a

---

[4] The relationship, when plotted in the form of a regular graph with angles along the
base line and errors on the vertical scale, appears as a trigonometric function, the
equation of which is

$7 = 2.1807 - 0.4561 \cos 2x + 0.3389 \sin 2x$

stylus, the individual (lateral) movements of the dot being 0.25, 0.50, and 1.00 in. [Ellson and Wheeler, 24]. The accuracy of the tracking operation was automatically recorded. A recap of some of the results is given below:

Length of stimulus, in.	Response		Standard deviation	
	Mean, in.	% of stimulus	In.	% of stimulus
0.25	0.32	127	0.040	16.0
0.50	0.52	104	0.040	8.0
1.00	0.96	95	0.045	4.5

Looking at the response means, we can see here in continuous tracking movements the range effect that we noted previously in positioning movements, that is, an overshooting of short movements and an undershooting of long movements. And we can also see from the standard deviation data, in the last columns above, that *relative* error (the percent column) is greater for the shortest movement than the the longest (although absolute errors, in inches, were about the same).

Manipulative Movements

The spectrum of possible manipulative movements is, of course, fairly broad. However, a fair share of the experimental data relating to them deals with the use of control devices and will be touched on in Chapter 11. In addition, a fair share of empirical data relating to different types of manipulative activities has been used as the basis for predetermined time systems [Karger and Bayha, 37] and will therefore not be covered here.

Repetitive Movements

The available information on repetitive movements is fairly skimpy, and is largely confined to such operations as tapping (e.g., sending Morse code). In such movements speed usually is more important than accuracy. A few generalizations and a few bits of data about tapping activities are given below:

- *Rate of tapping:* The maximum rate varies from about 5 to 14 taps per second, with a mean of about 8.4; preferred rates, however, are about 1.5 to 5 taps per second [Miles, 48].
- *Use of various fingers:* Rates of tapping vary considerably for separate fingers, as follows [Dvorak et al., 22]:

	Left hand				Right hand			
	4	*3*	*2*	*1*	*1*	*2*	*3*	*4*
Taps per 15 sec	48	57	63	66	70	69	62	56
Taps per sec	3.2	3.8	4.2	4.4	4.7	4.6	4.1	3.7

The rates for the right-hand fingers are systematically higher than for the left-hand fingers, and the rates decrease in order from the index finger, 1, to the little finger, 4.

· *Timing of stimuli:* If individual tapping responses are made to individual signals, the signals should be more than 0.5 sec apart; otherwise interference occurs [Vince, 73]. If responses are to be more frequent than about every 0.5 sec, it is preferable to have one signal trigger a series of taps, rather than to have a separate signal for each response.

· *Maintenance of tapping movements:* Repeated separate movements can be maintained at a constant rate much more accurately and consistently than continuous movements [Gottsdanker, 30].

Sequential Movements

In most instances sequential movements are of the same general kind, varying in some differentiating feature, as in operating a keyboard. In some instances, however, a potpourri of types of movements may occur in sequence. Most of the research on sequential movements relates to the first type (movements of the same kind), especially in the use of keyboards. Further discussion of keyboards, however, will follow in Chapter 11.

Studies with Universal Motion Analyzer. In connection with sequential movements, however, we should touch on the studies of Smith and his associates [18, 32, 59, 63, 64, 74, 77] using the Universal Motion Analyzer shown earlier in Figure 10-6. Certain of these studies were primarily concerned with time used in making a series of specified hand movements from one position on the panel to another, performing some manipulation at each position such as turning a switch or pulling out pins from holes, etc. Electrical recording devices made it possible to record the time for each component of the movements. In executing different patterns of complex movements, it was found [18, 59] that neither *travel* time nor *manipulation* time (the time, in these studies, to turn a switch at each position) varied with the serial pattern of movements from position to position. However, it was found that although practice did improve manipulation time, it did not reduce the travel-time component; presumably this is largely a function of the already developed motor skills of people.

More in line with our current interests, however, was the fact that the *type* of manipulation performed at each position has an influence on travel time *between* positions. This interaction effect was demonstrated in a comparison of travel times when the manipulations were a switch-turning operation versus a pin-pulling operation [77] and three variations in the angle of turn of a switch, 40, 80, and 120° [32], as well as in an assembly task of placing washers on pins either side up versus a specified side up [63].

Interactions of motions.   Interactions among components of sequential activities (such as discussed above in connection with the Universal Motion Analyzer) are the kinds of phenomena that could cause time-study engineers to grow old before their time, since it implies that the times required for individual motions to be executed in a sequence cannot be added up to derive estimates of the total time to be allowed for the complete sequence. In a review of research relating to this, Schappe [61] concludes that travel time (of a body member such as the hand) is influenced by the manipulation time at the terminals of the travel movement and by the precision requirements of the manipulation activity, and that both of these are influenced by perceptual factors. In summary, he questions seriously the assumption of the *additivity* of times of the individual components. As indicated earlier, however, some predetermined time systems do, to some degree, take such interaction into account in deriving total time allowances for sequences of movements.

Static Reactions

In static reactions, certain sets of muscles typically operate in opposition to each other to maintain equilibrium of the body or of certain portions of it. Thus, if a body member, such as the hand, is being held in a fixed position, the various muscles controlling hand movement are in a balance that permits no net movement one way or the other. The tensions set up in the muscles to bring about this balance, however, require continued effort, as most of us who have attempted to maintain an immobile state for any length of time can testify. In fact, it has been stated that maintaining a static position produces more wear and tear on people than some kind of adjustive posture [Harston, 33]. It was reported, for example, that holding a weight was three to six times more fatiguing than lifting it up and down.

Deviations from static postures are of two types: those called *tremor* (small vibrations of the body member) and those characterized by a gross drifting of the body or body member from its original position.

Tremor in maintaining static position.   Tremor is of particular importance in work activities in which a body member must be maintained in a precise and immovable position (as in holding an electrode in place when welding or in holding a needle while threading it). An interesting aspect of tremor, incidentally, is that the more a person tries to control it, the worse it usually is [Young, 79]. There are certain ways in which tremor can be reduced, as reported by Craik [16] and other sources. The following are four conditions that help to reduce tremor:

1. Use of visual reference.
2. Support of body in general (as when seated) and of body member involved in static reaction (as hand or arm).
3. Hand position. (There is less hand tremor if the hand is within 8 in. above or below the heart level.)
4. Friction. (Contrary to most situations, mechanical friction in the devices used can reduce tremor by adding enough resistance to movement to counteract in part the energy of the vibrations of the body member.)

DISCUSSION

Many of the sketches and blueprints of systems on the drawing board predetermine the nature of the physical activities that will be required later in the use of the systems, including their energy costs, the range of motions, and their strength, endurance, speed, and accuracy requirements. Some timely consideration given to these affairs during the system design processes frequently will pay handsome dividends in later system performance, and may even relieve someone's aching back or help to keep his bowling arm in good shape.

SUMMARY

1. Human motor abilities have implications for the design of control devices, hand tools, and related devices; the handling of materials; the physical work layout; and work methods and procedures.
2. Human motor activities depend essentially upon the skeletal structure, the skeletal muscle system, neural control, and body metabolism.
3. There are numerous measures of physical activity, including physiological criteria (heart rate, etc.) and operational criteria (some of which relate directly to body activity and others to measuring and recording devices).
4. The energy expenditure in activities can be measured by different units, including the kilocalorie. Work that causes the expenditure of more than 4 or 5 kcal/min usually cannot be continued over a regular day's activity without rest.
5. Biomechanics is concerned with aspects of physical movements of the body and body members, including the range, strength, endur-

ance, speed, and accuracy of movements. Sets of biomechanical data include information on these and other aspects of biomechanics for various samples of people.

6. When an individual is seated, the relative strength of arm movements in different directions are, in descending order, push and pull (about equal), up, down, in, and out; but the strength of the movements is related to the arm position.

7. In hand-turning action, exertible force is related to hand position and direction of turn. For a pronation (turning-in) direction of turn, force is greatest when the hand is at a turned-out position; for a supination (turning-out) direction of turn, force is greatest when the hand is at a turned-in position.

8. In elbow movements, a flexion action (bending) is considerably stronger than an extension action (straightening). The flexion force is maximum with the elbow bent at about a right angle.

9. A shoulder action is generally stronger than an elbow action. In a seated position, shoulder extension (an upward push) is more forceful than flexion (a downward pull).

10. Reaction time is a function of several variables. Simple reaction time is around 0.20 sec, and the time increases with increasing numbers of alternatives from which to choose. Unexpected signals increase reaction time. Average reaction time that includes movement time (such as activation of controls) may be 0.5 to 1 sec or more.

11. Long movements can be made in proportionately less time in relation to length than short movements.

12. Movements terminated by mechanical devices take less time than those which are terminated exclusively by visual cues.

13. One-handed visually controlled positioning movements are performed best in combined accuracy and speed when at about a 60° angle right from a straight-ahead position and when they are reasonably short.

14. Two-handed positioning movements made simultaneously are slightly faster at approximately 30° angles right and left from a straight-ahead position and are distinctly most accurate straight ahead (0° angle).

15. In executing blind positioning movements, people tend to overshoot short distances and to undershoot long distances. With vertical movements from top to bottom there is a tendency to overshoot both short and long movements.

16. Blind positioning movements in different directions from the body are made most accurately when they are to be made straight ahead and are to be made below shoulder height.

17. Continuous movements in a horizontal plane are most accurate in the directions represented by the 1:30 and 7:30 o'clock positions.
18. While repetitive finger-tapping movements can be made at rates from 5 to 14 taps per second, most people prefer rates below 5 per second.
19. The tapping rates of different fingers are greatest for the index finger and decrease systematically for the second, third, and fourth fingers.
20. In serial movements consisting of travel and manipulation actions, travel time is affected by the nature of the manipulation task.
21. Maintaining static posture is generally more fatiguing than some kind of adjustive posture.
22. Tremor can be reduced by (1) use of visual reference, (2) support of body or body member, (3) having hand near heart level, and (4) having moderate friction in devices used.

REFERENCES

1. Barany, J. W.: The nature of individual differences in bodily forces exerted during a simple motor task, *Journal of Industrial Engineering*, 1963, vol. 14, no. 6, pp. 332–341.
2. Barany, J. W., and J. H. Greene: The force platform: instrument for selecting and training employees, *American Journal of Psychology*, 1961, vol. 74, no. 1, pp. 121–124.
3. Barnes, R. M.: *An investigation of some hand motions used in factory work*, University of Iowa, Iowa City, Studies in Engineering, Bulletin 6, 1936.
4. Barnes, R. M., and M. E. Mundell: *A study of simultaneous symmetrical hand motions*, University of Iowa, Iowa City, Studies in Engineering, Bulletin 17, 1939.
5. Barter, J. T., I. Emanuel, and B. Truett: *A statistical evaluation of joint range data*, USAF, WADC, Technical Note 57–311, 1957.
6. Bedale, E. M.: *Comparison of the energy expenditure of a woman carrying loads in eight different positions*, Medical Research Council (Great Britain), Industrial Fatigue Research Board Report 29, 1924.
7. Briggs, S. J.: *A study in the design of work areas*, unpublished doctoral dissertation, Purdue University, Lafayette, Ind., August, 1955.
8. Brouha, L.: *Physiology in industry*, Pergamon Press, New York, 1960.
9. Brouha, L., and H. Krobath: Continuous recording of cardiac and respiratory functions in normal and handicapped people, *Human Factors*, 1967, vol. 9, no. 6, pp. 567–572.
10. Brown, J. S., E. B. Knauft, and G. Rosenbaum: The accuracy of positioning reactions as a function of their direction and extent, *American Journal of Psychology*, 1948, vol. 61, pp. 167–182.
11. Brown, J. S., and A. T. Slater-Hammel: Discrete movements in the horizontal plane as a function of their length and direction, *Journal of Experimental Psychology*, 1949, vol. 39, pp. 84–95.

12. Brown, J. S., E. W. Wieben, and E. B. Norris: *Discrete movements toward and away from the body in a horizontal plane*, ONR, USN, SDC, Contract N5ori–57, Report 6, September, 1948.

13. Caldwell, L. S.: Relative muscle loading and endurance, *Journal of Engineering Psychology*, 1963, vol. 2, pp. 155–161.

14. Christensen, E. H.: "Physiological valuation of work in the Nykroppa iron works," in W. F. Floyd and A. T. Welford (eds.), *Ergonomics Society Symposium on Fatigue*, Lewis, London, 1953, pp. 93–108.

15. Corrigan, R. E., and W. J. Brogden: The trigonometric relationship of precision and angle of linear pursuit-movements, *American Journal of Psychology*, 1949, vol. 62, pp. 90–98.

16. Craik, K. J. W.: *Psychological and physiological aspects of control mechanisms with special reference to tank gunnery. Part I*, Medical Research Council (Great Britain), Military Personnel Research Committee, B.P.C. 43/254, August, 1943.

17. Damon, A., H. W. Stoudt, and R. A. McFarland: *The human body in equipment design*, Harvard University Press, Cambridge, Mass., 1966.

18. Davis, R. T., R. F. Wehrkamp, and K. U. Smith: Dimensional analysis of motion: I. Effects of laterality and movement direction, *Journal of Applied Psychology*, 1951, vol. 35, pp. 363–366.

19. Dempster, W. T.: The anthropometry of body action, *Annals of the New York Academy of Sciences*, 1955, vol. 63, pp. 559–585.

20. Deupree, R. H., and J. R. Simon: Reaction time and movement time as a function of age, stimulus duration, and task difficulty, *Ergonomics*, 1963, vol. 6, no. 4, pp. 403–411.

21. Drillis, R. J.: The use of gliding cyclograms in biomechanical analysis of movement, *Human Factors*, April, 1959, vol. 1, no. 2, pp. 1–11.

22. Dvorak, A., N. I. Merrick, W. L. Dealey, and G. C. Ford: *Typewriting behavior*, American Book Company, New York, 1936.

23. Edholm, O. G.: *The biology of work*, World University Library, McGraw-Hill Book Company, New York, 1967.

24. Ellson, D. G., and L. Wheeler: *The range effect*, USAF, Air Materiel Command, Wright-Patterson Air Force Base, TR 4, Apr. 22, 1947.

25. Fisher, M. B., and J. E. Birren: Standardization of a test of hand strength, *Journal of Applied Psychology*, 1946, vol. 30, pp. 380–387.

26. Fitts, P. M.: "A study of location discrimination ability," in P. M. Fitts (ed.), *Psychological research on equipment design*, Army Air Force, Aviation Psychology Program, Research Report 19, 1947.

27. Goldman, J., and D. K. Ross: Measuring human work performance, *Electronics*, Mar. 10, 1961.

28. Gilbreth, F.: *Motion study*, D. Van Nostrand Company, Inc., New York, 1911.

29. Gordon, E. E.: The use of energy costs in regulating physical activity in chronic disease, *A.M.A. Archives of Industrial Health*, November, 1957, vol. 16, pp. 437–441.

30. Gottsdanker, R. M.: The continuation of tapping sequences, *Journal of Psychology*, 1954, vol. 37, pp. 123–132.

31. Greene, J. H., W. H. M. Morris, and J. E. Wiebers: A method for measuring physiological cost of work, *Journal of Industrial Engineering*, vol. 10, no. 3, May–June, 1959.

32. Harris, S. J., and K. U. Smith: Dimensional analysis of motion: VII. Extent and direction of manipulative movements as factors in defining motions, *Journal of Applied Psychology*, 1954, vol. 38, pp. 126–130.

33. Harston, L. D.: Contrasting approaches to the analysis of skilled movements, *Journal of General Psychology*, 1939, vol. 20, pp. 263–293.

34. Hick, W. E.: On the rate of gain of information, *Quarterly Journal of Experimental Psychology*, 1952, vol. 4, pp. 11–26.

35. Hilgendorf, L.: Information input and response time, *Ergonomics*, 1966, vol. 9, no. 1, pp. 31–37.

36. Hunsicker, P. A.: *Arm strength at selected degrees of elbow flexion*, USAF, WADC, TR 54–548, August, 1955.

37. Karger, D. W., and F. H. Bayha: *Engineered work measurement*, 2d ed., The Industrial Press, New York, 1965.

38. Kroemer, K. H. E.: *Maximal static force versus stress measurements as criteria for establishing optimal work conditions*, USAF, AMRL, TR 67–32, 1967.

39. Kroemer, K. H. E., and J. M. Howard: *Human strength: measurement, interpretation, and application of data*, paper presented at meetings of the Human Factors Society, Chicago, Ill., Oct. 30, 1968.

40. Lauru, L.: The measurement of fatigue, *The Manager*, 1954, vol. 22, pp. 299–303 and 369–375.

41. LeBlanc, J. A.: Use of heart rate as an index of work output, *Journal of Applied Physiology*, 1957, vol. 10, pp. 275–280.

42. Lehmann, G.: Physiological measurements as a basis of work organization in industry, *Ergonomics*, 1958, vol. 1, pp. 328–344.

43. Lundervold, Arne: Electromyographic investigations during typewriting, *Ergonomics*, 1958, vol. 1, pp. 226–233.

44. Malhotra, M. S., S. S. Ramaswamy, G. L. Dua, and J. Sengupta: Physical work capacity as influenced by age, *Ergonomics*, 1966, vol. 9, no. 4, pp. 305–316.

45. Malhotra, M. S., and J. Sengupta: Carrying of school bags by children, *Ergonomics*, 1965, vol. 8, no. 1, pp. 55–60.

46. Margaria, R., F. Mangili, F. Cuttica, and P. Cerretelli: The kinetics of the oxygen consumption at the onset of muscular exercise in man, *Ergonomics*, 1965, vol. 8, no. 1, pp. 49–54.

47. Maynard, H. B.: *Industrial engineering handbook*, 2d ed., McGraw-Hill Book Company, New York, 1963.

48. Miles, D. W.: Preferred rates in rhythmic response, *Journal of General Psychology*, 1937, vol. 16, pp. 427–469.

49. Monod, Par H.: La validité des mesures de fréquence cardiaque en ergonomie, *Ergonomics*, 1967, vol. 10, no. 5, pp. 485–537.

50. Müller, E. A.: The physiological basis of rest pauses in heavy work, *Quarterly Journal of Experimental Physiology*, 1953, vol. 38, p. 205.

51. Murrell, K. F. H.: *Human performance in industry*, Reinhold Publishing Corporation, New York, 1965.

52. Nadler, G.: *Motion and time study*, McGraw-Hill Book Company, New York, 1955.
53. Nadler, G., and J. Goldman: The UNOPAR, *Journal of Industrial Engineering*, 1958, vol. 9, no. 1, p. 58.
54. Passmore, R.: Daily energy expenditure by man, *Proceedings of the Nutrition Society*, 1956, vol. 15, pp. 83–89.
55. Passmore, R., and J. V. G. A. Durnin: Human energy expenditure, *Physiological Reviews*, 1955, vol. 35, pp. 801–875.
56. Pattie, C.: *Response times in use of toggle switches and push buttons in various locations around vehicle steering wheel*, unpublished paper, Purdue University, Lafayette, Ind., 1968.
57. Provins, K. A.: Effect of limb position on the forces exerted about the elbow and shoulder joints on the two sides simultaneously, *Journal of Applied Physiology*, 1955, vol. 7, pp. 387–389.
58. Provins, K. A., and N. Salter: Maximum torque exerted about the elbow joint, *Journal of Applied Physiology*, 1955, vol. 7, pp. 393–398.
59. Rubin, G., P. von Trebra, and K. U. Smith: Dimensional analysis of motion: III. Complexity of movement pattern, *Journal of Applied Psychology*, 1952, vol. 36, pp. 272–276.
60. Salter, N., and H. D. Darcus: The effect of the degree of elbow flexion on maximum torques developed in pronation and supination of the right hand, *Journal of Anatomy*, 1952, vol. 86, pp. 197–202.
61. Schappe, R. H.: Motion element synthesis: an assessment, *Perceptual and Motor Skills*, 1965, vol. 20, pp. 103–106.
62. Schmidtke, H., and F. Stier: Der aufbau komplexer bewegungsabläufe aus elementarbewegungen, *Forschungsberichte des landes Nordrhein-Westfalen*, 1960, no. 822, pp. 13–32.
63. Simon, J. R., and R. C. Smader: Dimensional analysis of motion: VIII. The role of visual discrimination in motion cycles, *Journal of Applied Psychology*, 1955, vol. 39, pp. 5–10.
64. Smader, R. C., and K. U. Smith: Dimensional analysis of motion: VI. The component movements of assembly motions, *Journal of Applied Psychology*, 1953, vol. 37, pp. 308–314.
65. Snook, S. H., and C. H. Irvine: The evaluation of physical tasks in industry, *American Industrial Hygiene Association Journal*, May–June, 1966, vol. 27, pp. 228–233.
66. Snook, S. H., and C. H. Irvine: Maximum acceptable weight of lift, *American Industrial Hygiene Association Journal*, July–August, 1967, vol. 28, pp. 322–329.
67. Spector, W. S.: *Handbook of biological data*, USAF, WADC, TR 56–273, October, 1956.
68. Swink, J. R.: Intersensory comparisons of reaction time using an electropulse tactile stimulus, *Human Factors*, 1966, vol. 8, no. 2, pp. 143–145.
69. Switzer, S. A.: *Weight lifting capabilities of a selected sample of human males*, AMRL, MRL, TDR 62–57, 1962.
70. Teeple, J. B.: Work of carrying loads, *Perceptual and Motor Skills*, 1957, vol. 7, p. 60.

71. Tuttle, W. W., C. D. Janney, and C. W. Thompson: Relation of maximum grip strength to grip strength endurance, *Journal of Applied Physiology*, 1950, vol. 2, pp. 663–670.

72. Tuttle, W. W., and B. A. Schottelius: *Textbook of physiology*, 16th ed., The C. V. Mosby Company, St. Louis, 1969.

73. Vince, M. A.: The intermittency of control movements and the psychological refractory period, *British Journal of Psychology*, 1948, vol. 38, pp. 149–157.

74. von Trebra, P., and K. U. Smith: Dimensional analysis of motion: IV. Transfer effects and direction of movement, *Journal of Applied Psychology*, 1952, vol. 36, pp. 348–353.

75. Wargo, M. J.: Human operator response speed, frequency, and flexibility: a review and analysis, *Human Factors*, 1967, vol. 9, no. 3, pp. 221–238.

76. Warrick, M. J., A. W. Kibler, and D. A. Topmiller: Response time to unexpected stimuli, *Human Factors*, 1965, vol. 7, no. 1, pp. 81–86.

77. Wehrkamp, R., and K. U. Smith: Dimensional analysis of motion: II. Travel-distance effects, *Journal of Applied Psychology*, 1952, vol. 36, pp. 201–206.

78. Williams, M., and H. R. Lissner: *Biomechanics of human motion*, W. B. Saunders Company, Philadelphia, 1962.

79. Young, I. C.: A study of tremor in normal subjects, *Journal of Experimental Psychology*, 1933, vol. 16, pp. 644–656.

80. *Manual lifting and carrying*, International Occupational Safety and Health Information Centre, Geneva, CIS Information Sheet 3, 1962.

81. *Meeting of experts on the maximum permissible weight to be carried by one worker*, International Labour Office, Geneva, MPW/1964/14, 1964.

82. *Teach them to lift*, Bureau of Labor Standards, U.S. Department of Labor, Bulletin 110, Mechanical and Physical Hazards Series, 1965.

chapter

eleven

CONTROLS,

TOOLS,

AND

RELATED

DEVICES

Archeological discoveries of recent times have linked prehuman primates of half a million years ago with stone tools [Washburn, 52]. Even the ancestors of man found that they could perform certain chores more effectively with devices which served essentially as extensions of their own upper extremities. The evolution of tools and related devices, in combination with mechanization and automation, has resulted in the present-day inventory of an amazing assortment of nonpowered and powered hand tools, control mechanisms that effect changes by controlling the action of machines and other mechanical equipment, and other devices that are used by people to achieve some desired change in the *environment* (i.e., some part of the situation in which people find themselves).

At the stage that a designer sets himself to design some type of device, usually two considerations are already crystallized, namely, its purpose and its gross nature (such as a nonpowered or powered hand tool or a control mechanism). At this point the design objective should be to develop a device that will perform its function effectively when used by the intended user(s) and that will minimize human wear and tear on the user. Usually these two objectives are compatible with the same design; for example, the pliers that perform their functions best are usually also the easiest for people to use. The dominant factors that contribute to good design of such objects are the psychomotor abilities of people and their closely related anthropometric characteristics. In considering such human characteristics, however, the designer may, in different circumstances, be concerned from two different points of view. In some circumstances he may be primarily interested in the range of *individual* differences in some particular characteristic, as in designing for, say, the *extreme* individuals (such as the smallest, the largest, the weakest, or the slowest). In other circumstances, however, he may be primarily interested in *group* differences, such as in the relative performance of groups of people in using one device versus another (such as a short lever versus a long one or a crank versus a lever).

In our discussion of controls, tools, etc., it will be our intent to illustrate how human performance and anthropometric characteristics can influence the use of such devices, rather than to present a comprehensive set of recommendations on specific design features. (See Appendix B for a summary of such recommendations on controls.)

FUNCTIONS OF CONTROLS

Controls are mechanisms which transmit "information" to the system and therefore can be characterized by the *type* of information given. In fact, since displays and controls frequently are used in combination, we can characterize the information related to certain types of control functions in terms of corresponding types of information related to displays as discussed in Chapter 5. The major classes of control functions and their corresponding types of information are given below:

Type of control function	*Type of related information*
Activation (usually on-off)	Status (dichotomous)
Discrete setting (at any separate, discrete position)	Status (discrete indications) Quantitative Check
Quantitative setting (individual settings of control at any position along quantitative continuum)	Quantitative
Continuous control	Quantitative Qualitative Tracking
Data entry (as in typewriters, computers, pianos)	Coded

It might be noted in passing that display and control relationships sometimes have a strange parallel to the chicken-and-the-egg conundrum, since in some circumstances the control response is made to a display indication and in other circumstances the display indication is the consequence of the control response. And some controls double in brass by also serving as displays, such as selector switches and volume controls.

TYPES AND USES OF CONTROLS

Although many different types of controls have been used for various purposes, the primary types are included in the following table; these are related to the types of control functions (mentioned above) for which they are generally used.

Type of control	Activation	Discrete setting	Quanti- tative setting	Con- tinuous control	Data entry
Hand push button	×				
Foot push button	×				
Toggle switch	×	×			
Rotary selector switch		×			
Knob		×	×	×	
Thumbwheel		×	×	×	
Crank			×	×	
Handwheel			×	×	
Lever			×	×	
Pedal			×	×	
Keyboard					×

Some indication of possible uses of most of these controls is shown in Figure 11-1. But although a general type of control might be considered most appropriate for a given purpose, the specific utility of a particular variant of that type for some specific application is influenced by such features (if relevant) as ease of identification, location, size, control-display ratio, resistance, lag, backlash, rate of operation, and distance of movement. A discussion of some such characteristics will follow.

IDENTIFICATION OF CONTROLS

Although the correct identification of controls is not really critical in some circumstances (as in operating a pinball machine), there are some operating circumstances in which their correct and rapid identification is of major consequence—even of life and death. For example, McFarland [37, pp. 605–608] cites cases and statistics relating to aircraft accidents that have been attributed to errors in identifying control devices. For example, confusion between landing gear and flap controls was reported to be the cause of over 400 Air Force accidents in a 22-month period during World War II. It is with these types of circumstances in mind that consideration of control identification becomes important.

The identification of controls is essentially a coding problem, the primary coding methods including shape, texture, size, location, operational method, color, and labels. (Since color and alphanumeric codes were discussed in Chapter 5, they will not be covered in this chapter.) The utility of these methods typically is evaluated by such criteria as the number of discriminable differences that people can make (such as the number of shapes they can identify), bits of information, accuracy of use, and speed of use.

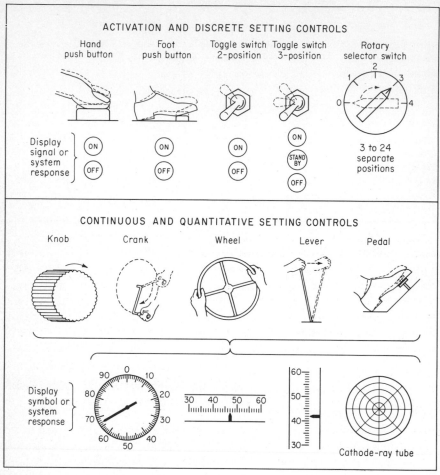

Figure 11-1 Examples of some types of control devices and their uses.

Shape Coding of Controls

The primary consideration in shape coding of controls is the accuracy of identification of the controls. Two factors contribute to such identification, namely, the tactual discrimination of various shapes and the symbolic association of such shapes (if this is feasible).

Discrimination of shape-coded controls. The discrimination of shape-coded controls is essentially one of tactual sensitivity. The procedure generally used in the selection of controls that are not confused with each other is illustrated by the study by Jenkins [25] in which he had 25 controls mounted on a rotating lazy Susan. Each subject, blindfolded, was presented with one knob which he touched for 1 sec. The experimenter then rotated the turntable to a predesignated point from which the subject went from knob to knob, feeling each in turn, until he found the one he thought was the one he had previously touched. It was then possible to determine which knobs were confused with which other knobs. While the statistical results will not be presented, it can be said that two sets of eight knobs were identified, such that the knobs within each group were rarely confused with each other. These two sets of knobs are shown in Figure 11-2.

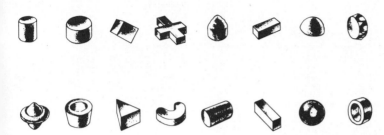

Figure 11-2 Two sets of knobs for levers that are distinguishable by touch alone. The shapes in each set are rarely confused with each other. [From Jenkins, 25.]

Following essentially the same tack as that mentioned above, the United States Air Force has developed 15 knob designs which are not often confused with each other. These designs are of three different types, each type being designed to serve a particular purpose [Hunt, 23]:

- *Class A: Multiple rotation.* These knobs are for use on controls (1) which require twirling or spinning, (2) for which the adjustment range is one full turn or more, and (3) for which the knob position is not a critical item of information in the control operation.
- *Class B: Fractional rotation.* These knobs are for use on controls (1) which do not require spinning or twirling, (2) for which the adjustment range usually is *less* than one full turn, and (3) for which the knob position is not a critical item of information in the control operation.
- *Class C: Detent positioning.* These knobs are for use on discrete setting controls.

The 15 knobs in these three classes are shown in Figure 11-3.[1] In connection with sizes of knobs in these three classes, Hunt suggests that

[1] A few of these knobs were confused with each other, and such combinations should not be used together if identification is critical. These combinations were *ab, co, cd#, do#, eg#, kp, ln, lo, np,* and *op#*. Those with a number sign (#) were confused only with gloves on and were not confused without gloves.

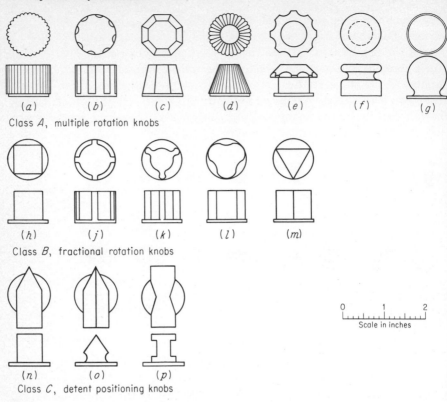

Class *A*, multiple rotation knobs

Class *B*, fractional rotation knobs

Class *C*, detent positioning knobs

Figure 11-3 Knob designs of three classes that are seldom confused by touch. [Adapted from Hunt, 23.]

they be not more than 4 in. in their maximum dimension and not less than ½ in. (except for class C, for which he suggests a ¾-in. minimum). In height they should not be less than ½ in., but need not be more than 1 in.

Symbolic associations of controls. If in addition to being individually discriminable by touch, the controls have shapes that are associated with their use, the learning of their use usually is simplified. They do not then require the learning of a new code. In this connection, the United States Air Force has developed a series of 10 knobs that have been standardized for aircraft cockpits. These standard knob shapes, besides being distinguishable from each other by touch, include some that also have symbolic meaning. In Figure 11-4, which includes these shapes, it will be seen, for example, that the landing-gear knob is like a landing wheel, the flap control is shaped like a wing, and the fire-extinguishing control resembles the handle on some fire extinguishers.

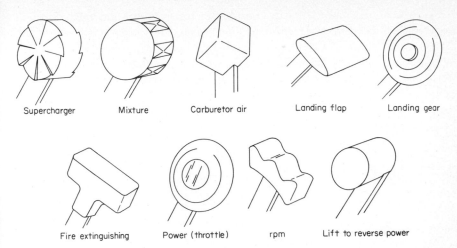

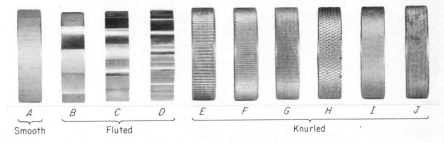

Figure 11-4 Standardized shape-coded knobs for United States Air Force aircraft. A number of these have symbolic associations with their functions, such as a wheel representing the landing-gear control. [*Personnel Subsystems*, 56, chap. 2, sec. 2D18, p. 3.]

Texture Coding of Controls

In addition to shape, control devices can be varied in their surface texture. This characteristic was studied (along with certain other variables) in a series of experiments with flat cylindrical knobs such as those shown in Figure 11-5 [Bradley, 8]. In one phase of the study, knobs of this type

Figure 11-5 Illustration of some of the knob designs used in study of tactual discrimination of surface textures. Smooth: *A*; fluted: *B* (6 troughs), *C* (9), *D* (18); and knurled: *E* (full rectangular), *F* (half rectangular), *G* (quarter rectangular), *H* (full diamond), *I* (half diamond), and *J* (quarter diamond). [From Bradley, 8.]

of 2-in. diameter were used, and subjects were presented with individual knobs through a curtained aperture and were asked to identify the particular design they felt. The results are shown as a "confusion matrix" in

Figure 11-6, this indicating the number of times each knob was identified correctly and incorrectly (and in such cases the knobs with which the one was confused). The smooth knob was not confused with any other, and vice versa; the three fluted designs were confused with each other, but not with other types; and the knurled designs were confused with each other, but not with other designs. It should be added that with gloved hands and with smaller-sized knobs (in a later phase of the study) there was some cross-confusion among classes, but this was generally

Knob that was felt

	Smooth	Fluted			Knurled					
	A	B	C	D	E	F	G	H	I	J
A	45									
B		42	6							
C		3	33	1						
D			6	44						
E					29	11	1	4	1	
F					8	8	7	8	5	6
G					1	7	27		15	9
H					6	7		27	1	2
I					1	7	6	5	2	18
J						5	4	1	21	10

Response (the knob that the "felt" knob was "identified" as)

Figure 11-6 "Confusion matrix," showing results of study in which knobs with different surface textures were presented through an aperture to 45 subjects. (The knobs are those shown in Figure 11-5.) The numbers in any given column are the numbers of times each knob was "identified" as the one that was felt by hand, when the one actually felt was the one identified at the top of the column. The numbers of *correct* identifications (out of 45) are those in the cells along the diagonal. [Adapted from Bradley, 8.]

minimal. The investigator proposes that three surface characteristics can thus be used with reasonably accurate discrimination, namely, smooth, fluted, and knurled.

Size Coding of Controls

Size coding of controls is not as useful for coding purposes as shape, but there may be some instances where it is appropriate. When such coding is used, the different sizes used should of course be such that they are discriminable one from the others. Part of the study by Bradley reported above [8] dealt with the discriminability of cylindrical knobs of varying diameters and thickness. It was found that knobs that differ by $\frac{1}{2}$ in. in diameter and by $\frac{3}{8}$ in. in thickness can be identified by touch very accurately, but that smaller differences between them sometimes result in confusion of knobs with each other. Incidentally, Bradley proposes that a combination of three surface textures (smooth, fluted, and knurled), three diameters ($\frac{3}{4}$, $1\frac{1}{4}$, and $1\frac{3}{4}$ in.), and two thicknesses ($\frac{3}{8}$ and $\frac{3}{4}$ in.) could be used in all combinations to provide 18 tactually identifiable knobs.

Aside from the use of size coding for individual control devices, size coding is part and parcel of ganged control knobs, where two or more knobs are mounted on concentric shafts with various sizes of knobs superimposed on each other like the layers of a wedding cake. When this type of design is dictated by engineering considerations, the differences in the sizes of superimposed knobs need to be great enough to make them clearly distinguishable, as illustrated in Figure 11-7 [Bradley and Stump, 9].

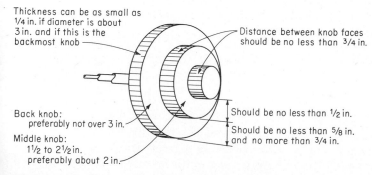

Thickness can be as small as 1/4 in. if diameter is about 3 in. and if this is the backmost knob

Distance between knob faces should be no less than 3/4 in.

Back knob: preferably not over 3 in.

Middle knob: 1½ to 2½ in. preferably about 2 in.

Should be no less than 1/2 in.

Should be no less than 5/8 in. and no more than 3/4 in.

Figure 11-7 Dimensions of concentrically mounted knobs that are desirable in order to allow human beings to differentiate knobs by touch. [Adapted from Bradley and Stump, 9.]

Location Coding of Controls

Whenever we shift our foot from the accelerator to the brake, feel for the light switch at night, or grasp for a machine control that we cannot see, we are responding to *location coding*. But if there are several similar controls from which to choose, the selection of the correct one may be difficult unless they are far enough apart that our kinesthetic sense makes it possible for us to discriminate. Some indications about this come from a study by Fitts and Crannell as reported by Hunt [23]. In this study blindfolded subjects were asked to reach for designated toggle switches on vertical and horizontal panels, the switches being separated by 1 in. The major results are summarized in Figure 11-8, which shows the percentage of reaches that were in error by specified amounts when the panels were in horizontal and vertical positions, left and right from cen-

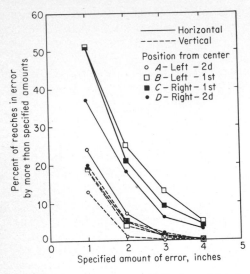

Figure 11-8 Accuracy of blind reaching to toggle switches (nine in a row on switch box) with switch box positioned horizontally and vertically at four locations. [Adapted from Fitts and Crannell, as presented by Hunt, 23.]

ter. The curves indicate quite clearly that accuracy was greatest when the toggle switches were arranged vertically. For the vertically arranged locations probably a 5-in. difference would be desirable for reasonably accurate positioning. Since errors drop quite low at about 2½ in. and since these errors are in both directions from the control, the central range of errors is double this distance, or 5 in. For horizontally arranged controls there should be 8 in. or more between them if they are to be recognized by location.

Operational Method of Coding Controls

In the operational method of coding controls, each control has its own unique method for its operation. For example, one control might be of a push-pull variety, and another of a rotary variety. Each can be activated *only* by the movement that is unique to it. It is quite apparent that this scheme would be inappropriate if there were any premium on time in operating a control device and where operating errors are of considerable importance. When such a method is used, it is desirable that compatibility relationships be utilized, if feasible. By and large, this method of coding should be avoided except in those individual circumstances in which it seems to be uniquely appropriate.

Discussion of Coding Methods

In the use of codes for identification of controls, two or more code systems can be used in combination. Actually, combinations can be used in two ways. In the first place, *unique combinations* of two or more codes can be used to identify separate control devices, such as the various combinations of texture, diameter, and thickness mentioned before [Bradley, 8]. And, in the second place, there can be completely *redundant codes*, such as identifying each control by a distinct shape *and* by a distinct color. Such a scheme probably would be particularly useful when accurate identification is especially critical. In discussing codes, we should be remiss if we failed to make a plug for standardization in the case of corresponding controls that are used in various models of the same type of equipment, such as automobiles and tractors. When individuals are likely to transfer from one situation to another of the same general type, the same system of coding should be used if at all possible. Otherwise, it is probable that marked "habit interference" will result [Weitz, 54] and that people will revert to their previously learned modes of response. In connection with the use of individual control coding methods, a few general principles can be set forth, as evolving from both research and experience. Some of these are given below, with the usual words of caution about the usual exceptions to general principles:

1. **Shape and texture**
   A. Desirable features. (1) Useful where illumination is low or where device may be identified and operated by feel, without use of vision; (2) can supplement visual identification; (3) useful in standardizing controls for identification purposes.
   B. Undesirable features. (1) Limitation in number of controls that can be identified (fewer for texture than for shape); (2) use of gloves reduces human discrimination.

2. **Location**
   A. Desirable features. (1) Same advantages as for shape and texture.
   B. Undesirable features. (1) Limitation in number of controls that can be identified; (2) may increase space requirements; (3) identification may not be as certain (may be desirable to combine with other coding scheme).

3. **Color**
   A. Desirable features. (1) Useful for visual identification; (2) useful for standardizing controls for identification purposes; (3) moderate number of coding categories possible.
   B. Undesirable features. (1) Must be viewed directly (but can be combined with some other coding method, such as shape); (2)

cannot be used under poor illumination; (3) requires people who have adequate color vision.

4. **Labels**

 A. Desirable features. (1) Large number can be identified; (2) does not require much learning.
 B. Undesirable features. (1) Must be viewed directly; (2) cannot be used under poor illumination; (3) may take additional space.

5. **Operational method**

 A. Desirable features. (1) Usually cannot be used incorrectly (control usually is operable in only one way); (2) can capitalize on compatible relationships (but not necessarily).
 B. Undesirable features. (1) Must be tried before knowing if correct control has been selected; (2) specific design might have to incorporate incompatible relationships.

CONTROL–DISPLAY RATIO

In continuous control tasks or when a quantitative setting is to be made with a control device, the ratio of the movement of the control device to the movement of the display indicator (i.e., the controlled element) is called the *control-display ratio* (C/D ratio). The movement may be measured in *distance* (in the case of levers, linear displays, etc.) or *angle* or *number of revolutions* (in the case of knobs, wheels, circular displays, etc.). When there *is* no display, the display movement is some measure of system response (such as angle of turn of an automobile). A very "sensitive" control is one which brings about a marked change in the controlled element (display) with a slight control movement; its C/D ratio would be low (a small control movement is associated with a large display movement). Examples of low and high C/D ratios are shown in Figure 11-9.

C/D Ratios and Control Operation

The performance of human beings in the use of control devices which have associated display movements is distinctly affected by the C/D ratio. This effect is not simple, but rather is a function of the nature of human motor activities when using such controls. In a sense, there are two types of human motions in such tasks. In the first place, there is essentially a gross adjustment movement (travel time or a slewing movement) in which the operator brings the controlled element (say, the display indicator) to the approximate desired position. This gross movement is followed by a fine adjustment movement, in which the operator makes an adjustment to bring the controlled element right to the desired location. (Actually, these two movements may not be individually identifiable, but there is typically some change in motor behavior as the desired position is approached.)

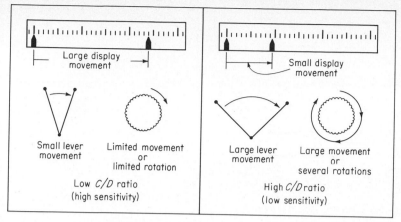

Figure 11-9 Generalized illustrations of low and of high control-display ratios (*C/D* ratios) for lever and rotary controls. The *C/D* ratio is a function of the linkage between the control and display.

Optimum C/D Ratios

The determination of an *optimum C/D* ratio for any given control-display setup needs to take into account these two components of the human motions in such a task. This was illustrated by the results of studies by Jenkins and Connor [27] in which a knob was used to control the movement of a pointer along a horizontal scale. The *C/D* ratio was varied by changing the linkage between the control knob and the display indicator. By an ingenious electrical recording device it was possible to record the *travel* time and the *adjust* time (the gross and fine movements mentioned above). The results, shown in Figure 11-10, illustrate the essential points that travel time drops off sharply with increasing *C/D* ratios and then tends to level off, and that adjustment time has the reverse pattern. The optimum, then, is somewhere around the point of intersection. Within this general range the combination of travel and adjust time usually would be minimized. The numerical values of the *C/D* ratios are of course a function of the physical nature and sizes of the controls and displays; the *C/D* ratio that would be applicable in controlling, say, a crane would be very different from that for a radar set. There are no formulas for determining what *C/D* ratio would be optimum for given circumstances. Rather, this ratio should be determined experimentally for the control and display being contemplated.

RESISTANCE IN CONTROL DEVICES

As pointed out by Burrows [12], there are many factors that impinge on the interaction between human beings and control devices, including the various forms of resistance in the controls. All mechanical devices have

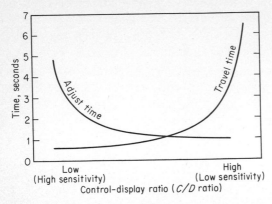

Figure 11-10 Relationship between C/D ratio and movement time (travel time and adjust time). While the data are from a study by Jenkins and Connor [27], the specific C/D ratios are not meaningful out of that context, so are omitted here. These data, however, depict very typically the nature of the relationships.

some resistance, although it may be extremely limited. The nature and amount of resistance in controls, however, are *not* factors over which the designer has *no* influence. Within limits he can build in those resistance characteristics that might be useful in human operation of controls. This can be done in a number of ways, such as with servomechanisms, hydraulic or other power, or mechanical linkages.

Types of Resistance in Controls

The primary types of resistance, as related to the human use of control devices, are:

- *Static and coulomb friction:* Static friction, the resistance to initial movement, is maximum at the initiation of a movement but drops off sharply. Coulomb (sliding) friction continues as a resistance to movement, but this friction force is not related to either velocity or displacement.
- *Elastic resistance:* Such resistance (as in spring-loaded controls) varies with the displacement of a control device (the greater the displacement, the greater the resistance). The relationship may be linear or nonlinear.
- *Viscous damping:* This is caused by a force operating opposite to that of the output, but proportional to the output velocity.

· *Inertia:* This is the resistance to movement (or change in direction of movement) caused by the mass (weight) of the mechanism involved. It varies in relation to acceleration.

The above relationships are illustrated in Figure 11-11. In the case of a spring-centered joy stick (i.e., a control with elastic resistance), the

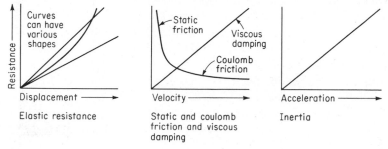

Figure 11-11   Illustration of certain types of resistance in control devices as related to movement variables.

amount of *displacement* of the stick is directly proportional to the amount of force applied (i.e., the resistance that is overcome). In a viscously damped control stick, the force applied will produce an angular *velocity* of the stick that is proportional to the force and, in turn, a displacement of the stick that is proportional to the time integral of the force. With a high inertia control, the *acceleration* of the stick will be proportional to the force applied; thus its displacement will be proportional to the second integral of the force applied. We can see, then, that certain control devices, by reason of their interaction with the forces applied to them, can bring about the effects of integration when using them.   These effects are summarized below [adapted from Birmingham and Taylor, 5]:

Type of joy stick	*No. of integrations*	*Joy stick displacement proportional to:*	*Input (force) proportional to:*
Spring-centered	0	Force applied	Displacement
Viscous-damped	1	Time integral of force	First derivation of displacement
High-inertia control	2	Second integral of force	Second derivation of displacement

Effects of Resistance on Performance

Although, as Burrows laments [12], there is much that is not known about the man-control interface, including the effects of resistance on human performance, there are at least a few gleanings from research and experience that impinge upon this matter.

Static and coulomb friction.   With a couple of possible exceptions, static and coulomb friction tend to cause degradation in human performance. This is essentially a function of the fact that there is no systematic relation between such resistance and any aspect of the control movement such as displacement, speed, or acceleration; thus it can not produce any meaningful feedback to the user of the control movement. The possible adverse effects of such friction were indicated by the results of a study by Jenkins, Maas, and Rigler [29] in which friction was added with a prony brake to the operation of a $2\frac{3}{4}$-in. knob. The braking was applied in varying degrees to require pulls of 100, 400, 700, 1000, and 1300 g at the periphery of the knob. Such resistance generally increased travel time, especially for longer movements, although it did not affect the adjust time. On the other hand, such friction can have the advantages of reducing the possibility of accidental activation and of helping to hold the control in place.

Elastic resistance.   A major advantage of elastic resistance (such as in spring-loaded controls) is that it usually (but not always) serves as a source of useful feedback, presumably because the displacement feedback has some systematic relation to resistance. By so designing the control that there is a distinct gradient in resistance at critical positions (such as near the terminal), additional cues can be provided. In addition, such resistance permits sudden changes in direction and automatic return to a neutral position.

Viscous damping.   Viscous damping generally has the effect of helping to execute smooth control, especially in maintaining a prescribed *rate* of movement [Bahrick, Fitts, and Schneider, 2]. The feedback from such resistance, however, being related to speed (velocity) and not displacement, probably is not readily interpretable by operators. Such resistance, however, does minimize accidental activation.

Inertia.   Inertia also aids in smooth control and minimizes the possibility of accidental activation. Because of the forces required to overcome inertia, however, it complicates the process of changing directions. Its use probably is particularly warranted when heavy friction is involved in a control. In such a situation, inertia presumably compensates in part for friction drag.

Control Resistance in Relation to Display Gain

The C/D ratio discussed above influences the sensitivity of a control device. A different facet of sensitivity is associated with the resistance in controls as this influences the force that must be applied to bring about a particular effect. In some controls both force and displacement are involved, but in others, such as pressure joy sticks, virtually no displacement occurs. An investigation of the sensitivity of pressure joy sticks was carried out by Tipton and Birmingham [50], in which the sensitivity required to cancel the maximum course displacement was varied over a range from 1 to 128 oz (8 lb). The average integrated tracking errors made by the eight subjects are shown in Figure 11-12. It will be noted

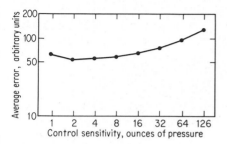

Figure 11-12 Tracking errors for eight variations in control sensitivity (control sensitivity is in ounces of pressure required to move display cursor its maximum displacement). [From Tipton and Birmingham, 50.]

that the curve has a fairly broad range of sensitivity values (from 1 to about 8) that are conducive to fairly minimal errors. It will be noted further that the magnitude of errors—even for the least-sensitive situation—does not vary greatly. This was interpreted as an indication that the human being is capable of varying his own gain (or sensitivity) to the demands of the situation in order to maintain a given overall system gain and thereby produce a nearly constant error performance.

Ability of People to Judge Resistance

Regardless of what variety of resistance is intrinsic to a control mechanism, if it is to be used as a source of feedback, the meaningful differences in resistance have to be such that the corresponding pressure (touch) cues can be discriminated. Such discriminations were the focus of an investigation by Jenkins [26], with three kinds of pressure controls, namely, a stick, a wheel, and a pedal (like a rudder control in a plane). After some training and practice in reproducing specified forces, a series of trials was made by each subject. Measures of actual pressure exerted were then compared with the pressures that the subjects attempted to reproduce. The difference limens by pounds of pressure for the various types of controls are shown in Figure 11-13. Since the difference limen is the average difference that can just barely be detected, two pressures have to differ by an amount greater than the limen to be detected as being different.

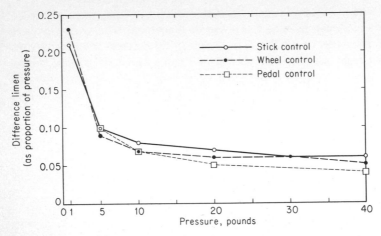

Figure 11-13   Difference limens for three control devices for various pressures that were to be reproduced. (Limen is the standard deviation divided by the standard pressure.) [From Jenkins, 26.]

The systematic drop-off of all the curves between 5 and 10 lb implies that if differences in pressure are to be used as feedback in operation of control devices of the types used, the pressures used preferably should be around or above these values (and perhaps more for pedals because of the weight of the foot). The experimenter suggested that if varying levels of pressure discrimination are to be made, the equipment should provide a wide range of pressures up to 30 or 40 lb. Beyond these pressures, the likelihood of fatigue increases, and also the likelihood of slower operation.

PRESSURE AND AMPLITUDE AS CONTROL FEEDBACK CUES

When control devices are to be used to control the *amount* of some variable (as in the use of speed and volume controls), the operator has three possible types of feedback, namely, (1) visual (such as observing the distance of movement of the device, or observing a corresponding display), (2) pressure (the sensation via the sense of touch of the pressure required to overcome the control resistance), and (3) displacement (i.e., amplitude or distance) of movement (as sensed through the kinesthetic sense). When vision is not to be used as a feedback source (or at least not as the dominant source), it would be helpful for the designer to know the relative usefulness of pressure and of amplitude cues, singly or in combination. The evidence regarding their combination, however, is not entirely consistent.

In reviewing the use of different types of controls in *tracking* tasks, Zeigler and Chernikoff [55] cite certain studies in which tracking errors were lower with pressure types of controls than with displacement (e.g., free-moving) controls. In their own study Zeigler and Chernikoff found a pressure type of lever to result in a somewhat lower error rate than a displacement type of lever, in a third-order tracking task. Similar results also were reported by Burke and Gibbs [11] in a position tracking (zero-order) task; however, in an analysis of errors in relation to *gain* (i.e., displacement-force ratios which were varied) it was found that errors were *fewer* with *low* ratios, that is, when *greater* forces were associated with a given displacement. This finding, of course, ties in with the results reported by Jenkins [26] as shown above in Figure 11-13, which indicate that differences between *small* forces cannot be differentiated very reliably with only pressure feedback.

On the other hand the results of certain other studies have shown that amplitude of movement as a source of feedback tends to be superior to pressure. Such studies, however, generally have dealt with *positioning* tasks (in which the control serves to move the controlled element to a particular position or location) rather than with tracking tasks. For example, in a study by Weiss [53] of blind positioning movements it was found that the distance the lever moved (amplitude) provided much better cues for positioning the lever accurately than pressure (force) cues did. Positioning errors were greater for short movements than for long ones, however. Another investigation [Bahrick, Bennett, and Fitts, 1] tended to confirm these inklings as illustrated in Figure 11-14. That figure shows that changes in torque (increases in torque as the movement continues) aided in accurate positioning for the short movements (the angular excursions of 17.5 and 35°) but not for the longer movements (70°).

In still another study it was found that changes in amplitude of movement significantly affected performance only when force cues were high—not when force cues were low; and changes in force cues affected performance only when amplitude was high—not when it was low [Briggs, Fitts, and Bahrick, 10]. It was proposed that the combination effect of amplitude and force cues is expressed by the following ratio:

$$\frac{\Delta F}{F(\Delta D)}$$

where ΔF is the force associated with a given displacement change ΔD and F is the terminal force required for attaining the displacement. In turn, Burrows [12] points to the very complicated effects of the dynamic characteristics of the controlled element on the feedback that the operator would receive (at least in the case of complex systems) as a potentially limiting factor in the use of such feedback in control operations.

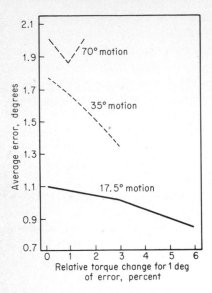

Figure 11-14 Average error in degrees in blind positioning of arm control lever in relation to torque changes from beginning to termination of movement. [From Bahrick, Bennett, and Fitts, 1.]

It is thus evident that there are numerous missing and confusing pieces of this jigsaw puzzle. In sifting through some of the evidence, however, a few conclusions seem to be warranted. To begin with, it seems evident that there is no clear and consistent superiority of one mode of feedback over the other, which suggests that each may have utility in certain kinds of control circumstances. For example, as indicated above, pressure controls may be more useful in certain types of continuous-tracking tasks than in positioning tasks (but this is not to say that all tracking tasks should be executed with a pressure type of control). And, in any event, pressure controls presumably should be used only when the pressure feedback cues have a tolerably systematic relationship to changes in the controlled variable. If a pressure-stick control is to be used, however, the force to be overcome needs to be of some consequence (that is, the gain, or displacement-force ratio, needs to be low) for the feedback to be of greatest use. Since such forces might induce fatigue (if the controls are used continuously), one might follow the suggestion of Burke and Gibbs [11] to use a nonlinear control with lower force required for large displacements.

On the other side of the coin, if moving controls are to be used (as opposed to strictly pressure controls), one needs to be forewarned that distance of movement is not a very reliable source of feedback when the amount of displacement is limited. All of this adds up to the notion that for some conventional positioning control purposes (as contrasted with tracking tasks) a combination of force and distance feedback may have some advantage, using dominantly pressure cues for limited displacements of the controlled element and movement cues for larger displacements. Still keeping in mind Burrows' observations [12] on the abysmal lack of knowledge about the effects of various types of control feedback, it probably can be said that, barring complicated pressure cues, neither type of feedback (pressure or movement) in combination with the other seems to affect performance adversely. Although there may still be some question as to whether, with long movements, the combination actually facilitates performance over either one independently, the reference of pilots and other vehicle operators to the *feel* of the control device seems to imply the desirability of at least moderate feel in such mechanisms.

RESPONSE LAG

Virtually inherent in any man-machine system is some lag or delay in the response to a changed input. This lag generally can be viewed as consisting of two components, namely, lag in the system itself and human reaction time. In practice, however, these certainly interact.

System Lag

As an operator makes a change in a control device, the system (as he perceives its response directly or by a display) usually does not respond immediately. In the design of some systems it may be useful to know what effect such lag has on operator performance. Probably the most important sources of such lag as far as the operator is concerned are *transmission* lag and *exponential* lag. Transmission lag refers to a situation in which there is a constant delay between input and output; the output is identical with the input, but simply follows it, temporally, after a constant time interval. Exponential lag, on the other hand, refers to the situation where the output follows essentially an exponential function following a *step* input. These two types of lags are illustrated in Figure 11-15. With exponential lag, the time delay (sometimes referred to as the *time constant*) is usually considered the time required for the output to reach 63 percent of the input change; in the figure illustrated there are two such time delays, t_1 and t_2.

Effects of system lag. The effects of system lag need to be viewed within the context of the *use* of a mechanical system by the *operator* of

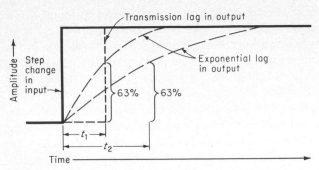

Figure 11-15 Illustration of transmission lag and exponential lag between control action (input) and display response (output). The input change shown is a step change. Transmission lag in output follows the input by a constant time interval, such as t_1. Exponential lag follows an exponential curve; time delay (time constant) is usually considered the time it takes to reach 63 percent of the input value. In this figure, time delays t_1 and t_2 are given for two exponential lags. As illustrated, the transmission lag and one exponential time delay are the same, t_1.

the system. With this point of view in mind, let us take a quick look at a few of the investigations that have dealt with such lags. Although some studies of the effects of lag on tracking tasks have indicated that lag in appreciable amounts brings about degradation in performance [Levine, 36; Wallach, 51], such effects are not universal. In fact, in some circumstances it has been found to have a facilitating effect, these differential effects being particularly related to the C/D ratio being used, as shown in Figure 11-16. This figure presents data on the relationship between time delay and average TOT (time on target) for eight subjects. This shows that with high C/D ratios (1:6 and 1:3, limited display movement relative to control movement), a greater C/D time lag apparently causes degradation in compensatory tracking performance, whereas with low C/D ratios (considerable display movement relative to control movement), longer time lags are not serious and, in fact, were even associated with improved performance (especially with a 1:30 ratio).

While there has perhaps been a tendency to consider any delay between control and display as undesirable, such results have suggested that for exponential time delay, a more appropriate principle would be: The optimum delay between control and display depends, among other things, on the magnitude of display change produced by a given control input [Rockway, 43].

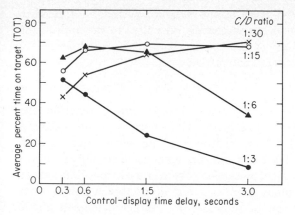

Figure 11-16 Relationship between control-display (*C/D*) time delay and average percentage of time on target (TOT) scores for subjects using various *C/D* ratios on a compensatory tracking task. Under long (as opposed to short) delays, high *C/D* ratios (1:3 and 1:6) result in performance degradation, whereas low *C/D* ratios (1:15 and 1:30) result in tracking improvement. [From Rockway, 43.]

Anticipation of input. In some circumstances, the operator can anticipate input changes. This would be possible in a tracking task, for example, where the operator can see the course to be followed, as in driving a vehicle. In other circumstances, where there may be no advance information available as such to the operator, he may be able to deduce the nature of future signals from past experience; this would be the case, for example, when a person learns the time interval between a warning signal and a subsequent signal to which he is to react, or where in a tracking task, there is a systematic input such as a sine wave. This type of circumstance has been referred to by Poulton [42] as *perceptual anticipation*.

There is evidence from a number of studies to the effect that people can compensate fairly well for lag if they are able to anticipate future inputs by either of these methods. This was shown, for example, by Poulton [42], Conklin [13], and Fenwick [20]. Conklin concluded that skilled performance in a perceptual motor task depends to a large extent upon the individual's ability to anticipate and thus predict system performance; he thus can compensate for his basic "intermittency" and behave as a continuous error-correction device. In turn, Fenwick found that while performance was somewhat poorer under lag conditions than under no-lag conditions, this difference was of less consequence than certain other variables (such as speed); and further, with some procedure for providing visual cues to aid in attending (visually) to the *appropriate* advance-information position of the moving course, it was possible almost to eliminate any adverse effects of lag.

Discussion

In considering the effects of lag on system output, human reaction time must of course be considered independently where it is the exclusive source of lag, and in combination with system lag where there is such. While human and system lag can (and do) have adverse effects on system performance in many circumstances (even in such mundane affairs as driving a car), it is evident that it is not invariably a goblin to be avoided at all costs. As seen in some of the studies discussed above, there are circumstances where its effects can be counteracted and even where it can contribute to the adequacy of system performance.

CONTROL BACKLASH AND DEADSPACE

Human performance with control devices can be influenced by still other control characteristics, such as backlash and deadspace; these are particularly relevant with continuous-control tasks.

Backlash

Backlash in a control system is a tendency for the system response to be reversed when a control movement is stopped; and typically it cannot be coped with very well by operators. This effect was illustrated by the results of an investigation by Rockway and Franks [45], using a control task under varying conditions of backlash and of display gain (which is essentially the reciprocal of C/D ratio). Figure 11-17 summarizes the results and shows that performance deteriorated with increasing backlash for all display gains, but was most accentuated for high gains. The implications of such results are that if a high display gain is strongly indicated (as in, say, high-speed aircraft), the backlash needs to be minimized in order to reduce system errors; or conversely, if it is not practical to minimize backlash, the display gain should be as low as possible—also to minimize errors from the operation of the system.

Deadspace

Deadspace in a control mechanism is the amount of control movement that results in no movement of the device being controlled. It is almost inevitable that some deadspace will exist in a control device. Deadspace of any consequence usually affects control performance, but here, again, the amount of effect is related to the sensitivity of the control system.

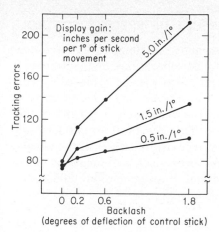

Figure 11-17 Relationship between backlash in a control system and tracking errors for various display gains (sensitivity). [Adapted from Rockway and Franks, 45.]

This is indicated in Figure 11-18 [Rockway, 44]. It can be observed that tracking performance deteriorated with increases in deadspace (in degrees of control movement that produced no movement of the controlled device). But the deterioration was less with the less-sensitive systems (higher C/D ratios) than with more sensitive systems (lower C/D ratios). This, of course, suggests that deadspace can, in part, be compensated for by building-in less-sensitive C/D relationships.

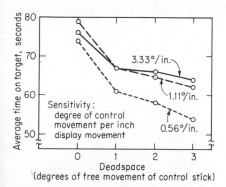

Figure 11-18 Relationship between deadspace in a control mechanism and tracking performance for various levels of control sensitivity. [Adapted from Rockway, 44.]

DESIGN ASPECTS OF SPECIFIC CONTROLS

The aspects of control devices discussed above predominantly have some general implications in control design (considerations of identification, resistance, lag, etc.), rather than a relation to specific types of controls (although some of the aspects dealt with are more relevant to tracking controls than to other types). It is neither relevant nor feasible to discuss or illustrate the many control mechanisms that ingenious people have concocted. It may be useful, however, to illustrate with examples the principle that the specific design features, sizes, and locations of certain types of such gadgets have a direct bearing on how adequately people can use them for their intended purpose.

Selector Switches

As a case in point, in one investigation a comparison was made of the use of four types of rotary selector switches for making settings of three-digit numbers, and, separately, for reading three-digit numbers already set into the four switches. One type of switch (1) was a fixed-scale model with a moving pointer, as shown in Figure 11-19; the others were moving-

Figure 11-19 Illustration of fixed-scale, moving-pointer selector switches used in study by Kolesnik [35]. See text for discussion. [Reproduced by permission of Autonetics Division, North American Rockwell Corporation.]

scale models with fixed pointers with 10, 3, or 1 of the 10-scale positions visible. Some of the results of the study are given below [Kolesnik, 35].

Type of switch	Average setting time, sec	Preference, rank order	Reading errors
Fixed scale	4.5	1	287
Moving scale, 10 digits shown	5.4	3	135
Moving scale, 3 digits shown	5.8	2	101
Moving scale, 1 digit shown	6.3	4	92

In the task of making *settings* of specified three-digit numbers, there was little difference in accuracy among the four styles, but the fixed scale resulted in lowest setting times and was first in preferences of the users. However, in the task of *reading* three-digit values at which the switches were set, it had the most errors. These differences, of course, need to be considered in the light of the *use* to which selector switches are to be put.

Cranks and Handwheels for Moving Objects

Cranks and handwheels frequently are used as a means of applying force to move something to a particular location, such as moving a carriage or cutting tool or lifting objects. Let us see to what extent performance with such control devices is affected by their size, friction, direction of rotation, location, and use of preferred versus nonpreferred hand.

Size of cranks and handwheels. In a study by Davis [18] several different sizes of cranks and handwheels were used to control the position of a pointer on a dial. In general, about one revolution of the crank was required in making the settings. The cranks and handwheels were mounted so that the plane of rotation was parallel to the frontal plane of the body. By using torques of 0, 20, 40, 60, and 90 in.-lb, it was possible to determine the time required to make settings under different friction-torque conditions that typically occur in the operation of such controls.

Figure 11-20 shows average times for certain torque levels by size of crank or handwheel. A couple of points are illustrated by this figure. In the first place, times for making settings under 0 torque were shortest for small sizes of cranks and wheels, whereas for torques of 40 and 90 in.-lb, the small cranks were definitely inferior to the larger ones. In the second place, the average times for cranks were, in general, lower than those for handwheels. In some cases, however, the time differences were not great, especially with the larger sizes.

Work output with different cranks. The work output of subjects using three sizes of cranks and five resistant torque loads was investigated by Katchmer [33], with a view to identifying fairly optimum combinations. The cranks were adjusted to waist height of each subject, and the subjects were instructed to turn the crank at a rapid rate until they felt they could no longer continue the chore (or until 10 min. had passed). A summary of certain data from the study is given in Figure 11-21. In total foot-pounds of work (shown by the bars), it is apparent that the 30 and 50 in.-lb of torque load resulted in higher values than the lower or higher loads did. The 5-in.- and especially the 7-in.-radius cranks were generally superior on this score to the 4-in. In average time that subjects could "take it," the lower loads of course were highest, but it is apparent that they could operate the 7-in. cranks longer under moderate loads than the shorter cranks. While average horsepower (hp) per minute was

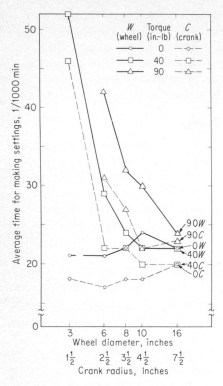

Figure 11-20 Average times for making settings with vertically mounted handwheels and cranks of various sizes under different torque conditions. [Adapted from Davis, 18.]

highest for the 90 in.-lb torque (average for the period actually worked), the durability of the subjects for this heavy load was short (about 1 min). These data illustrate emphatically the point that *different* design features might be preferable if different criteria are used (in this case the different criteria being time tolerance, i.e., average minutes; speed, i.e., rpm; and average horsepower per minute when working).

Rotary Controls in Tracking Operations
Where rotary controls are used for tracking operations, some determination needs to be made regarding the location of the controls and their size.

Plane of rotation in tracking. In a series of studies, Spragg et al. [40, 41, 49] investigated different aspects of tracking, including the posi-

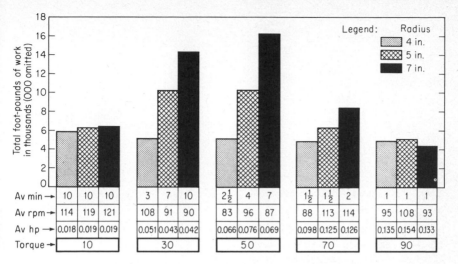

Av min →	10	10	10	3	7	10	2½	4	7	1½	1½	2	1	1	1
Av rpm →	114	119	121	108	91	90	83	96	87	88	113	114	95	108	93
Av hp →	0.018	0.019	0.019	0.051	0.043	0.042	0.066	0.076	0.069	0.098	0.125	0.126	0.135	0.154	0.133
Torque →	10			30			50			70			90		

Figure 11-21   Average performance measures for subjects using 4-, 5-, and 7-in.-radius cranks under 10, 30, 50, 70, and 90 in.-lb torque resistance. [Adapted from Katchmer, 33.]

tion of cranks for two-handed cranking. In all these studies two cranks controlled the movement of a *cursor* in tracking a moving target, one crank controlling the cursor in a left-and-right direction and the other in a front-and-back direction.

In one of the studies nine combinations of positions of left-and-right cranks were tried out, all these being combinations of three different crank positions, namely, horizontal, vertical facing body, and vertical at right angles to body. The combinations that were superior (in TOT) were those in which (*a*) the two cranks faced the body and (*b*) the left crank faced the body and the right crank was at right angles to the body at the right side of the equipment.

For a single-crank tracking operation for tracking in only one direction, a side crank resulted in fewer tracking errors than a front crank [Helson, 22]. It was also found that, for such a tracking task, two side cranks, one on either side, operated in a symmetrical double-handed movement, reduced errors below the level of a single side crank.

Crank radius in tracking.   In one of the above studies, cranks of different radii were tried out, these being 1, 2, 3, 4, and 5.5 in. [49]. Times on target for these cranks are shown in Figure 11-22 for crank combinations 1 and 4. For both these combinations, TOT was greatest for the 2- and 3-in. cranks, 4-in. cranks being next best. Thus, the optimum crank radius for two-handed tracking operations seems to be in the range of about a 2- to 3-in., or possibly 4-in., radius.

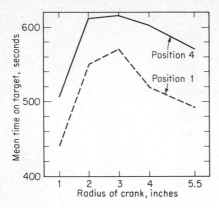

Figure 11-22 Accuracy of pursuit tracking with various sizes of cranks. The two curves show the relationship between radius of crank and time on target (TOT) for two combinations of crank positions (1, both cranks vertical and facing body; 4, left crank same as 1 but right crank vertical and at right angles to body). [Adapted from Swartz, Norris, and Spragg, 49, fig. 1.]

Rate of cranking in tracking. In a tracking operation, the ratio of crank rotations to rate of movement of the indicator or target can be juggled around by mechanical or electrical means. Is there, then, some rate of rotation that is most desirable for tracking accuracy? Some information on this point has been provided by Helson [22]. In a compensatory tracking task, he determined the relationship between speed of cranking and tracking errors. The results are summarized in Figure 11-23 for two sizes of cranks. Errors were high for slow speeds, but improved toward the higher speeds, then leveled off at around 50 rpm. For the long-radius crank (4.50 in.), errors increased again at 200 rpm. There seems to be a range of speeds at which errors are minimum, and this extends from around 50 rpm up to 100 or 200 rpm depending on the radius of the crank.

Stick-type Controls

As indicated earlier, there is evidence that the ratio of lever movement to display-indicator movement apparently is a fairly critical feature of stick control devices. In one study it was found that length of joy stick (12, 18, 24, or 30 in.) was relatively unimportant in both speed and accu-

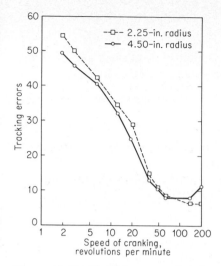

Figure 11-23 Tracking errors in relationship to speed of cranking for two crank radii. [Adapted from Helson, 22.]

racy of making settings, as long as the C/D ratio was around 2.5 or 3.0 [Jenkins and Karr, 28].

In a tracking task (as opposed to a task of making a particular setting), however, it was found that joy-stick length had at least a moderate effect upon tracking performance [Hartman, 21]. Within the range of stick lengths from 6 to 27 in., those around 18 in. were relatively optimum, although the advantage of that length over others was not great— about 10 percent. It was proposed that, where it would not interfere with other more critical design requirements, the length of such sticks be about 18 in. It was further proposed that especially long sticks (say, 27 in.) be avoided since they tend to interfere with comfortable positioning of the operator.

Control devices for cranes. An interesting series of applied studies dealing with human factors has been carried out by the Ergonomics Research Section of the British Iron and Steel Institute. One of these studies was concerned, in part, with the design of *master controllers* for cranes used in steel mills [Box and Sell, 7; Sell, 47]. In the pertinent part of the study, lever-type hand controls with the following features were used experimentally over 3 weeks by three subjects:

· Speed steps: the number of different speeds for which specific settings could be made by the lever, specifically 3, 4, and 6 speeds on either side of the off position

· Arc of movement of the handle: 20, 35, and 50° in each direction
· Handle length: 9, 12, and 15 in.

Granting that the investigation was highly situational, and used only three subjects, the following conclusions may still be of interest:

1. Performance was significantly better when using four speed steps than when using either three or six.
2. With either three or four speed steps, handle lengths of 9 to 12 in. were better than the larger one (15 in.); there was no difference with the 15-in. handle for six speed steps.
3. With six speed steps an arc of 20° movement in each direction was significantly worse than arcs of 35 or 50°.

Design of Pedals

There are many variations in the operational requirements of pedals considering such aspects as the degree of activation (whether the pedal is applied all the way or only to some degree), the frequency of operation (i.e., continuous, frequent, or infrequent), and the appropriate criterion (such as speed, force, precision). The design that would be best for one situation might not be best for another. We shall discuss briefly the results of two studies dealing with the use of pedals. In one study, five different pedal designs were tried out with 15 subjects, the criteria of performance being the number of strokes attempted and the number of cycles completed within a given time. The designs used and the results are given in Figure 11-24. We can see that pedals *a* and *b* were operated

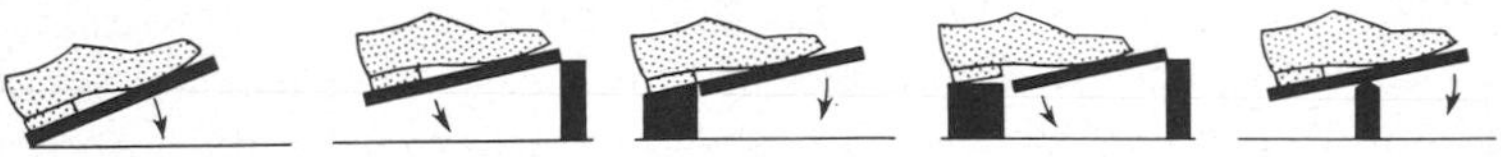

Strokes per minute	187	178	176	140	171
Percent cycles completed	81	83	70	74	74
	(*a*)	(*b*)	(*c*)	(*d*)	(*e*)

Figure 11-24 Strokes per minute and percentage of cycles completed in operation of five pedal designs (20 in.-lb pressure). [Adapted from Barnes, Hardaway, and Podalsky, 4.]

most rapidly, and had the highest percents of cycles completed. The biomechanical characteristics of the foot and leg presumably lend themselves more readily to the foot actions used with these pedals than with the others.

In turn Box and Sell [7] analyzed the use of pivotal pedals in a steel mill, in which control is maintained by toe operation (as when increasing a controlled variable) and by heel operation (as when reducing the control value) and came to the following conclusions:

1. The pivot point should be about 5 in. in front of the back of the pedal (a bit in front of the tibial axis, incidentally) and at a vertical position relative to the foot ranging from about 1 in. above the pedal to about 2 in. below.
2. An operating torque (measured at the front of the pedal) in the range of 30 to 60 lb-in. is preferable to one of 72 lb-in.
3. A range of angular separation of the pedal positions of 40 to 50° is slightly preferable to a more limited range of 30°.
4. Where settings of various discrete pedal positions are to be made, it seems evident that people can use two or three more adequately than they can four or more.

Data-input Devices

Various kinds of devices, especially keyboards, are used to enter digital information (usually alphabetical or numerical) into the many mechanical and electronic contraptions of our present world that have an insatiable appetite for the "information" that is part and parcel of this world. In most such devices, individual mechanisms are used, separately or in combination, to enter unique items of information. In some devices the operator applies the power for activation (such as with manual typewriters and pianos); in other cases the power is furnished by the equipment (such as with electric typewriters and certain calculating machines). The usual input in the operation of data-input devices consists of coded visual or auditory stimuli (alphanumeric, symbolic, etc.) in a sequence; in some instances the input is retrieved from memory in its original or reorganized form. Most such mechanisms are operated by the fingers (although organs have foot keyboards). The time probably will come, however, when the activation of data input devices will be by voice, or possibly by eye movements that are sensed by optical or other instruments. The typical requirement in the use of such devices is the execution of responses that correspond with input stimuli in the exact sequence in which the stimuli occur.

The type of machine, however, can influence the effectiveness with which given types of data can be entered. This was illustrated, for example, by a study in which numerical data were entered with four different types of input devices [Miner and Revesman, 38], with the following results:

Type of device	Average time to enter 10 digits, sec	Percent of error
10-key keyboard, as on hand calculators	12	0.6
10 by 10 matrix keyboard	13	1.2
10-lever device, one for each digit	17	2.3
10 rotary knobs, one for each digit	18	2.3

Recognizing the whole gamut of data input devices, we shall here simply touch on a few examples.

Sequential and chord keyboards.   There are two basic schemes for using keyboards in data input tasks. One is a sequential procedure, in which a separate device is associated with every possible stimulus such as a letter or digit, as with a typewriter. In the other a *chord* keyboard is used on which some stimuli require the simultaneous activation of two or more keys, such as with stenotype machines, pianos, and certain mail-sorting machines. Figure 11-25 shows the Burroughs chord key-

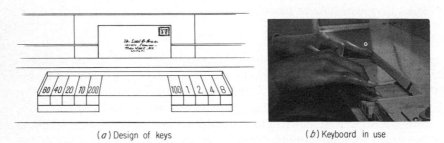

(*a*) Design of keys   (*b*) Keyboard in use

Figure 11-25   Burroughs chord keyboard used in letter sorting in some post offices. Individual numbers can be reproduced by simultaneous activation of certain combinations of keys, following the binary system. For example, 3 is activated by pressing 1 and 2, and 5 by pressing 1 and 4. [Courtesy of the Burroughs Corporation.]

board that is used in a few post offices for semiautomatic mail sorting. The theoretical ceilings of both types of keyboards can be expressed in terms of information bits per stroke. As pointed out by Seibel [46], for any given number of keys, there is an approximate ratio of about 2:1 in terms of such information, in favor of the chord keyboard. The actual limits, however, are influenced by the coding system and are constrained by human performance capabilities. For example, the information rates of stenotyping versus conventional typing are in the ratio of 2:1, and the stroke rates per second for good operators are in the ratio of 3:8.3 per second, which gives an advantage to stenotyping of 5.6:1 (in terms of information input).

An actual comparison of the two types of keyboards was made by Bowen and Guinness [6] in the context of a simulated semiautomatic mail-sorting task. In this study the *encoding* of mail by memory into various classes (each with a three-digit numerical code) was done by some subjects with a regular typewriter and by other subjects with a small (12-key) chord keyboard (requiring the use of 1 to 4 keys simultaneously) and a large (24-key) chord keyboard (requiring use of 1 to

3 keys). The digitation patterns used represented the fingering difficulty and frequency that would be used for a 500-separation mail-sorting scheme. Some of the results are summarized below.

Type of keyboard	No. sorted per minute	
	Correct	Incorrect
Sequential, typewriter	40.4	4.2
Chord, small	55.3	7.8
Chord, large	49.0	3.3

Both chord-keyboard tests resulted in higher numbers of items sorted correctly than did the typewriter tests. The superiority of the large over the small chord keyboard in reduced error probably can be attributed to the fact that it did not require such difficult finger patterns. Some generally confirming evidence comes from a study by Conrad and Longman [14], who report that for subjects working over a period of weeks, those using a chord keyboard had higher keystroke per minute rates than those using a typewriter. Bowen and Guinness [6] suggest that the apparent advantage of chord over sequential encoding in the context of their study may be attributable to the fact that when using a learned memory code, one pairs it immediately with a *unit* response—not a response *spread out over time* as called for in sequential keying.

Keyboards in specific contexts. The above discussion certainly does not mean that we should spread the gospel of using chord keyboards everywhere. Many data-entry processes usually would require keying mechanisms and coding systems that would be specially designed for the process in question. (A piano obviously requires a different type of keyboard than a computer.) We must, therefore, agree with Hillix and Coburn [24] when they warn us that the fact that a keyboard is useful in one context should not be accepted as valid evidence that it would also be good in another context. Some reflection will bring to mind a number of different features of keyboard design that might have a bearing on the effectiveness with which keyboards might be used by people in different situations—such features as the coding system, the keyboard configuration, the keying process, and the nature of the feedback. We shall illustrate research on keysets using telephone push buttons as an example.

Telephone push buttons. Previous reference was made to a study by Deininger [19] dealing with the arrangement of telephone keysets, specifically in connection with compatibility. Another phase of that study dealt with the size and spacing of keys. The specific sizes and spacings are shown in Figure 11-26, along with both performance measures (keying time and errors) and subjective preference of 15 people used as subjects. While the variation in keying time for the different designs was not great, there was a wide range in keying errors (from 1.3 to 7.1 percent). In both performance and preferences, the middle-size push buttons (½ in., sets 2, 3, 4) were better than either the smaller (⅜ in., set 1) or larger (¾ by ½ in., set 5). Set 4, with the letters on the plate rather than on

Button arrangement	Three rows of three; one in center of fourth row		Two rows of five each		
Inches between centers	¾ in.	¾ in.	¾ in.	¾ in.	27/32 in.
Button and lettering size	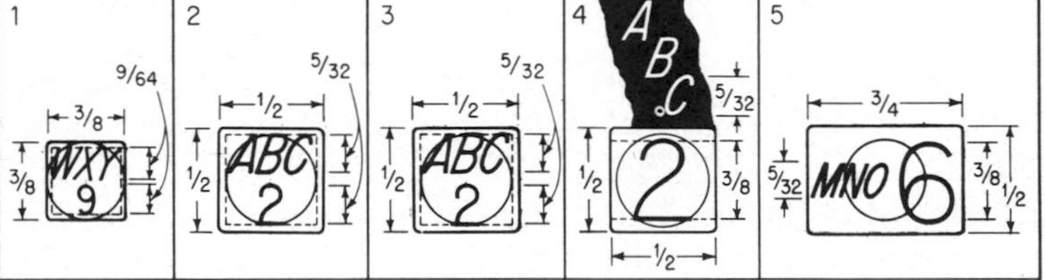				
Keying time, sec	6.35	5.83	5.75	5.77	6.07
Error, %	7.1	1.3	2.0	3.3	5.3
Votes for	2	1	4	6	2
Votes against	9	1	1	4	0

Figure 11-26 Variation in telephone push-button size, spacing, arrangement, and lettering used experimentally, with average keying times, percentage of errors, and votes for and against. [Adapted from Deininger, 19, by permission of copyright owner, American Telephone and Telegraph Company.]

the key itself, was preferred by more subjects than any other set, but some other subjects disliked this idea. In related studies, various arrangements of keys were investigated, with the result that the push-button telephone that was ultimately produced had three rows of three numerals, with a zero push button at the bottom, as shown in Figure 11-27.

Remote Controls

Present-day technology has generated the requirement that remote-control manipulators be used for vehicular control and for handling materials, tools, etc., in certain circumstances, especially in environments with hazards which preclude proximity of the individual to the materials or tools to be handled; handling radioactive materials and performing certain operations in space are examples of such situations. An illustration of

Figure 11-27 Illustration of push-button telephone design adopted for use. Most of the features of this design were based on the research by Deininger [19]. [Reproduced through the courtesy of the American Telephone and Telegraph Company.]

one remote-handling control is shown in Figure 11-28. Also, Figure 11-29 shows an experimental remote-handling setup used in a series of studies at the United States Air Force Behavioral Sciences Laboratory.

Remote-control feedback.   As one example of the human factors aspects of such controls, let us consider briefly the problem of feedback. In the use of such devices the individual is bereft of his normal direct physical contact with (and, therefore, such feedback from) the object(s)

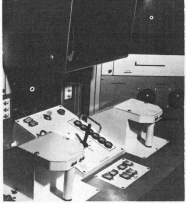

Figure 11-28 An example of a remote control system, the Mobile Remote Manipulating Unit (MRMU), showing the vehicle to be controlled and the remote-control console. [Courtesy U.S. Atomic Energy Commission, Idaho Falls.]

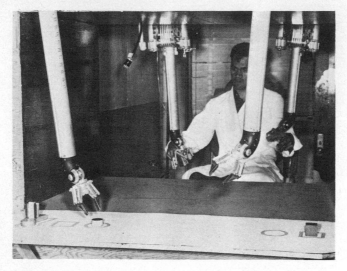

Figure 11-29 Experimental remote-handling equipment used in various studies. [Courtesy USAF, AMRL, Behavioral Sciences Laboratory.]

being handled or controlled, and he may also lose direct visual feedback. However, as reported by Kama [32], one does not really need direct viewing of objects being manipulated in order to control a remote-handling unit (specifically a movable cart); in fact, he found that TV images with augmented depth cues (by shadows or converging lights) assisted the operator in the control of the unit. Somewhat in line with this, Knowles [34] expresses the opinion that the normality or naturalness of the situations, movements, and feedback information is in all probability irrelevant. (After all, we mortals frequently perform activities in ways that are not natural, such as unplugging the kitchen drain with a plumber's snake.) Knowles points out that what *is* needed for remote-handling devices is a situation in which *usable inputs* are combined with *usable control devices*—or what has been referred to as the "verified body image"; in simpler terms this means that the operator needs to have some meaningful *identification* or *integration* with his equipment.

Perhaps it is because of this need for integration with the equipment that operators of remote-handling controls tend to do better with those that operate on a pantograph principle (in which the operator's response is translated directly into corresponding three-dimensional movements) as contrasted with some *rectilinear* models (in which electric-powered manipulators must be controlled independently in each of the three dimensions). Probably another manifestation of this principle comes from the study by Crawford [15] in which he found that in the remote handling of disks to be placed in holes in a form board (by using a variation of the device shown in Figure 11-29), subjects did better with a joy stick than with multiple-lever controls.   These results are shown in Figure 11-30. Manipulations of the single joy stick actually controlled six different control movements, simulating shoulder pivot, shoulder rotation, elbow pivot, wrist pivot, wrist rotation, and grip. This single joy stick provides a higher degree of integration of the control with the operator than the independent use of six separate multiple levers. (This probably is, in a sense, another manifestation of the principle of compatibility.)

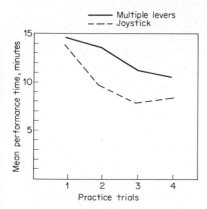

Figure 11-30 Comparison of re-mote-handling performance as a function of practice with a single joy stick (which governed six control functions) versus six multiple levers. Although practice resulted in improvement with both types of controls, performance with the joy stick was consistently superior. [From Crawford, 15.]

As another indication of the need for integration in remote handling, when remote viewing is required (with the use of TV cameras), it has generally been found that a major alteration of the relations between the visual image and the control output will render an otherwise satisfactory control system ineffective [Knowles, 34].[2]

And as still another aspect of the same basic theme, it has been found that position control (zero order) generally results in better per-

[2] The reader interested in further discussion of remote handling is referred to Baker [3], Crawford and Kama [16], Johnson and Corliss [30], Kama [31 and 32], and Knowles [34].

formance than rate control (first order) [Crawford and Kama, 16; Kama, 32]. Presumably position control contributes to the integration of the input-output relationships as perceived by the operator.

Mechanical Men

Many of the fantasies of the comic strips and of the Jules Vernes' of the past are now everyday realities. Thus, the notion of a giant mechanical man stalking the face of the moon would hardly cause an eyebrow to be raised today, as it would have many years ago. It may be consoling to the human ego, however, to realize that (at least to date) the control of such devices can best be done by human beings. The current concept of such devices is predicated on the symbiosis of man and machine, of an integration of the two into a mutually complementing system to achieve certain objectives. In this integration, certain as-yet-not-to-be-outdone human sensory skills and adaptive decision-making abilities would serve in the control of devices that mimic the human motor responses but that apply forces beyond the levels of human limits. Such a device would, in effect, serve as an extension of the man.

HardiMan. Figure 11-31 illustrates this concept. This particular device, called *HardiMan*, is what is known as an *exoskeleton* device, since its various members are attached to the corresponding body members and reproduce the same movements as the body members of a human operator. (Other species of mechanical men do not have this same human form, but are equally controlled by human responses.) In its present (and still somewhat experimental) form, HardiMan can perform such load-handling tasks as walking, lifting, climbing, pushing, and pulling with a lift capacity of 1500 lb that, to a man, would feel like 60 lb [Mosher, 39]. The intricacies of his human factors features cannot be described here, but it might be pointed out that, aside from the obvious aspects of anthropometric fit, important features are those of spatial correspondence and near-natural kinesthetic-force feedback as aids in the human control of HardiMan. Thus, the complete ensemble is much like the chambered nautilus that maneuvers himself around within his own shell on the floor of the sea.

TOOLS AND OTHER DEVICES

There are probably a couple of interrelated considerations (criteria, if you will) that are (or should be) dominant in the design or selection of the many hand tools and other hand devices that people use (ranging from eyebrow tweezers to carving knives, to billiard sticks, to hacksaws). Certainly they need to be capable of performing their function (such as pulling eyebrows), but in addition they need to be usable by people, and

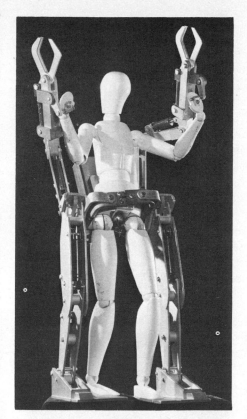

Figure 11-31 A model of HardiMan, a walking machine that mimics the control responses of a human operator and that is capable of amplifying those responses by a ratio of 25:1 up to a force of 1500 lb. [Courtesy of the General Electric Company Research and Development Center, Schenectady, N.Y.]

this means that they need to be compatible with the anthropometric and biomechanical performance characteristics of people. (And in some cases aesthetic considerations may also become relevant, for better or worse; if the purpose of a knife is to cut the steak, there is nothing like a plain butcher knife, but if the purpose is to impress one's guests, a solid-silver place setting probably is indicated.) To illustrate the relevance of anthropometric and biomechanical considerations (if not the aesthetic aspects), let us cite two or three different types of examples.

Pliers

The Western Electric Company has been a forerunner in the application of biomechanic principles to the design of tools and work places. As an example, in a certain wiring operation it was found that the incidence of tenosynovitis (inflammation of the tendon sheaths of the fingers) among the operators was extremely high [Damon, 17]. This operation called for the use of a conventional set of pliers in the wiring process which required the operator to work with her wrist bent in such a manner as to place excessive pressure of the ulnar side of the wrist against the tip of the ulna bone, as shown in Figure 11-32a. In addition, the pressure on the nerves of the palm sometimes caused pain in the wrist and even in the elbow and shoulder. An analysis of the operation indicated that the design of the pliers was at fault, and a redesigned set of pliers was ultimately produced that (1) eliminated the need to bend the wrist, (2) allowed the operator to hold the pliers without producing pressure on the palm, and (3) did not materially modify the existing motion pattern. The ultimate design is shown in Figure 11-32b.

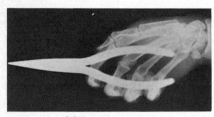 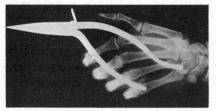

(a) Conventional design   (b) Redesigned pliers

Figure 11-32 X-rays of hand using conventional pliers in a wiring operation, a, and in using a redesigned model, b. The redesigned model eliminated the need to bend the wrist (which eliminated the pressure on the tip of the ulna bone) and reduced the pressure on the palm. [From Damon 17; photographs courtesy of Western Electric Company, Kansas City.]

Devices for Handling Silicon Wafers

As another example of the human factors aspects of hand tools, let us summarize briefly the results of a comparison of the use of four types of small tools in the handling of silicon wafers, which are small, very fragile parts used in the production of microelectronic devices [Springer and Harris, 48]. Four designs were used in one study, namely, (a) thin-blade tweezers, (b) wide-bill tweezers, (c) wide-bill tweezers with a *stop* (a small metal stop on the inside of the tweezers), and (*d*) a vacuum spatula (to take hold of the wafer by vacuum). The handling errors in the use of these were about 15, 16, 5.5, and 6.5, respectively, and indicated significant superiority in the use of the wide-bill tweezers with a stop and the vacuum spatula. These four designs are shown in Figure 11-33, along with a fifth design (an improved variation of *c*) that actually was adopted for use.

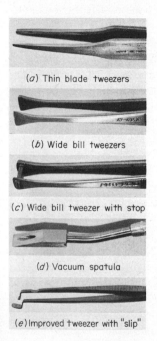

(*a*) Thin blade tweezers

(*b*) Wide bill tweezers

(*c*) Wide bill tweezer with stop

(*d*) Vacuum spatula

(*e*) Improved tweezer with "slip"

Figure 11-33 Illustration of five types of tools used in the handling of fragile silicon wafers. Models *c* and *d* were markedly superior to *a* and *b* in handling errors during an experiment. Model *e* was later developed and adopted for use. [From Springer and Harris, 48; reproduced by permission of Autonetics Division, North American Rockwell Corporation.]

SUMMARY

1. The principal types of functions of controls are (a) activation, (b) discrete setting, (c) quantitative setting, (d) continuous control, and (e) data entry. Each type, in effect, transmits information to the mechanism being controlled, the information generally corresponding to the control functions.
2. Certain types of controls are preferable for each of the various control functions.
3. Controls can be coded in many ways, such as by shape, texture, size, location, operational method, color, and labels.
4. With certain devices for continuous control and quantitative setting, a corresponding display is used. The control-display (C/D) ratio is the ratio of the movement of the control to the movement of the display indicator. This ratio is *low* with a small control movement and large display movement, and is *high* with the reverse relationship.
5. Where a quantitative setting is to be made, the most desirable C/D ratio usually is a value that optimizes the travel time (gross adjustment) and adjust time (fine adjustment) of the control operation.
6. Resistance in control devices can be of different types: elastic resistance (as in spring-loaded controls), viscous damping (a force opposite to that of input and proportional to output speed), static and coulomb friction, and inertia. While the effects of different types of resistance on human performance are complex, it is generally true that static and coulomb friction are undesirable (except possibly to prevent accidental activation). Within reasonable bounds, elastic resistance, viscous damping, and inertia seem to facilitate certain aspects of control operation.
7. There is no clear-cut evidence regarding the relative advantages of feedback cues received in the operation of controls from force (tactual sense) and from amplitude of movement (kinesthetic sense). For short movements, however, force (pressure) cues seem to be particularly useful, and in general some "feel" of control devices seems to be desirable.
8. The effects of lag in a system depend in part upon the type of system; lag causes more degradation in compensatory tracking, for example, than in pursuit tracking. Within reasonable bounds, people can learn to anticipate lag and to correct for it.
9. Backlash and deadspace in control systems generally are undesirable for human performance. Such effects can be minimized, however, with high C/D ratios.

10. The type, design, and location of control devices are related to the effectiveness with which they are used by human beings in terms of the accuracy, speed, and force that can be applied. A number of studies relating to this are summarized.

11. The use of remote-control handling devices imposes special problems for operators, especially because usual feedback is missing. However, normal feedback presumably is not essential for satisfactory control operation, if *usable* inputs and control devices are available.

REFERENCES

1. Bahrick, H. P., W. F. Bennett, and P. M. Fitts: Accuracy of positioning responses as a function of spring loading in a control, *Journal of Experimental Psychology*, 1955, vol. 49, pp. 437–444.

2. Bahrick, H. P., P. M. Fitts, and R. Schneider: Reproduction of simple movements as a function of factors influencing proprioceptive feed back, *Journal of Experimental Psychology*, 1955, vol. 49, pp. 445–454.

3. Baker, D. F.: *Survey of remote handling in space*, USAF, AMRL, TDR 62–100, September, 1962.

4. Barnes, R. M., H. Hardaway, and O. Podalsky: Which pedal is best? *Factory Management and Maintenance*, January, 1942, vol. 100, pp. 98–99.

5. Birmingham, H. P., and F. V. Taylor: *A human engineering approach to the design of man-operated control systems*, USN, NRL, Report 4333, Apr. 7, 1954.

6. Bowen, H. M., and G. V. Guinness: Preliminary experiments on keyboard design for semiautomatic mail sorting, *Journal of Applied Psychology*, 1965, vol. 49, no. 3, pp. 194–198.

7. Box, A., and R. G. Sell: Ergonomic investigations into the design of master controllers, *Journal of the Iron and Steel Institute*, October, 1958, vol. 90, pp. 178–187.

8. Bradley, J. V.: Tactual coding of cylindrical knobs, *Human Factors*, 1967, vol. 9, no. 5, pp. 483–496.

9. Bradley, J. V., and N. E. Stump: *Minimum allowable dimensions for controls mounted on concentric shafts*, USAF, WADC, TR 55–355, December, 1955.

10. Briggs, G. E., P. M. Fitts, and H. P. Bahrick: Effects of force and amplitude cues on learning and performance in a complex tracking task, *Journal of Experimental Psychology*, 1957, vol. 54, pp. 262–268.

11. Burke, D., and C. B. Gibbs: A comparison of free-moving and pressure levers in a positional control task, *Ergonomics*, 1965, vol. 8, no. 1, pp. 23–29.

12. Burrows, A. A.: Control feel and the dependent variable, *Human Factors*, 1965, vol. 7, no. 5, pp. 413–421.

13. Conklin, J.: Effect of control lag on performance in a tracking task, *Journal of Experimental Psychology*, 1957, vol. 53, pp. 261–268.

14. Conrad, R., and D. J. A. Longman: Standard typewriter versus chord keyboard—an experimental comparison, *Ergonomics*, 1965, vol. 8, no. 1, pp. 77–88.

15. Crawford, B. M.: Joy stick vs. multiple levers for remote manipulator control, *Human Factors*, 1964, vol. 6, no. 1, pp. 39–48.
16. Crawford, B. M., and W. N. Kama: Remote handling research and potential space applications, in *National Conference on Space Maintenance and Extravehicular Activities*, sponsored by USAF Aero Propulsion Laboratory, Mar. 1–3, 1966, AF APL, Conf 66–8.
17. Damon, F. A.: The use of biomechanics in manufacturing operations, *The Western Electric Engineer*, October, 1965, vol. 9, no. 4.
18. Davis, L. E.: Human factors in design of manual machine controls, *Mechanical Engineering*, October, 1949, vol. 71, pp. 811–816.
19. Deininger, R. L.: Human factors studies of push-button characteristics and information processing in keyset operation, *Bell System Technical Journal*, July, 1960, vol. 39, no. 4, pp. 995–1012.
20. Fenwick, C. A.: *Effects of simulated display lag on performance in pursuit tracking with advance course information*, Ph.D. thesis, Purdue University, Lafayette, Ind., January, 1962.
21. Hartman, B. O.: *The effect of joystick length on pursuit tracking*, U.S. Army Medical Research Laboratory, Fort Knox, Ky., Report 279, November, 1956.
22. Helson, H.: Design of equipment and optimal human operation, *American Journal of Psychology*, 1949, vol. 62, pp. 473–479.
23. Hunt, D. P.: *The coding of aircraft controls*, USAF, WADC, TR 53–221, August, 1953.
24. Hillix, W. A., and R. Coburn: *Human factors in keyset design*, NEL, Report 1023, Mar. 22, 1961.
25. Jenkins, W. O.: "The tactual discrimination of shapes for coding aircraft-type controls," in P. M. Fitts (ed.), *Psychological research on equipment design*, Army Air Force, Aviation Psychology Program, Research Report 19, 1947.
26. Jenkins, W. O.: The discrimination and reproduction of motor adjustments with various types of aircraft controls, *American Journal of Psychology*, 1947, vol. 60, pp. 397–406.
27. Jenkins, W. L., and M. B. Connor: Some design factors in making settings on a linear scale, *Journal of Applied Psychology*, 1949, vol. 33, pp. 395–409.
28. Jenkins, W. L., and A. C. Karr: The use of a joy-stick in making settings on a simulated scope face, *Journal of Applied Psychology*, 1954, vol. 38, pp. 457–461.
29. Jenkins, W. L., L. O. Maas, and D. Rigler: Influence of friction in making settings on a linear scale, *Journal of Applied Psychology*, 1950, vol. 34, pp. 434–439.
30. Johnson, E. G., and W. R. Corliss: *Teleoperators and human augmentation*, NASA SP–5047, December, 1967.
31. Kama, W. N.: *Human factors in remote handling: a review of past and current research at the Aerospace Medical Research Laboratories*, USAF, AMRL, TR 64–122, July, 1964.
32. Kama, W. N.: *Effect of augmented television depth cues on the terminal phase of remote driving*, USAF, AMRL, TR 65–6, April, 1965.

33. Katchmer, L. T.: *Physical force problems: 1. Hand crank performance for various crank radii and torque load combinations*, USA, Human Engineering Laboratory, Aberdeen Proving Ground, Technical Memorandum 3–57, March, 1957.

34. Knowles, W. B.: *Human engineering in remote handling*, USAF, MRL, TDR 62–58, August, 1962.

35. Kolesnik, P. E.: *A comparison of operability and readability of four types of rotary selector switches*, Autonetics Division, North American Aviation, Inc., T5–1187/3111, June, 1965.

36. Levine, M.: *Tracking performance as a function of exponential delay between control and display*, USAF, WADC, TR 53–236, October, 1953.

37. McFarland, R. A.: *Human factors in air transport design*, McGraw-Hill Book Company, New York, 1946.

38. Minor, F. J., and S. L. Revesman: Evaluation of input devices for a data setting task, *Journal of Applied Psychology*, 1962, vol. 46, pp. 332–336.

39. Mosher, R. S.: *Handyman to Hardiman*, paper presented at Automotive Engineering Congress, Detroit, June, 1967, SAE paper 670088.

40. Norris, E. B., and S. D. S. Spragg: Performance on a following tracking task as a function of the planes of operation of the controls, *Journal of Psychology*, 1953, vol. 35, pp. 107–117.

41. Norris, E. B., and S. D. S. Spragg: Performance on a following tracking task as a function of the relations between direction of rotation of controls and direction of movement of display, *Journal of Psychology*, 1953, vol. 35, pp. 119–129.

42. Poulton, E. C.: Perceptual anticipation in tracking with two-pointer and one-pointer displays, *British Journal of Psychology*, 1952, vol. 43, pp. 222–229.

43. Rockway, M. R.: *The effect of variations in control-display ratio and exponential time delay on tracking performance*, USAF, WADC, TR 54–618, December, 1954.

44. Rockway, M. R.: *Effects of variations in control deadspace and gain on tracking performance*, USAF, WADC, TR 57–326, September, 1957.

45. Rockway, M. R., and P. E. Franks: *Effects of variations in control backlash and gain on tracking performance*, USAF, WADC, TR 58–553, January, 1959.

46. Seibel, R.: Data entry through chord, parallel entry devices, *Human Factors*, 1964, vol. 6, no. 2, pp. 189–192.

47. Sell, R. G.: Letter to the editor: Ergonomic investigations into the design of master controllers, *Journal of the Iron and Steel-Institute*, February, 1960.

48. Springer, R. M., and D. H. Harris: Human factors in the production of microelectronic devices, *Technical Digest, 28C72, 8th Annual IEEE Symposium, Human Factors in Electronics*, Palo Alto, Calif., May 3–5, 1967.

49. Swartz, P., E. B. Norris, and S. D. S. Spragg: Performance on a following tracking task as a function of radius of control cranks, *Journal of Psychology*, 1954, vol. 37, pp. 163–171.

50. Tipton, C. L., and H. P. Birmingham: The influence of control gain in a first-order man-machine control system, *Human Factors*, 1959, vol. 1, no. 3, pp. 69–71.

51.   Wallach, H. C.: *Performance of a pursuit tracking task with different time delays inserted between the control mechanism and the display cursor,* USA Ordnance Human Engineering Laboratories, Technical Memorandum 12–61, OMS Code 50 10.11.841 A, August, 1961.

52.   Washburn, S. L.: Tools and human evolution, *Scientific American,* September, 1960, vol. 203, no. 3.

53.   Weiss, B.: *Building "feel" into controls: the effect on motor performance of different kinds and amounts of feedback,* USN/ONR, SDC, TR 241–6–11, Oct. 14, 1953.

54.   Weitz, J.: "The coding of airplane control knobs," in P. M. Fitts (ed.), *Psychological research on equipment design,* Army Air Force, Aviation Psychology Program, Research Report 19, 1947.

55.   Ziegler, P. N., and R. Chernikoff: A comparison of three types of manual controls on a third-order tracking task, *Ergonomics,* 1968, vol. 11, no. 4, pp. 369–374.

56.   *Personnel subsystems,* USAF, AFSC design handbook, Series 1–0, General, AFSC DH 1–3, 1st ed., Jan. 1, 1969, Headquarters, AFSC.

part
five
WORK
SPACE
AND
ARRANGEMENT

The comfort, physical welfare, and performance of people frequently are influenced, for better or worse, by the extent to which the physical facilities they use are designed to fit people in terms of relevant features of the human body, such as body dimensions and the range of normal body movements. Such anthropometric and biomechanical features of the body can have a bearing on such physical facilities as the work space, the arrangement of the work space, seating, and specific items of equipment including personal equipment.

ANTHROPOMETRY

The design of the physical facilities that fit people is reflected even in children's stories such as the one about the poppa, mamma, and baby bears (and the beds and chairs that fit each one) and the one about Cinderella and her slipper (that fit only her foot). Anthropometry deals with the measurement of physical features of the body, including linear dimensions, weight, and volume. As indicated in Chapter 10, biomechanics generally deals with the range, strength, speed, and other aspects of physical movements. In practice, however, these variables are also dealt with by the anthropometrists. Although we cannot, of course, reproduce here the anthropometric data that people have accumulated over the years, we shall at least illustrate some such data.[1] Before illustrating some aspects of anthropometry as they might have a bearing on the design and arrangement of physical space and equipment, we should distinguish between two facets of this field, namely, static and dynamic.

Static Anthropometry

Traditionally, anthropometry has been concerned with the measurement of physical features or characteristics of the static body. For various purposes, many different features may be measured. In a survey by Hertzberg, Daniels, and Churchill [13], for example, 132 different features

[1] The interested reader is referred particularly to the excellent compilation of such data by Damon, Stoudt, and McFarland [4], Hansen and Cornog [10], Hertzberg, Daniels, and Churchill [13], and Public Health Service Publication No. 1000, series 11, no. 8 [26].

were measured of the 4000 Air Force flying personnel in the sample. Measurements of different body features could have some specific application, whether in designing chest protectors for baseball umpires, earphones, or pince-nez glasses. However, measurements of certain body features probably have rather general utility, and summary data on some of these features will be presented for illustrative purposes. These data come from a survey by the United States Public Health Service [26] of a representative sample of 6672 adult males and females. The specific body features measured are shown in Figure 12-1; for each of these (plus weight) data on the 5th, 50th, and 95th percentile are given in Table 12-1. It should be pointed out that these values cover ages from 18 to 79 and that most of the measurements given did vary somewhat by age, particularly weight and height, as illustrated in Figure 12-2. Further, corre-

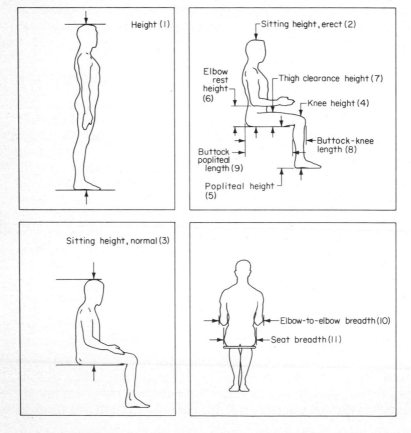

Figure 12-1 Diagrams of body features measured in National Health Survey of anthropometric measurements of 6672 adults [26]. See Table 12-1 for selected data based on the survey.

Table 12-1   Selected Anthropometric Features of Adults*

	Male, percentile			*Female, percentile*		
Body feature	*5th*	*50th*	*95th*	*5th*	*50th*	*95th*
1. Height	63.6	68.3	72.8	59.0	62.9	67.1
2. Sitting height, erect	33.2	35.7	38.0	30.9	33.4	35.7
3. Sitting height, normal	31.6	34.1	36.6	29.6	32.3	34.7
4. Knee height	19.3	21.4	23.4	17.9	19.6	21.5
5. Popliteal height	15.5	17.3	19.3	14.0	15.7	17.5
6. Elbow-rest height	7.4	9.5	11.6	7.1	9.2	11.0
7. Thigh-clearance height	4.3	5.7	6.9	4.1	5.4	6.9
8. Buttock-knee length	21.3	23.3	25.2	20.4	22.4	24.6
9. Buttock-popliteal length	17.3	19.5	21.6	17.0	18.9	21.0
10. Elbow-to-elbow breadth	13.7	16.5	19.9	12.3	15.1	19.3
11. Seat breadth	12.2	14.0	15.9	12.3	14.3	17.1
12. Weight, lb	126	166	217	104	137	199

*Note: measurements are in inches, excluding item 12.
SOURCE: From *Weight, height, and selected body dimensions of adults: 1960–1962.* Data from National Health Survey, USPHS Publication 1000, series 11, no. 8, June, 1965.

sponding data from surveys of other samples can vary from data from this survey. And as still another word of caution, measurements of personnel wearing special gear, such as arctic clothing or heavy work clothing, can add inches to their space requirements [for example, see Kobrick, 15].

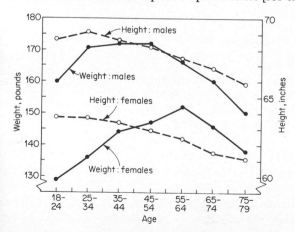

Figure 12-2   Average weight (left scale) and height (right scale) of 6672 adults, showing changes by age. Other physical characteristics also are related somewhat to age. [Based on data from National Health Survey, 26, tables 1 and 2.]

Dynamic Anthropometry

Data on static measurements can effectively serve certain design purposes that relate to circumstances in which people do in fact tend to stay put, such as in some chairs and seats, or in which mobility is relatively minimal. In most circumstances in life, however, people are not inert (not even when sleeping). Rather, in most work and nonwork situations people are functional—whether operating a steering wheel, assembling a mouse trap, or reaching across the table for the salt. The focus of dynamic anthropometry is on operational measurements of the human body while its inhabitant is performing some function. Figure 12-3 illustrates the *dynamic* aspects of an operation, that of driving a vehicle, as contrasted with the less realistic *static* aspects of vehicular cab design (which tend to emphasize the physical *clearance* between the driver and the features of his physical environment).

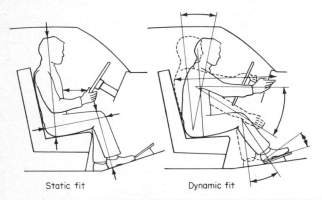

Static fit   Dynamic fit

Figure 12-3   Illustration of a static versus dynamic fit in the context of vehicular cab design. A static approach tends to focus on clearances of body members with the surroundings, whereas a dynamic approach tends to focus on the functions or operations involved. [Adapted from Damon et al., 4.]

Probably a central postulate of dynamic anthropometry relates to the fact that in performing physical functions, the individual body members normally do not operate independently, but rather in concert. The practical limit of arm reach, for example, is not the sole consequence of arm length; it is also affected in part by shoulder movement, partial trunk rotation, possible bending of the back, and the function that is to be performed by the hand. These and other variables make it difficult, or at least very risky, to try to resolve all space and dimension problems on the basis of static anthropometric data. An impression of the manner

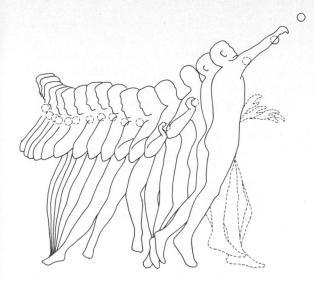

Figure 12-4 Pattern of body movements in the shot put, showing rhythm, sequence of posture, and a pattern of force application directed toward getting the maximum throw and maintaining body control to stay within the throwing ring. [From Dempster, 6.]

in which the body members interact in performing some function is shown in Figure 12-4, this particular example showing the pattern of body movements involved in the shot put.

THE USE OF ANTHROPOMETRIC DATA

As indicated above, anthropometric data can have a wide range of applications in the design of physical equipment and facilities. In the use of such data, however, the same caution mentioned in Chapter 10 regarding biomechanical data is repeated here, namely, that the designer should use data from samples of people who are reasonably similar to the ones who will actually use the facilities in question. Although we shall not deal with many of the considerations involved in the design of facilities in terms of specific body features, we shall discuss certain general considerations.[2]

Principles in the Application of Anthropometric Data

In the application of anthropometric data there are certain principles that may be relevant, each one being appropriate to certain types of design problems.

[2] For a discussion of design considerations that relate to *specific* body features, the reader is referred to Damon et al. [4].

Design for extreme individuals. In the design of certain aspects of physical facilities there is some "limiting" factor that argues for a design that specifically would accommodate individuals at one extreme or the other of some anthropometric characteristic, on the grounds that such a design *also* would accommodate virtually the entire population. A *minimum* dimension, or other aspect, of a facility would usually be based on an *upper* percentile value of the relevant anthropometric feature of the sample used, such as the 90th, 95th, or 99th. Perhaps most typically a minimum dimension would be used to establish clearances, such as for doors, escape hatches, and passageways. If the physical facility in question accommodates large individuals (say, the 95th percentile), it also would accommodate all those smaller in size. The minimum weight carried by supporting devices (a trapeze, rope ladder, or other support) is another example. On the other hand, *maximum* dimensions of some facility would be predicated on *lower* percentiles (say, the 1st, 5th, or 10th) of the distribution of people on the relevant anthropometric feature. The distance of control devices from an operator is an example; if those with short functional arm reach can reach a control, persons with longer arm reach generally could also do so. In setting such maximums and minimums it is frequently the practice to use the 95th and 5th percentile values, if the accommodation of 100 percent would incur trade-off costs out of proportion to the additional benefits to be derived. To take an absurd case, we do not build 8½-ft doorways for the rare 8 footers, or dining-room chairs for the rare 400-lb guest. There are circumstances, however, in which designs that accommodate all people can be achieved without appreciable trade-off costs.

Design for adjustable range. Certain features of equipment or facilities preferably should be adjustable in order to accommodate people of varying sizes. The forward-backward adjustments of automobile seats and the vertical adjustments of typists chairs are examples. In the design of adjustable items such as these, it is fairly common practice to do so for the range of cases from the 5th to the 95th percentiles. The example given in Figure 12-5 illustrates seat adjustability requirements (i.e., the range of adjustment that should be provided for in a seat) to accommodate different segments of the population according to their sitting height. This illustrates the point that the amount of seat adjustment required to accommodate the extreme cases (such as below the 5th percentile and above the 95th) is disproportionate to the additional numbers of individuals who would be accommodated.

In this connection, as in other contexts, trade-off considerations may be in order. The military practice of rejecting persons who are extremely short or tall is dictated, in part, by the fact that the requirement for smaller or larger items of clothing, shoes, etc., would impose an additional

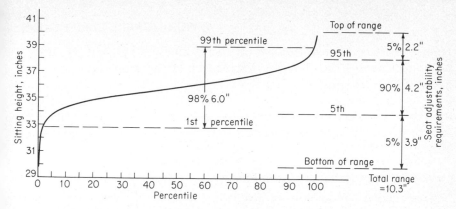

Figure 12-5 Illustration of the relationship between percentiles of cases and anthropometric measurements. This figure shows, specifically, the cumulative percentiles (along the base line) of people whose sitting height is at or below the values on the left vertical scale. The right vertical scale shows the corresponding seat adjustability requirements for various percentile groups. The differences in seat adjustability requirements are disproportionate at the extremes to the additional percentiles that would be accommodated. [From Hertzberg, 12.]

administrative load in supplying such items; this is trading off the possible utility of such men in the service for a certain degree of administrative simplicity.

Design for the average. While we frequently hear of the "average" or "typical" man, this is, in one sense, an illusive, will-of-the wisp concept. In the domain of human anthropometry there probably are few, if any, people who would really qualify as average—average in each and every respect. In connection with this, Hertzberg [11] indicates that in a survey of over 4000 Air Force personnel there were *no* men who fell within the (approximately) 30 percent central (average) range on all 10 of a series of measurements. Since the concept of the average man is then something of a myth, there is some rationale for the common proposition that physical equipment should not be designed for this mythical individual. Recognizing this, however, we would like to make a case here for the use of "average" values in the design of *certain* types of equipment or facilities, specifically those in which, for legitimate reasons, it is not appropriate to pitch the design at an extreme value (minimum or maximum) or feasible to provide for an adjustable range. As an example, the checkout counter of a supermarket built for the average customer probably would discommode customers less in general than one built either for the circus midget or for Goliath. This is not to say that it would be optimum for all people, but that, collectively, it would cause less inconvenience and difficulty than one which might be lower or higher.

WORK–SPACE DIMENSIONS

Human *work space* can consist of many different physical situations, including that of the plumber working under a stopped-up sink, the astronaut in his capsule, the assembler at his position on the assembly line, the flagpole painter, and the minister in his pulpit. Since we cannot here work out the space problems of the plumbers or flagpole sitters, we shall consider certain of the more conventional work locations.

Work Space for Stationary Work Places

There probably are millions of people whose physical work activities are carried out while seated in a fixed location. The space within which such an individual works is sometimes referred to as the *work-space envelope*. This envelope preferably should be circumscribed by the functional arm reach of the individuals affected, and most of the things they need to handle should be arranged within this envelope. The fact of individual differences, as well as differences in the activities to be performed, argue for approaching such design problems on a strictly situational basis, but a couple of anthropometric studies will be brought in to illustrate the concept of the work-space envelope as it would be predicated on data from such studies.

Functional arm reach in grasping. One example comes from a study carried out with a fairly representative sample of 20 United States Air Force personnel [Kennedy, 14]. In a seated position, each subject was presented with a vertical rack of measuring staves, each pointing toward the approximate joint center of the right shoulder, each with a knob at the end; the subject grasped each rod between the thumb and forefinger and moved it out until the arm was fully extended without pulling the shoulder away from the seat back. This was done with the rack of staves in various positions around him, actually at positions 15° apart around an imaginary vertical reference line beginning at a seat reference point behind the subject. As each subject positioned each stave, its distance from the vertical reference line was recorded, and these distances served as the basic data. Figure 12-6*a* illustrates the general seat arrangement, and *b* presents the resulting curves for the 5th and 95th percentiles of the subjects for each of certain horizontal "slices" of the three-dimensional space. (The constraints of two-dimensional pages make three-dimensional representations a bit unrealistic.) The 5th percentile values for certain angular positions of these and other horizontal slices are given in Table 12-2. Data such as these naturally cannot be extrapolated to different types of people, to other seating arrangements, or to markedly different manual tasks.

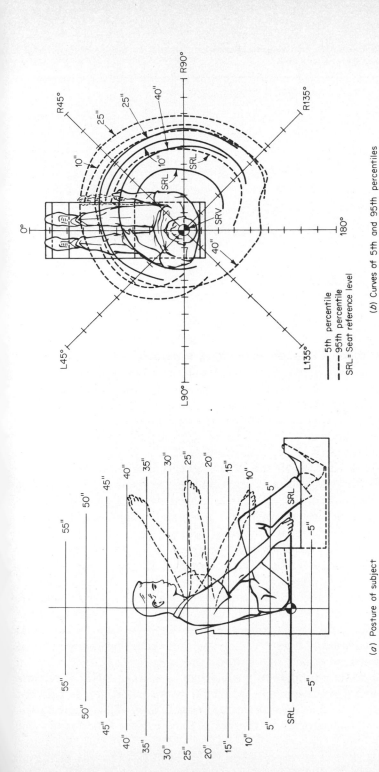

(a) Posture of subject

(b) Curves of 5th and 95th percentiles

——— 5th percentile
— — — 95th percentile
SRL = Seat reference level

Figure 12-6 Part *a* illustrates the physical arrangement used in an anthropometric study of the three-dimensional space envelope of seated subjects (male United States Air Force personnel); the grasping reach was measured at different positions relative to the seat reference level (SRL) and at every 15° around the subject. Part *b* presents the curves of the 5th and 95th percentiles for each of four horizontal "slices" of the space, namely, at the seat reference level and at 10, 25, and 40 in. above that level. [Adapted from Kennedy, 14.]

Table 12-2   Fifth Percentile of Grasping Reach, in Inches, to Selected Horizontal Planes above Seat Reference Level*

	Inches above seat reference level (SRL)						
Angle	*SRL*	10	20	25	30	40	45
L 135							7.75
L 90						12.25	7.25
L 45			19.50	20.00	19.00	14.00	8.50
L 30			21.50	22.50	21.50	15.50	9.50
0			25.50	26.25	25.50	19.00	12.75
R 30	17.50	27.00	30.00	30.25	29.00	22.75	17.50
R 45	19.50	28.25	31.00	31.00	30.25	24.75	19.00
R 90	19.50	29.25	32.25	32.25	31.25	26.25	21.00
R 135	16.50	26.25					20.00
180							12.75

* Sample: 20 male USAF personnel.
SOURCE: Adapted from Kennedy [14].

Effects of manual task on work-space envelope.   The confounding effects upon the work-space envelope of the nature of the manual task being performed are illustrated by the results of some anthropometric research by Dempster [5] and Dempster, Gabel, and Felts [7]. Some of their research involved the analysis of photographic traces of contours of the hand as it moved over a series of frontal planes spaced at 6-in. intervals. Eight different hand grasps were used, in which the hand, grasping a handlelike device, was in one of the eight fixed orientations (supine, prone, inverted, and at specified angles); but the hand was free to be moved over the plane in question. The mean data for 22 male subjects were summarized in several ways and presented in different forms in order to characterize different functional areas of the three-dimensional space of the subjects. *Kinetospheres* were developed for each type of grasp, showing graphically the mean contours of the tracings as photographed from each of three angles, top (transverse), front (coronal), and side (sagittal). Although the kinetospheres for the several types of grip will not be illustrated, they were substantially different. These were, however, combined to form *strophospheres* as shown in Figures 12-7 and 12-8. The shaded areas in Figure 12-7 define the region that is common to the hand motions made with the various hand grips.

Effects of backrest angle on work space.   Another factor that can influence the limits of the convenient manual work space of seated personnel is the angle of the backrest of the seat that is used. That angle naturally alters the relationship of the body (particularly functional arm

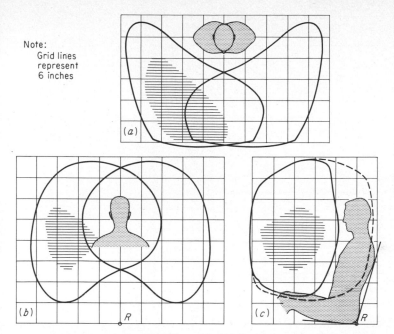

Note:
  Grid lines
  represent
  6 inches

Figure 12-7 Strophosphere resulting from superimposition of kinetospheres of range of hand movements with a number of hand-grasp positions in three-dimensional space. The shaded areas depict the region common to all hand motions (prone, supine, inverted, and several different angles of grasps), probably the optimum region, collectively, of the different types of hand manipulations. [Adapted from Dempster, 5.]

reach) to the space immediately in front. This effect is illustrated in Figure 12-9, which shows the generally optimum location of control devices for three angles of backrests (0, 10, and 20°); (these configurations would accommodate about 90 percent of an adult male population).

Minimum Requirements for Restricted Spaces

People sometimes find themselves working in, or moving through, some restricted and sometimes awkward spaces (such as an astronaut crawling through an escape hatch). For certain types of restricted spaces dynamic anthropometric data have been derived that provide minimum values. Some such data are given in Figure 12-10 for illustration. Note that the dimensions given include those applicable to individuals with heavy clothing. In most cases such clothing adds 4 to 6 in., and in the case of a vertical escape hatch it adds 10 in. to the requirements.

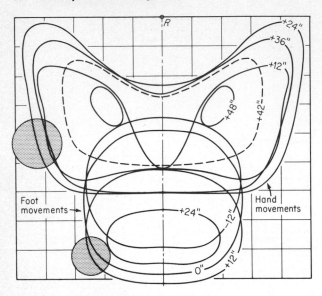

Figure 12-8 Floor plan of work space shown in Figure 12-7 relative to the standard seat, shown by 12-in. contours. In general, these contours represent the average shape and dimensions of space required in several specified hand and foot movements used in collection of the data. The 0-in. contour is at the seat reference level; −12-in. is below, and +12, +24, +36, etc., are above that level. The radii of the two shaded circles represent widths to be added to, or subtracted from, the different curves to accommodate the 5th and 95th percentiles of hand and foot movements, respectively. [From Dempster, 5.]

WORK SURFACES

Within the three-dimensional envelope of a work space, more specific considerations of work-area design relate to horizontal (dimensions, contours, height, etc.), vertical, and sloping work surfaces (dimensions, positions, angles, etc.). These features of the work situation also preferably should be determined on the basis of anthropometric considerations of the people who are to use the facilities in question.

Horizontal Work Surface

Many types of manual activities are carried out on horizontal surfaces such as work benches, desks, tables, and kitchen counters. For such work surfaces, the *normal* and *maximum* areas have been proposed by Barnes

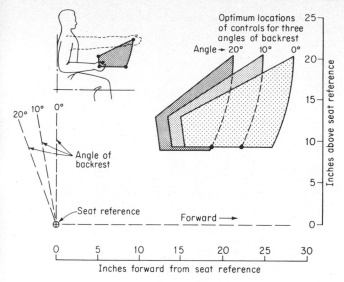

Figure 12-9 Optimum manual-control areas in relation to angle of backrest. The shaded areas show optimum manual-control areas for use with backrest angles 0, 10, and 20°. Areas are scaled from seat reference point at lower corner of seat. (See inset at upper left for illustration of areas relative to body.) [Adapted from Morgan et al., 18.]

[1] based on the measurements of 30 men. These two areas are shown in Figure 12-11 and have been described as follows:

1. *Normal area:* This is the area that can be conveniently reached with a sweep of the forearm, the upper arm hanging in a natural position at the side.
2. *Maximum area:* This is the area that could be reached by extending the arm from the shoulder.

Related investigations by Squires [23], however, have served as the basis for proposing a somewhat different work-surface contour that takes into account the dynamic interaction of the movement of the forearm as the elbow also is moving. The area that is so circumscribed[3] is superimposed over the area proposed by Barnes in Figure 12-11. The fact that

---

[3] The resulting curve is known as a *prolate epicycloid.* The following data characterize certain points of this curve; the first value of each pair given is the distance in inches from the center (right or left), and the second value, in parentheses, is the distance in inches from the table edge of the curve at that position: 0 (10.04), 1.8 (11.57), 6.2 (12.49), 9.6 (12.86), 13 (12.65), 16.2 (11.88), 19.2 (10.57), 21.2 (8.67), and 24.5 (6.59).

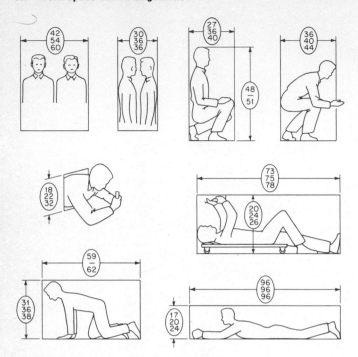

Figure 12-10 Clearances for certain work spaces that individuals may be required to work in or pass through. *Note:* The three dimensions given (in inches) are (from top to bottom in each case) minimum, best (with normal clothing), and with heavy clothing (such as arctic). [Adapted from Rigby, Cooper, and Spickard, 20.]

the normal work area proposed by Barnes has gained wide acceptance probably indicates that it is quite adequate, although the somewhat shallower area proposed by Squires probably tends to correspond somewhat better with dynamic anthropometric realities.

Work-surface Height: Seated

The wide range of tasks performed by seated personnel at tables, desks, work benches, etc., plus the range of individual differences, obviously precludes the establishment of any single, universal height that would be appropriate for such surfaces. However, considering body structure and biomechanics, at least one can set forth a guiding principle that should be applied, namely, that the work surface (or, really, the location of the devices or objects to be used continuously) should be at such a level that the arms can hang in a reasonably natural, relaxed position from the

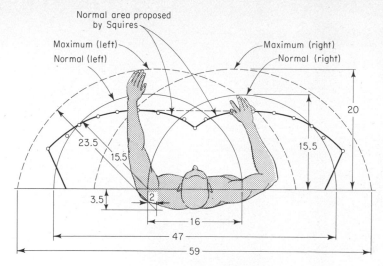

Figure 12-11 Dimensions in inches of normal and maximum working areas in horizontal plane proposed by Barnes, with normal work area proposed by Squires superimposed to show differences. [From Barnes, 1, and Squires, 23.]

shoulder, with the elbow having, as Floyd and Roberts [9] refer to it, a "satisfactory" relationship to the working surface. Generally this would mean that the forearm would be approximately horizontal or sloping down slightly when performing most simple manual tasks; when the work surface is higher than the elbow, the need to keep the arm raised can generate stresses and strains, including those in the shoulder.

The work surface that would be most applicable for an individual (or people generally), however, is closely tied in with seat height (to be discussed later) and the thickness of the surface, and with the thickness of the thigh. In fact, if a fixed facility is to be used, it virtually becomes impossible to design a setup that would be satisfactory for both a short and a tall person. Thus, where feasible, some adjustable features should be provided, such as seat height, foot position (as by the use of a foot-rest), or even work-surface height. In this connection, an interesting innovation has been developed by the Western Electric Company, as illustrated in Figure 12-12. The work surface is adjustable in height by an electrically activated control. The seat also has adjustments for height and backrest (height, angle, degree of rigidity, etc.). (The frame on the surface holds the chassis of the electrical equipment to be wired by an operator, and can be rotated to the desired position.)

In connection with fixed desk and table heights, it is fairly common practice to use those around 30 in. But most of the recommendations

Figure 12-12 Illustration of adjustable work place and chair. The height of the work surface can be adjusted by use of an electrically controlled device. The chassis (to be wired by the operator) can be rotated on its frame to any desired position. Various features of the chair also can be adjusted. [Photograph courtesy of Western Electric Company, Kansas City.]

summarized by Kroemer and Robinette [16] recommend heights somewhat below this. Burandt and Grandjean [3], for example, suggest a range from about 27 to 30 in. (69 to 77 cm), and propose that with higher desks and tables, such as the widely used height (in Switzerland) of 78 cm (30.7 in.) there should be available a footrest (of up to 4 in.). They point out that the most critical aspect of table height, however, is in the relation of the seat to the surface, 11 in. (28 cm) being the mean value preferred by a sample of office workers; they propose an adjustable seat-level range of about 1 in. (± 2 cm) to provide some adaptation from this mean value. (Actually, the mean elbow-rest heights of men and women are 9.5 and 9.2 in., as shown earlier in Table 12-1, so even an 11-in. difference might preclude a *relaxed* elbow position for many persons.) Figure 12-16 later in this chapter shows certain of their recommendations regarding table, seat, and footrest dimensions.

Work-surface Height: Standing

Some experimental evidence relating to work-surface height for persons working in a standing posture comes from a study by Ellis [8]. Using a manipulation test of turning over wooden disks, he varied the work-surface height in relation to the distance from floor to fingertip of each subject. Six different levels were used for each subject, his performance being measured by number of disks turned over during a 3-min trial, these distances in inches being as follows: 25.9, 31.3, 36.6, 42.0, 47.4, and 52.7. The fourth height (42 in.) was optimum for speed of performance, and the third (36.6 in.) was nearly as good. These provided average distances below elbow height of 2.8 and 8.2 in., respectively, for the 42 and 36.6 in. heights, and lead to the conclusion (which is supported from other investigations and by experience) that for standing, a work surface preferably should be a few inches below elbow height (Barnes proposes 2 to 4 in.) for at least light assembly or similar manipulatory tasks. If work surfaces are fixed, platforms can, of course, be used to adjust the work-surface height to the elbow height of individuals.

Vertical Work Surfaces

Vertical work surfaces typically include visual displays and manual controls. Where visual displays alone are mounted, the distance of the surfaces is not critical if the displays are large enough to be seen clearly. When vertical surfaces include controls to be used, however, the dynamic anthropometric arm reach defines the generally acceptable distance of such panels. The area that can conveniently be reached is, of course, related to distance from the surface. Some useful data on this point are provided by a study in which subjects sat erect in a chair in front of a wall chart and reached out at various specified angles on the chart. The distance of the fingertips in each position was marked with a pin [Sandberg and Lipshultz, 22]. The distance from the eyes to the wall was varied, these distances being 10, 15, and 20 in.

Average circles were then derived for the eight subjects. Subsequently, a comparison was made of the arm reach of the subjects in this study with the arm lengths of nearly 3000 United States Air Force cadets; these comparative data came from another source [Randall et al., 19]. Estimates were then made of the area that could be reached by 95 percent of the base group of 3000 men. The result of the study was the preparation of Figure 12-13, which shows, respectively, the span that can be reached by the three groups indicated. For most purposes, of course, the smaller area A would be most suitable for the 20-in. distance, but for longer distances (such as the common 28-in. viewing distance), even that would be too large an area to use.

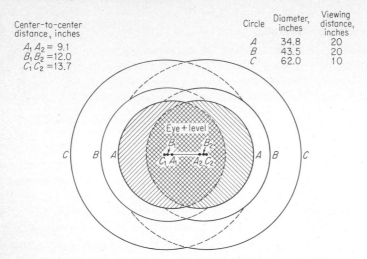

Center-to-center
distance, inches

$A_1 A_2 = 9.1$
$B_1 B_2 = 12.0$
$C_1 C_2 = 13.7$

Circle	Diameter, inches	Viewing distance, inches
A	34.8	20
B	43.5	20
C	62.0	10

Figure 12-13   The limits of maximum working areas on vertical surface for various sizes of personnel and viewing distances. *A*, very conservative, includes 95 percent of personnel; *B*, average mean, includes 50 percent of personnel; *C*, largest mean, includes 5 percent of personnel. [From Sandberg and Lipshultz, 22.]

THE SCIENCE OF SEATING

Whether at work, at home, at horse races, on busses, or elsewhere, the members of the human race spend a major fraction of their lives sitting down. As we know from experience, the chairs and seats we use cover the gamut of comfort; they can also vary in their influence on the performance of people who use them when carrying out some types of work activities.

Principles of Seat Design

The relative comfort and functional utility of chairs and seats are, of course, the consequence of their physical design in relationship to the physical structure and biomechanics of the human body. The uses of chairs and seats (from TV lounge chairs to stadium bleachers) obviously require different designs, and the range of individual differences complicates the design problem. Granting that compromises sometimes are necessary in the design of seating facilities, there nonetheless are certain general guidelines that may aid in the selection of designs that are sufficiently optimum for the purposes in mind. Some such guidelines have been expressed by Floyd and Roberts [9] and Kroemer and Robinette [16], including most of those discussed below.

Weight distribution.   Various seating studies have led to the conclusion that people are generally most comfortable when the weight of the body is borne primarily by the ischial tuberosities [Lay and Fisher,

17; Swearingen et al., 24]. These are the bony structures of the buttocks and in their anatomical features seem well suited to their weight-bearing responsibilities. Some indications of the distribution of pressures in sitting are given below (although we shall not go into the details of the method of measurement). These data are presented for each of four areas of the total sitting area, the areas differing in the average pressure over them [Swearingen et al., 24, tables 4 and 5]:

	Sitting area			
Location	Under ischial tuberosities (1)	Near tuberosities (2)	Adjacent to periphery (3)	Peripheral (4)
Range of pressure, lb/in.²	10–60	4–9	½–3	⅛–½
Total sitting area, %	3.5	4.5	34.0	58.0
Average weight, lb	51.5	27.2	76.8	25.9
Average pressure, lb/in.²	11.5	4.6	1.8	0.4

It can be seen that nearly half of the body weight is concentrated on about 8 percent of the sitting area, namely, that under and near the tuberosities (areas 1 and 2). Further evidence of the very unequal distribution of weight in sitting comes from the study by Lay and Fisher [17].

Seat height.   To avoid excessive pressure on the thigh (toward the front of the seat) the front of the seat should be no higher than the distance from the floor to the thigh when seated (i.e., popliteal height). This dimension generally should be selected to accommodate, say, all individuals from the 5th percentile up. Referring back to Table 12-1, the 5th percentiles for males and females are 15.5 and 14.0 in., respectively. Fixed seat heights of such values, however, can complicate the mechanics of sitting for the taller members of the human clan by a chain reaction starting with the knee angle that can cause such an individual to sit with his lumbar back area in a convex rather than concave posture. By taking into account the fact that heels typically add an inch or more to the 5th percentile values (and more for females), it has become fairly common practice to use seat heights of around 17 in. Since this may still be too high for short individuals, it has been proposed by the Engineering Equipment Users Association [25] that where feasible, an adjustable range of from 15 to 19 in. be provided. As an aside, common practice also provides for a slightly angular seat, such angles ranging from, say, 3 to 5° for work seats up to, say, 8° for lounge chairs and seats.

Seat depth and width.   Seat depth is another feature of seats that generally should be set in terms of a maximum value (i.e., adapted to, say, the 5th percentile) to provide adequate clearance for the calf of the leg and to minimize thigh pressure.   For work seats (such as chairs for typists and bench workers) values of 14 to 16 in. are generally satisfactory, although these values may be increased for more "comfy" types of seats such as airline seats and lounge chairs. On the other hand, seat width should be set to minimum values (adapted to, say, the 95th percentile). Although a seat breadth of 16 to 17 in. does the trick for individual seats, if people are to be lined up in a row, or seats are to be adjacent to each other, elbow-to-elbow breadth values need to be taken into account, with even 95th percentile values around 19 and 20 in. producing a moderate sardine effect. (And for bundled-up football observers the sardine effect is amplified further.) In any event these are approximate minimum values for chairs with arms on them (and for your well-fed friends you should have even wider lounge chairs).

Trunk stabilization.   As indicated above, the primary weight of the body should be borne by the area around the ischial tuberosities. By proper seat design this concentration of pressure can be achieved without excessive muscular stresses and strains or improper posture such as slumping. The seat back and angle play an important role in such stabilization of the trunk with minimal effort expenditure. Probably the critical area of support is in the lumbar region (about 8 in. or so above the seat level). Thus, if a small backrest is to be provided (such as for certain work chairs), it should avoid the sacral region (below the lumbar region) and the shoulder blades. Chairs with complete backrests should be so designed as to provide particular support in the lumbar region, as shown later in Figure 12-16. If a full backrest is provided, it should be angled to provide for greatest comfort. In this connection Lay and Fisher [17] report that the average angle preferred by their subjects was 111.7° from the horizontal (being comprised of an average seat angle of 6.4° plus an average angle between the seat and back of 105.3°).

Trunk stability can also be aided by the use of arm rests and even by resting the arms on desks or work-surface areas, *but* these should also be at levels that would make it possible for the arms to hang freely, and for the elbows to rest in a natural position.

Postural changes.   Although some seats have been tested by the postural changes that people make in them (such as the number of fidgits), this does not mean that the goal of seat design should be to reduce mobility to zero. Generally a chair or seat should permit moderate mobility and changes in posture.

Seat Designs and Dimensions

As indicated earlier, the specific features of seats need to be adapted to the purpose in mind. In this connection, Ridder [21] characterizes three main types of sitting positions preferred by adults, namely, (1) an erect position preferred near a table, a desk, or other surface while dining, writing, etc.; (2) a less erect, more relaxed, position preferred mainly for conversation, listening, or viewing; and (3) a relaxed position preferred while reading, watching TV, or for informal conversations. In an investigation with 72 men and 90 women, she obtained data on the profiles of seats that people preferred for these three purposes. In this she used the experimental chair shown in Figure 12-14 with rubber-tipped plungers

Figure 12-14 Experimental chair used in study of preferred contours of chairs for three purposes (see text). The rubber-tipped plungers could be adjusted to form, collectively, the contour preferred by each subject. [From Ridder, 21, fig. 1.]

that could be adjusted to each person's own preference. The extensive data so collected were used to develop, for each of the three types of postures, the dimensions that represented, collectively, the preferred contours of the subjects. Figure 12-15 shows the side contours of the three types, to illustrate the variability in seat designs for different purposes. To represent, further, such variations the results of a couple of specific studies will be cited.

Office and factory seats. As indicated earlier, when seats are to be used in combination with desks, tables, etc., the dimensions of the two need to be worked out together. For office seats, Burandt and Grandjean [3], on the basis of a substantial amount of data from their study, have proposed the design features and dimensions shown in Figure 12-16.

Train seats. As another example of a study of seats for a given purpose, Branton and Grayson [2] carried out a comparative evaluation of two train seats which were substantially the same in their linear dimensions [see Figure 12-17] but differed particularly in their internal characteristics. Type 1 was a traditional spring and upholstered seat (with a relatively *soft* feel), whereas type 2 had a glass-reinforced plastic

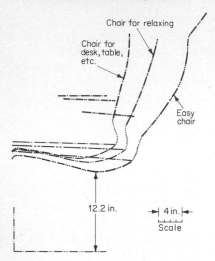

Figure 12-15 Side view of contours of seat support preferred by adults for three purposes. (The details of the contours of the seats and backs, characterized by 1.5-in. intervals, are not shown, but are presented in the original report.) [Adapted from Ridder, 21, fig. 13.]

shell and urethane foam cushion of high ·density (with a relatively *firm* feel). In the evaluation, time-lapse camera recordings (one frame every 10 sec) were made of passengers in the two seats on the 4-hr train run between London and Edinburgh. The resulting 30,000 frames were then classified by type of posture, and the data were analyzed by duration of postures and the number of fidgits, (a fidgit being any discernible change in posture from one frame to the next). In brief, it was found that the passengers maintained their dominant postures longer with type 2 (the firm type) than with type 1, especially after the first hour of sitting (55 percent for type 2 versus 30 percent for type 1), and fidgited less in type 2. In addition, however, some interesting differences bobbed up between tall and short passengers in the number of fidgits; the short people fidgited more than the tall in type 1, but the tall passengers fidgited more in type 2, although the reason for this difference was not evident.

Seating of Groups

The seating of people in groups, such as when working together, in schoolrooms, and auditoriums, becomes very much a problem in applied physical anthropometry. To illustrate the application of anthropometry

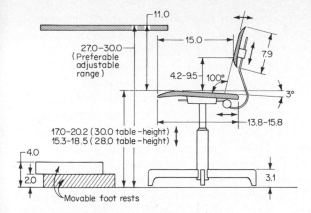

Figure 12-16 Recommended dimensions (in inches) of adjustable features of office seat. Note ranges of adjustability of seat height in relation to two table heights (30 and 28 in.). To maintain approximately 11 in. between seat height and work surface, the seat should have a range of adjustability that would depend on the work surface height, as indicated. With short persons, footrests may be required if work-surface height is high. (Data in inches converted from centimeters.) [Adapted from Burandt and Grandjean, 3.]

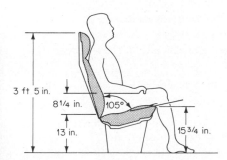

Figure 12-17 Illustration of train-seat proportions adopted by British Railway and used in seat evaluation by Branton and Grayson [2]. In the evaluation, two types of internal characteristics were compared, one being a spring and upholstered seat and the other having urethane cushions of high density.

to such a problem, let us take as an example the problem of arranging rows of seats, with seats directly in back of each other (rather than staggered). Let us assume that the person in one row needs to see over the head of the person in front of him (and assume that that person is not a woman with a huge flowered hat; in this case all bets are off); let us also assume random seating with respect to height. If the viewing angle (that is, the object to be viewed, as a screen, a speaker, or a display) is horizontal to a given individual (say, in the second row), his seat must be raised enough to permit him to see over the head of the person in front. The eye height of the second person and the sitting height of the first person then become the critical values to deal with in determining the rise of the second row. Following are selected data relating to these values, based on male flying personnel [Hertzberg, Daniels, and Churchill, 13]:

Percentile	Eye height, in.	Sitting height, in.
5th	29.4	33.8
50th	31.5	36.0
95th	33.5	38.0

Taking the values shown in Figure 12-18a, we could figure out that the rise would have to be 8.6 in. for a small person (5th percentile) in the second row to see over the head of a tall person in the first row (95th percentile). In attacking this problem more systematically, one can use an adaptation of a formula from Morgan et al. [18] to derive a value for the rise *in addition* to that required for the row ahead (in this case, 8.6 in.):

$$R_{2-1} \text{ (difference in rise between any two rows)} = \frac{[(D + F) - A]C}{B}$$

in which the ingredients are those shown in Figure 12-18b. As an example let us put in these values in inches: A, 43; B, 60; C, 40; D, 38; and F, 17. Applying these values, we arrive at a value for R_{2-1} of 8.0 in., which, added to our 8.6 in. above, gives a total rise X of 16.6 in. The rise for succeeding rows can be derived in the same manner. Appropriate adaptations of this type of approach can be used in other circumstances, such as with standing individuals, with other samples of subjects, and with other assumptions (such as the use of percentile ranges different from the 5th and 95th used in this case).

For audience seating, it is sometimes the practice simply to use the *average* differences in eye height and seating (head) height (which is about 5 in.) as a rule of thumb in raising the row above the one ahead, or even for every two rows, recognizing that viewing will be affected for some

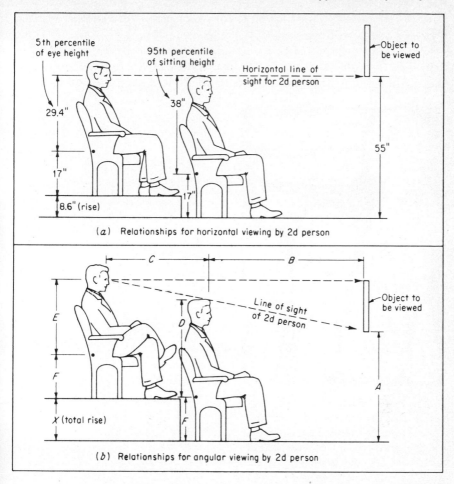

Figure 12-18 Illustration of relationships of individuals seated behind others, for use in deriving estimates of the rise X required for permitting adequate viewing by one person behind another. [Adapted from Morgan et al., 18.]

people. Incidentally, the growing practice of staggering seats is, of course, a very sensible scheme.

While the above elaboration regarding seating deals with only one (out of many) anthropometric problems, it will perhaps illustrate how anthropometric data can be utilized in practical situations.

USE OF MANIKINS AND BODY FORMS

For some work-space and seating-design problems, human manikins or body-form models are used. Two-dimensional manikins usually are built to scale of cardboard, plywood, or other flat firm materials to represent certain body sizes. Generally such forms represent certain percentiles, probably most commonly the 5th, 50th, and 95th percentiles of some sample, such as the set of three manikins developed by Dempster [6] to represent the 5th, 50th, and 95th percentiles of an Air Force sample. One of these is illustrated in Figure 12-19. In use, such a manikin is ad-

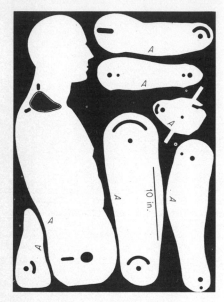

Figure 12-19 Pattern of body segments for drafting-board manikin. This particular example is one for an average man; others (not shown) represent the 5th and 95th percentiles. Note line on thigh segment which indicates scale. [From Dempster, 5.]

justed to a particular working posture over a scale drawing of the work place, and the effects of hand, foot, knee, elbow, and other locations are marked on the drawing. These and other manikins should of course be used in situations in which static anthropometric data are appropriate, and in such circumstances, one should use the particular manikin or manikins that are most appropriate (such as the 5th, 50th, or 95th percentile models).

DISCUSSION

It must be granted that tables of anthropometric data do not make for very enthralling bedtime reading (although they might help those who have insomnia). But when we consider the application of such data to the design of the physical facilities and objects we use, we can see that such intrinsically uninteresting data play a live, active role in the real, dynamic world in which we live and work.

SUMMARY

1. Static anthropometry is concerned with the measurement of (static) features of the body and of body members. Dynamic anthropometry is concerned with the measurement of body features while people are in the process of performing some function; it is based on the assumption that the body members interact with each other. Sets of data from both static and dynamic anthropometry have been developed.
2. There are three primary principles in the application of anthropometric data:
    (a) Design for extreme individuals. This principle usually should be followed where there is some limiting consideration that would affect the convenient use of a facility for those whose measurement on a given physical characteristic is above or below the value that is used as the design base. For example, in the case of a control panel the limiting variable is length of arm reach, and it should be designed for those with short reach. On the other hand, door heights should accommodate tall people.
    (b) Design for adjustable range. Where a facility can be adjusted (such as height of desk chairs and positions of automobile seats), it should be designed to accommodate a reasonable range of individuals, usually from the 5th to the 95th percentiles.
    (c) Design for the average. Certain types of facilities preferably should be designed for the mythical average individual on the grounds that this will discommode people less than a facility designed for extremes.
3. In the design of work space or other facilities, various conditions need to be taken into account, for example, the type of data that would be most appropriate (static data, dynamic data, etc., including the activity to be performed), the types of people who are to use the facility as related to the types of people for whom appropriate data are available, the possible need for trade-off of one advantage for another, and the clothing and other personal gear to be used.
4. Among the principles relevant to the design of seating are the following: (a) weight distribution (the weight of the body should be borne primarily by the ischial tuberosities, the bony structures of the buttocks); (b) seat height (the front of the seat should be no higher than the distance from the floor to the thigh when seated, to avoid pressure on the underside of the thigh); (c) seat depth (for seats to be

used by many persons, the depth should not exceed that suitable for comparatively short people; for work seats, typing chairs, etc., this would be about 14 to 16 in.); (*d*) seat width (width should accommodate those who are broad-of-beam, not less than 16 to 17 in. and preferably more); (*e*) trunk stabilization (the seat back should have a modest angle, such as 105°, and the seat should slope a bit, such as 5 or 6°, in order to minimize the need to maintain an erect posture as would be the case with vertical backs; there should be particular back support for the lower lumbar area); (*f*) postural changes (most seats should permit reasonable changes in posture).

5. Audience seating should provide for adequate rise to permit viewing over the heads of people in front.

REFERENCES

1. Barnes, R. M.: *Motion and time study*, 5th ed., John Wiley & Sons, Inc., New York, 1963.
2. Branton, P., and G. Grayson: An evaluation of train seats by observation of sitting behavior, *Ergonomics*, 1967, vol. 10, no. 1, pp. 35–51.
3. Burandt, V., and E. Grandjean: Sitting habits of office employees, *Ergonomics*, 1963, vol. 6, no. 2, pp. 217–228.
4. Damon, A., H. W. Stoudt, and R. A. McFarland: *The human body in equipment design*, Harvard University Press, Cambridge, Mass., 1966.
5. Dempster, W. T.: *Space requirements of the seated operator*, USAF, WADC, TR 55–159, July, 1955.
6. Dempster, W. T.: *The anthropometry of body action*, USAF, WADD, TR 60–18, January, 1960.
7. Dempster, W. T., W. C. Gabel, and W. J. L. Felts: The anthropometry of the manual work space for the seated subject, *American Journal of Physical Anthropology*, December, 1959, vol. 17, no. 4, pp. 289–317.
8. Ellis, D. S.: Speed of manipulative performance as a function of work-surface height, *Journal of Applied Psychology*, 1951, vol. 35, pp. 289–296.
9. Floyd, W. F., and D. F. Roberts: Anatomical and physiological principles in chair and table design, *Ergonomics*, 1958, vol. 2, no. 1, pp. 1–16.
10. Hansen, R., and D. Y. Cornog: *Annotated bibliography of applied physical anthropology in human engineering*, USAF, WADC, TR 56–30, May, 1958.
11. Hertzberg, H. T. E.: *Some contributions of applied physical anthropology to human engineering*, USAF, WADD, TR 60–19, January, 1960.
12. Hertzberg, H. T. E.: Dynamic anthropometry of working positions, *Human Factors*, August, 1960, vol. 2, no. 3, pp. 147–155.
13. Hertzberg, H. T. E., G. S. Daniels, and E. Churchill: *Anthropometry of flying personnel–1950*, USAF, WADC, TR 52–321, September, 1954.
14. Kennedy, K. W.: *Reach capability of the USAF population: Phase I. The outer boundaries of grasping-reach envelopes for the shirt-sleeved, seated operator*, USAF, AMRL, TDR 64–59, 1964.

15. Kobrick, J. L.: *Quartermaster human engineering handbook series: I. Spatial dimensions of the 95th percentile arctic soldier*, Quartermaster Research and Development Center, Natick, Mass., TR EP–39, September, 1956.

16. Kroemer, K. H. E., and J. C. Robinette: *Ergonomics in the design of office furniture: A review of European literature*, USAF, AMRL, TR 68–80, 1968.

17. Lay, W. E., and L. C. Fisher: Riding comfort and cushions, *Journal of the SAE* (Transactions), November, 1940, pp. 482–496.

18. Morgan, C. T., J. S. Cook, III, A. Chapanis, and M. W. Lund (eds.): *Human engineering guide to equipment design*, McGraw-Hill Book Company, New York, 1963.

19. Randall, F. E., A. Damon, R. S. Benton, and D. I. Patt: *Human body size in military aircraft and personal equipment*, USAF, TR 5501, June 10, 1946.

20. Rigby, L. V., J. I. Cooper, and W. A. Spickard: Guide to integrated system design for maintainability, USAF, ASD, TR 61–424, October, 1961.

21. Ridder, Clara A.: *Basic design measurements for sitting*, Agricultural Experiment Station, University of Arkansas, Fayetteville, Bulletin 616, October, 1959.

22. Sandberg, K. O. W., and H. O. Lipshultz: *Maximum limits of working areas on vertical surfaces*, USN ONR SDC, Report 166–1–8, reprint, April, 1952.

23. Squires, P. C.: *The shape of the normal work area*, Navy Department, Bureau of Medicine and Surgery, Medical Research Laboratory, New London, Conn., Report 275, July 23, 1956.

24. Swearingen, J. J., C. D. Wheelwright, and J. D. Garner: *An analysis of sitting areas and pressures of man*, Civil Aeronautical Research Institute, Federal Aviation Agency, Oklahoma City, Report 62–1, January, 1962.

25. *Factory seating*, Engineering Equipment Users Association, London, Handbook 16, 1962.

26. *Weight, height, and selected body dimensions of adults: United States, 1960–1962*. Data from National Health Survey, USPHS Publication 1000, series 11, no. 8, June, 1965.

thirteen

ARRANGEMENT

AND

UTILIZATION

OF

PHYSICAL

SPACE

Human activities are carried out in many kinds of physical situations, including those that are man made (such as specifically designed work stations or work-space envelopes as discussed in the last chapter, i.e., vehicle cabs, warehouses, and kitchens) and those that are part of a more natural environment (such as certain agricultural and other outdoor activities). When human activities are carried out in man-made environments, there typically are some degrees of freedom in arranging the physical objects and facilities (i.e., displays, controls, tools, machines, work materials, etc.) in order to facilitate their use by people.

CONCEPT OF OPTIMUM LOCATION OF COMPONENTS

In this juggling process there is a basic premise that we should set forth now, namely, the premise of *optimum locations* of physical components. It is reasonable to hypothesize that any given type of activity using a physical component could be carried out best if the component is in a satisfactorily optimum general location as far as relevant human sensory, anthropometric, and biomechanical characteristics are concerned (i.e., reading a visual display, activating a foot push button, etc.). Preferably, of course, components should always be placed in their optimum locations, but since this frequently is not possible, *priorities* sometimes must be established. These priorities, however, do not descend from heaven like manna, but must be otherwise determined, usually on the basis of some factors such as those mentioned below.

GUIDING PRINCIPLES OF ARRANGEMENT

Before touching on a few methods that are used in trying to figure out what should go where, however, let us set down a few general guides (in addition to the idea of optimum location) that may be helpful. Depending on the circumstance, these guidelines can deal with either or both of two separate, but interrelated, phases, as follows: that concerned with the *general location* of components (such as specific components within a fixed work space or usually larger components that might be located in a more general work area such as an office), and that concerned with the *specific arrangement* of components. (We shall use the term *component* in this discussion to cover any kind of physical object, i.e., displays, controls, equipment, materials, etc.)

Importance Principle

This principle deals with operational importance, that is, the degree to which the performance of the activity with the component is vital to the achievement of the objectives of the system. The determination of importance is largely a matter of judgment.

Frequency-of-use Principle

As implied by the name, this concept applies to the frequency with which some component is used. An additional twist to this idea places it in the frame of reference of information theory [28, part B, chap. 4, p. B.4–12A], in which both frequency of use and the number of possible components from which a selection is to be made are taken into account. (This will be discussed later.)

Functional Principle

The functional principle of arrangement provides for the grouping of components according to their function, such as the grouping together of displays, or controls, that are functionally related in the operation of the system.

Sequence-of-use Principle

In the use of certain items, there are sequences or patterns of relationship that typically or frequently occur in the operation of the equipment. In applying this principle, the items would be so arranged as to take advantage of such patterns; thus, items used in sequence would be in close physical relationship with each other.

Discussion

In putting together, as pieces of a jigsaw puzzle, the various components of a system, it is manifest that no single guideline can, or should, be applied consistently, across all situations. But, in a very general way, and in addition to the optimum premise, the notions of importance and frequency probably are particularly applicable to the more basic phase of locating components in a general area in the work space; in turn, the sequence-of-use and functional principles tend to apply more to the arrangement of components within a general area. In all of this, the need for trade-offs is ever with us.

METHODS OF ACTIVITY ANALYSIS

Although the process of arranging work spaces or other facilities for human use generally cannot be reduced to a completely objective approach, there are nonetheless certain methods for the collection of relevant data and for its use that can aid the process. In some situations it may be worth the effort to obtain data on the activities or operations that are to be carried out (such as the visual, psychomotor, movement, and other activities). To apply any of the guidelines discussed above in any given situation, it might be useful then to have information on the frequency of various activities, their relative importance, the time devoted to them, the sequence of activities, the interrelationships among activities, etc. The methods for collecting such activity data, however, would be quite different for an entirely new system than for one that is a modification or a new generation of an existing system.

Methods with Existing Systems

When a modification of an existing system is to be developed, it is possible to carry out activity analyses with the current version or to take advantage of experience with it in developing the new model (recognizing, of course, some of the risks involved in such extrapolation). In this connection, a number of the procedures or methods of analysis used in the field of industrial engineering can be of direct use in developing the layout and arrangement of work areas.[1] It is not intended here to describe or illustrate these; rather, we shall simply discuss certain of the more general approaches to the analysis of activities in an existing system.

1. *Film analysis:* Conventional motion-picture films (usually 8 or 16 mm) can be used to record and later analyze overt physical activities of individuals or groups.
2. *Eye-movement recordings:* For recording eye movements, eye cameras or other special devices are available.
3. *Observation:* For certain purposes, the activities of individuals can be observed and recorded in predetermined categories. Sampling procedures are used in some instances, whereas in other cases continuous observations are made. Timing devices may be used in conjunction with the process to determine the length of time for certain activities.

[1] The reader is referred to such texts as R. M. Barnes, *Motion and time study: design and measurement of work*, 6th ed., John Wiley and Sons, Inc., New York, 1968.

An example of the results of the observation method is given in Figure 13-1. In this example, the percentage of time devoted to each activity was obtained from the frequency with which the activity was observed during a random sampling of activities.

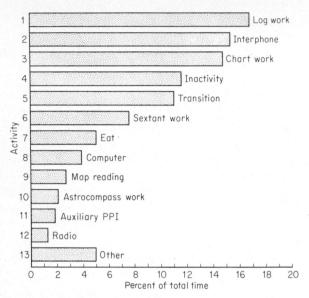

Figure 13-1   Distribution of first navigator's time on three aircraft missions. [Adapted from Christensen, 2.]

4. *Interview:* The interview is used with personnel experienced in the operation of the type of equipment or facility in question, in order to obtain factual information, or opinions, about the activities involved, such as frequency or importance of use of certain components; such opinions can be obtained with rating scales or by having the individuals rank various activities or components in terms of the factor in question (e.g., frequency or importance).

5. *Questionnaire:* This has the same possible uses as the interview, except that the individuals furnish the information on a questionnaire of some type.

Methods with New Systems

In the case of new systems (without current counterparts) the frequency, sequence, and other activity parameters must be inferred from whatever tentative drawings, plans, or concepts are available. In some cases (especially during early design stages) the designer (or others who are knowledgeable about the system being designed) may have to rely on some rather gross estimates, but as plans become further crystallized, it may be possible to use a more systematic method in the analysis of the activities that *would* be involved in the use of the system. Such analyses may, in turn, aid in rearranging the features of the system until, through an iterative process, a satisfactory arrangement is developed. Such a tentative arrangement sometimes can be further tested by the use of prototypes or physical models (if these are developed).

SUMMARIZING AND USING ACTIVITY DATA

The activity data collected, by whatever method, need to be summarized in order to assist in their interpretation and use. Frequently it is sufficient to derive simple sums (such as frequency of use of a component) or simple averages (such as the average ratings of importance). In other circumstances, some special manipulations in summarizing such data may be handy. A couple of these will be discussed below, as illustrations.

Indexes of Priorities

In figuring out the priorities in locating components (mentioned above), the designer has to decide what the appropriate basis for the priorities (such as rated importance, frequency of use, etc.) should be. When two or more such factors seem relevant, however, it is possible to combine them into a single index, either by adding the ratings on, say, importance and frequency, or by multiplying them. The table below shows, for each of five displays (*A*, *B*, *C*, *D*, and *E*) the average ratings on these two factors (a rating of 3 being high and 1 being low); in addition, the sums and the products of the two ratings are given for each display.

	Display				
	A	*B*	*C*	*D*	*E*
1. Av. rating, importance	3.0	1.7	1.6	2.5	1.1
2. Av. rating, frequency	3.0	1.2	1.7	1.0	2.7
3. Sum of 1 and 2	6.0	2.9	3.3	3.5	3.8
4. Product of 1 and 2	9.0	2.0	3.7	2.5	3.0

The sums, or products, reflect the relative priorities of the displays, for consideration in deciding where to place them. Since the *relative* values of the sums and the products can be different (as in the above example), the manner in which the ratings are combined (including any weighting system) should be given careful thought.

Another example, with a slightly different twist, actually relates to the establishment of *coding* priorities of control devices, but the method probably could be adapted to other situations as well [29, chap. 2, sec. 2D3, pp. 3–4]. The scheme provides for combining the rank orders of each of three different variables for each control into a priority index, these being:

- *Amount of information:* This is the number of bits of information, based on the probability of occurrence of an event (actually the number of controls available for selection and the number of times, relative to other controls, that the particular control will be used).
- *Penalty rate:* The controls are ranked according to the seriousness of the effects of slow or improper control selection or operation.
- *Operational importance:* The controls are ranked on the basis of their relative importance to the achievement of system objectives.

An example of this particular approach (slightly adapted) is given below (rank orders are from 1, low, to 8, high):

	Control							
	A	*B*	*C*	*D*	*E*	*F*	*G*	*H*
Amount of information	8	7	2	1	3	4	6	5
Penalty rate	1	7	3	4	7	7	2	5
Operational importance	6	7	2	3	1	8	5	4
Total rank value	15	21	7	8	11	19	13	14
Priority rank	6	8	1	2	3	7	4	5

Links as Indexes of Interrelationships

The operational relationships between individual components usually can be expressed in terms of *link* values. Actually link indexes can be developed for a wide range of such relationships, although they fall generally into two classes, namely, functional and sequential. Although we could consider the connections between items, or components, of equipment as links, for our current purposes we will think in terms of those relationships that involve human beings. Some such versions are:

Functional communication links

1. Visual (man to man or equipment to man)
2. Auditory, voice (man to man, man to equipment, or equipment to man)
3. Auditory, nonvoice (equipment to man)
4. Touch (man to man or man to equipment)

Functional control links

5. Control (man to equipment)

Sequential movement links (movements from one location to another)

6. Eye movements
7. Manual and/or foot movements
8. Body movements

Link indexes can be used as aids in connection with the general location of components or with their relative arrangements. In some circumstances they can be used as the basis for assignment of priorities.

Derivation of link values. In most cases *functional* link values are based on either the frequency or the importance of the connections between men, or between men and the equipment components in question. And, actually, there can be two or more types of links in some cases, such as voice communications and visual links between men. The method of deriving link values will of course depend upon their nature. The frequency of some links can be obtained objectively by film or observation, but can also be estimated on the basis of interviews or questionnaires. Importance, however, almost of necessity is evaluated on the basis of judgments, as obtained by interview or by questionnaire. When link ratings are derived for both importance and frequency, it is usually the practice to compute a composite link value by multiplying or adding the importance and frequency values of the individual links, in much the same manner as illustrated above in the case of priority indexes. In turn, *sequential* links, of course, relate to human movements between equipment components and therefore reflect frequency of such relationships.

Link values in operational procedures sometimes can be derived by a graphic approach in which the sequential steps in an operation are recorded. Subsequently the functional links and the sequential links (showing relations in operation between all pairs of components) can be tallied in the manner presented by Haygood et al. [10] and illustrated in Table 13-1. This table actually presents data for only half of the "panels" that were considered as components, but these will at least illustrate the nature of the results. Incidentally, the purpose of this investigation was that of comparing a computerized approach to the development of link values with a more conventional graphic approach. It was found that both methods produced substantially the same results in link values, but that the computerized approach was more economical of time and cost.

Use of Activity Data in Arranging Components

In the use of activity data for developing a reasonably optimal arrangement of components, the typical approach is through some form of physical simulation. However, in some circumstances, more systematic, quantitative methods can be used.

Arrangement by physical simulation. In the use of a physical simulation approach, the various components are juggled around on paper

Table 13-1 Graphic Link Analysis of Certain Panels of Flight System Checkout Console

Panel	Functional links		Sequential links among panels						
	Visual	Control	2	3	4	5	6	7	8
1. Master selector	175	884	348	97	44	7	1	3	0
2. Programmer	31	637		32	11	10	3	3	3
3. Recorder	51	81			0	1	2	9	0
4. Meter panel	63	0				8	0	0	0
5. Power supply	5	49					0	0	0
6. Oscilloscope	12	12						0	0
7. Preamplifier	0	17							0
8. Signal generator	0	5							

SOURCE: Adapted from Haygood et al. [10], tables 1 and 2.

(in graphic form) or in the form of models or mock-ups, until an arrangement is achieved that is judged to be reasonably optimum for whatever considerations are relevant (e.g., the optimum location of components, their priorities, or their functional or sequential link values). An example of this approach is shown in Figure 13-2, this example consisting of a combination of three men and five machines. The symbols for men (circles) and equipment units (squares) are in this case connected with three types of functional links, namely, control, visual, and auditory. In each case a composite link value is shown (the frequency rating times the importance rating). It should be noted, however, that equal numerical values of control, visual, and auditory links are not necessarily *equal* as far as arrangement is concerned. The primary consideration with each type of link is different, as follows:

Type of link	Primary consideration
Control	Ease of physical access
Visual	Ability to make visual discriminations
Auditory	Ability to hear relevant sounds or voice

The lower part of Figure 13-2 shows a revised arrangement of the men and machines, this particular arrangement, having been developed by essentially graphic methods, was about the best of several that were tried out.

Another example of an arrangement developed by essentially graphic methods comes from a study of the layout of aircraft instruments [Jones, Milton, and Fitts, 11]. The basic data for the study came from the use of an eye camera that records the reflection of the light from the cornea of the eye. Recordings were made of the eye movements of 36 pilots dur-

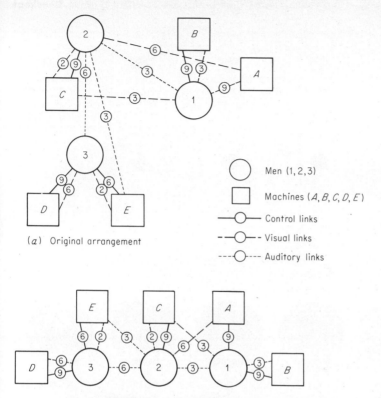

Figure 13-2   Example of rearrangement of men and machines on basis of linkages between them. (Link values are shown in small circles on link connections.)

ing various maneuvers. Figure 13-3 shows the sequential link values between the various instruments (expressed as percentages of eye shifts between instruments) for one particular maneuver. Figure 13-4, in turn, shows the average length of eye fixations, the number of fixations per minute, and the proportion of time spent on each instrument. These, in a sense, are functional links.

Data such as shown in these figures were used to develop a standard instrument arrangement (basically the one shown) which has been accepted by the U.S. Air Force, the U.S. Navy, U.S. airlines, the Royal Canadian Air Force, and the Royal Air Force (U.K.).

Quantitative solutions to arrangement problems.   Especially in simple problems of arranging components, it would be like gilding the lily to apply sophisticated quantitative methods. But with complex systems that have many components, some quantitative attack may well be

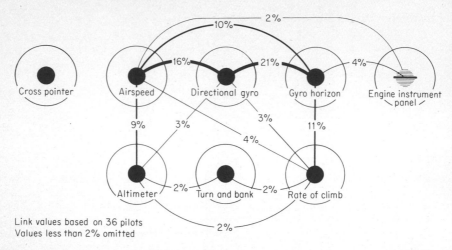

Figure 13-3 Eye-movement link values between aircraft instruments during climbing maneuver with constant heading. [Courtesy of USAF, AMRL, Behavioral Sciences Laboratory.]

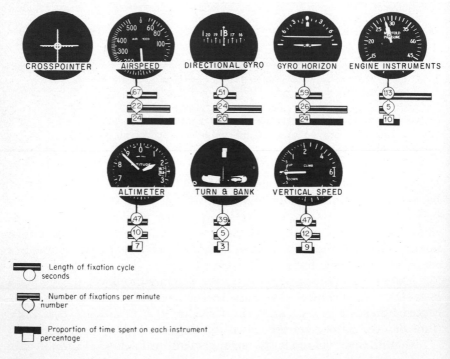

Figure 13-4 Length of eye fixations, number of fixations, and proportion of time spent on instruments during climbing maneuver with constant heading. [Courtesy of USAF, AMRL, Behavioral Sciences Laboratory.]

justified. One such method is that of linear programming. This is a statistical method that results in the optimizing of some criterion or dependent variable by manipulation of various independent variables. The optimum in some cases would be the minimum, and in other cases the maximum—whichever is the desired value in terms of the criterion.

An example of this technique will be given primarily to illustrate the applicability of the technique to a practical problem. The example in question draws upon data from the following two sources: (1) data on frequencies with which a pilot made task responses, with eight controls, in flying simulated cargo missions in a C-131 aircraft, over 139 one-minute periods, given in the column headings of Table 13-2 [Deininger, 3]; and

Table 13-2   Results of Linear Programming Applied to the Problem of Developing Optimum Arrangement of Eight Control Devices

Mean accuracy score	*Area*	*Controls: abbreviation* and frequency*							
		AP	*I*	*E*	*T*	*C*	*AC*	*M*	*G*
		115†	*40*	*30*	*27*	*20*	*15*	*13*	*12*
2.14	5	246.10†	85.60	64.20	57.78	42.80	32.10	27.82	25.68
2.46	2	282.90	98.40†	73.80	66.42	49.20	36.90	31.98	29.52
2.76	8	317.40	110.40	82.80†	74.52	55.20	41.40	35.88	33.12
3.03	7	348.45	121.20	90.90	81.81†	60.60	45.45	39.39	36.36
3.07	6	353.05	122.80	92.10	82.89	61.40†	46.05	39.91	36.84
3.13	4	359.95	125.20	93.90	84.51	62.60	46.95†	40.69	37.56
3.21	3	369.15	128.40	96.30	86.67	64.20	48.15	41.73†	38.52
3.49	1	401.35	139.60	104.70	94.23	69.80	52.35	45.37	41.88†

* Abbreviations of controls: auto pilot panel, *AP*; intercom panel, *I*; elevator trim, *E*; throttle, *T*; cross-pointer set, *C*; auto compass panel, *AC*; mixture, *M*; and gyro-magnetic compass set, *G*.

† Optimal solution.

SOURCE: Based on data from Deininger [3] and Fitts [6] as presented by Freund and Sadosky [8].

(2) data on the accuracy of manual blind-positioning responses in various areas, based on the study by Fitts [6]; for this particular purpose accuracy data for 8 of the 20 target areas were used, as shown in Figure 13-5. The accuracy scores are average errors in inches in reaching to targets in the 8 areas. Linear programming, when used in the analysis of the data by Freund and Sadosky [8] involved the derivation of a *utility cost* rating for each of eight controls in each of the eight areas shown in Figure 13-5. These utility-cost values are given in Table 13-2. Each one was computed by multiplying the *frequency* of responses involving each control (from Deininger) by the *accuracy* of responses in each area (from Fitts); for example, the first value given for control *AP* in area 5 (246.10) is the

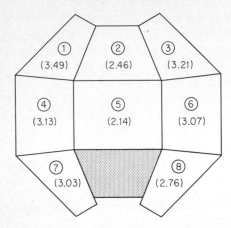

Figure 13-5 Perspective drawing of the eight areas used in application of linear programming to arrangement of eight aircraft control devices [Freund and Sadosky, 8]. Mean-accuracy scores of blind-positioning movements are given in parentheses for the eight target areas [based on data from Fitts, 6].

product of the *frequency* value for AP (115) and the mean *accuracy* score for area 5 (2.14). For any possible arrangement of controls, one can derive a *total cost*, this being the sum of the utility costs of the controls in their individual locations (areas) of that arrangement. By linear programming it was possible to identify the particular arrangement whose total cost was minimum. The locations of the various controls in this optimum arrangement are identified by daggers in Table 13-2, the dagger for each control being in the *row* for the area in which it would be located. In the use of linear programming sometimes two or more combinations have relatively similar total costs; in this particular application, for example, a couple of other arrangements would be about as optimum as the one indicated.

Although this particular exercise in linear programming deals with the location of controls, this and other quantitative techniques can also be applied to the arrangement of displays of items of equipment or of people.

GENERAL LOCATION OF COMPONENTS

The premise of optimum locations of components mentioned several times before is predicated essentially on sensory, anthropometric, and biomechanical considerations (depending on what kinds of components one is talking about). To the extent that circumstances preclude the placement of all components in their own optimum locations, we need a system of priorities as discussed above (such as based on importance and frequency of use). Successively *less* optimum locations presumably should then be earmarked for components with correspondingly lower priorities. Although it is not feasible or appropriate to catalogue the optimum locations for each and every type of component, we shall at least bring in some examples.

Visual Displays

The normal line of sight of individuals is about 15° below the horizon. Limited eye and head movements permit a fairly convenient visual scanning of an area roughly 15° around the normal line of sight—up and down and sideways. This area, then, defines the approximate optimum location for visual displays; it is represented later in Figure 13-11, which depicts various features of a console. However, the neighborhood areas are also reasonably satisfactory for visual displays, such as the angle from the horizontal down to about 35 to 45° and a total lateral range of about 45 to 60° from a straight-ahead direction (i.e., of about 22.5 to 30° on either side); these areas, then, preferably should be used for displays of *second* priority. Visual areas beyond these fringes have a couple of strikes against them for displays that need to be monitored very frequently. In the first place, continual scanning of such areas would require considerable head movement, and in the second place, peripheral vision normally would not pick up changes in conventional displays. However, peripheral vision is more sensitive in the case of light signals (as contrasted with conventional displays). This is indicated by the results of a study by Kobrick [13] in which intentional response time (IRT) was recorded for subjects who responded with a push button when a light flashed on briefly, the lights being positioned at 32 positions in the hemispherical visual field. Figure 13-6 shows the area within which the IRT was unaffected. That figure suggests that flashing-light indicators can be detected with approximately equal effectiveness across almost the entire horizontal line of sight (about 90° to either side) and about 30° above and below that line (but with some restriction to the left and right in the area above the horizontal).

Angle of visual displays.   Wherever visual displays are located—even if outside the normal span of vision—they should preferably be so placed that the eyes look straight at the display and therefore the display is perpendicular to that line of sight. However, the movement of the head to bring about a straight-ahead line of sight to a display shifts the position of the eyes down and forward, or up and back, as the case may be. This shift in the relative position of the eyes is shown in Figure 13-7. This illustrates how the displays should be positioned in the direct line of sight

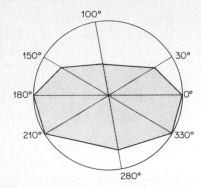

Figure 13-6 Visual field in which intentional response times (IRT) to flashing lights are approximately equal. The center is at the eye level of the subject and straight ahead. [From Kobrick, 13.]

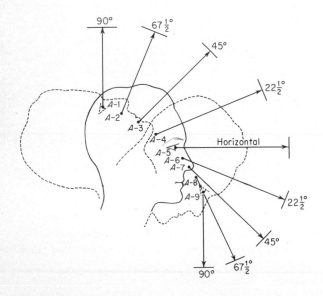

Figure 13-7 Diagram showing eye positions for various sighting angles. Note change in vertical and lateral positions of eye as sighting angle is changed. Lines perpendicular to ends of arrows show preferable angles of displays for viewing angles. [Adapted from Randall et al., 22.]

from the position from which the eyes will actually be looking and also shows the desirable angles of the displays in those locations (these angles being perpendicular to the lines of sight).

Hand Controls

The optimum location of hand control devices is, of course, a function of the type of control; the mode of operation, and the appropriate criterion of performance (accuracy, speed, force, etc.). Certain preceding chapters have dealt with some tangents to this matter, such as the discussion of the work-space envelope in Chapter 12, including Figures 12-6, 12-7, and 12-8, in that chapter.

Controls that require force. Many controls are easily activated, so a major consideration in their location is essentially one of ease of reach. Controls that require at least a moderate force to apply (such as certain control levers and hand brakes in some tractors), however, bring in another factor, that of force that can be exerted in a given direction with, say, the hand in a given position. Investigations by Dupuis [4] and Dupuis, Preuschen, and Schulte [5] dealt with this question, specifically the pulling force that can be exerted, when seated, when the hand is at various distances from the body (actually, from a seat reference point). Figure 13-8, which illustrates the results, shows the serious reduction in

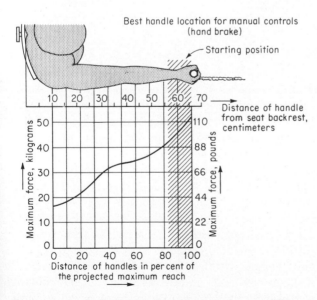

Figure 13-8 Relationship between maximum pulling force (such as on a hand brake) and location of control handle. [From Dupuis, Preuschen, and Schulte, 5.]

effective force as the arm is flexed when pulling toward the body. The maximum force that can be exerted by pulling is about 57 to 66 cm forward from the seat reference point, and this span, of course, defines the optimum location of a lever control (such as a hand brake) if the pulling force is to be reasonably high.

Since most manual controls do not require maximum force in their use, however, Konz and Day [14] suggest that it may be more desirable to consider the *distribution* of forces exerted when an operator is using a control. For this purpose they used a force platform in measuring the forces required in operating a push-pull device that was positioned at eye, chest, waist, hip, and knee heights. Recordings were made of the frontal, lateral, and vertical force components when operating the device. The frontal and lateral vectors of force did not vary much by location of the device, but the vertical, and total, forces did differ by location, as indicated below:

Location of control handle	Mean force, lb	
	Vertical	Total
Eye	4.6	9.2
Chest	3.4	7.8
Waist	4.2	8.5
Hip	7.4	12.4
Knee	5.7	10.4

The operation of the device when level with the chest area resulted in the minimal level of effort expended, with the waist area next. As an aside, it might be added that the forces varied somewhat with the orientation of the handle used, with positions ranging from 9 o'clock (with a prone position of the hand) to 12 o'clock requiring lower forces than those involving a more supine position of the hand.

Control location and response time.   For still other types of control devices, response time (that is, the time to activate them) may be the dominant consideration in their location. In the discussion of reaction time in Chapter 10 some mention was made of this. For present illustrative purposes, let us pull out certain of the results from a study by Sharp and Hornseth [24] in which seated subjects operated each of 3 types of controls (knobs, toggle switches, and push buttons) at each of 12 locations in each of 3 consoles (far, middle, and close). The controls of the close console were positioned for convenience of reach of individuals of small build (about the 5th percentile of males), and only data for this console will be given here. Two further points should be made. First, the controls were positioned to the *left* of a straight-ahead position for use only by the *left* hand (on the perhaps arguable grounds that such devices might be used most by the left hand, since the right hand normally is used for other controls, such as vehicular controls); and second, the time recorded for the knob was measured until it was touched, whereas the time recorded for the toggle switch and push button included activation of the control. With these points now in mind, let us look at some of the results as shown in Figure 13-9. Part *a* shows that,

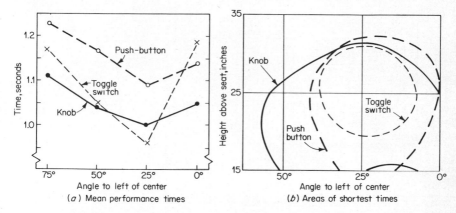

Figure 13-9   Data on time to activate push button and toggle switch and time to reach knob, when in various positions. Data are for *left* hand, and for close console. Part *a* gives mean times for controls positioned 25 in. above seat reference level. Part *b* gives contours of the area for which times were *shortest* for the particular control (e.g., times within 5 percent of the minimum time for the control). [Adapted from Sharp and Hornseth, 24.]

in the case of all three types of controls, the location about 25° left of center resulted in shortest times, and other data (not shown) indicated that mean times were lowest around 25 in. above the seat reference level (as contrasted with 15 or 35 in.). Part *b* in turn shows the areas of the shortest performance times for each type of control (e.g., the area within which times were *within 5 percent* of *minimum* times for the control in question); the smaller area for the toggle switch suggests that the selection of a location for such devices may be more critical than for the other devices if time is of the essence. However, as suggested by the investigators, a well-designed system preferably should not impose response-time requirements on operators in which differences of 0.1, 0.2, or 0.3 sec are crucial.

Console Design

Many hand controls, of course, are positioned on consoles, such consoles usually including displays as well. In connection with the arrangement of controls on a vertical console surface, one proposal has been made to segment that surface into what we might think of as priority areas. This arrangement, shown in Figure 13-10, is based on dynamic anthropometric considerations (although for essentially a military population) and indicates the outer bounds of four areas, decreasing in their order of preference, for control locations.

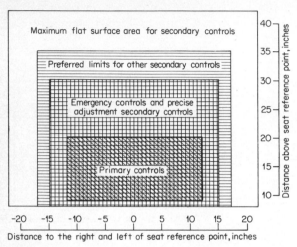

Figure 13-10 Preferred vertical surface areas and limits for different classes of manual controls. [Adapted from *Personnel subsystems*, 29, chap. 2, sec. 2D6, p. 7.]

A more generalized sketch of an angled console is shown in Figure 13-11, this particular design providing for both controls and displays. It should be emphasized that although the dimensions and relations given are quite compatible with human anthropometric and biomechanical characteristics, reasonable variations from these specific values would also be within acceptable bounds.

When all the control devices and associated displays cannot be placed within convenient range on a flat console, a sectional or curved arrangement may be useful, as shown in Figure 13-12. Although the curved design probably has some nominal advantages over the sectional variety, the curved design may introduce construction difficulties. The sectional design, however, is probably about as satisfactory, assuming appropriate dimensions and angles. In this connection, an evaluation of the angular orientation of such side panels was carried out by Siegel and Brown [25] in which subjects, using a 48-in. front panel with side panels at 35, 45, 55, and 65°, followed a sequence of verbal instructions to use the controls on the panels. A number of criteria were obtained, including objective criteria of average number of seat movements, average seat

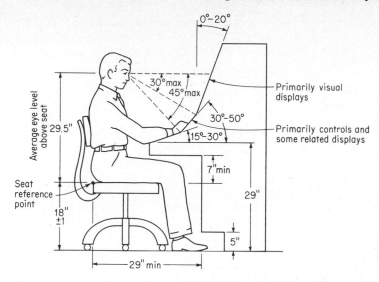

Figure 13-11 Generalized schematic drawing of a console in which some separation could be made between a primarily visual display area and a primarily control area (with limited visual displays). The dimensions and angles shown are intended to be suggestive of generally desirable ranges of such values.

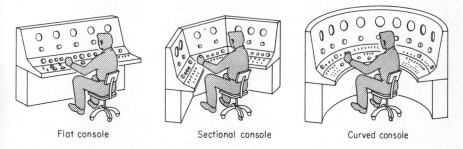

Figure 13-12 Sketches of flat, sectional, and curved consoles. Where considerable panel space is required for displays and controls, the sectional or curved arrangements are definitely superior to flat consoles.

displacement, average body movements (number and extent), average number of arm extensions (part and full), subjective criteria based on the subjects' responses of degree of ease or difficulty, judgments that the panels should be wider apart or closer together, and preference ranking for the four angles.

Only some of the data will be presented, but they will characterize the results generally. Figure 13-13 shows data for four of the criteria for the four angles. The criterion scales have been converted here to

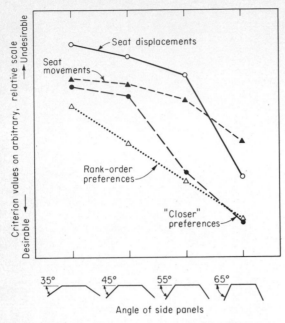

Figure 13-13 Representation of criteria from study relating to angles of side panels of console. The criterion values are all converted to an arbitrary scale for comparative purposes, but they all indicate the desirability of the 65° angle panels over the others. [Adapted from Siegel and Brown, 25, courtesy of Applied Psychological Services, Wayne, Pa.]

fairly arbitrary values, and only the "desirable" and "undesirable" directions are indicated. It can be seen, however, that all four criteria were best for the 65°-angle side panels. The consistency across all criteria (these and the others) was quite evident.

Foot Controls

Since only the most loose-jointed among us can put their feet behind their heads, the location of foot controls generally needs to be in the fairly conventional areas, such as those depicted in Figure 13-14. These areas, differentiated as optimal and maximal, for toe-operated and heel-operated controls, have been delineated on the basis of dynamic anthropometric data. The maximum areas indicated require a fair amount of thigh or leg movement or both, and preferably should be avoided as locations for frequent or continual pedal use. Incidentally, that figure is predicated on the use of a horizontal seat pan; with an angular seat pan (and more angled back rest) the pedal locations need to be manipulated accordingly (although such adjustments are published, they will not be given here).

The areas given in Figure 13-14 generally apply to foot controls that do not require substantial force. For applying considerable force, a

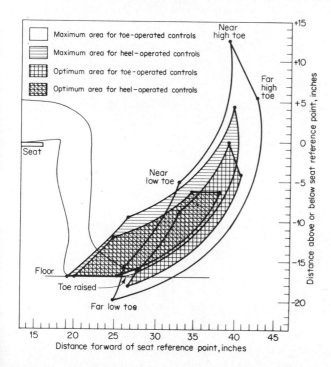

Figure 13-14 Optimum and maximal vertical and forward pedal space for seated operators. [Adapted from *Personnel subsystems*, 29, chap. 2, sec. 2D7, p. 3.]

pedal preferably should be fairly well forward. This point is illustrated by some data obtained at the Max Planck Institutes in Germany [5], as illustrated in Figure 13-15. That figure shows the sharp loss in *foot power* when the pedal requires more of a downward than forward thrust, or a thrust off to the side.

SPECIFIC ASPECTS OF ARRANGEMENT

The discussions above deal with some of the general approaches and considerations in the jig-saw process of arranging the components of some systems. There are, however, special considerations that are relevant

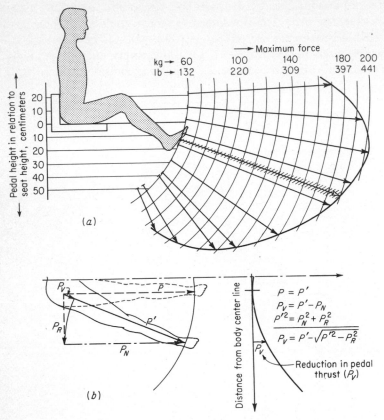

Figure 13-15 (a) Maximum pedal thrust in various directions relative to seat height, and (b) reduction in pedal thrust at various lateral distances from body center line. [Adapted from Dupuis, Preuschen, and Schulte, 5.]

to specific arrangement problems, as, for example, the compatibility between displays and controls (as discussed in Chapter 8) and the need to consider human psychomotor skills (as discussed in Chapter 10). For our present purpose we shall discuss, in particular, certain points on the spacing of control devices.

Spacing of Control Devices

Although we have talked about minimizing the distances between components, such as the sequential links between controls, there are obvious lower-bound constraints that need to be respected, such as the physical space required in the operation of individual controls to avoid touching other controls.

Spacing of knobs. In this particular connection, Bradley and Stump [1] investigated the spacing between knobs on a control panel, with particular reference to inadvertent errors in touching other knobs. They varied the diameter of the knob to be operated and the spacing between the knob and four others around it. The errors in touching the surrounding knobs are shown in Figure 13-16. From this and other results

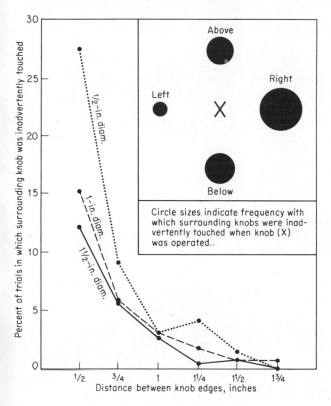

Figure 13-16 Frequency of inadvertent touching errors for knobs of various diameters as a function of the distance between knob edges. Areas of circles in inset indicate relatively the frequency with which the four surrounding knobs were inadvertently touched when knob x was being operated. [From Bradley and Stump, 1.]

of the study it was found that performance (in both touching errors, as shown, and time) improves quite rapidly with increasing distance between knob edges, up to a distance between edges of 1 in.; beyond that distance performance improves at a much slower rate. When comparisons are made between knob centers, rather than edges, however, performance is more nearly error-free for ½-in.-diameter knobs than for the larger knobs. This suggests that when panel space is at a high premium, the smaller diameter knobs are to be preferred. By referring again to Figure 13-16, it can be seen that touching errors were greatest for the knob to the right of the one to be operated, and were minimal for the one on the left.

Spacing of adjacent controls. On the basis of studies of control devices, Morgan et al. [19] have set forth certain recommended distances (preferred and minimal) between pairs of similar devices, as given in Figure 13-17.

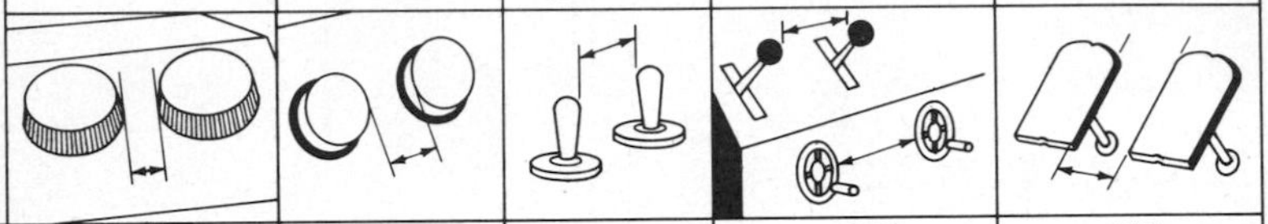

Number of body members and type of use	Knobs	Push buttons	Toggle switches	Cranks, levers	Pedals
1, randomly	2(1)	2(1/2)	2(3/4)	4(2)	6(4)
1, sequentially		1(1/4)	1(1/2)		4(2)
2, simultaneously	5(3)			5(3)	
2, randomly, sequentially	1/2 (1/2)		3/4 (5/8)		

Figure 13-17 Recommended separations, in inches, between adjacent controls. Preferred separations are given for certain types of use, with corresponding minimum separations in parentheses. [Adapted from Morgan et al., 19, table 7-38, p. 313.]

Spacing for blind location of control devices. Control devices that are to be reached and used without visual reference should have a separation between them great enough that devices are not confused with each other. This is essentially a matter of location coding, and the accuracy of location discrimination is based on the spatial differences that can be identified on an absolute basis. While coding was discussed in Chapter 4, it might be added that, on the basis of a study on distance discrimination [Fitts and Crannell, reported by Hunt, 7], it has been proposed that a "safe" distance between such devices as switches is about 5 in. when they are at shoulder height and within easy reach. Human discriminations at higher or lower levels, and in less easily accessible locations, probably are less precise; this suggests the desirability of greater spacing in such locations.

EXAMPLES OF PHYSICAL-ARRANGEMENT PROBLEMS

Recognizing that few systems are ideal in every respect, let us now illustrate a few items of equipment or facilities that pose some human factors problems in the location or arrangement of their components.

Vehicle-cab Areas

The cabs of vehicles serve to illustrate the need for integration of space relations in work situations. Since truck and bus drivers spend at least a fourth of their lives in the cabs of their vehicles, the design of their cabs is of some consequence to their comfort, and probably also to their safety.

Undesirable features of some cabs. In the evaluation of truck and bus cabs, McFarland and his associates at Harvard University [16, 17, 18] have unearthed certain features of some cabs that certainly contribute to difficulty of operation, to operator fatigue, and probably to accidents. Among these undesirable features were the following: hand brakes that required the driver to lean forward; a gear shift that was in such a position that the driver could not get his knee between it and the steering wheel when trying to apply the brake; and steering wheels that were so positioned that tall drivers could not get their knees under the wheel and thus had to apply the brake from an angle.

A study of cab areas. Seating is, of course, an important problem in designing a vehicle-cab area, but seating must be considered in relation to the situation. In a study pertaining to this problem, a mock-up of a truck cab was used in which a variety of adjustments could be made, as shown in Figure 13-18. Three groups of male subjects were used: 10 tall men, 10 average men, and 10 short men. They represented, approxi-

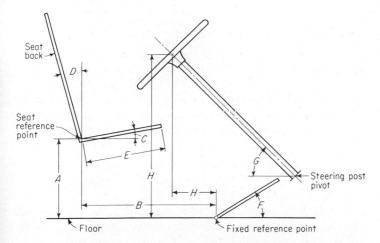

Figure 13-18 Schematic diagram of mock-up of truck cab used in studying cab areas. *A*, vertical seat height; *B*, horizontal seat position; *C*, angle of seat bottom; *D*, angle of seat back; *E*, length of seat bottom; *F*, angle of toe pan; *G*, angle of steering wheel; *H*, position of steering wheel. [Adapted from Kephart and Dunlap, 12.]

mately, the 95th 50th, and 5th percentiles of army truck drivers [Kephart and Dunlap, 12]. Each subject adjusted the seating, the steering mechanism, and the toe pan until they suited him. He then performed a "driving" task by manipulating the steering wheel to keep a pointer on a "road," which was a winding strip on a moving paper tape. Time and frequency "off the road" were recorded automatically. For comparative purposes, each subject also performed this driving task when the adjustments were set to conform to two standard cab arrangements.

Two important results came out of this study: In the first place, the study led to the recommendations for the design of truck cabs shown in Table 13-3. These recommendations take into account both human and

Table 13-3   Recommendations for Design of Truck Cabs

Feature	Recommendation
A. Vertical seat height	Adjustable: 14.69–17.39 in.
B. Horizontal seat position	Adjustable: 22.97–29.25 in.
C. Angle of seat bottom	Adjustable: 9.20–11.38°
D. Angle of seat back	Adjustable: 8.55–24.07°
E. Length of seat bottom	Fixed: 16.06 in.
F. Angle of toe pan	Fixed: 31.96° (28.75–35.17° acceptable)
G. Angle of steering wheel	Fixed: 45.34° (42.51–48.17° acceptable)
H. Position of steering wheel (center)	Fixed: 33.4 in. high, 8.5 in. in front of fixed reference point*

* Variation in the position of the steering wheel would be acceptable with the wheel center within the outlines of a parallelogram described as follows (horizontal distances from fixed reference point, vertical distances above floor): horizontal of 4.8 in. with vertical from 28.0 to 31.4 in.; horizontal of 12.25 in. with vertical from 35.4 to 38.8 in.

engineering considerations. The first four features are those for which adjustments are feasible from an engineering point of view, and the ranges are those which would accommodate about 90 percent of army truck drivers. The other features are not as amenable to adjustment and are set at approximately average values, or at values that would meet the operational needs of most of the population. Such values, of course, represent some compromise with the ideal design.

The second result of this study related to the performance of the subjects on the driving task. Their performance was significantly better for the cab that conformed to their own preferred adjustments than for two standard cab arrangements. While there is no assurance that performance on this task is related to actual driving ability, it is reasonable to suppose that driving performance would be related to comfort.

Cab of lift truck.   In the development of a sit-down lift truck for use in narrow aisles, a number of human factors problems were clearly obvious or came to light during the development process. Without pinpointing the specific problems nor describing the evolution of the design, we shall simply illustrate the final model in Figure 13-19.

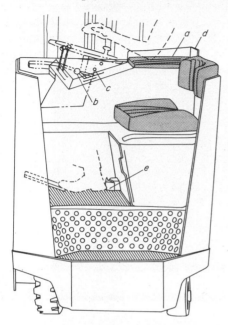

Figure 13-19   Rear view and view of floor controls of final Raymond Model 820 narrow-aisle sit-down lift truck, resulting from a series of human factors and engineering design modifications. Letter identifications: *a*, padded elbow rest, *b*, lift control, *c*, lower control, *d*, arm shield, *e*, heel-operated "butterfly" forward-reverse switch. [From Stevens et al., 26. Courtesy Raymond Corporation, Greene, N.Y.]

BART rapid-transit system.   The Bay Area Rapid Transit District, in the development of the BART car for use in the San Francisco Bay area, set forth certain human factors design criteria for the system, some of these relating to the comfort and convenience of the passengers, and others relating to the operation of the car. In the latter connection, the system will generally be controlled electronically by a central computer system, but there will be an attendant on each train for various purposes, especially for overriding the central computer system in case of emergency. Actually the attendant will be in a detachable control *pod* that can be positioned at the front of the first car. In the design of this pod, a prime requirement was that of ensuring adequate forward visibility to observe signals, the track ahead, and around curves, and backward visibility to observe the interior of the car. Figure 13-20 shows, for a prototype of the pod, the fields of view for the attendant, both when standing and sitting. In turn, Figure 13-21 shows the control console of the prototype pod.

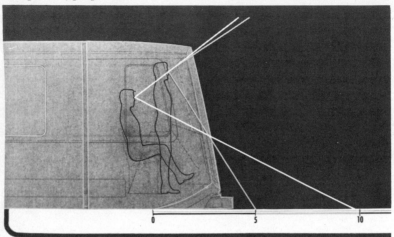

Figure 13-20   Fields of view from a prototype of the attendant's pod of the BART (The San Francisco Bay Area Rapid Transit District) car. The fields of view are for the attendant when standing and when seated and are shown from the top and side of the prototype pod. [From Sundberg and Ferar, 27, fig. 5.]

Crane Controls and Cabs

In the iron and steel industries, as well as certain others, traveling overhead cranes are used to move materials from one place to another. The operational requirements of this moving control would suggest the need for unobstructed vision, easily operated controls that respond with acceptable fidelity to the operator's responses, and a relatively comfortable posture. The British Iron and Steel Research Association has interested itself in the design and arrangement of equipment for use in iron and steel operations, including this one. On the basis of some research, plus the application of sound human factors principles, a modification was made of an original crane cab along with its associated controls. The before and after designs are given in Figure 13-22. The improved visibility, better arrangement of the controls, and posture of the operator can be seen from these photographs.

Figure 13-21   Attendant's control console for the prototype of the pod of the BART rapid-transit car described by Sundberg and Ferar [27]. [Photograph courtesy of Sundberg-Ferar, Southfield, Mich.]

Office Arrangement

Most of us who live our working hours in offices probably have become oblivious to the many inefficiencies of our conventional arrangement. As Propst [21] points out, although office work has undergone a revolution in recent decades, the functional format of offices has remained substantially static. But there are some stirrings of interest in the design of offices and office facilities with a view to creating facilities that would more adequately fulfill their current-day requirements. This reexamination would force one to analyze the functions that are to be carried out in offices, and to try to so design offices and their facilities that these functions can be performed more adequately and more efficiently. Admittedly research in office design poses a sticky problem, in part because relevant criteria are hard to come by.

However, some probing efforts are being made along these lines. For example, Herman Miller, Inc., has developed the concept of the Action Office [Propst 20, 21] that is predicated on certain principles such as the following: (1) enclosure and access (in line with the expression, "an office should neither closed nor open be," an objective should be that of achieving an appropriate degree of enclosure); (2) the vertical function of space (this principle deals with the effective use of vertical space, including walls); (3) work-station generation (providing for highly varied work

(a) Original crane cab

(b) Original crane controls

(c) Improved crane cab

(d) Improved crane controls

Figure 13-22 Illustration of original and improved cab and controls for overhead traveling crane used in steel mill. [From Laner, 15, and Sell, 23. Courtesy The British Iron and Steel Research Association.]

stations, adaptable to the needs of the individual); (4) the arena effect (providing an *arena* in which the individual is free to turn to separate work surfaces or to converse with visitors); (5) conversation and conference controls (providing for options in contacts with others, such as for person-to-person conversation, stand-up conversations, and conferences); (6) display (providing *display* elements such as shelves, blackboards, and racks, as necessary); (7) paper handling, storage, and retrieval (providing facilities that aid in the active use of information); (8) the telephone and office machines (providing for flexibility of location and use); (9) privacy and security; (10) portability and change (providing for movement and rearrangement of the various facilities); (11) traffic (providing for convenient traffic access of functionally interactive individuals and groups, and for insulation for diversionary traffic); and (12) social climate (providing a physical arrangement that generates a satisfactory social climate). An example of an office that illustrates the "arena" principle is shown in Figure 13-23.

Figure 13-23   An illustration of the concept of the Action Office, developed by Herman Miller Research Corporation. This particular example illustrates the arena principle of office arrangement in which an individual can have ready access to facilities that are associated with different functions. [From Propst, 21; courtesy The Business Press, Elmhurst, Ill., and Herman Miller Research Corporation, Ann Arbor, Mich.]

A modest experiment with an Action Office is reported by Fucigna [9] in which a few people used such facilities over a period of a few months. Although activity analyses of the inhabitants did not show material differences from the activities carried out in their previous offices, there was a rather universally favorable reaction to the Action Office on the part of the individuals, including a concensus that the arrangement facilitated various aspects of their activities.

SUMMARY

1. In the arrangement of components (displays, controls, equipment, etc.) two determinations need to be made, namely, their *general location* and their specific *arrangement* relative to others.

2. It is reasonable to hypothesize that most displays, controls, and other components people use have an *optimum* location in terms of relevant human sensory, anthropometric, and biomechanical characteristics. (In practice, of course, it is frequently not possible to place all such items in their optimum locations.)

3. Guidelines, based on research, have been developed for the optimum location of various types of components (visual displays, specific control devices, etc.). On the basis of such data, recommendations have been made about the design of consoles, work surfaces and areas, etc. Further, such guidelines can aid in the design of specific work stations and equipment.

4. Determinations about the location and arrangement of components sometimes can be made on the basis of such considerations as these: (*a*) importance (placing important components in their optimum locations), (*b*) frequency of use, (*c*) functional relationships, and (*d*) sequence of use.

5. In analyzing existing systems with a view to modifications of location and arrangement of components, different methods of analysis can be used, such as film analysis, eye-movement recordings, observation, interviews, and questionnaires. In the case of completely new systems being designed, the activities to be performed must be inferred from tentative design data and in some instances by the use of prototypes or models.

6. Usually some summary of data from such analyses is useful (such as frequency of use of components or average ratings of importance of components). Such summaries can serve as indexes for subsequent use in making determinations about location and arrangement of components.

7. The operational relationships between components sometimes can be expressed in terms of *link* values, these being either functional (essentially frequency) or sequential. Link values can characterize communication, control, or movement relationships.

8. The arrangement of related components preferably should be based on activity data such as reflected by link values, usually with the objective of optimizing the arrangement in terms of link values; this can be done graphically, by judgment, or by quantitative procedures.

REFERENCES

1. Bradley, J. V., and N. E. Stump: *Minimum allowable knob crowding*, USAF, WADC, TR 55–455, December, 1955.
2. Christensen, J. M.: *Aerial analysis of navigator duties with special reference to equipment design and work-place layout: II. Navigator and radar operator*

activities during three Arctic missions, USAF, Air Material Command, Engineering Division, Memorandum Report MC REXD 694–15A, Feb. 2, 1948 (restricted).

3. Deininger, R. L.: *Process sampling, workplace arrangements, and operator activity levels*, unpublished report, Engineering Psychology Branch, USAF, WADD, 1958.

4. Dupuis, H.: *Farm tractor operation and human stresses*, paper presented at the meeting of the American Society of Agricultural Engineers, Chicago, Dec. 15–18, 1957.

5. Dupuis, H., R. Preuschen, and B. Schulte: *Zweckmäbige gestaltung des schlepperführerstandes*, Max Planck Institutes für Arbeitsphysiologie, Dortmund, Germany, 1955.

6. Fitts, P. M.: "A study of location discrimination ability," in P. M. Fitts (ed.), *Psychological research on equipment design*, Army Air Force, Aviation Psychology Program, Research Report 19, 1947.

7. Fitts, P. M., and C. W. Crannell: "Studies in location discrimination," in D. P. Hunt, *The coding of aircraft controls*, USAF, WADC, TR 53–221, August, 1953.

8. Freund, L. E., and T. L. Sadosky: Linear programming applied to optimization of instrument panel and work-place layout, *Human Factors*, 1967, vol. 9, no. 4, pp. 295–300.

9. Fucigna, J. T.: The ergonomics of offices, *Ergonomics*, 1967, vol. 10, no. 5, pp. 589–604.

10. Haygood, R. C., K. S. Teel, and C. P. Greening: Link analysis by computer, *Human Factors*, 1964, vol. 6, no. 1, pp. 63–78.

11. Jones, R. E., J. L. Milton, and P. M. Fitts: *Eye fixations of aircraft pilots: IV. Frequency, duration, and sequence of fixations during routine instrument flight*, USAF, AF TR 5975, 1949.

12. Kephart, N. C., and J. W. Dunlap: *Human factors in the design of vehicle cab areas*, Department of Psychology, Purdue Department of Psychology, Purdue University, Lafayette, Ind., December, 1954.

13. Kobrick, J. L.: Effects of physical location of visual stimuli on intentional response time, *Journal of Engineering Psychology*, 1965, vol. 4, no. 1, pp. 1–8.

14. Konz, S. A., and R. A. Day: Design of controls using force as a criterion, *Human Factors*, 1966, vol. 8, no. 2, pp. 121–127.

15. Laner, S.: *Ergonomics in the steel industry*, The British Iron and Steel Research Association, Report 19/61, List 120, November–December, 1961.

16. McFarland, R. A., et al.: *Human body size and capabilities in the design and operation of vehicular equipment*, Harvard School of Public Health, Boston, Mass., 1953.

17. McFarland, R. A., et al.: *Human factors in the design of highway transport equipment*, Harvard School of Public Health, Boston, Mass., 1953.

18. McFarland, R. A., and A. L. Moseley: *Human factors in highway transport safety*, Harvard School of Public Health, Boston, Mass., 1954.

19. Morgan, C. T., J. S. Cook, III, A. Chapanis, and M. W. Lund (eds.): *Human engineering guide to equipment design*, McGraw-Hill Book Company, New York, 1963.

20. Propst, R.: The Action Office, *Human Factors*, 1966, no. 4, pp. 299–306.
21. Propst, R.: *The office: a facility based on change*, The Business Press, Elmhurst, Ill., 1968.
22. Randall, F. E., A. Damon, R. S. Benton, and D. I. Patt: *Human body size in military aircraft and personal equipment*, USAF, TR 5501, June 10, 1946.
23. Sell, R. G.: The ergonomic aspects of the design of cranes, *Journal of the Iron and Steel Institute*, 1958, vol. 190, pp. 171–177.
24. Sharp, E., and J. P. Hornseth: *The effects of control location upon performance time for knob, toggle switch, and push button*, AMRL, TR 65–41, October, 1965.
25. Siegel, A. I., and F. R. Brown: An experimental study of control console design, *Ergonomics*, 1958, vol. 1, pp. 251–257.
26. Stevens, P. H., D. O. Chase, and A. W. Brownlie: Industrial design of a narrow aisle sit-down lift truck, *Human Factors*, 1966, no. 4, pp. 317–325.
27. Sundberg, C. W., and M. Ferar: Design of rapid transit equipment for the San Francisco Bay Area rapid transit system, *Human Factors*, 1966, vol. 8, no. 4, pp. 339–346.
28. *Handbook of instructions for aerospace personnel subsystem design* (HIAPSD), USAF, AFSC Manual, 80–3, 1967.
29. *Personnel subsystems*, USAF, AFSC design handbook, Series 1–0, General, AFSC DH 1–3, 1st ed., Jan. 1, 1969, Headquarters, AFSC.

part
six
ENVIRONMENT

In Chapter 3 we discussed some of the variables that are related to the seeing process, such as luminance, contrast, luminance ratio, time to view whatever is to be seen, and illumination. Because these variables are so intertwined with each other, our present discussion of illumination obviously will have to include some consideration of some of them. In our discussion of illumination it is not our intent to cover the engineering aspects or to present recommendations of illumination standards for the many circumstances in which people find themselves. Rather, the primary focus will be on the interaction of illumination and related variables on human visual performance and visual comfort.

AMOUNT OF ILLUMINATION

Probably the most important consideration related to the human aspects of illumination is that of the amount required for various human tasks and activities. In this connection, it is pertinent first to raise the question, as related to the determination of the appropriate level of illumination *for a given visual task:* On the basis of what set of standards, or by what criterion or criteria, should the illumination level for the task be established?

Criteria of Adequate Illumination Levels

In the years gone by, considerable research effort has been devoted to the matter of illumination, especially by such individuals as Luckiesh and his associates [22–24, 27] and Tinker [34–37]. In the research of these and others, various criteria were used, including visual acuity (actually some measure of the ability to make visual discriminations under different illumination conditions); heart rate; blink rate; muscular tension; opinions; the critical level (the critical level as proposed by Tinker [37] is that level of illumination beyond which there is no appreciable increase in efficiency of visual performance); and certain optical systems for measuring visibility, such as the Luckiesh-Moss visibility meter and the Blackwell Visual Task Evaluator (VTE) (these systems generally take into account considerations of size, luminance, and contrast of visual tasks and result in estimates of illumination requirements for different tasks by comparison with certain standards). The historical note should be added that spirited controversies once raged about illumination research, particularly with regard to the appropriate criteria to use in establishing illumination standards. These controversies have cooled down in more recent years, and the Illuminating Engineering Society (IES) has adopted, as the basis is for establishing illumination standards, a procedure developed by Blackwell [2] that involves the use of the VTE.

Visibility Research of Blackwell

Because Blackwell's research has been accepted by the IES as the primary base for specifying interior illumination levels for tasks and activities, a summary of some of his work will be in order.[1]

Laboratory procedures and results. Blackwell's research was concerned primarily with the study of basic parameters of visibility. His laboratory setup consisted of a large cubicle painted white and illuminated by concealed lighting from the side. The subjects viewed one side of this that consisted of a translucent screen, the luminance of which could be varied from 0.001 to 800 fL. The target to be viewed consisted of a disk of light which could be projected onto that screen. The disk could be varied in size (1 minute of arc to 64), ratio of contrast with background (0.01 to 300), and time of exposure (0.001 to 1.0 sec). A given disk was projected during any one of four time intervals, these being delineated by a buzzer sound at the beginning of each interval. As a disk was presented on the screen (during one of the four intervals delineated by the buzzer), the subject was to identify the interval and press a button that identified the interval in question. Some of the results are summarized in Figure 14-1. This shows, for various target sizes (1, 2, 4, and 10 minutes of arc), the relationship between luminance of the background and the contrast between the background and the target disk required for 50 percent accuracy of identification of the disk. For estimating values that would be required for 99 percent accuracy, the contrast value should be multiplied by a factor of about 2 [Crouch, 9]. One can here see the beginnings of a scheme for utilizing such data for determining illumination requirements for various tasks.

Before pursuing the further details of such a scheme, however, there are a couple of additional aspects of the research that should be mentioned. One of these deals with the use of a *static* viewing condition versus a dynamic *moving-eye* condition. For studying dynamic conditions, a *field task simulator* was constructed. This consisted of a 7-ft wheel with 50 circular plaques, 4 in. in diameter, around the circumference, as illus-

---

[1] The details of his research and of the procedures for specifying illumination levels for various tasks and activities are available in other sources [Blackwell, 2, 4; Crouch, 9; 46, p. 425; 47, p. 428; *IES lighting handbook*, 42].

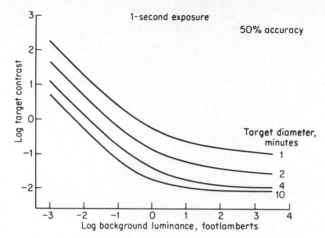

Figure 14-1 Relationship between luminance of the background and contrast required for 50 percent accurate discriminability of targets of different sizes. [From Blackwell, as presented by Crouch, 9.]

trated in Figure 14-2. A few of the plaques (from 1 to 6) had defects, and the subjects were to identify the defective ones as the wheel rotated. The amount of illumination was varied systematically for different experimental testing sessions. By a process that need not be described here, it was possible to derive a *field factor* value that, in effect, would make it possible to equate the visibility requirements of static and of moving targets. On the basis of updated data [*IES lighting handbook*, 42, pp. 2–14], this field factor is established as 6.67, meaning that, for equal visibility, moving targets would require the light necessary for the same size test object with a contrast of 1/6.67 that of a static (laboratory) task.

The other point that should be mentioned relates to the concept of *visual capacity* advanced by Blackwell [2]. Various studies of eye fixation have indicated that, under normal, unhurried circumstances, the average visual fixation is about $\frac{1}{4}$ to $\frac{1}{5}$ sec. Such fixations give rise to the notion of *assimilations*, which, in turn, can be time related. Thus, four or five assimilations per second would occur under normal visual-fixation conditions. With more assimilations per second than this, the contrast would have to be increased to bring about equal visual performance.

Comparison of actual visual tasks with standard laboratory task. All this is leading up to the derivation of a *field-use curve*, which, in turn, serves as part of a bridge between the extensive laboratory data, on the one hand, and the use of such data in establishing illumination standards, on the other hand [Crouch, 9]. This curve, shown in Fig. 14-3, shows the

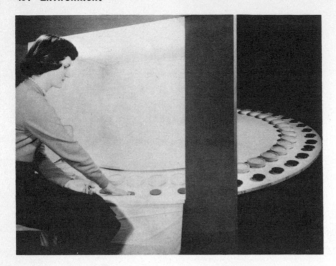

Figure 14-2 Illustration of field task simulator used for studying visibility requirements of moving objects. The subjects were to identify the defective plaques as the wheel rotated to bring them into view. [From Blackwell, 2.]

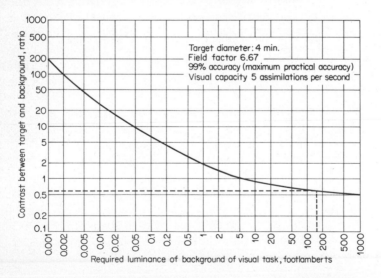

Figure 14-3 Field-use curve used in derivation of illumination standards with Visual Task Evaluator (VTE). All unknown field tasks may be related by the VTE to this curve, for a circular disk of 4 minutes of visual arc, which has been weighted to represent moving-eyes field conditions and maximum field accuracy. See text for discussion. (Dotted line shows how to determine luminance for given contrast.) [Based on updated data presented in *IES lighting handbook*, 42, fig. 2–26.]

relationship of luminance of background and of the contrast between the target and background required for the following conditions: 99 percent accuracy, target diameter of 4 minutes of arc, a field factor of 6.67 (the general correction for seeing moving objects under field conditions), and visual capacity of 5 assimilations per second.

The bridge, however, requires another process, namely, equating the visual task for which one wishes to determine illumination standards with the now standardized laboratory visual task for which data are given in Figure 14-3. This is done with the Visual Task Evaluator (VTE). With this device the material to be discriminated in the visual task in question is viewed under controlled optical and illumination conditions. The contrast of that task material is reduced by placing over it a "veiling" brightness which, as it is increased, begins to wash out the details and brings the configuration to the threshold of being seen. By a subsequent operation, with another optical device, the visibility of the task in question is equated with that of the standard 4-minute disk target. The next step is that of determining (from Figure 14-3) the luminance required to see the disk under field-use conditions and, by implication, the luminance required to see the previously equated task material to be viewed.

Determining Illumination Level Recommendations

The research summarized above and the use of the VTE provide a basis for determining illumination levels for various tasks, but do not provide convenient procedures for so doing. Since the scheme requires a number of manipulations and intervening operations, one might wonder whether there would be enough "slippage" to limit its general application. In a number of subsequent validation studies summarized by Blackwell [5] and Blackwell and Blackwell [6], there is substantial evidence to indicate that the visual task evaluation procedures proposed actually do result in assessments of task difficulty that jibe well enough with real life conditions of viewing. In one comparison [Blackwell and Blackwell, 6] a very high degree of correspondence was reported between the relative *contrast sensitivity* as a function of luminance for the disk targets under normal experimental conditions, and in a field task simulator, as shown in Figure 14-4. Relative contrast sensitivity is the inverse of the minimum contrast required for threshold detection. On the basis of this figure and other data, it appears that one can summarize the effects of the quantity of illumination upon visual performance of many visual tasks by a single curve relating contrast sensitivity to task background luminance. Toward this end, five categories of difficulty of seeing tasks have been established by the Illuminating Engineering Research Institute [9, p. 422], as shown in Table 14-1. These are expressed in terms of guide brightness in *foot-lamberts*. The *footcandles* required to produce a given level of guide bright-

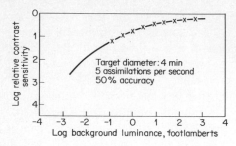

Figure 14-4 Relative contrast sensitivity as a function of luminance for disk targets (1) under normal conditions (shown by curve), and (2) in the field task simulator (shown by points). [From Blackwell and Blackwell, 6, fig. 4.]

Table 14-1 Categories of Difficulty of Seeing Tasks with Guide Brightness and with Footcandles for Specified Reflectance Conditions

Category of seeing task	Guide brightness, fL	Footcandles for specified reflectance conditions		
		90%	*50%*	*10%*
A. Easy	Below 18	Below 20	Below 36	Below 180
B. Ordinary	18–42	20–45	36–84	180–420
C. Difficult	42–120	45–133	84–240	420–1200
D. Very difficult	120–420	133–455	240–840	1200–4200
E. Most difficult	420 up	455	840	4200

SOURCE: Adapted from *Illuminating Engineering* [46, p. 422].

ness in footlamberts, however, is a function of the (average) reflectance of the object being viewed. (An illustration of low reflectance and low contrast is shown in Figure 14-5.) The illumination required for a task is derived as follows:

$$\text{Required illumination, fc} = \frac{\text{required luminance, fL}}{\text{reflectance}}$$

Table 14-1 also shows the footcandles required for three reflectance conditions (90, 50, and 10 percent) for the five categories. As a further phase in the procedures for deriving the illumination levels required for specific visual tasks, Blackwell used the VTE to derive illumination standards for 56 bench-mark tasks, these in turn being classified in the five "categories" shown in Table 14-1 [46, p. 423]. A few examples of these are given in Table 14-2.

Figure 14-5 Example of a visual task object that has very low reflectance and low contrast between the detail to be discriminated and its background. A spot on gray cloth such as this would require 1100 fc in order to be discriminated adequately under operational conditions. With similar contrast but higher reflectance the footcandle requirements would be lower. [From Crouch, 9.]

The specific procedures in the estimation of illumination requirements will not be spelled out [see *Illuminating Engineering*, 46, p. 424], but in general they include (1) the selection of the most difficult seeing task and the measurement of its average reflectance; (2) the selection, from the 56 bench-mark tasks, of the one that is most nearly comparable to the task in question and the determination of its corresponding luminance in footlamberts from the table [46, table 2, p. 423; in this text Table 14-2 shows a few examples]; (3) dividing this luminance level by the average reflectance of the task in question to determine the range of footcandles.

The Committee on Recommendations for Quality and Quantity of Illumination of the IES has presented recommendations based on the application of this procedure, these being for industrial areas, institutions, offices, stores, residences, and marine transportation [46, pp. 425–432]; other sets of recommendations have been published elsewhere. A few examples from these are given in Table 14-3.

Table 14-2 Examples of Visual Tasks Classified by Difficulty of Seeing

Task number and description	Required brightness, fL	Reflectance, %	Required illumination, fc
A. Easy tasks:			
23 Light stitching on blue serge cloth, vertical stitching	0.09	0.018	5.05
7 12-point text type	0.42	0.70	0.60
17 Gray line defect on black cloth	0.66	0.024	27.5
18 Darned blemish on gray cloth	7.41	0.10	74.1
B. Ordinary tasks:			
41 Small skip defect, viewed from 30 in.	35.6	0.19	187
C. Difficult tasks:			
21 Orange chalk on light-brown tweed	47.9	0.18	266
25 Dark-brown raised threads on silk	58.9	0.035	1680
8 Sample of shorthand copy with No. 3 pencil	60.5	0.79	76.5
D. Very difficult tasks:			
42 Chip grain, viewed from 30 in., no specular	195	0.19	1030
13 Thermofax copy, poor quality	371	0.63	589
E. Most difficult tasks:			
26 Blue-gray raised threads on silk	447	0.046	9650
10 Typed original, extremely poor ribbon	1950	0.62	3140
50 Brown-thread stitching on brown silk tweed	>3000	0.12	10,000

SOURCE: From *Illuminating Engineering* [46, p. 423].

Discussion

Although the IES procedures for deriving illumination standards for various tasks and activities are predicated on the very extensive and impressive research by Blackwell, there are still a few qualifications, loose ends, and dangling questions relating to this matter. For example, most of the current recommendations are higher than previous standards and probably are thus at odds with some earlier research such as that of Tinker [37]. Further, it has been suggested that lighting can weaken information cues [Logan and Berger, 21] by suppressing the "visual gradients" of the pattern density of the objects being viewed. In other words, high levels of illumination might tend to minimize the (perceived) differences in the features of the object, such as by reducing shadows which characterize features of the object. Thus, to the extent that this might be the case, more footcandles might reveal less than fewer footcandles.

Table 14-3 Illumination Standards Recommended by the IES for Several Selected Types of Situations and Tasks

Situation or task	*Recommended illumination, fc*
Assembly:	
Rough easy seeing	30
Rough difficult seeing	50
Medium	100
Fine	500
Extra fine	1000
Machine shops:	
Rough bench and machine work	50
Medium bench and machine work	100
Fine bench and machine work	500
Extra-fine bench and machine work, grinding—fine work	1000
Storage rooms or warehouses: Inactive	5
Offices:	
Cartography, designing, detailed drafting	200
Accounting, bookkeeping, etc.	150
Regular office work	100
Corridor, elevators, stairways	20
Residences:	
Kitchen, sink area	70
Kitchen, range and work surfaces	50

SOURCE: Examples selected from *Illuminating Engineering* [46, pp. 425–432].

In addition, people whose visual acuity is noticeably below par (such as aged individuals) require more illumination than those with normal acuity [Fortuin, 11; Guth and Eastman, 13; and Guth et al., 14]. In order to compensate for the usual reduction in visual acuity with age, Fortuin [11] proposes that additional illumination be provided, and he has developed equations to use in estimating the required luminance for individuals of advanced ages.

Some of the work of Tinker, the argument posed by Logan and Berger, the qualifications and cautions mentioned by Blackwell [2], and other cues suggest that there still are many remaining questions to this problem.

Illumination Level in Work Activities

In all this, there is an implicit assumption that the level of illumination for a work task may have some bearing on the performance and other criteria of people on the task in question. The possible effect of illumination on work performance undoubtedly would depend in part upon the criticalness to the job of visual discriminations and upon the difficulty of those discriminations (size, contrast, movement, etc.). Although there have been many laboratory studies of visual performance, oddly enough there have been relatively few documented studies of actual work performance under various levels of illumination. The results of a few such surveys are given in Table 14-4. In each case the change in illumination

Table 14-4　Results of Surveys Showing the Change in Work Output Following Improvement of Illumination of Work Areas

Type of work activity	Illumination, fc		Change in work output, %	Source
	Original	New		
Metal-bearing manufacturing	4.6	12.7	15	Luckiesh and Moss [26]
Steel machining	3.0	11.5	10	Luckiesh and Moss [26]
Carburetor assembly	2.1	12.3	12	Luckiesh and Moss [26]
Iron manufacturing	0.7	13.5	12	Luckiesh and Moss [26]
Buffing shell sockets	3.8	11.4	9	Luckiesh and Moss [26]
Letter sorting	3.6	8.0	4	Luckiesh and Moss [26]
Piston-ring manufacturing	1.2	6.5	13.0	Magee, as described by
	1.2	9.0	17.9	Luckiesh [22]
	1.2	14.0	25.8	
Inspecting roller bearings	2.0	6.0	4	Hess and Harrison [16]
	2.0	13.0	8	
	2.0	20.0	12.5	
Iron-pulley finishing	0.2	4.8	35	Viteles [38, p. 301]
Spinning	1.5	9.0	17	Viteles [38, p. 301]
Weaving worsted cloth	13.3	29.0	5.3	Weston [39]
Spinning wool yarn	11	42	9.6	[44]
Weaving automobile cloth	14–17	32	4.7	[44]
Card punching	28	49	6.7	[43]
Mail handling (Richmond Post Office)	10	45–50	8.0	[40]

level is indicated, and the change in work output is reported. In connection with such surveys, it should be kept in mind that other variables in addition to illumination conceivably can change, as, for example, the psychological atmosphere.

A two-step illumination-improvement plan in one heavy-manufacturing company was followed by a decrease in the accident rate [48]. First, the illumination in the erection shop of this company was increased from 5 to 20 fc by the use of a combination of mercury-vapor and incandescent lamps. Later the shop was painted in order to get better light usage and a more favorable luminance ratio. Accidents in the shop dropped 32 percent after the lights were changed and dropped an additional 11 percent after the painting, a total drop of 43 percent.

GLARE

Glare is produced by brightness within the field of vision that is sufficiently greater than the luminance to which the eyes are adapted to cause annoyance, discomfort, or loss in visual performance and visibility. *Direct* glare is caused by light sources in the field of view, and *reflected* or *specular* glare is caused by reflections of high brightness from polished or glossy surfaces that are reflected toward an individual. Most of the research relating to glare has been concerned with direct glare, with a view to the creation of illumination conditions that provide adequate levels of illumination but minimize such glare.

Effects of Glare on Visual Performance

The effects of glare on the visual performance of people are illustrated by the results of a study in which the subjects viewed test targets with a glare source of a 100-watt inside-frosted tungsten-filament lamp in various positions in the field of vision [25]. The test targets consisted of parallel bars of different sizes and contrasts with their background. The glare source was varied in position in relation to the direct line of vision, these positions being at 5, 10, 20, and 40° with the direct line of vision as indicated in Figure 14-6. The effect of the glare on visual performance is shown as a percentage of the visual effectiveness that would be possible

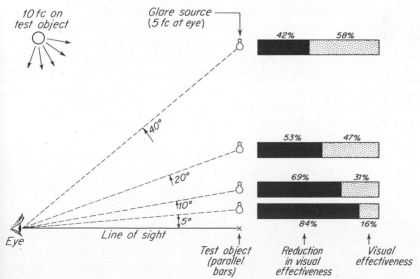

Figure 14-6 Effects of direct glare on visual effectiveness. The effects of glare become worse as the glare source gets closer to the line of sight. [After Luckiesh and Moss, 25.]

without the glare source. It will be seen that with the glare source at a
40° angle, the visual effectiveness is 58 percent, this being reduced to
16 percent at an angle of 5°.

Glare and Visual Comfort

Visual discomfort from glare is, unfortunately, a common experience, and
a major concern in the design of luminaires and lighting installations
should be to minimize such discomfort. Toward this end, there has been
considerable research in the past couple of decades or so relating to glare
and its effects on the subjective sensations of visual comfort and discomfort. Much of this research has been carried out at the Lamp Division,
General Electric Company, Cleveland, Ohio [Allphin, 1; Guth, 12; Guth
and Eastman, 13; Guth and McNelis, 15; Luckiesh and Guth, 24]. As a
direct result of such research, the IES has adopted a standard procedure
for computing *discomfort glare ratings* (DGR) for luminaires and for
tentatively planned interior lighting situations [49]. Involved in this
formulation is the notion of the *borderline between comfort and discomfort*
(BCD), which is essentially a subjective threshold, or limen [Luckiesh and
Guth, 24].

Derivation of Discomfort Glare Ratings (DGR). The DGR lends
itself to various purposes, such as the preparation of *glare tables* for
typical or specific types of luminaires or for deriving ratings of tentative
light layouts. Although we shall not present the details of the procedures
[see Reference 49], it may be useful to know what variables affect visual
comfort and to understand certain aspects of the procedures. The calculation of such ratings for specific lighting layouts takes into account most
of the situational factors that affect visual comfort, as follows: (1) room
size and shape; (2) room surface reflectances; (3) illumination level; (4)
luminaire type, size, and light distribution; (5) number and location of
luminaires; (6) luminance of entire field of view; (7) observer location and
line of sight; and (8) equipment and furniture. In addition, the procedures
can take into account a ninth variable, if desired, namely, differences in
individual glare sensitivity.

The scheme for estimating the DGR for any specific lighting layout
is basically the derivation of an index of sensation for each source, M,[2]

---

[2] The basic equation for M (index of sensation for each source) is

$$M = \frac{L s Q}{P F^{0.44}}$$

in which L_s = luminance of each source, fL
$\quad\quad\ Q$ = function of the solid angle (derived by a formula based on the solid
$\quad\quad\quad\quad$ angle, in steridians, of the two-dimensional area subtended by the
$\quad\quad\quad\quad$ light source)
$\quad\quad\ P$ = position index of each source
$\quad\quad\ F$ = average luminance of entire field of view, fL

and subsequently deriving the DGR for the entire ensemble of luminaires. Bypassing the several intervening calculations, the DGR is derived from the sum of the indices M of the several individual sources of illumination and the number of such sources, as shown in the nomograph in Figure 14-7. Aside from the derivation of the DGR from this nomograph, it is

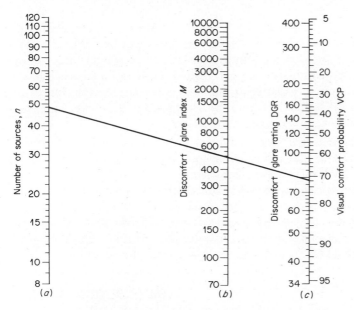

Figure 14-7   Nomogram for computing discomfort glare ratings (DGR) and visual comfort probabilities (VCP). Procedure: Draw a line from the number of sources, n on axis a through the discomfort glare index M on axis b to axis c, from which the DGR and VCP are obtained. [*Illuminating Engineering,* 49, fig. 4, p. 655.]

also possible to take into account the factor of individual differences mentioned above, specifically by a direct conversion of the DGR into a visual comfort probability (VCP), which is the percentage of individuals who generally would be expected to be visually comfortable (i.e., to be above the BCD). Aside from the research underlying this general formulation Guth [12] has, in effect, validated the procedure in a specific interior lighting installation. Subsequently McGowan and Guth [28] developed procedures for extending the application of this procedure to a wider variety of lighting situations.

In the application of the procedure, it has been stated by the IES Subcommittee on Direct Glare [49, p. 643] that direct glare will not be a problem in a lighting installation if three specific conditions are fulfilled. One of these is that the VCP be 70 or more.[3]

The Reduction of Glare

The methods of glare reduction depend, of course, upon the nature of the glare.

- *To reduce direct glare:* (1) Select luminaires with low DGR; (2) reduce the luminance of the light sources (for example, by using several low-intensity luminaires instead of a few very bright ones); (3) position luminaires as far from line of sight as feasible; (4) increase luminance of area around any glare source, so the luminance (brightness) ratio is less; and (5) use light shields, hoods, and visors where glare source cannot be reduced.

- *To reduce reflected glare:* (1) Keep luminance level of luminaires as low as feasible; (2) provide good level of general illumination (such as with many small light sources and use of indirect lights); (3) use diffuse light, indirect light, baffles, window shades, etc.; (4) position light source or work area so reflected light will not be directed toward the eyes; and (5) use surfaces that diffuse light, such as flat paint, nongloss paper, and crinkled finish on office machines; avoid bright metal, glass, glossy paper, etc.

DISTRIBUTION OF LIGHT

As described earlier, the luminance ratio is the ratio of the luminance of a given area (usually the work area) and a surrounding area. There are bits of evidence that visual performance is generally enhanced if the ratio between the work area and the surrounding area is reasonably limited.

Luminance Ratios

Based on the fragmentary research evidence, but perhaps more on experience, it has become general practice to provide adequate general illumination for work tasks, with somewhat higher levels of illumination at the site of the visual task. The ratios recommended by the IES for various areas relative to the visual task, for both office and industrial situations, are given in Table 14-5. Certain other recommendations, however, depart somewhat from these ratios; Kahler [18], for example, proposes luminance ratios about double those of the IES.

---

[3] The others are that the ratio of maximum-to-average luminaire luminance should not exceed 5:1 (preferably 3:1) at 45, 55, 65, 75, and 85° from the nadir crosswise and lengthwise, and maximum luminances of luminaires should not exceed values of 2250, 1605, 1125, 750, and 495 fL, respectively, at these angles.

Table 14-5 Recommended Luminance Ratios for Offices and Industrial Situations

Areas	Recommended maximum luminance ratio	
	Office	Industrial
Task and adjacent surroundings	3:1	
Task and adjacent darker surroundings		3:1
Task and adjacent lighter surroundings		1:3
Task and more remote darker surfaces	10:1	10:1
Task and more remote lighter surfaces	1:10	1:10
Luminaires (or windows, etc.) and surfaces adjacent to them	20:1	20:1
Anywhere within normal field of view	40:1	40:1

SOURCE: From *IES lighting handbook*, 42, fig. 11–11, p. 11–7, and fig. 14–2, p. 14–3.

In this connection, there is at least a hint that, for some visual tasks, higher concentrations of illumination at the center of the work—and lower levels of general illumination—may be justified. Hopkinson and Longmore [17], for example, offer evidence that, in some operations, the greater concentration of local lighting tends to focus visual attention on the most important work areas. In part this means that the luminaires must be appropriately positioned in order that the concentration of light is at the location where there should be the most visual attention. This is illustrated in Figure 14-8 for a wood planer.

Figure 14-8 Effects of change of location of luminaires on wood planer. Original position (left) caused reflection from polished surface, which distracted attention from cutters. Relighting (right) gives lower brightness reflection, which makes cutter the center of attraction. [From Hopkinson and Longmore, 17, by permission of the Controller of H. M. Stationery Office.]

The fact that, at least in some circumstances, more local lighting and less general lighting have been found to be desirable does not necessarily mean that this should hold true for all, or many, work situations.

Rather, it serves as an additional reminder that human performance is complex, subject to influence by numerous variables. Simple, pat, all-embracing answers to some human factors problems are few and far between. In the case at hand, the challenge to research is that of further delineating those *kinds* of situations in which more local illumination would be in order.

Reflectance

But the distribution of light within a room is not only a function of the amount of light and the location of the luminaires. It is also a function of the reflectance of the walls, ceilings, and other surfaces in the room, that is, the percentage of light reflected from a surface (sometimes called a *reflectance factor*). And, incidentally, chromatic surfaces of equal reflectance with neutral (gray) surfaces will produce greater illumination from the *interflections* of the light as it bounces around the room from one surface to another and another. Tied in with reflectance is the concept of the utilization coefficient, which is the percentage of light that is reflected, collectively, by the surfaces in a room or area. Table 14-6 shows

Table 14-6 Effect on Illumination of Various Ceiling, Wall, Floor, and Furniture Combinations

Ceiling		*Walls*		*Floor*		*Furniture*		*Utilization coefficient,*
Color	*RF**	*Color*	*RF*	*Color*	*RF*	*Color*	*RF*	*%*
Cream	65	White and gray	40	Dark red	12	Dark oak	20	29
Cream	85	White and gray	40	Dark red	12	Dark oak	20	33
Cream	85	Green	72	Dark red	12	Dark oak	20	45
Cream	85	Green	72	White	85	Dark oak	20	56
Cream	85	Green	72	White	85	Blond	50	57
Cream	85	Green	72	White and russet	70	Blond	50	55

* RF = reflectance factor (percentage of light reflected).
SOURCE: A. A. Brainard and R. A. Massey, Salvaging waste light for victory, *Edison Electric Institute Bulletin*, 1942, vol. 10, pp. 341–343, 355.

the utilization coefficient for each of several combinations of reflectance of ceilings, walls, floor, and furniture. The last three combinations of conditions just about double the utilization coefficient over the first one.

In order to contribute to the effective distribution and utilization of light in a room, it is generally desirable to use rather light walls, ceilings, and other surfaces. However, areas of high reflectance in the visual field can become sources of reflected glare. For this, and other reasons (including practical considerations), the reflectances of surfaces in a room (such as an office) generally increase from the floor to the ceiling. Figure 14-9 illustrates the IES recommendations on this score, indicating for

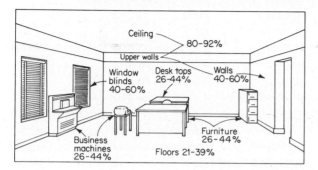

Figure 14-9 Reflectances recommended for room and furniture surfaces in offices. [From *IES lighting hand-book*, 42, fig. 11-10, p. 11–7.]

each type of surface the range of acceptable reflectance levels. Although that figure applies specifically to offices, essentially the same reflectance values can be applied to other work situations, as in industry [42, fig. 14-3].

TYPE OF LUMINAIRE

Luminaires come in many varieties, the most common types being incandescent (most of which have tungsten filament) and fluorescent. For special purposes other types are used, such as mercury and sodium lamps.

Visual Performance with Different Types of Luminaires

For interior lighting purposes, there is no consistent and systematic superiority of one type of luminaire over others in visual performance. In one study, for example, a comparison was made of incandescent, sodium, clear mercury, and color-improving mercury lamps [Eastman and McNelis, 10]. There was no significant difference among these either in the threshold contrast required for equal visibility or in glare (as indicated by the BCD). It should be noted, however, that—while not statistically significant—the BCD for clear mercury lamps was about one-third less that for the other three, which suggests (but does not demon-strate conclusively) that clear mercury lamps may be slightly more glar-ing. There are, however, at least some types of tasks in which the type of luminaire does have some impact on task performance. For example, a comparison was made of performance on four laboratory tasks carried out with tungsten filament and fluorescent lighting [Lion, 19], with the following results:

Task and criterion	Tungsten	Fluorescent
1. Grading ball bearings, no. sorted	306.2	322.1*
1. Grading ball bearings, no. errors	30.7	34.1
2. Needles, no. threaded	149.7	156.2*
3. Number reading, no. read	91.6	91.7
4. Rod measuring, time, sec	718.3	695.8*
4. Rod measuring, no. errors	13.4	13.3

* Statistically significant.

On the three manipulative tasks (1,2,4) the subjects worked more quickly under fluorescent light without any detrimental effect on accuracy of performance; it was suggested that this might be the consequence of lower glare with fluorescent than with incandescent tungsten lamps. In a subsequent investigation [Lion et al., 20] it was found in the inspection of links (black plastic disks with white links engraved on one side) that fewer faulty links were missed under fluorescent light (22.7) than under tungsten light (26.3), although there was no appreciable difference in the inspection of buttons for off-center holes. The investigators suggested that the link inspection task, being more of a visual acuity task than the button-sorting task, is at a disadvantage under a *point source* of incandescent tungsten illumination, as contrasted with the more diffused fluorescent illumination. Thus, at least as compared with tungsten lamps, it seems that fluorescent lamps may have an edge for fairly demanding visual tasks.

However, fluorescent lamps have some possible disadvantages in their use, in particular in the flicker effect caused by the on-off cycle (usually 60 Hz). In this connection Rey and Rey [33] report that in performing an intricate, mechanized test, performance on the test and on three *fatigue* tests given before and after the test (flicker fusion frequency, reaction time, and rhythmical irregularity) showed deterioration under a fluorescent lamp of 50 Hz as contrasted with one functioning at 100,000 Hz (which is more comparable to a continuous light). With the use of pairs of fluorescent tubes (with opposite phase) and the use of many luminaires in a given room, however, the flicker effect presumably is not of major general consequence, even though it might be relevant in the case of certain demanding visual tasks.

Spectral Quality of Luminaires

The spectral characteristics of different types of lamps vary considerably. For example, most filament lamps tend to enhance the reds, oranges, and yellows and to subdue the greens and blues. While fluorescent lamps vary considerably in spectral qualities, they tend more to accentuate the blues, greens, and somewhat the yellows. The color which people perceive is the consequence of the spectral characteristics of the lamp and of the spectral absorption characteristics of the colorant of the object in question. Thus, blues tend to look greenish under yellow lamps. Such color distortion effects, however, generally do not have any adverse effects on visual performance except in those tasks where some color discrimination is involved. In such a case, the color of the illuminant should be carefully selected in order to facilitate the color discriminations required.

ILLUMINATION FOR SPECIAL SITUATIONS

There are, of course, many circumstances that require special illumination installations in order to make the necessary visual discriminations possible. Two examples will be given to illustrate the human factors questons that bob up in illumination problems.

Roadway Illumination

It is not being proposed that every street, road, or lane be illuminated, but highway engineers have provided illumination for some particularly well-traveled or particularly hazardous stretches of streets, highways, and intersections. The before-and-after accident records in a few of these situations provide extremely persuasive evidence that this pays off. Following is a summary of the data compiled by the Street and Highway Lighting Bureau of Cleveland for 31 thoroughfare locations throughout the country showing traffic deaths for the year before they were illuminated and for the year after [45, pp. 585–602]:

		Reduction	
Year before illumination	Year after	No.	Percent
556	202	354	64

Such evidence is fairly persuasive, and argues for the expansion of highway illumination programs.

In connection with road and highway illumination, Christie [8] makes the point that in the absence of direct measures of "visual safety" (or even of visibility), it is customary to focus attention on the attributes of the lighting which are thought to have the greatest effect on visibility, as follows: the general level of road luminaires, the degree of patchiness (i.e., the variability of road luminance), and the amount of disability glare (i.e., the effect of glare on visibility), and sometimes discomfort glare. Some of the situational variables that can affect visibility and visual comfort include the luminaires (their power, elevation, number, and spacing), road surface, etc. Some practices that have been recommended (and that to some extent are being implemented) are discussed below [adapted in part from Rex, 29].

Provision of transition lighting.   Where roadway luminaires are to be used at some particular location (such as an intersection) it is desirable to provide some transition, as by extending the lighting system in each approach and exit direction, using approximately the same spacing and size of luminaire, but graduating the size of lamp used.

Better system geometry.   There is a trend toward improving the geometry of highway lighting systems, especially with respect to the mounting height of luminaires; in some installations the mounting height is 35 ft or more, and in some European systems installations at heights of 40 ft or more are increasing. Such heights (while requiring larger lamps) help to increase the *cutoff* distance, that is, the distance from the light source at which the top of the windshield cuts off the view of the luminaire from the driver's eyes. Under average conditions this cutoff is about 3.5 times the mounting height (MH) of the luminaire.

The effect of the geometry of the light system upon visibility is illustrated in Figure 14-10. This shows the relative visibility (from the driver's seat) at various positions along a lighted roadway, the visibility being the consequence of the locations of the luminaires, and the cutoff distances (the cutoff from each luminaire is somewhere around 3.5 times the MH, and visibility tends to be greatest at such distances).

Luminaire design.   Through appropriate design of luminaires, the distribution of candlepower can be controlled to some extent. In particular, by improved design it is possible to control the cutoff of luminaire candlepower at driver approach distances that are greater than the top of the windshield cutoff and thus improve visual comfort and visibility. This was illustrated by the results of a survey of 121 drivers who rated each of two stretches of illuminated roadway, one having an installation of *noncutoff* luminaires and the other of *cutoff* luminaires [Christie, 8]. The mean ratings on glare and visibility are given below (based on a scale from very poor to very good, ranging from -2 to $+4$):

	Type of luminaire	
	Cutoff	*Noncutoff*
Glare rating, mean	1.8	-0.9
Visibility rating, mean	1.3	0.4

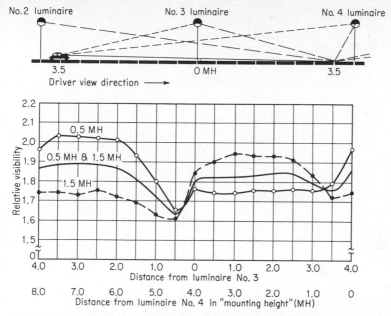

Figure 14-10 Relative visibility produced along roadway in relation to location of luminaires. Distances are given in units of mounting height (MH) of the luminaires. Visibility is shown for three transverse locations, that is, the relationship of the road lane to the position of the luminaires off to the side of the road, these being expressed in terms of MH, namely, 0.5 MH, 1.5 MII, and the average of these two (0.5 and 1.5). Normally, visibility increases as the view of the luminaire is blocked by the windshield, this cutoff distance usually being about 3.5 MH. [Adapted from Rex; 30.]

While on the subject of roadway luminaires, it might be relevant to illustrate the differences in the amount of illumination produced by luminaires of different types. Figure 14-11 shows the minimum and average illumination (in footcandles) and of pavement illumination (in footlamberts) produced by high-intensity sodium, iodine mercury, and clear mercury lamps (400-watt) at two mounting heights [Rex, 32]. Translating such data into economic terms, it is proposed by Rex that the high-intensity sodium lamp is the best investment now in units of illumination per mile.

Roadway surface. Difference in the reflectance characteristics of the surface of roadways are illustrated by the fact that asphalt pavement surface (about 6 years old) reflects approximately 8 percent of the light, whereas concrete pavement has a surface reflectance of about 20 percent [Rex, 31]. It has been estimated that the amount of illumination (in foot-

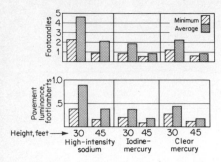

Figure 14-11 Minimum and average illumination (footcandles) and pavement luminance (footlamberts) produced by 400-watt luminaires of three types of roadway lights at two mounting heights (MH) above ground. [Adapted from Rex, 32.]

lamberts) required for equal brightness of asphalt pavements, compared with concrete, is of approximately the ratio of 1.92 : 1 [Blackwell, Prichard, and Schwab, 7]; this is roughly a 2 : 1 ratio. With respect to levels of illumination, the United States of America Standards Institute proposes 0.75 and 1.0 fc for reflective surfaces of 20 percent and 10 percent, respectively [41]. Making further allowances for the fact that seeing time may be abbreviated (and is not static) and adjusting to the 2 : 1 ratio mentioned above, values of 1.5 and 3.0 fc are derived [Rex, 31]. These compare with 1.2 and 2.4 fc proposed by Blackwell [3].

Aside from the possibility of increasing the amount of illumination in the case of less-reflective surfaces to achieve some desired brightness level, it is of course possible to increase the reflectivity of the surface, as by the use of a top surface treatment of white or light-gray aggregate or by the use of epoxy plastic.

Inspection Processes

The inspection of products in processing or manufacturing establishments is another kind of situation that sometimes requires special lighting treatment. A particularly difficult visual inspection process is that of inspecting empty glass jars when they pass in single file on a conveyor belt. Since it is necessary to focus on each jar in succession, the time available and the illumination level are important factors in making the necessary visual discriminations. An interesting solution to the illumination problem is illustrated in Figure 14-12. The luminous horizontal and vertical panels are illuminated from behind with fluorescent lamps that produce

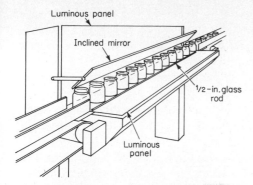

Figure 14-12 Booth for inspection of glass jars, with horizontal and vertical translucent glass panels illuminated from behind with fluorescent lamps. [*IES lighting handbook*, 42, fig. 14-45.]

300 to 400 fL. The inclined mirror above the containers aids in viewing the interior of the jars.

SUMMARY

1. Various standards or criteria have been proposed as a basis for evaluating the adequacy of illumination for different human activities. These include (*a*) visual acuity, (*b*) heart rate, (*c*) blink rate, (*d*) muscular tension, (*e*) opinions, (*f*) critical level, and (*g*) others.

2. The research of Dr. Blackwell has led to the development of the Visual Task Evaluator (VTE). This device, and its associated procedures, have been accepted by the IES as the basis for establishing illumination standards. The *guide* brightness used in this procedure is in footlamberts; the footcandles specified for a particular task are derived by dividing the appropriate guide brightness by the percentage reflectance of the visual task.

3. While general practice is tending toward increasing levels of illumination, there remain certain unresolved aspects of illumination practice. For example, there is some evidence to suggest that high levels of illumination (in some tasks) may suppress the *visual gradients* and reduce visual information.

4. Both laboratory and work-situation studies reveal the influence of illumination on human performance on visual tasks.

5. While experience and research generally argue for fairly high levels of general illumination (relative to that at the work area), some evidence suggests that, in some circumstances at least, more local lighting and less general light may be preferable.

6. Individuals whose visual acuity is poor (such as the aged) need higher levels of illumination than other people.

7. Procedures have been developed for deriving discomfort glare ratings (DGR) for luminaires and possible interior lighting situations.

8. To reduce direct glare (*a*) select luminaires with low DGR, (*b*) reduce the luminance of light sources, (*c*) position luminaires as far from line of sight as possible, (*d*) increase luminance around glare source, and (*e*) use light shields, hoods, etc., if glare cannot be reduced.

9. To reduce reflected glare (*a*) keep luminance of luminaires low, (*b*) provide good general illumination, (*c*) use diffuse light (indirect light, etc.), (*d*) position luminaires so reflected light will not be directed toward eyes, and (*e*) use surfaces that diffuse light.

10. There usually should not be marked differences between the illumination on the task and that on the surrounding area.

11. Although the type of luminaire (tungsten, fluorescent, etc.) does not affect visual performance on many tasks, for some tasks certain types are preferable.

12. The reflectance of walls, ceilings, and other surfaces in rooms determine the *utilization coefficient* of the light sources. Fairly high reflectances are generally recommended for ceilings and upper walls, and somewhat lower values for other areas.

13. Roadway and street lighting has been found to reduce accidents. Such installations can be designed to optimize visual comfort and visibility by appropriate *geometry* of lighting systems, by providing transition lighting (such as at approaches and exits of lighted areas), by improved design of luminaires, and by the use of more highly reflective road surfaces.

REFERENCES

1. Allphin, W.: BCD appraisals of luminaire brightness in a simulated office, *Illuminating Engineering*, 1961, vol. 56, p. 31.

2. Blackwell, H. Richard: Development and use of a quantitative method for specification of interior illumination levels on the basis of performance data, *Illuminating Engineering*, 1959, vol. 54, pp. 317–353.

3. Blackwell, H. R.: *Proceedings of the 1960 research symposium*, Illuminating Engineering Research Institute, New York.

4. Blackwell, H. R.: Development of visual task evaluators for use in specifying recommended illumination levels, *Illuminating Engineering*, 1961, vol. 56, pp. 543–544.

5. Blackwell, H. R.: Further validation studies of visual task evaluation, *Illuminating Engineering*, 1964, vol. 59, no. 9, pp. 627–641.

6. Blackwell, H. R., and O. M. Blackwell: The effect of illumination quantity

upon the performance of different visual tasks, *Illuminating Engineering*, 1968, vol. 63, no. 3, pp. 143–152.

7. Blackwell, H. R., B. S. Prichard, and R. N. Schwab: *Illumination requirements for roadway visual tasks*, Highway Research Board Bulletin 255, Publication 764, 1960.

8. Christie, A. W.: Visibility in lighted streets and the effect of the arrangement and light distribution of the lanterns, *Ergonomics*, 1963, vol. 6, no. 4, pp. 385–391.

9. Crouch, C. L.: New method of determining illumination required for tasks, *Illuminating Engineering*, 1958, vol. 53, pp. 416–422.

10. Eastman, A. A., and J. F. McNelis: *An evaluation of sodium, mercury and filament lighting for roadways*, paper presented at National Technical Conference of the IES, Dallas, Texas, Sept. 9–14, 1962.

11. Fortuin, G. J.: Age and lighting needs, *Ergonomics*, 1963, vol. 6, no. 3, pp. 239–245.

12. Guth, S. K.: Computing visual comfort ratings for a specific interior lighting installation, *Illuminating Engineering*, 1966, vol. 61, no. 10, pp. 634–642.

13. Guth, S. K., and A. A. Eastman: Lighting for the forgotten man, *American Journal of Optometry*, 1955, vol. 32, pp. 413–421.

14. Guth, S. K., A. A. Eastman, and J. F. McNelis: Lighting requirements for older workers, *Illuminating Engineering*, 1956, vol. 51, pp. 656–660.

15. Guth, S. K., and J. F. McNelis: A discomfort glare evaluator, *Illuminating Engineering*, 1959, vol. 54, pp. 398–406.

16. Hess, D. P., and W. Harrison: The relation of illumination to production, *Transactions of the Illuminating Engineering Society*, 1923, vol. 18, pp. 787–800.

17. Hopkinson, R. G., and J. Longmore: Attention and distraction in the lighting of work-places, *Ergonomics*, 1959, vol. 2, pp. 321–334.

18. Kahler, W. H.: Visual comfort in the plant, *Industrial Medicine and Surgery*, 1958, vol. 27, pp. 556–557.

19. Lion, J. S.: The performance of manipulative and inspection tasks under tungsten and fluorescent lighting, *Ergonomics*, 1964, vol. 7, no. 1, pp. 51–61.

20. Lion, J. S., E. Richardson, and R. C. Browne: A study of the performance of industrial inspectors under two kinds of lighting, *Ergonomics*, 1968, vol. 11, no. 1, pp. 23–34.

21. Logan, H. L., and E. Berger: Measurement of visual information cues, *Illuminating Engineering*, 1961, vol. 56, pp. 393–403.

22. Luckiesh, M.: *Light and work*, D. Van Nostrand Company, Inc., Princeton, N.J., 1924.

23. Luckiesh, M.: *Light, vision, and seeing*, D. Van Nostrand Company, Inc., Princeton, N.J., 1944.

24. Luckiesh, M., and S. K. Guth: Brightness in visual field at borderline between comfort and discomfort, *Illuminating Engineering*, 1949, vol. 44, pp. 650–670.

25. Luckiesh, M., and F. K. Moss: "The new science of seeing," in *Interpreting the science of seeing into lighting practice*, vol. 1, 1927–1932, General Electric Co., Cleveland.

26. Luckiesh, M., and F. K. Moss: *Seeing*, The Williams & Wilkins Company, Baltimore, 1931.

27. Luckiesh, M., and F. K. Moss: *The science of seeing*, D. Van Nostrand Company, Inc., Princeton, N.J., 1937.

28. McGowan, T. K., and S. K. Guth: *Extending and applying the IES visual comfort rating procedure*, paper presented at National Technical Conference of the IES, Phoenix, Ariz. Sept. 9–12, 1968.

29. Rex, C. H.: New developments in the field of roadway lighting, *Institute of Traffic Engineers*, 1959, vol. 30, pp. 15–25.

30. Rex, C. H.: Computation of relative comfort and relative visibility factor ratings for roadway lighting, *Illuminating Engineering*, 1959, vol. 54, pp. 291–314.

31. Rex, C. H.: *Effectiveness ratings for roadway lighting*, paper presented at National Technical Conference of the IES, Dallas, Texas, Sept. 9–14, 1962, preprint 36.

32. Rex. C. H.: Roadway lighting for the motorist, *Illuminating Engineering*, 1967, vol. 62, no. 2, pp. 98–110.

33. Rey, P. P., and Jean-Pierre Rey: Les effects comparés de deux éclairages fluorescents sur une tâche visuelle et des tests de "Fatigue," *Ergonomics*, 1963, vol. 6, no. 4, pp. 393–401.

34. Tinker, M. A.: Illumination standards for effective and comfortable vision, *Journal of Consulting Psychology*, 1939, vol. 3, pp. 11–20.

35. Tinker, M. A.: Effect of visual adaptation upon intensity of light preferred for reading, *American Journal of Psychology*, 1941, vol. 54, pp. 559–563.

36. Tinker, M. A.: Illumination intensities for reading newspaper type, *Journal of Educational Psychology*, 1943, vol. 34, pp. 247–250.

37. Tinker, M. A.: Trends in illumination standards, *Transactions of the American Academy of Ophthalmology and Otolaryngology*, March–April, 1949, pp. 382–394.

38. Viteles, M. S.: *The science of work*, W. W. Norton & Company, Inc., New York, 1934.

39. Weston, H. C.: *The effects of conditions of artificial lighting on the performance of worsted weavers*, Medical Research Council (Great Britain), Industrial Health Research Board, Report 81, 1938.

40. A revolution in post office lighting, *The Magazine of Light*, 1953, vol. 22, no. 4.

41. *American standard practice for street and highway lighting*, IES, New York, Feb. 28, 1953.

42. *IES lighting handbook*, 4th ed., IES, New York, 1966.

43. *Influence of lighting, eyesight, and environment upon work production*, General Services Administration, Public Buildings Service, Washington, D.C., 1949.

44. Lighting for woolen and worsted mills, *Illuminating Engineering*, 1949, vol. 44, pp. 364–373.

45. Public lighting needs, *Illuminating Engineering*, 1966, vol. 61, no. 9, pp. 585–602.

46. Recommendations for quality and quantity of illumination and new footcandle tables, *Illuminating Engineering*, 1958, vol. 53, pp. 422–432.

47. Recommended levels, interior and exterior, *Illuminating Engineering*, 1959, vol. 54, pp. 428–433.

48. *See better–work better bulletin*, no. 1, General Electric Co., Lamp Division, Cleveland, 1953.

49. Visual comfort ratings for interior lighting: Report 2 (prepared by Subcommittee on Direct Glare, Committee on Recommendations for Quality and Quantity of Illumination, IES), *Illuminating Engineering*, 1966, vol. 61, no. 10, pp. 643–666.

Since the current model of the human organism is the result of evolutionary processes over millions of years, it has developed substantial adaptability to the environmental variables within the world in which we live, including its atmosphere. There are, however, certain natural environmental conditions that are outside the repertoire of the human being's range of adaptability. In addition to the environments to which man is exposed in his everyday activities, science and technology are busy developing new kinds of environments for him, including capsules for outer space, as well as in more mundane situations such as blast furnaces, cold-storage warehouses, x-ray rooms, and underwater caissons for use in building bridges. It should be added that people are becoming concerned about changes that are occurring in our natural environment that are the unintentional by-products of civilization—such as smog, air pollution from industrial processes, gradual increase in carbon monoxide in the air, the harmful effects from the increasing use of insecticides, and radiation from nuclear explosions and x-rays.

THE HEAT–EXCHANGE PROCESS

The human body is continually generating heat as the consequence of metabolic activity. In a state of rest, an adult male generates a little over 1 kcal/min; some sedentary activities cause expenditures of from 1.5 to 2.0 kcal/min, and physical activities range up to about 5.0 kcal/min for moderate activity, and up to 10 or even 20 kcal/min for extremely heavy work. Since the metabolic activity is continuous, the body is continually in the process of trying to maintain thermal equilibrium with its environment.

Body Changes during Thermal Adjustment

When the body changes from one thermal environment to another, certain physical adjustments are made by the body, especially the following:

- *Changes from optimum environment to a cold one:* (1) The skin becomes cool; (2) the blood is routed away from the skin and more to the central part of the body; (3) "goose flesh" occurs on the skin, increasing the insulation power of the skin (a rough surface loses less heat than a smooth one); (4) rectal temperature rises slightly; (5) shivering may occur; (6) the cooled blood flowing back from the skin area is warmed by the fresh blood flowing out to the skin area; and (7) the body may eventually stabilize in the cold environment with a lowered blood pressure and large areas of the skin receiving very little blood (although this increases the tendency to frostbite).

· *Changes from cool environment to a warm one:* (1) The skin surface is warmed; (2) more blood is routed to the surface of the body; (3) rectal temperature falls; (4) shivering may occur; (5) sweating may begin; (6) the body may eventually stabilize in the warm environment, with continual sweating, increased blood flow to the surface of the body, and increase in skin temperature.

The question of the sensory mechanisms that serve to regulate these adjustments in heat control has been a matter of some dispute. In particular, Benzinger [6], believes that the hypothalamus (at the base of the brain) is the controlling mechanism, whereas Hertzman [29] has shown that thermoreceptors in the skin are capable of eliciting sweat-gland activity and therefore take part in the regulation of body and skin temperature. It is probable that both these mechanisms participate in the temperature-control process in a manner that is as yet not completely understood.

Acclimatization to heat and cold. Acclimatization consists of a series of physiological adjustments that occur when an individual is habitually exposed to extreme thermal conditions, hot or cold as the case may be. An illustration of the changes that occur during acclimatization to heat is shown in Figure 15-1. The men involved in the study worked each of 9 days for 100 min at an energy expenditure of 300 kcal/hr in a hot climate [Lind and Bass, 36]. The figure shows changes in rectal temperature, pulse rate, and sweat loss. The changes in these physiological indices are very distinct. Although there is some question about the time required to become fully acclimatized, it is evident that much acclimatization to heat occurs within 4 to 7 days, and reasonably complete acclimatization occurs usually in 12 to 14 successive days of heat exposure [Leithead and Lind, 33, p. 21]. Quite a bit of acclimatization to cold occurs within a week of exposure, but full acclimatization may take months or even years [McBlair, et al., 39]. Even complete acclimatization, however, does not fully protect an individual from extreme heat or cold, although such an individual can tolerate extremes better than his unacclimatized brother. A comment should be added about ethnic and other groups of people who have lived for generations in a given climate such as the arctic or tropics. Through succeeding generations such groups tend to develop some hereditary characteristics that make them physiologically more

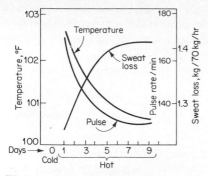

Figure 15-1 Changes in rectal temperature, pulse rate, and sweat loss of a group of men during 9 days of exposure to a hot climate (dry-bulb and wet-bulb temperatures of 120 and 80°F), during which they worked for 100 min at an energy expenditure of 300 kcal/hr. Comparative temperature and pulse values are given for a preceding control day (day 0) during which the men worked in a cool climate. [Adapted from Lind and Bass, 36.]

suitable to the environment. While their increased tolerance would not become complete, they usually can tolerate living and working in their native environment more effectively than those from other environments, and vice versa.

Methods of Heat Exchange

Given the heat produced by the metabolic process, primary sources of heat gain and heat loss to and from the body are (1) convection (heat gain or loss by contact with the air), (2) radiation (heat gain or loss depending on skin temperature and the temperature of surrounding areas), and (3) evaporation (evaporation of sweat can only result in heat loss). These variables can be measured by a method called *partial calorimetry* [Leithead and Lind, 33, chap. 3]; the basic formula is as follows:

M (metabolism) $\pm$ C (convection) $\pm$ R (radiation) $-$ E (evaporation)
$$= \pm S \text{ (storage)}$$

The S factor is the amount of heat gained or lost; if the body is in a state of heat balance, S becomes zero. It should be added that there are other methods of heat exchange, although they are typically so nominal that they do not find their way into the common formula above. They include conduction and the ingestion of hot or cold fluids.

Factors That Affect Heat Exchange

The discussion above has given hints of the environmental variables that affect the heat-exchange process, but it may be useful here to pin these down specifically.

- *Temperature* changes affect heat exchange primarily through the convection process. Air temperature below body temperature results in body cooling, and above body temperature it results in body warming.
- *Humidity* has little effect on heat exchange for normal temperatures, but with extreme temperatures humidity has an important bearing on comfort and physiological tolerance. When temperatures are high and heat exchange depends more on evaporation, high humidity will reduce the possible heat exchange. At low extremes, also, discomfort is greater with high humidities.
- *Circulation of air* exposes the body to more air than would be possible in still air. The cooling or heating of the body by circulation will depend largely on air temperature (whether below or above body temperature) and humidity (as mentioned above).
- *Temperature of objects in the environment* (walls, ceiling, and other objects) comes into play through the radiation process. Whether there is a net heat loss or gain from radiation depends on the relative temperature of these objects in relation to the body surface that is exposed.

Heat loss under various conditions. While there are a number of ways that heat from the body may be exchanged with its environment, their relative importance is markedly different under different atmospheric conditions. Some examples are shown in Figure 15-2. This figure shows, for each of five conditions of air and wall temperature, the percentage of heat loss by evaporation, radiation, and convection. In particular, conditions *d* and *e* illustrate the fact that, with high air and wall temperatures, convection and radiation cannot dissipate much body heat, and the burden of heat dissipation is thrown on the evaporative process. But, as we all know, evaporative heat loss is limited by the humidity; there is indeed truth to the old statement that "it isn't the heat—it's the humidity." To see the limiting effect of humidity on evaporation, let us look at Figure 15-3. This shows the upper limits of tolerance in relation to temperature and relative humidity for working and for resting nude subjects. For any one of the three curves in Figure 15-3, combinations of temperature and humidity to the right of the curve represent

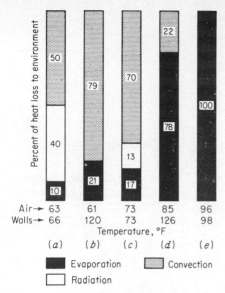

Figure 15-2 Percentage of heat loss to environment by evaporation, radiation, and convection under different conditions of air and wall temperature. [Adapted from Winslow and Herrington, 56.]

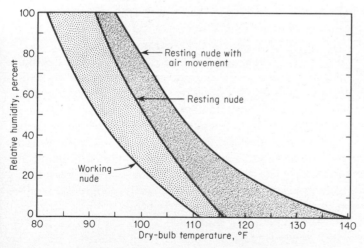

Figure 15-3 Approximate upper limits of tolerance for heat loss by evaporation. For any of the three conditions, temperature and humidity combinations to the right of the curve prevent evaporation. [Derived from data by Winslow et al., 55.]

conditions which, if accentuated enough and long enough, could result in some physiological consequences ranging up to heatstroke or death. A comparison of the two curves at the right shows that air movement usually helps to make conditions more tolerable by exposing the surface of the body to more air.

Composite Indexes of Environmental Factors

Since there are different environmental factors that affect the heat-exchange process, it would be desirable if there were available some composite index that would take them all into account as far as their effect upon human beings is concerned. While it has not been possible to include all these factors in a single index, there are a few indexes that do take into account certain combinations of the factors.

Effective temperature.   One of these indexes is *effective tempera-ture* (ET), which is defined by the *ASHRAE handbook of fundamentals* [59, p. 117] as an empirical sensory index, combining into a single value the thermal effect of temperature, humidity, and movement of air upon the human body. The ET scale was developed through a series of studies in which subjects compared the relative warmth of various air conditions in adjoining rooms by passing back and forth from one to the other. In this way, combinations of temperature and humidity were identified which gave the same average sensation of warmth.

Figure 15-4 shows these relationships for air movement of 15 to 25 ft/min (other figures, not shown here, give variations for other rates of air movement). In the main part of the figure the combination of a given dry-bulb temperature such as 75° (horizontal scale) and 100 per-cent relative humidity (the top diagonal line from southwest to north-east) defines a specific ET (such as 75°); all combinations of dry-bulb temperature and humidity, along the northwest-to-southeast line begin-ning at that point, have the same ET, that is, they were found to produce equal sensations of warmth. (The vertical wet-bulb temperature scale is, in effect, a combination of dry-bulb temperature and 100 percent relative humidity.) The upper left and lower right charts show the percent of individuals who feel comfortable in summer and winter at different ETs.

The basic figure also has three superimposed lines characterizing conditions that generated average ratings of slightly cool, comfortable, and slightly warm on the part of moderately clothed subjects during a 3-hour stint in an environmentally controlled room [Nevins et al., 46]. We should note that these three curves of relative comfort tend to be more vertical than the ET lines, which suggests that relative humidity may not be as important a factor in sensations of *comfort* as in sensations of equal *warmth* (as represented by the ET lines). This, however, brings up two points. The first is that several factors serve to complicate the

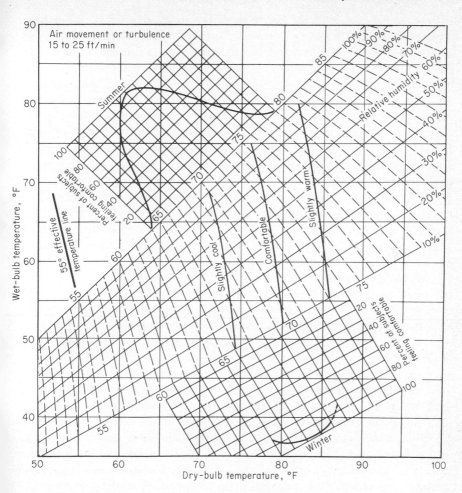

Figure 15-4 Scale of effective temperature (ET) [basic scale reprinted by permission from *ASHRAE handbook of fundamentals*, 59]. In addition, the figure shows lines representing conditions subjects rated as slightly cool, comfortable, and slightly warm after a 3-hr exposure [adapted from Nevins et al., 46, fig. 7, p. 289.]

effects of the existing environmental variables on people, such as the nature of the activity, clothing, acclimatization of the individual, and exposure time. With respect to exposure time, the ET lines were based on responses of subjects after a momentary exposure to temperature and humidity conditions, whereas the superimposed curves of Nevins were based on responses after 3 hours of exposure; this difference in exposure very possibly contributed to these differences. The second point concerns the relative effects of the separate environmental conditions. Questions have been raised, for example, about the relative importance of humidity as depicted by the ET lines of Figure 15-4. For example, Leithead and Lind [33, p. 61], on the basis of evidence from various studies, state that it is clear that there is an inherent error in the ET scales if they are used as an index of physiological effect, especially under severe environmental conditions; but they conclude that, for practical purposes, the scales are reasonably accurate in assessing heat stress for low and moderate degrees. Despite this and other criticisms and qualms about the ET scale, it nonetheless is the best known and most commonly used composite scale of environmental conditions.

Although the ET scale does not take radiation into account, subsequent studies have indicated that a change of about 1°F in wall temperature corresponds roughly to about a 0.5°F change in ET, although such a correction should be accepted with reservations, especially if there is a marked disparity between wall and air temperature.

Operative temperature. Since the ET scale does not account for radiation to or from wall surfaces, Winslow, Herrington, and Gagge [57] developed a scale of *operative temperature.*[1] This scale takes into account air temperature and wall temperature, but not humidity or air flow.

A recent adaptation of the operative temperature scale has been developed, referred to as a *comfort* equation, T_o, specifically to assess the physiological effects of high-intensity radiant-beam heating [Gagge, 19].

Heat-stress index. This index, originally developed by Belding and Hatch [5] has been modified by Hatch [26] and Hertig and Belding [28]. The index expresses the heat load in terms of the amount of perspiration that must be evaporated to maintain heat balance; this is referred to as E_{req} (the *required evaporation* heat loss). In turn, it is possible to determine the maximum heat that can be lost through evaporation E_{max}, from assumptions of body size, weight, and temperature and by taking into account water-vapor pressure of the environment and air velocity. While the details of these derivations will not be repeated, the *difference* between these two values (expressed in Btu per hour) indicates the load

---

[1] Operative temperature is the sum of a radiation constant multiplied by the mean wall temperature, and of a convection constant multiplied by the mean air temperature, divided by the sum of the two constants, as derived from this formula.

$$T_o = \frac{K_R T_W + K_C T_A}{K_R + K_C}$$

in which T_o = operative temperature
$\quad\quad K_R$ = radiation constant
$\quad\quad T_W$ = temperature of walls
$\quad\quad K_C$ = convection constant
$\quad\quad T_A$ = temperature of air

that must be reduced or dissipated otherwise. The *otherwise* can take various forms, such as further reduction of convection or radiation sources, reduction of task demands by reducing physical requirements or by rest pauses, and by proper clothing.

To illustrate this general approach, Hertig and Belding [28] present a hypothetical example of a task as follows:

Source of heat load	Btu/hr
Metabolism (based on type of activity)	800
Radiation	2800
Convection	60
E_{req}	3660
E_{max} (computed by formula)	2530
Difference	1130

In this particular case, heat by radiation is the major source of the heat load and would be the primary aspect of the situation to do something about.

In connection with the heat-stress index, McKarnes and Brief [41] have worked up a set of nomographs based on the modifications reported by Hertig and Belding, and have also elaborated on the theme by setting up a formula for estimation of allowable exposure time (AET) and minimum recovery time (MRT).

Predicted 4-hr sweat rate (P4SR). In a series of investigations McArdle et al. [38] found that sweat loss was the physiological measurement that correlated best with the severity of experimental environments. They then developed a scale of physiological stress based on sweat loss, the basic reference value being the amount of sweat produced in 4 hr by young, fit, acclimatized men. The derivation of the index is rather intricate and will not be spelled out, but it takes into account the environmental variable of globe or dry-bulb temperature, wet-bulb temperature (which in part takes humidity into account), and air speed; but it also adds in adjustments for energy expenditure and clothing worn. Leithead and Lind [33, pp. 62–68] pull together evidence that convinces them that it is the most accurate scale of (climatic) heat stress presently available. It should be added, however, that it is not a measure of physiological strain of individuals [Belding and Hatch, 5]. But, for that matter, none of the other indexes discussed above is a measure of physiological strain.

Oxford index. The Oxford index, or WD index, is a simple weighting of wet-bulb and dry-bulb temperature, as follows [Leithead and Lind, 33, p. 82]:

$$WD = 0.85w \text{ (wet-bulb temperature)} + 0.15d \text{ (dry-bulb temperature)}$$

It has been found to be a reasonably satisfactory index to equate climates with similar tolerance limits.

HEAT STRESS

Having discussed the heat-exchange process and the various indexes of environmental conditions, let us now turn to the effects of heat stress.

Physiological Effects

One of the most direct effects of heat stress is on the temperature of the body. Body temperature measurements include measures of *core* temperature (oral, rectal, etc.) and of *shell* temperature (usually skin temperature). The various core and shell temperature measurements are not the same and, in their individual ways, are influenced by thermal conditions and are manifestations of physiological strain.

As an opener, Figure 15-5 shows data summarized from different studies by Leithead and Lind [33] on the relation between ET and skin

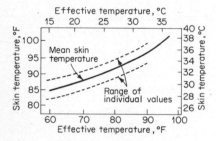

Figure 15-5 Relationship between effective temperature (ET) and skin temperature as consolidated from various studies. [Adapted from Leithead and Lind, 33, fig. 6.]

temperature. The upswing of the curves probably reflects the consequence of the flow of blood toward the surface of the skin to increase the dissipation of surplus heat with increasing effective temperatures. As a follow-up, Figure 15-6 represents the relation between ET and rectal temperature for one individual engaged in three levels of physical activity and for the average of three individuals at a single level of activity. In each instance the long flat sections of the curves represent what Lind [34] refers to as the *prescriptive zone* for whatever activity is involved. As a prelude to a later discussion of the effects of work load, we can see the differential upswings for the three levels of work activity represented,

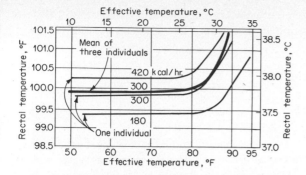

Figure 15-6 Rectal temperature as related to effective temperature (ET) for one individual working at three levels of work activity and for three individuals working at the same level (300 kcal/hr). [Adapted from Lind, 34.]

the upswing starting at successively lower points with increasing work load. The middle, light line (that represents a single individual working at a level of 300 kcal) parallels the heavy line (that represents the average for three individuals), which indicates considerable stability of the patterns of the curves.

The disproportionate physiological *costs* of work at higher temperatures are reflected in another way by some data from Brouha [7, p. 61] on the average heart-rate recovery time from bicycling under various temperature conditions, shown below as related to a 60°F condition:

Change in environment, °F	*Average recovery time, %*
60–60	100
60–72	123
60–90	968
60–100	1222

Levels of work in combination with environmental conditions that bring about the rise in core temperature (as shown in Figure 15-6) also induce other corresponding physiological changes, which, if continued, can cause hypothermia, a condition that renders normal heat loss more difficult [Hunter, 31]. Dehydration, such as from sweat, is another possible consequence of heat stress. In the case of 51 operating engineers working during mild summer weather in California's Central Valley, the average weight loss from water depletion per day was about 5 lb 2 oz. Such a deficit, of course, needs to be replaced during the evening and night.

If adverse physiological effects are permitted to accumulate, serious heat strain and prostration can occur, and if the process is not reversed, death will follow.

Heat Stress and Performance

The physiological effects of heat stress are paralleled by degradation in performance in many kinds of human activity. Brief reference will be made to a few studies that illustrate such effects.

Physical work. One investigation of the effects of heat stress (heat and humidity) on physical activity involved 16 army enlisted men who performed a marching task in a controlled room on each of several days, marching at 3 mph for 4 hr with a 20-lb pack [Eichna et al., 15]. Observation of these men made it possible to characterize their performance as (1) relatively easy, (2) difficult, and (3) impossible. The results, shown in Figure 15-7, emphasize the importance of humidity in combination with temperature as a factor that limits performance on such a task.

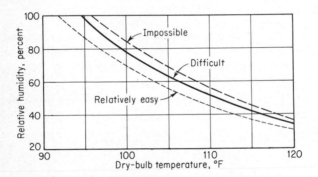

Figure 15-7 Relative difficulty of performing a marching task under various temperature and humidity conditions. The three lines show the conditions under which the work was easy, difficult, and impossible, respectively. [From Eichna, Ashe, Bean, and Shelley, 15.]

A number of studies of performance under high temperatures were carried out by Mackworth [see Mackworth, 45, for summary]. In one of these, the subjects performed on a pull test that required raising and lowering a 15-lb weight by bending and straightening the arm to a metronome that beat every second until the men could not lift the weight again, the task being performed under different ETs. The results, shown in Figure 15-8, indicate that performance (the amount of work done)

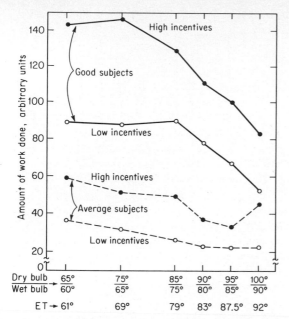

Figure 15-8 Relationship between effective temperature (ET) and performance on a weight-lifting test for good and average subjects under high and low incentives. [From Mackworth, 45.]

deteriorated with higher ETs, both for good and average subjects, under both low- and high-incentive conditions. A similar pattern of deterioration occurred in performance in a heavy pursuit meter task [Mackworth, 45] and in the case of gold miners filling mine cars in South Africa [Wyndham et al., 58]; and Pepler [in Leithead and Lind, 33, chap. 12] summarizes the results of several industrial surveys that further nail down the point that heat stress takes its toll in performance in physical work activities.

But, for most practical purposes, one would like to have some inkling of the tolerance and comfort of people under different *levels* of stress, for different *levels* of energy expenditure, and for different periods of *time*. Although there is probably no simple scheme or set of data that completely fulfills this order, Figure 15-3 (discussed earlier) shows some such interrelationships (as related to tolerance), and these are further illustrated in Figure 15-9. And if we are interested in the sensations of comfort of people working at different levels of activity, Fanger [16] provides some data, as shown in Figure 15-10; this shows that the comfort lines

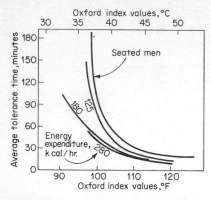

Figure 15-9 Average tolerance times of men seated, and of men working at three levels of energy expenditure, in relationship to Oxford index values. Safe tolerance values should be taken to be no more than about 75 percent of the times shown. [Adapted from Lind, 35, in Hardy, 25, as based on data from other studies.]

for working at the different levels of activity are shifted down the scale by noticeable increments.

Mental activities. The effects of heat stress on performance in mental activities are entwined with related factors such as the type of task, the duration of the task, the degree of acclimatization of the individuals, and the level of their training. Figure 15-11 shows, as one example, the pattern of relationship between ET and errors in Morse code receiving, as related to time [Mackworth, 43]; the error curve for the 3d hour is clearly higher than for the 1st hour. A generalized pattern of the temperature-duration function related to mental performance has been developed by Wing [54] on the basis of a very thorough analysis of the results of 15 previous studies. In this analysis he identified, from the several studies, the lowest temperature at which a statistically significant performance decrement occurred. These points, plotted at their respective exposure durations, were then used in drawing the curve shown in Figure 15-12. In addition, that figure shows, for comparative purposes, curves of tolerable and marginal physiological limits based on earlier studies by other investigators. Wing suggests that the thresholds for at least some mental tasks might be between the lower curve and the tolerable physiological limit curve in the case of fully acclimatized or highly practiced individuals.

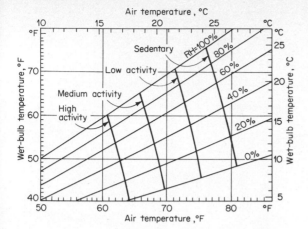

Figure 15-10 Comfort lines (air temperature versus wet-bulb temperature) for three levels of work activity. [From Fanger, 16, fig. 3, pt. 3, p. 4.11. Reprinted by permission of ASHRAE. ASHRAE does not necessarily agree with statements or opinions advanced at its meetings or printed in its publications. Please note that all articles published in ASHRAE publications are accepted on an exclusive basis and are copyrighted by the Society.]

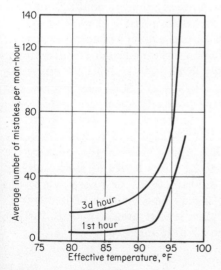

Figure 15-11 Average number of mistakes in Morse code receiving in relation to effective temperature. [After Mackworth, 43.]

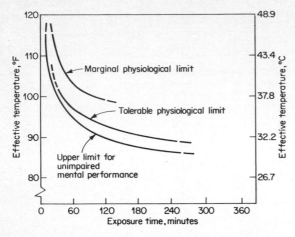

Figure 15-12 Tentative upper limit of effective temperature for unimpaired mental performance as related to exposure time; data are based on an analysis of 15 studies. Comparative curves of tolerable and marginal physiological limits are also given. [Adapted from Wing, 54, fig. 9. See Wing's report for original sources.]

Discussion

The discussion and illustrations above probably tend to confirm our own experiences—that the effects of heat stress are related to what activities we are engaged in, the duration of exposure, the degree of acclimatization, how much insulation we wear, etc. These intertwined relationships pretty well preclude the setting of any single level of tolerable heat stress for all circumstances. In this connection, Mackworth [45] does propose the concept of a *critical region* on the atmospheric temperature scale above which most acclimatized men dressed in shorts will not work effectively indoors, this region being between ETs of 83 and 87.5°F. This probably represents something of a ceiling. The upper ends of the prescriptive zones shown in Figure 15-6 [Lind, 34] for three levels of work activity were 80.4, 81.3, and 86.1°F, which indicate that, for heavy work, the critical region proposed by Mackworth may be a bit high. The comfortable conditions (as opposed to tolerance levels) are, of course, further down the scale, as reflected by Figure 15-10 [Fanger, 16].

COLD

While civilization is generally reducing the requirement for many people to work in cold environments, there still are some circumstances where people must work, and live, in cold environments. These situations include outdoor work in winter, arctic locations (especially in military and exploration activities), cold-storage warehouses and food lockers. As in the case of heat exposure, there are a number of interlaced factors that affect the tolerance, comfort, and performance of people in cold environments; these include the level of activity, degree of acclimatization, duration, and insulation.

Indexes Related to Effects of Cold

First, let us mention a couple of indexes that are related to the effects of cold.

Wind chill.   One of these is a wind-chill index [Siple and Passel, 48]; it provides a means for making a quantitative comparison of combinations of temperature and wind speed. The quantitative value corresponds to a calorie scale (actually kilocalories per square meter per hour) but is converted into a sensation scale ranging from hot (about 80), to pleasant (about 200), cool (400), cold (800), bitterly cold (1200), and even colder values (including that at which exposed flesh freezes in 1 min or less). By the use of a nomograph with temperature and wind velocity, one can note (as an example) that the same sensation of very cold (1000) would be produced under the following conditions: $-50°F$ and 0.1 mph wind velocity; $-15°F$ and 1.0 mph; and $+24°F$ and 10 mph.

Clo unit: a measure of insulation.   The clo unit is a measure of the thermal insulation necessary to maintain in comfort a sitting, resting subject in a normally ventilated room at 70°F temperature and 50 percent relative humidity [Gagge et al., 20].[2] Since the typical individual in the nude is comfortable at about 86°F, one clo unit would be required to produce an equal sensation at about 70°F; a clo unit has very roughly the amount of insulation required to compensate for a drop of about 16°F, and is approximately equivalent to the insulation of the clothing normally worn by men. To lend support to the old adage that there is nothing new under the sun, the Chinese have for years described the weather in terms of the number of suits required to keep warm, such as a "one-suit" day (reasonably comfortable), a "two-suit" day (a bit chilly), up to a limit of a "twelve-suit" day (which would be really bitter).

Physiological Effects of Cold

With inadequate protection, exposure to the cold brings about a reduction of both core and shell temperature. Continued exposure, of course, can bring about frostbite and other effects, and ultimately death. As a case in point, the mean body and skin temperatures for four subjects exposed to

---

[2] A clo unit is defined as follows:

$$\text{Clo unit} = \frac{°F}{\text{Btu}/(\text{hr})(\text{ft}^2 \text{ of body area})}$$

conditions of $-26°F$ after about 75 min are given below, for the conditions indicated [Veghte and Clogston, 51].

	Temperature, °F	
Condition	Body	Skin
Walking, 1 mph (160 kcal/hr)	95.6	89
Resting, 4.2 clo units	93.5	84
Resting, 2.3 clo units	91.8	79

Normal skin temperature is about 92°F, and the critical mean skin temperature—excluding the hands and feet—is about 76°F [Barnett, 4]. This is the level at which extreme discomfort generally occurs.

Performance in Cold

Since most activities performed in the cold are of a physical nature, the possible effects of cold on such tasks is of particular concern. An impression of the decrement in performance on such activities is shown in Figure 15-13, which shows results from McCleary [40] of a Brush Assembly Breakdown (BAB) test, performed under several temperatures; this test is something like a line-maintenance job. In addition, this figure shows

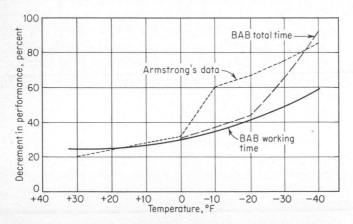

Figure 15-13 Percentage of decrement in performance on two tasks under different temperature conditions. The BAB (brush assembly breakdown) data are for working time and for total time (including working time and warming-up time). Armstrong's data relate to operational efficiency of flying personnel in open cockpit. [After McCleary, 40, and Armstrong, 2.]

the results of an earlier study by Armstrong [2] in which he determined the loss in efficiency of flying personnel under operational conditions of flight in an open cockpit. The noticeable difference between the two sets of data for temperatures around −10 and −20°F can be accounted for by an explanation that need not be given here. Discounting that discrepancy, however, there is an amazing degree of similarity between the two sets of data.

For strictly manual tasks, Fox [17] puts together a fairly persuasive body of evidence to the effect that hand-skin temperature (HST) is a critical factor in performance. The critical region is apparently between about 55 and 60°F, as illustrated in Figure 15-14. Fine finger dexterity

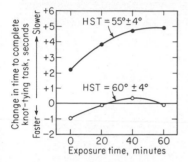

Figure 15-14 Changes in manual performance on knot-tying test as function of hand-skin temperature (HST) and duration of cold exposure. The differences for the 60°F HST are not significant. Those for the 55°F temperature are. [From Clark, 11.]

is more susceptible to adverse temperature conditions than grosser types of manual activities [Dusek, 14]. There have been some inklings that manual performance can be maintained if hand-skin temperature is above the critical level of about 60°F [Clark, 11]; more recent evidence, however, indicates that this is not universally the case, and that, for certain types of tasks, HST must be maintained at reasonable levels to preclude work decrement [Lockhart, 37].

In outdoor activities, wind chill is a factor to be contended with. Among its effects are increased numbness [Mackworth, 44] and increased reaction time [Teichner, 50]. In this connection acclimatization somewhat increases tolerance to wind chill but does not eliminate the effects of cold on manual performance. This was shown, for example, with a group of 22 men who lived in a room with temperatures of a brisk $-20°F$ for a period of 8 to 14 days [Horvath and Freedman, 30]. It was found that such continuous exposure resulted in deteriorating performance on both a gear-assembly test and a code test with the use of a pencil. Two other interesting points also were discovered, namely, that neither mental performance (as measured by the code test) nor visual performance (as measured by speed or precision in responding) was particularly affected by the cold.

Maintenance of Heat Balance

Our sensations of chilliness provide a reasonably valid cue of heat loss (negative storage). Some indication of the lower bounds of subjective tolerance in different atmospheric conditions is shown in Figure 15-15.

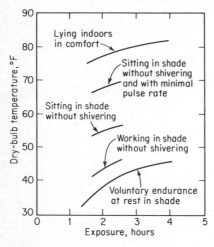

Figure 15-15 Lower bounds of tolerance limits (for indicated criteria) for seminude men under different circumstances. [Adapted from Adolph and Molnar, 1, from *American Journal of Physiology*.]

The men, seminude, were exposed to a number of air temperatures, wind velocities, and solar intensities under such circumstances as sitting, standing, working, and lying down. Their tolerances are expressed as comfort and discomfort, voluntary endurance, and goose flesh and shivering.

As indicated earlier, however, heat balance is influenced by such factors as insulation, energy expenditure, and duration of exposure. An indirect indication of the protection provided by insulation for a low activity level is shown in Figure 15-16. This shows the combination of exposure time and temperature that was tolerable for each of four insulation levels; the difference in tolerance between 1 and 4 clo units is upward of 60°F or more. This figure demonstrates the trade-off effects, in terms of maintaining heat balance, of exposure time and insulation.

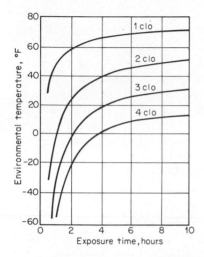

Figure 15-16 Exposure time and temperature that are tolerable for different levels of insulation (clo units). [Adapted from Burton and Edholm, 8, as presented in Webb, 53, p. 125.]

THE MANAGEMENT OF TEMPERATURE PROBLEMS

It is neither feasible nor appropriate here to try to set forth the engineering and other specifications for the resolution of all kinds of existing or anticipated temperature problems. However, let us touch very briefly on the major aspects of the management of such problems.

Atmospheric Control

One possible solution is of course the control of the atmosphere. While Mark Twain once said that people talk a lot about the weather but never do anything about it, he apparently did not anticipate the ingenuity of the heating and air-conditioning engineers who have been able to create practically any indoor climate you might specify. This is done, of course, through heating; air conditioning; circulation of air; shielding against possible sources of radiant energy; humidity control; selection of construction and other materials for the reflectivity, convection, and insulation properties; awnings; insulation; water spraying or water layer on roof in dry weather; plantings of grass and trees around buildings; and other means.

Clothing and Other Protective Gear

The effects of insulation have been illustrated above, and certainly argue for the use of clothing to provide thermal protection for people in cold environments, be it the polar ice cap or the cold-storage warehouse. Some such apparel also needs to provide protection against wind and water and at the same time permit reasonable physical activity. At the hot end of the scale, clothing needs to be light and loose to permit evaporation and convection heat loss, and to minimize absorption of radiant energy from the sun. The rather common use in the tropics of loose-fitting, white apparel probably is not happenstance, but rather is the result of experience which has proved such apparel more suitable than other types; white fabric, for example, would absorb less energy from the sun than dark fabric.

Management of Personnel

Where it is not possible or practical to modify extreme environmental conditions to bring them into the bounds of normal, continuous human tolerance, there are certain actions which can be taken in the management of personnel who are to work or live in the environment that may make it tolerable. Some such actions are as follows [adapted in part from Machle, 42]: (1) selection of personnel who can tolerate the condition (sometimes by tryout for 4 or 5 days); (2) permitting people to become gradually acclimatized; (3) establishing appropriate work and rest schedules; (4) rotating personnel; (5) modifying the work (such as by reducing energy requirements); and (6) maintaining hydration by seeing that people drink enough water to replace that lost (thirst is not an adequate indication of water requirements).

AIR POLLUTION

We humans have inherited from our natural environment a number of environmental hazards to contend with. But our science and technology have resulted in the creation of a whole host of new environmental hazards, especially pollutants including smoke, exhaust fumes, toxic vapors and gases, insecticides, herbicides, and ionizing radiation. Some of these contaminants are the by-products of industrial processes and tend to be confined to their industrial environments, whereas others escape into the general atmosphere; and our automobiles, household furnaces, and other nonindustrial sources add their bit to the ever-increasing level of pollution of the air we breathe. Although the constraints of pages preclude thorough discussion of this topic, we should be remiss if we did not at least take cognizance of this important problem.

General Atmospheric Pollution

We are now suddenly becoming aware of the specter that this problem of general air pollution poses to us and our descendants, along with the related problems of soil and water pollution and of waste disposal. Commoner [12], in reflecting about these and other unwanted by-products of our technology, likens our situation to that of the sorcerer's apprentice (who, you may recall, in the absence of the sorcerer, produced floods of water, but had not the ability to halt the torrents he had turned on).

Insecticides and herbicides. As one manifestation of this potentially catastrophic problem, there is evidence that the use of insecticides and herbicides is having an effect on the ecology of nature, that is, on the delicate balance of animal and plant life [Carson, 9]. And, aside from the effects of our unintentional but real tinkering with nature, there is also the risk to health and to life of those who use such materials. In aerial dissemination, for example, the potential danger for pilots and for ground personnel has been clearly demonstrated [Smith, 49]. It has been reported that "empty" 5-gal drums of parathion can contain after months of exposure to the weather as much as 10 gm, an amount sufficient to kill many adults or large farm animals.

Automobile exhaust fumes. As another example, Commoner [12] calls our attention to the effects of automobile exhaust fumes. Aside from the problem of the toxic material itself (which presumably will be somewhat abated over time with the enforced use of antipollution control devices), there is evidence that the world temperature is being gradually raised because of such fumes; one possible long-range consequence of this could be the gradual melting of the Antarctic icecap which, over a period of centuries, could raise the water level of the oceans substantially. Commoner raises the question as to whether we can long continue to take such risks with the future, and suggests that we should consider the use of alternative modes of transportation that do not carry such risks (and, we may hope, will not have any others!).

Smog. If we want still other examples, reflect on the occasional events in New York, London, Los Angeles, and Donora, Pennsylvania, in which atmospheric conditions have caused smog to settle for periods of several days, with unhappy consequences including higher incidences of respiratory ailments and of deaths.

Discussion. Aside from the possible direct affects of all the forms of pollution on people, there can be indirect effects on the climate [69]. The spectrum of environmental problems that can emerge from Pandora's box is of an order of magnitude many times that of the question of the temperature at which we should set the thermostat in the living room. But the field of human factors seems inevitably to be extending its horizons, and one aspect of such an expanded focus should be that of the effects of our technology on the quality of our general atmosphere as it relates to human life.

Local Atmospheric Impurities

In the more circumscribed environments of industrial plants, garages, submarines, laboratories, politicians' smoke-filled rooms, and other places where people work and live, toxic materials sometimes occur in quantities that are potentially dangerous. As with other environmental variables, the effects of some toxic materials are related to quantity and exposure time, such as in the case of carbon monoxide, shown in Figure 15-17. The tolerance levels and any legal or other standards that have been set are, of course, specific to the particular contaminant and will not be included here.[3]

[3] For sets of data on the toxicity of many contaminants, see Roth [47, sec. 13], and Webb [53, sec. 2].

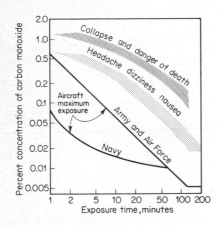

Figure 15-17 Effects of carbon monoxide on man as functions of concentration and exposure time. Shaded areas show conditions that are dangerous and lethal (heavy shading) and that are milder (light shading). The solid lines represent exposure limits set by the military services for aircraft. [Adapted from Webb, 53, p. 23, as based on various sources; for original sources, see Webb.]

The specific methods of protection of people from the effects of contaminants also depend on the nature of the particular toxic material and on the situation.[4] In very general terms, these methods include the following: (1) correcting the condition that creates the contaminant, (2) providing proper ventilation and exhaust equipment, (3) using equipment that is properly designed to minimize exposure, (4) providing physical separation of personnel from contaminants, (5) specifying and enforcing safety procedures, and (6) providing personal protection equipment when desirable.

Ionizing Radiation

Although few of us will have to worry about ionizing radiation in outer space, the increased use of nuclear energy and the use of radioactive materials in laboratories do pose a threat to some of us earthlings. Ionizing radiation is the form of radiation that penetrates matter, and, in its interaction with matter, causes pairs of positive and negative ions to be formed. The different types of ionizing radiation vary in their penetration abilities and thus in their potentially damaging effects. For example, gamma and neutron radiation have greater penetration ability than alpha and beta radiation [Glasstone, 23]. The field of ionizing radiation (and the establishment of guidelines for protection of personnel) comprises a highly specialized area that we can only touch on lightly in this text.[5]

Units of measurement of radiation. The most generally known measure of radiation is the roentgen (r), a measure of gamma radiation.[6] Another index is the radiation-absorbed dosage (rad). This is the measure of the energy imparted to matter (and retained) by ionizing radiation per unit mass of irradiated material; it is 100 ergs/g in any medium.

The biological effects of types of radiation, however, are different. Therefore, the biological effect, say, of a given number of rads of gamma radiation would be different from, say, an equal amount of alpha radiation. The relative biological effectiveness factor (RBE) is used to equate for such differences. In turn, the roentgen equivalent man (rem) is an index of the human biological dose as a result of exposure to one or many types of ionizing radiation. It is the absorbed dose in rads, times the RBE factor of the type of radiation being absorbed; thus the rem is the amount of RBE dose.

Effects of radiation. The gamut of biological effects of excessive radiation is not fully known, but it includes leukemia, skin disorders, loss

[4] For further treatment of certain phases of this subject the reader is referred to the following references: Dubois and Geiling [13], Henderson and Haggard [27], and von Oettingen [52].
[5] Readers are referred to other sources for further discussion, such as Gitlin and Lawrence [22], Langham [32], and Roth [47, sec. 3], two reports [61 and 62], and a manual [66].
[6] See *Radiological health handbook*, U.S. Public Health Service, 1960 [67] for a description of roentgen and other related terms.

of hair, change in blood composition, ulcerations of the digestive tract, cataracts, effects on central nervous system, sterility, cancer, and, in severe cases, death.

With acute doses (as in accidental exposure), the probable effects in terms of possible survival are given below [adapted from *The effects of nuclear weapons*, 63]:

Amount of radiation, r

0–50	No obvious effect, except possibly minor blood changes
100–250	Survival probable, recovery likely in about 3 months
300–550	Survival possible, death in most serious cases, mortality 50 percent for 450 r
700 or more	Survival improbable

And, aside from the direct effects of radiation listed above, there is also reason to believe that life expectancy is related to the degree of continuous exposure, the effects of exposure presumably being somewhat comparable to the normal aging process.

Level of exposure.   Since ionizing radiation is cumulative (and not very healthful for people), it is obvious that the less exposure the better, whether from natural environments, x-rays, laboratories, or otherwise. Because there is much that is not known about its effects, recommended permissible doses need to be accepted with some reservations. The maximum permissible radiation doses in adult radiation workers as recommended by the Federal Radiation Council [60] are as follows:

	Maximum dose accumlated, rem	
Type of exposure	*per year*	*per 13 weeks*
Whole body, head, trunk, gonads, eyes	5 times years beyond 18	3
Skin of whole body	30	10
Hands, forearms, feet, ankles	75	25
Other organs	15	

IONIZED AIR

In recent years there has been a flurry of interest in the possible effects upon human beings of ionized air. Earlier research, especially with laboratory animals, had suggested that high concentrations of negative ions in the air have a generally facilitating effect on the respiratory process (along with other possible beneficial effects) and that positive ions have a generally adverse effect on the respiratory process. If such effects also occurred in human beings, it would of course be possible that human performance, and other aspects of behavior, might be influenced (indirectly) by concentration of atmospheric ions.

Although there have been some indications that ionization has some effect on certain aspects of human behavior [as reviewed by Halcomb and Kirk, 24], these indications are far from universal [Chiles et al., 10; Frey, 18]. A hint from Halcomb and Kirk, however, may suggest a possible explanation for some of the apparently inconsistent experimental results. In their study, separate groups of subjects carried out a vigilance task over 4 hr, performing in environments with a high concentration of positive and of negative ions. The reaction times of the two groups gradually spread apart over the 4-hr, with the negative-ion group maintaining a shorter average reaction time than the positive-ion group. This suggests the possibility that a high concentration of positive ions might take its toll over a period of time—rather than have an immediate effect. However, one probably would be inclined to agree with Frey [18] that, on the basis of current readings of available research, the relative concentration of negative and positive ions is not of major consequence in human performance or general physical welfare (although further research might justify a reversal of this evaluation).

AIR PRESSURE

In the mundane lives of most mortals, the atmospheric variables that we complain about most are temperature and humidity. In less normal environments, however, other variables may play first fiddle. For people in high altitudes (e.g., mountainous areas and aircraft) and for people below sea level (e.g., diving and underwater construction work) air pressure and associated problems can be of paramount importance to well-being and to human performance.

The Atmosphere around Us

The atmosphere of the earth consists primarily of oxygen (21 percent) and nitrogen (78 percent), but also includes a bit of carbon dioxide (0.03 percent), plus other odds and ends. The density of the atmosphere, however, is reduced at higher altitudes; thus, at an altitude of, say, 20,000 ft, there is less air in a given volume than at sea level. Further, because of the weight of the atmosphere, the pressure decreases with altitude. Air pressure is frequently measured in pounds per square inch (psi) or in millimeters of mercury (mm Hg). At sea level the air pressure is 14.71 psi (760 mm Hg). Figure 15-18 shows the pressures at

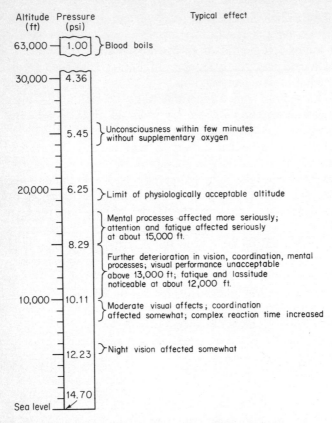

Altitude Pressure
(ft) (psi)   Typical effect

63,000 — | 1.00 | } Blood boils

30,000 — 4.36

5.45 } Unconsciousness within few minutes
without supplementary oxygen

20,000 — 6.25 } Limit of physiologically acceptable altitude

8.29 { Mental processes affected more seriously;
attention and fatigue affected seriously
at about 15,000 ft.

Further deterioration in vision, coordination, mental
processes; visual performance unacceptable
above 13,000 ft; fatigue and lassitude
noticeable at about 12,000 ft.

10,000 — 10.11 { Moderate visual affects; coordination
affected somewhat; complex reaction time increased

12.23 } Night vision affected somewhat

14.70
Sea level

Figure 15-18 General effects of hypoxia at various altitude levels and equivalent pressure levels. [Adapted in
part from Roth, 47, vol. 3, sec. 10; *Your body in flight,* 70;
Flight surgeon's manual, 64; *Handbook of human engineering data,* 65. part 7, chap. 2, sec. 14, tables 4-1 to 4-4;
and Balke, 3.]

other altitudes. A reduction of pressure by some ratio would mean that
the volume occupied by a given amount of air would be increased in-
versely (such as doubling the volume if the pressure is reduced by half).

Air Pressure and Oxygen Supply

A primary function of the respiratory system is transporting oxygen
from the lungs to the body tissue and picking up carbon dioxide on the
return trip and carting it back to the lungs where it is exhaled. Under
normal circumstances (including near-sea-level pressure) the blood (actually the red blood cells) carry oxygen up to about 95 percent of their

capacity. As air pressure is reduced, however, the amount of oxygen that the blood will absorb is reduced. For example, at approximately 10 psi (equivalent to about a 10,000-ft altitude) the blood will hold about 90 percent of its potential capacity; at about 7.3 psi (18,000 ft) the percentage drops to about 70. Incidentally, at 1.0 psi (63,000 ft) the pressure is so low that the blood actually boils, like water in a tea kettle. In addition, reduced pressure results in the inhalation of a smaller volume of air.

Hypoxia

If the oxygen supply is reduced, a condition of hypoxia (also called *anoxia*) can occur, the effects varying with the degree of reduction. Some indication of these is given in Figure 15-19 as related to the amount of oxygen

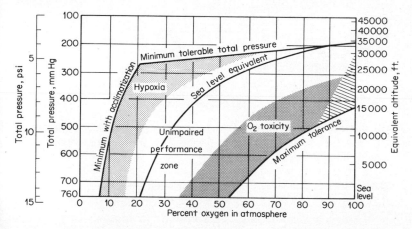

Figure 15-19 Physiological relationships between the percentage of oxygen in the atmosphere and pressure (and equivalent altitude). The white band following the sea-level equivalent curve shows the conditions in which human performance normally is unimpaired. [Adapted from Roth, 47, fig. 1-2, based on data compiled by U. C. Luft and originally drawn by E. H. Green of the Garrett Corporation.]

that would be available at various altitudes (or their equivalents). Generally speaking, the effects below about 8000 ft are fairly nominal, but above that level, or at least above about 10,000 ft, the effects become progressively more serious as shown in the figure. In connection with the effects of hypoxia, however, two qualifications are in order: (1) There are marked individual differences and (2) acclimatization does increase tolerance somewhat.

Use of oxygen. At altitudes where the hypoxia effects would normally be of some consequence, the use of oxygen masks can stave off the onset of hypoxia or minimize its degree. This is illustrated, for example, in Figure 15-20, which shows the relationship between altitude and percentage of oxygen capacity for subjects breathing air versus pure oxygen. This also shows the approximate degree of handicap for various percentage values. It can be seen, for example, that while breathing pure air reduces the percentage of capacity to about 70 at around 18,000 ft, the use of oxygen delays this amount of reduction up to about 42,000 ft.

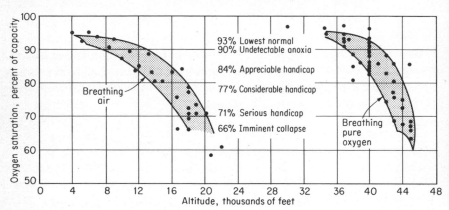

Figure 15-20   Relationship between altitude and oxygen saturation (percentage of capacity) for subjects breathing air and those breathing pure oxygen. [From *Flight surgeon's manual*, 64.]

Pressurization.   The ideal scheme for avoiding hypoxia at high altitudes is the use of a pressurized cabin that maintains the atmospheric conditions of some lower altitude. This is done in high-altitude civilian planes and in some military planes. The Air Force generally requires the use of oxygen equipment for aircraft that will operate at altitudes of 10,000 ft or above, or at 8000 ft on flights of 4 hr or longer [*System safety design handbook*, 68, chap. 3, sec. 3P], and at higher altitudes prescribes minimum differentials in pressure for different types of planes and flights [70]. Pressure suits usually are prescribed for high-altitude aircraft, in part as protection against decompression.

Decompression

It has been indicated that the volume of gas expands or contracts in proportion to the pressure applied to it (this is Boyle's law). The atmosphere within the human body is not immune from this law. Thus, as an individual changes from one pressure to another the atmosphere in body tissues and cavities expands or contracts. In this connection, as the external air pressure is increased, that within the body follows suit fairly closely, and (aside from some discomfort here and there) there are no serious consequences. But when the external air pressure is reduced suddenly, there can be some unhappy consequences, generally referred to as *decompression sickness*. Actually there are different kinds of physiological reactions. Among these is the formation of nitrogen bubbles in the blood, body tissue, and around the joints; the effect is somewhat like taking the top off a bottle of soda water. The manifestations of the physiological effects include various symptoms, such as those referred to familiarly as the *bends* and the *chokes*, and various skin manifestations. The bends consist of generalized pains in the joints and muscles. The chokes are characterized by breathing (choking) difficulty, coughing, respiratory distress, and accompanying chest pains; the skin sensations include hot and cold sensations and itching of the skin, and a mottling of the skin surface sometimes occurs. In extreme cases, the above symptoms become more severe, and in some cases shock, delirium, and coma occur; fatalities may occur in such cases.

Protection from decompression. Decompression sickness occurs primarily in underwater construction work (in sealed caissons), in underwater diving operations, in submarines, and in aircraft (as in the case of rapid ascent to, say, 30,000 ft or more, or in rapid decompression from a rupture in a sealed cabin). Where rapid decompression has occurred (say, by accident), it is usually the practice to subject the person to a higher pressure (near that to which he was originally exposed) and then gradually bring the pressure back to that of the earth.

SUMMARY

1. The body exchanges heat with its environment primarily by convection, radiation, and evaporation.
2. The primary environmental factors that affect the heat-exchange process are (*a*) *air temperature;* (*b*) *humidity of the air;* (*c*) *air circulation;* and (*d*) *temperature of objects* in the environment (walls, ceilings, etc.).
3. The relative importance of the various methods of heat exchange vary considerably, depending upon the environmental factors.
4. Indexes have been developed for combining the effects of two or more heat-exchange variables. These include: *effective temperature* (ET) which combines the thermal effects of temperature, humidity, and air movement; *operative temperature* which takes into account air temperature and wall temperature; the heat-stress index that expresses the heat *load* as influenced by various environmental factors and body metabolism; the P4SR, which is a scale of physiological stress; and the Oxford index, which is a simple weighting of wet-bulb and dry-bulb temperatures.

5. Heat stress occurs if the body cannot dissipate the heat generated by the metabolic process.

6. For equal comfort, individuals doing heavy work have to be in a cooler environment than individuals doing light or sedentary work.

7. Work performance is usually affected adversely in environments with ET above about 81 to 83°F (for heavy work) to about 86 to 88°F (for light or sedentary work).

8. Acclimatization to heat usually is reasonably complete within two weeks; substantial acclimatization to cold occurs within a week or so, but full acclimatization may take months or even years.

9. Despite acclimatization to heat or cold, the physiological costs of performing work under either extreme are greater than for doing similar work under more normal conditions.

10. The effects of cold on manual performance are primarily associated with hand-skin temperature (HST); HST of around 55° generally brings about deterioration in performance.

11. The amount of insulation (in clo units) required to maintain heat balance under cold conditions is a function of the level of physical activity.

12. The conditions under which people feel comfortable vary by individuals, age, clothing worn, and other variables. However, most people feel comfortable in the summer with ETs from about 69 to 74°F and in the winter from about 65 to 70°F.

13. The management of temperature problems includes (a) atmospheric control; (b) use of clothing and other protective gear; (c) management of personnel including selection of personnel, acclimatization, proper work schedules, personnel rotation, modification of work, and hydration.

14. Our environments include man-generated elements that can affect human welfare, such as many forms of air pollution including automobile exhaust fumes, ionizing radiation, and toxic materials. Some such elements pose potentially serious threats to human health.

15. The effects of radiation depend largely upon the dosage and include both pathological and genetic effects.

16. Hypoxia is a condition of oxygen shortage in the blood. Most frequently it occurs at high altitudes (which are characterized by lowered atmospheric pressure). The effects are of limited consequence under 10,000 ft (although there may be mild visual disturbances). The physiological, behavioral, and performance effects increase with higher altitudes and are of very serious consequence over about 18,000 ft.

17. In high-altitude flight, pressurized cabins or oxygen masks must be used to avoid hypoxia.

18. Sudden decompression (such as from accidents in underwater caissons and sudden loss of pressure in aircraft) causes nitrogen bubbles to form in the blood, body tissue, and around the joints. The manifestations of decompression sickness include the bends (pains in the joints and muscles), the chokes (respiratory difficulty), and other symptoms; in severe cases death occurs.

19. Where people are working in pressurized conditions (such as caissons), the pressure should be reduced gradually to avoid decompression sickness.

REFERENCES

1. Adolph, E. F., and G. W. Molnar: Exchanges of heat and tolerances to cold in men exposed to outdoor weather, *American Journal of Physiology*, 1946, vol. 146, pp. 507–537.

2. Armstrong, H. G.: The loss of tactical efficiency of flying personnel in open cock-pit aircraft due to cold temperature, *Military Surgeon*, 1936, vol. 79, pp. 133–140.

3. Balke, B.: *Human tolerances*, Civil Aeromedical Research Institute, Federal Aviation Agency, Aeronautical Center, Oklahoma City, Report 62–6, April, 1962.

4. Barnett, P.: *Field tests of two anti-exposure assemblies*, Arctic Aeromedical Laboratory, Fort Wainwright, Alaska, AAL TDR 61–56, 1962.

5. Belding, H. S., and T. F. Hatch: Index for evaluating heat stress in terms of resulting physiological strains, *Heating, Piping and Air Conditioning*, August, 1955, pp. 129–136.

6. Benzinger, T. H.: The human thermostat, *Scientific American*, 1961, vol. 204, pp. 134–147.

7. Brouha, L.: *Physiology in industry*, Pergamon Press, New York, 1960.

8. Burton, A. C., and O. G. Edholm: *Man in a cold environment: physiological and pathological effects of exposure to low temperatures*, Edward Arnold (Publishers), Ltd., London, 1955.

9. Carson, Rachel: *Silent spring*, Riverside Press, New York, 1962.

10. Chiles, W. D., R. E. Fox, and D. W. Stilson: *Effects of ionized air on decision making and vigilance performance*, USAF, MRL, Technical Data Report 62–51, May, 1962.

11. Clark, R. E.: *The limiting hand skin temperature for unaffected manual performance in the cold*, USA Quartermaster Research and Engineering Center, TR EP–147, February, 1961.

12. Commoner, B.: *Science and survival*, The Viking Press, Inc., New York, 1966.

13. Dubois, K. P., and E. M. K. Geiling: *Textbook of toxicology*, Oxford University Press, Fair Lawn, N.J., 1959.

14. Dusek, E. R.: *Manual performance and finger temperature as a function of skin temperature*, USA Quartermaster Research and Engineering Center, TR EP–68, October, 1957.

15. Eichna, L. W., W. F. Ashe, W. B. Bean, and W. B. Shelley: The upper limits of environmental heat and humidity tolerated by acclimatized men working in hot environments, *Journal of Industrial Hygiene and Toxicology*, 1945, vol. 27, pp. 59–84.

16. Fanger, P. O.: Calculation of thermal comfort: introduction to a basic comfort equation, *ASHRAE Transactions*, 1967, vol. 73, pt. 3, pp. 4.1–4.20.

17. Fox, W. F.: Human performance in the cold, *Human factors*, vol. 9, no. 3, pp. 203–220.

18. Frey, A. H.: Human behavior and atmospheric ions, *Psychological Review*, 1961, vol. 68, pp. 225–228.

19. Gagge, A. P.: Final progress report: RP–41—physiological effects of high intensity radiant beam heating, *ASHRAE Journal*, April 1968, vol. 10, no. 4, pp. 86–89.

20. Gagge, A. P., A. C. Burton, and H. C. Bazett: A practical system of units for the description of the heat exchange of man with his environment, *Science*, 1941, vol. 94, pp. 428–430.

21. Gaydos, H. F.: Effect on complex manual performance of cooling the body while maintaining the hands at normal temperature, *Journal of Applied Physiology*, 1958, vol. 12, pp. 373–376.

22. Gitlin, J. N., and P. S. Lawrence: *Population exposure to x-rays U.S. 1964*, U.S. Department of Health, Education, and Welfare, USPHS Publication 1519.

23. Glasstone, S. (ed.): *The effects of nuclear weapons*, U.S. Department of Defense, 1962.

24. Halcomb, C. G., and R. E. Kirk: Effects of air ionization upon the performance of a vigilance task, *Journal of Engineering Psychology*, 1965, vol. 4, no. 4, pp. 121–125.

25. Hardy, J. D. (ed.): *Temperature—its measurement and control in science and industry*, Reinhold Publishing Corporation, New York, 1963.

26. Hatch, T. F.: "Assessment of heat stress," in J. D. Hardy (ed.), *Temperature— its measurement and control in science and industry*, vol. 3, pt. 3, pp. 307–318, Reinhold Publishing Corporation, New York, 1963.

27. Henderson, Y., and H. W. Haggard: *Noxious gases*, Reinhold Publishing Corporation, New York, 1943.

28. Hertig, B. A., and H. S. Belding: "Evaluation and control of heat hazards," in J. D. Hardy (ed.), *Temperature—its measurement and control in science and industry*, vol. 3, pt. 3, pp. 347–355, Reinhold Publishing Corporation, New York, 1963.

29. Hertzman, A. B.: *Vasomotor control of the peripheral circulation and regulation of body temperature*, paper read at American Institute of Physics Symposium on Temperature—Its Measurement and Control in Science and Industry, Columbus, Ohio, March, 1961.

30. Horvath, S. M., and A. Freedman: The influence of cold upon the efficiency of man, *Journal of Aviation Medicine*, 1947, vol. 18, pp. 158–164.

31. Hunter, J. C.: Effects of environmental hyperthermia on man and other mammals: a review, *Military Medicine*, 1961, vol. 126, pp. 273–281.

32. Langham, W. H. (ed.): *Radiobiological factors in manned space flight,* NAS, NRC, Publication 1481, 1967.

33. Leithead, C. S., and A. R. Lind: *Heat stress and heat disorders,* Cassell & Co., Ltd., London, 1964.

34. Lind, A. R.: A physiological criterion for setting thermal environmental limits for everyday work, *Journal of Applied Physiology,* 1963, vol. 18, pp. 51–56.

35. Lind, A. R.: "Tolerable limits for prolonged and intermittent exposures to heat," in J. D. Hardy (ed.), *Temperature—its measurement and control in science and industry,* vol. 3, pt. 3, pp. 337–345, Reinhold Publishing Corporation, New York, 1963.

36. Lind, A. R., and D. E. Bass: The optimal exposure time for the development of acclimatization to heat, *Federal Proceedings,* 1963, vol. 22, no. 3, pp. 704–708.

37. Lockhart, J. M.: Extreme body cooling and psychomotor performance, *Ergonomics,* 1968, vol. 11, no. 3, pp. 249–260.

38. McArdle, B., et al.: *The prediction of the physiological effects of warm and hot environments,* Medical Research Council (Great Britain), H.S. 194 Royal Naval Personnel Research Committee, Report R.N.P. 47/391, October, 1947.

39. McBlair, W., D. Rumbaugh, and J. Fozard: *Ventilation, temperature, humidity,* San Diego State College Foundation, Contract Nonr–1268 (01), December, 1955.

40. McCleary, R. A.: *Psychophysiological effects of cold,* USAF School of Aviation Medicine, Project 21–1202–0004, Report 1, January, 1953.

41. McKarnes, J. S., and R. S. Brief: Nomographs give refined estimate of heat stress, *Heating, Piping and Air Conditioning,* January, 1966, vol. 38, no. 1, pp. 113–116.

42. Machle, W.: Control of heat in industry, *Occupational Medicine,* 1946, vol. 2, pp. 350–359.

43. Mackworth, N. H.: Effects of heat on wireless telegraphy operators hearing and recording Morse messages, *British Journal of Industrial Medicine,* 1946, vol. 3, pp. 143–158.

44. Mackworth, N. H.: Cold acclimatization and finger numbness, *Proceedings of the Royal Society,* 1955, vol. B, 143, pp. 392–407.

45. Mackworth, N. H.: *Researches on the measurement of human performance,* Medical Research Council (Great Britain), Special Report Series 268, 1950. Reprinted in H. W. Sinaiko (ed.), *Selected papers on human factors in the design and use of control systems,* Dover Publications, Inc., New York, 1961.

46. Nevins, R. G., F. H. Rohles, W. Springer, and A. M. Feyerherm: Temperature-humidity chart for thermal comfort of seated persons, *ASHRAE Transactions,* 1966, vol. 72, pt. 1, pp. 283–291.

47. Roth, E. M. (ed.): *Compendium of human responses to the aerospace environment,* NASA CR–1205, vols. 1–4, November, 1968.

48. Siple, P. A., and C. F. Passel: Measurement of dry atmospheric cooling in subfreezing temperatures, *Proceedings of the American Philosophical Society,* 1945, vol. 89, pp. 117–199.

49. Smith, P. W.: *Toxic hazards in aerial application*, Federal Aviation Agency, Civil Aeromedical Research Institute, Report 62–8, April, 1962.
50. Teichner, W. H.: Reaction time in the cold, *Journal of Applied Psychology*, 1958, vol. 42, pp. 54–59.
51. Veghte, J. A., and J. R. Clogston: *A new heavy winter flying clothing assembly*, Arctic Aeromedical Laboratory, Fort Wainwright, Alaska, AAL TN 61–4, 1961.
52. von Oettingen, W. F.: *Poisoning*, W. B. Saunders Company, Philadelphia, 1958.
53. Webb, P. (ed.): *Bioastronautics data book*, NASA SP–3006, 1964.
54. Wing, J. F.: *A review of the effects of high ambient temperature on mental performance*, USAF, AMRL, TR 65–102, September, 1965.
55. Winslow, C. E. A., et al.: Physiological influence of atmospheric humidity: second report of the ASHVE Technical Advisory Committee on Physiological Reactions, *Transactions of the ASHVE*, 1942, vol. 48, pp. 317–326.
56. Winslow, C. E. A., and L. P. Herrington: *Temperature and human life*, Princeton University Press, Princeton, N.J., 1949.
57. Winslow, C. E. A., L. P. Herrington, and A. P. Gagge: Physiological reactions of the human body to varying environmental temperatures, *American Journal of Physiology*, 1937, vol. 120, pp. 1–22.
58. Wyndham, C. H., N. B. Strydom, H. M. Cook, and J. S. Mavitz: *Studies on the effects of heat on performance of work*, Applied Physiology Laboratory Reports, 1–3/59, 1959, Transvaal and Orange Free State Chamber of Mines, Johannesburg, S. Africa.
59. *ASHRAE handbook of fundamentals (1967)*, ASHRAE, New York.
60. *Background material for development of radiation protection standards*, Federal Radiation Council, Staff Report 1, May 13, 1960.
61. *The biological effects of atomic radiation: a report to the public*, NAS, NRC, 1960.
62. *The biological effects of atomic radiation: summary reports*, NAS, NRC, 1960.
63. *The effects of nuclear weapons*, U.S. Government Printing Office, May, 1957.
64. *Flight surgeon's manual*, USAF Manual 160–5, July, 1954.
65. *Handbook of human engineering data*, 2d ed., Tufts University, Medford, Mass., 1952.
66. *Manual of industrial radiation protection: Part III. General guide on protection against ionising radiation*, International Labour Office, Geneva, 1963.
67. *Radiological health handbook*, USPHS, PB121784R, 1960.
68. *System safety design handbook*, USAF, AFSC DH 1–6, 2d ed., July 20, 1968.
69. *Weather and climate modification: Problems and prospects*, NAS, NRC, Publication 1350, 1966.
70. *Your body in flight*, USAF Pamphlet 160–10–3, Jan. 1, 1960.

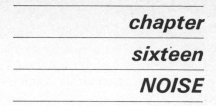

Before the days of machines and mechanical transportation equipment, mankind's noise environment consisted of noises such as those of household activities, domestic animals (and maybe of a few blood-curdling wild ones), horse-drawn vehicles, hand tools, and weather. But man's ingenuity changed all this through his creation of machines, motor vehicles, subways, radios, guns, bombs, fire sirens, jet aircraft, and New Year's Eve horns. The human problems associated with noise include working situations, the home, and the community [Cohen, 10–12]. City noise has been characterized as a form of *ear pollution,* and has moved from the annoyance stage to the level of a serious health hazard. It has been estimated that city noise is increasing at about 1 decibel (dB) per year [Jones and Cohen, 26]; such a rate of increase would mean that the intensity 20 years from now would be 100 times as high as the present level!

Noise usually is considered to be unwanted sound, but this concept has certain unsatisfying aspects, particularly the question of the basis on which the judgment of unwanted is to be made. A somewhat more definitive concept is the one proposed by Burrows [7], in which noise is considered in an information-theory context, as follows: Noise is "that auditory stimulus or stimuli bearing no informational relationship to the presence or completion of the immediate task." This concept applies equally well to attributes of task-related sounds that are informationally useless, as well as to sounds that are not task related. A tying together of these two concepts would seem to be reasonable, by considering unwanted that sound that has no informational relationship to the task or activity at hand.

NOISE AND LOSS OF HEARING

Of the different possible effects of noise, one of the most important is hearing loss. There are really two primary types of deafness. One is called *nerve* deafness and most frequently is caused by a condition of the nerve cells of the inner ear that reduces sensitivity. The other is *conduction* deafness and is caused by some condition of the outer or middle ear that affects the transmission of sound waves to the inner ear.

The hearing loss in nerve deafness is typically uneven; usually the hearing loss is greater in the higher frequencies than in the lower ones.

515

Normal deterioration of hearing through aging is usually of the nerve type, and continuous exposure to high noise levels also typically results in nerve deafness. Once nerve degeneration has occurred, it can rarely be corrected. Conduction deafness is only partial, never complete, since airborne sound waves strike the skull and may be transmitted to the inner ear by conduction through the bone. It may be caused by different conditions such as adhesions in the middle ear that prevent the vibration of the ossicles, infection of the middle ear, wax or other substance in the outer ear, or scars from a perforated eardrum.

People with this type of damage sometimes are able to hear reasonably well, even in noisy places, if the sounds to which they are listening (for example, conversation) are at intensities above the background noise. This type of deafness can sometimes be arrested, or even improved. Hearing aids are more frequently useful in this type of deafness than they are in nerve deafness.

Measuring Hearing

In order to review the effects of noise on hearing, we should first see how hearing (or hearing loss) is measured. There are two basic methods of measurement, namely, the use of simple tests of hearing and the use of an audiometer.

Simple hearing tests.   For some purposes simple hearing tests are used. These include a voice test, a whisper test, a coin-click test, and a watch-tick test. In the voice and whisper tests, the tester (out of sight) speaks or whispers to the testee, and the testee is asked to repeat what was said. This may be done at different distances and with different voice intensities. The primary shortcoming of such tests is that they usually lack standardization. If reasonable standardization can be achieved (such as using a particular person's voice or a particular watch), such tests may serve certain rough hearing-test purposes.

Audiometer tests.   Audiometers are of two types, the most common being an instrument that is used to measure hearing at various frequencies. It reproduces, through earphones, pure tones of different frequencies and intensities. As the intensity is increased or decreased, the testee is asked to indicate when he can hear the tone or when it ceases to be audible. It is then possible to determine for each frequency tested the lowest intensity that can just barely be heard; this is the *threshold* for the frequency. Examples of three audiograms are shown in Figure 16-1. For each individual the line shows the auditory threshold for sounds of various frequencies. It will be seen that these three patterns of hearing loss differ rather markedly from one another.

Another type of audiometer is a speech audiometer. Direct speech, or a recording of speech, is reproduced to earphones or to a loudspeaker,

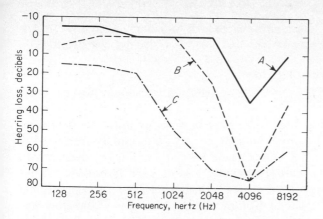

Figure 16-1 Examples of three audiograms. In each
· case the line represents the minimum sound intensity
of the frequencies that can just barely be heard by the
individual (the auditory threshold). [Adapted from
Carhart, 8.]

and intensity is controlled. Depending upon the type of material presented, the testee may be asked to repeat the words, or to answer simple
questions that would indicate whether he had heard the material, or to
explain something related to the material spoken. The speech intelligibility tests used with audiometers to measure hearing loss for speech have
been carefully constructed in order that the words selected for them
include the gamut of speech sounds. Probably the most commonly used
tests of this type are the tests of phonetically balanced words (called
PB lists). These tests include 20 lists of 50 one-syllable words [Egan, 15],
each of which includes words which collectively incorporate approximate
representation of speech sounds that occur in normal speech.[1] Another
type of list includes *spondees*, or two-syllable words in which both syllables
have equal emphasis [Hudgins et al., 21].[2] In the use of lists of words such
as those mentioned above, it is the usual practice to determine the thresh-

[1] Following are the words for one of these PB lists; as presented here, the words
are in alphabetical order, but in actual use they are randomized: are, bad, bar, bask,
box, cane, cleanse, clove, crash, creed, death, deed, dike, dish, end, feast, fern, folk,.
ford, fraud, fuss, grove, heap, hid, hive, hunt, is, mange, no, nook, not, pan, pants,
pest, pile, plush, rag, rat, ride, rise, rub, slip, smile, strike, such, then, there, toe,
use, wheat.
[2] Following are the words from one of the spondee lists: airplane, armchair, back-
bone, bagpipe, baseball, birthday, blackboard, bloodhound, bobwhite, bonbon,
buckwheat, coughdrop, cowboy, cupcake, doorstep, dovetail, drawbridge, earth-
quake, eggplant, eyebrow, firefly, hardware, headlight, hedgehog, hothouse, inkwell,
mousetrap, northwest, oatmeal, outlaw, playground, railroad, shipwreck, shotgun,
sidewalk, stairway, sunset, watchword, whitewash, wigwam, wildcat, woodwork.

old for speech, that is, the minimum intensity level at which a given percentage of the words can be identified correctly, usually 50 percent.

Normal Hearing and Hearing Loss

Before we see what effect noise has on hearing, we should first see what normal hearing is like.

Surveys of hearing loss. Surveys have been made to determine the hearing abilities of people. In such surveys individuals are tested at various frequencies to determine their loss of hearing at each of the tested frequencies. The average hearing loss at each frequency is then determined. Usually such data are developed separately for separate age groups. The USASI Sectional Committee on Acoustics, Vibration, and Mechanical Shock [48] has consolidated the results of three such surveys into smoothed curves showing average hearing loss by frequencies for various ages for both men and women, as shown in Figure 16-2. This

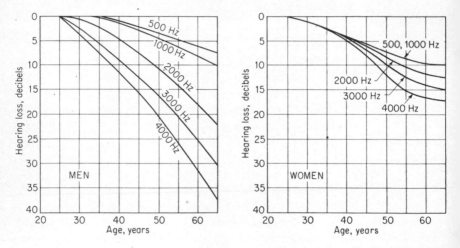

Figure 16-2 Average hearing loss by frequency for men and women of various ages. [From *The relations of hearing loss to noise exposure*, USASI, 48.]

shows, for each of five frequencies, the average hearing loss that typically occurs through age. There are certain points in particular that are revealed by this figure. We can see, for example, that older groups have greater hearing loss than younger groups. Further, hearing loss is generally greater in the higher frequencies than in the lower frequencies. In general, women have greater loss in the lower frequencies than men, and men, in turn, have greater loss in the higher frequencies than women. In interpreting such data, certain points need to be kept in mind, such as the fact that averages can be rather treacherous since they can mask wide ranges of individual differences (the samples used, although large, are not necessarily representative), and the probability that some people tested had experienced some hearing loss from noise exposure and not necessarily from age. But, despite such possible limitations, this figure probably gives us the best available picture of the hearing loss that occurs through age.

Hearing Loss from Continuous Noise Exposure

There is abundant evidence that continuous exposure to high noise levels contributes to hearing loss. There are, however, many unanswered questions with regard to what noise (frequency, intensity, duration of exposure, etc.) will produce what hearing loss (frequency, decibels, etc.). Noise levels of working or living environments vary a great deal. Octave-band spectrums of three industrial operations in an aircraft plant are shown in Figure 16-3. These three spectrums show, for the three noisy operations, how spectrums of noise can vary.

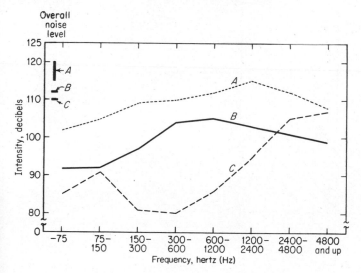

Figure 16-3   Octave-band spectrums of noises of three industrial operations: *A*, riveting (overall noise level 115–120 dB); *B*, speed hammers (112 dB); and *C*, compressed air drying (110 dB). [Adapted from MacLaren and Chaney, 33.]

Evidence of hearing loss from noise.   There have been a number of specific studies, and summaries thereof, that provide persuasive testimony about the effects of noise on hearing [Berrien, 2; Cohen, 10; Kryter, 27; and LaBenz et al., 31]. In this connection, a committee under the aegis of the American Industrial Hygiene Association has teased out and condensed, from many sources, data that show the incidence of hearing impairment, in a consolidated figure, for each of several age groups of individuals who have been exposed (in their work) to noise intensities of different levels. This consolidation is shown in Figure 16-4. Actually this figure shows, for any group, the probabilities of individuals having hearing impairment (specifically, an average hearing threshold in excess of 15 dB at 500, 1000, and 2000 Hz). Although the probabilities of impairment (so defined) are not much above those of the general population for individuals exposed to 85 dB, the curves shoot up sharply at higher levels, except for the youngest group (but their time will come!).

Although Figure 16-4 shows us something about the *odds* of having impairment, it does not depict the nature of the impairment in terms of frequencies. For this purpose, let us refer to the very thorough longitudinal survey carried out by a subcommittee of the USASI [48]. Audiograms were prepared on a selected group of 200 workers (1) who had been exposed continuously (during the working day) to a single type of

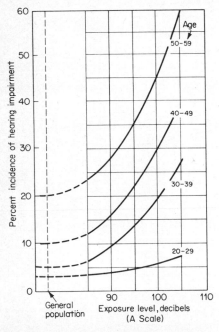

Figure 16-4 Incidence of hearing impairment in the general population and in selected populations by age group and occupational noise exposure; impairment is defined as a hearing threshold level in excess of an average of 15 dB at 500, 1000, and 2000 Hz. [From *Industrial Noise Manual*, 43, fig. 1, p. 420.]

noise for a period of time—the time ranging from 2 to 44 years; (2) who had no history of previous exposure to intense noise; and (3) whose noise environment in the plant was known (sound spectrums were developed for 30 different plant locations, with an octave-band analyzer).

By the use of an ingenious procedure that need not be described here, the investigators developed estimates of the hearing loss brought about by varying lengths of exposure to noises of various levels. These estimates were made of hearing loss at each of three frequency levels, namely, 1000, 2000, and 4000 Hz. The estimates were corrected for normal hearing loss attributable to age. The results are shown in Figure 16-5. The three parts of this figure, a, b, and c, represent the estimated-average-trend curves for net hearing loss at 1000, 2000, and 4000 Hz, respectively. The decibel level for each contour is the noise level for a *specific octave, not* the *overall* level. Without belaboring the point, it can be said that the octave selected for use with any one of the three frequency levels is the octave for which the noise level was found to be most nearly correlated with hearing loss at the frequency in question. This does not necessarily mean, however, that the noise in this selected octave caused the hearing loss of the frequency in question.

Three implications can be drawn from this figure: (1) The amount of hearing loss is related to level of noise to which exposed; the greater the exposure intensity, the greater the hearing loss. (2) Hearing loss is greater in the 4000-Hz range than in the 1000- and 2000-Hz ranges; this has been found in numerous studies. (3) Hearing loss is associated with exposure time for higher exposure intensities, though to a limited extent, or not at all, for lower exposure intensities.

Noncontinuous Noise

The gamut of noncontinuous noise includes intermittent (but steady) noise (such as machines that operate for short, interrupted periods of time), impact noise (such as that from a drop forge), and impulsive noise (such as from gun fire). In heavy doses, such noise levies its toll in hearing loss, but the combinations and permutations of intensity, noise spectrum, frequency, duration of exposure, and other parameters preclude any simple, pat descriptions of the affects of such noise.[3] In the case of impact and impulsive noise, however, it might be noted that the toll sometimes is levied fairly promptly. For example, 35 drop-forge operators showed a noticeable increase in hearing threshold within as little as 2 years [48], and 45 gunnery instructors averaged 10 percent hearing loss over only 9 months, even though most of them had used hearing protection devices [Machle, 32].

[3] A thorough treatment of exposure to intermittent noise is presented by Kryter et al. [30].

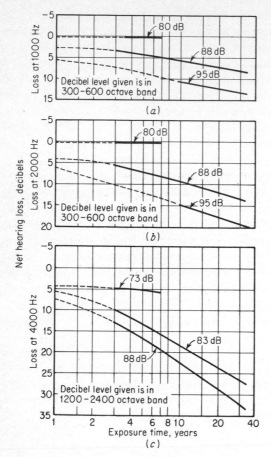

Figure 16-5 Estimated average trend curves for net hearing loss at 1000, 2000, and 4000 Hz, after continuous exposure to steady noise. Data are corrected for age but not for temporary threshold shift. Dotted portions of curves represent extrapolation from available data. [Adapted from USASI, 48.]

Temporary versus Permanent Hearing Loss

A daily quota of a fairly loud noise generally brings about some degree of temporary hearing loss, which usually is largely dissipated by the next morning. However, depending on a number of factors (intensity, period of exposure, etc.), the daily recovery may become less and less and the temporary hearing loss may become more permanent. This was illustrated, for example, by the fact that the *recovery* of temporary hearing loss in the case of textile workers with 10 years of exposure was very limited, as contrasted with groups exposed for 19 or 27 months [USASI, 48]. In fact, evidence indicates that the extent of temporary threshold changes following a day's exposure to continuous noise is surprisingly close to the magnitudes of permanent hearing loss after several years of such exposure. Thus, temporary hearing loss can be used as a reasonably valid secondary yardstick for permanent threshold shifts due to exposure to noise [Kryter, 28].

PHYSIOLOGICAL EFFECTS OF NOISE

Hearing loss is of course the consequence of physiological damage to the mechanisms of the ear. Aside from such damage, one might wonder whether, *concurrent with* noise exposure, there are any temporary physiological effects. Studies dealing with this facet generally indicate that initial exposure (especially sudden exposure) induces an initial physiological reaction, but that typically such responses settle down to normal or near-normal levels after a period of time.

EFFECTS OF NOISE ON PERFORMANCE

By selection of relevant studies, one can "prove" that noise (1) produces a decrement in human performance, (2) has no effect on such performance, or (3) produces an increment in performance. This confusing state of affairs is reflected in reviews [Berrien, 2; Cohen, 10; Kryter, 27; Plutchik, 35; and Teichner et al., 38] and probably somewhat supports the statements made in the *Industrial noise manual* [43] to the effect that the behavioral effects of noise are as complex and ill-defined as the noise itself and that a categorical statement that noise exposure has no ill effects on human behavior cannot be made at this time. In contemplating this unsatisfying state of affairs, Teichner et al. [38] chide those investigators who glibly attribute performance decrement under noise to "distraction," performance increment to motivational "compensation," and no effect to "lack of sensitivity" of the task. In a couple of studies they attempted to test a theoretical approach to the effects of noise on performance that takes into account the factors of distraction, habituation, auditory adaptation, and bodily arousal. Without bringing in the details, their controlled experiments suggested that such an approach is reasonable. If this be so, the results of various (and apparently conflicting) studies might be attributable to the way in which these factors combine to affect the performance in question.

Without citing any of the apparently conflicting studies, we probably should reject the notion that noise *generally* brings about degradation in human performance. There is accumulating evidence, however, that noise may adversely affect human performance under certain circumstances. This evidence comes particularly from Broadbent and his associates [5, 6] and from Jerison and his associates [22–25]. Without giving the specific results, we can state that these investigators found significant decrements in performance on vigilance tasks under noise conditions as contrasted with quiet conditions. Jerison, however, found such decrement only after a period of time, about $1\frac{1}{2}$ hr [22]. In addition he found a significant adverse effect of noise on a mental counting task. In interpreting such results in the light of several investigations in which negative results have been found, Jerison suggests that adverse performance effects (such as he found) might be expected where noise might serve as a source of psychological stress. If this hypothesis is valid, the implication would be that for short, spurtlike efforts no performance decrements would be expected from noise, but that decrement might be expected when sustained performance is required and where the task is not intrinsically challenging; under such conditions noise might serve to change the motivational level or emotional balance, and thus serve as a source of psychological stress.

Aside from vigilance tasks, it is very possible that performance on certain other types of tasks might be vulnerable to the effects of noise, but, at the present stage of affairs, hard, definitive evidence about such tasks is hard to come by. However (with fingers crossed) it seems that performance on the following kinds of tasks is most likely to be affected by noise: vigilance tasks (as discussed above); certain types of complex mental tasks; tasks calling for skill and speed [Roth, 37]; and tasks that demand a high level of perceptual capacity, such as in some time-shared tasks that press one's perceptual abilities [Boggs and Simon, 3].

In view of the ambiguity relating to the effects of noise on performance, Roth's observations may be consoling [37]. He points out that the level of noise required to exert a measurably degrading effect on task performance (such as a sound-pressure level of 90 dB) is considerably higher than the highest levels that are acceptable by other criteria, such as hearing loss and effects on speech communications. Thus, if noise levels are kept within reasonable bounds in terms of, say, hearing loss considerations, the probabilities of serious effects on performance probably would be relatively nominal.

ANNOYANCE AND WELL–BEING

In discussing the annoying qualities of noise, Kryter [29] uses the term *perceived noisiness* as being somewhat synonymous with unwantedness, unacceptableness, annoyingness, objectionableness, and disturbingness. You and I know full well what he is talking about, and the noise characteristics that do bother us are intensity, bandwidth, spectral content, and duration. Taking frequency to illustrate this point, Fenwick [16], using 32 different sound spectra equated for subjective loudness, found that subjects judged *most annoying* those spectra with highest frequencies.

Annoying Aspects of Community Noise

In a survey carried out in Boston, Los Angeles, and New York for the Federal Housing Administration [46], traffic was generally reported by people to be the most "bothersome" source of urban noise, with noise from planes, industry, children and neighbors, and sirens and horns coming in for frequent complaint, depending in part on location (as, for example, near airports) and distance (such as from streets). As an example, a 10-min sample of noise of heavy traffic in New York averaged 81 dB at 15 ft and 76 dB at 50 ft. As another illustration, a properly muffled diesel truck produces about 77 dB at 100 ft, and this value can increase by about 5 dB during its acceleration; the absence of a muffler can add 15 dB more. In three schools within 1½ miles of an airport there were between 40 and 60 interruptions per day from aircraft, which resulted in a loss of 10 to 20 min/day in each classroom [Cohen, 12]. The combinations of noise sources in some cities probably cannot be passed off simply as "annoying;" they probably approach the levels of a health hazard [Cohen, 12; Jones and Cohen, 26] because of the possible effects in disturbing privacy, rest, relaxation, and sleep (as well as hearing) and because all these are essential to an individual's well-being.

There is, of course, some habituation to the annoying aspects of noise, as reported, for example, by Culbert and Posner [14] on the basis of a 3-week adaptation period. Although some adaptation does occur, however, the extent of this adaptation is not yet known, and it undoubtedly does not make a person impervious to the effects of a roaring subway or a nearby jackhammer.

Procedure for rating annoyance. The level of annoyance of community (especially residential) noise probably is reflected in the actions that people in the community take. A number of individuals and organizations have systematically analyzed community reactions to various noise conditions in order to be able to predict such reactions [Kryter, 29; Rosenblith and Stevens, 36; see also *Land use planning with respect to aircraft noise*, 44]. One such procedure is that developed by the International Organization for Standardization (ISO) [47]. This procedure is essentially an adaptation of the one developed earlier by Rosenblith and Stevens [36] and takes into account such factors (aside from the noise level itself) as the nature of the community, the time of day, season of the year, and the initial effects (as in the case of a new noise source in the community). The procedures are outlined below:

1. Develop a spectrum of the noise (or the noise that is expected to be generated, in the case of a new facility).
2. Superimpose this spectrum over Figure 16-6, and determine the highest noise rating curve N that the spectrum exceeds.

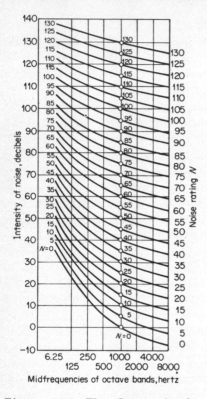

Figure 16-6 The International Organization for Standardization (ISO) noise rating curves. Such curves can be used in the estimation of public reaction to noise sources (such as new sources) and for establishing noise ceilings for specific purposes (see text for discussion). [From 47.]

3. Apply the corrections shown in Table 16-1; add algebraically these corrections, and add or subtract from the noise rating N derived above. This is the *corrected noise rating number*.
4. Refer to Table 16-2 to ascertain the estimated public reaction to the noise.

Table 16-1 Correction to Noise Rating Number (Primarily Applicable to Residential Cases)

Influencing factor	Possible conditions	Corrections to uncorrected N
Noise spectrum character	Pure tone components	+5
	Wide-band noise	0
Peak factor	Impulsive	+5
	Nonimpulsive	0
Repetitive character (about ½ min noise duration assumed)	Continuous exposure to one-per-minute	0
	10–60 exposures per hour	−5
	1–10 exposures per hour	−10
	4–20 exposures per day	−15
	1–4 exposures per day	−20
	1 exposure per day	−25
Adjustment to exposure	No previous conditioning	0
	Considerable previous conditioning	−5
	Extreme conditioning	−10
Time of day and season	Only during daytime	−5
	At night	+5
	Winter	−5
	Summer	0
Allowance for local conditions	Neighborhood:	
	Rural	+5
	Suburban	0
	Residential, urban	−5
	Urban near light industry	−10
	Industrial area, heavy industry	−15

SOURCE: ISO [47].

Table 16-2 Public Reaction to Noise in Residential Districts

Corrected noise rating number	Estimated public reaction
Below 40	No observed reaction
40–50	Sporadic complaints
45–55	Widespread complaints
50–60	Threats of community action
Above 65	Vigorous community action

SOURCE: ISO [47].

Such a procedure, although admittedly rough, probably would be particularly useful when the creation of a new facility that could be a potential source of noise is being considered for a community. If, for example, a noise rating for a facility being planned would turn out to be about 50, one could expect "widespread complaints" about it.

ACCEPTABLE LIMITS OF NOISE

In trying to figure out the upper ceiling of noise that would be acceptable in a given situation, the question of the criterion of acceptability immediately bobs up. Criteria in terms of speech communications were discussed in Chapter 7 [such as the speech interference level (SIL) and noise criteria (NC) curves], and the above discussion of residential noise levels dealt in part with the criterion of annoyance.

Hearing-loss Criteria

In connection with hearing loss, various damage risk criteria and other standards have been developed by several organizations. Such standards typically are expressed in terms of specific factors such as intensity and duration of each exposure and years of exposure. As yet, however, there is probably no single set of standards that has been widely accepted, although the USASI and the ISO are actively working in this direction. It is neither feasible nor desirable to work in a compendium of all the specific sets of standards that have been developed, but certain examples would be in order.[4]

Continuous noise. Actually, the ISO noise rating curves given above in Figure 16-6 serve such purposes in connection with hearing loss that might occur from broad-band continuous noise during the working day (5 hr or more). In such continuous exposure, an N of 85 dB is suggested as an upper limit for conservation of hearing. An impression of several other specific sets of noise criteria for steady-state noise is given in Figure 16-7. The several criteria included in that figure are proposed as ceilings for 5 to 8 hr daily exposure to steady-state noise. Although there are obvious differences among these sets, they also have quite a bit in common in their general patterns. The U.S. Air Force [39] has established a simple standard, using only the four octave-band pressure levels having center frequencies of 500, 1000, 2000, and 4000 Hz; this standard provides that the sound pressure level should not exceed 85 dB for *any* of these four octave bands, for conventional daily exposure of 8 hr (although there is provision for permitting 3 dB increases for each *halving* of the daily exposure duration).

---

[4] For further discussion of these standards the reader is referred to such sources as Beranek [1], Botsford [4], Cohen [10], Coles et al. [13], Glorig et al. [19], Kryter [28], Kryter et al. [30], *Guide for conservation of hearing in noise* [41], Guidelines for noise exposure control [42], and *Industrial noise manual* [43].

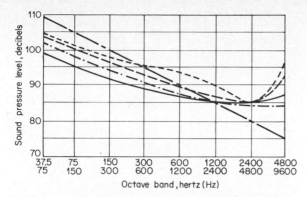

Figure 16-7 Proposed hearing-conservation criteria proposed by various investigators for 5 to 8 hr daily exposures to steady-state noise. For any given criterion, the noise spectrum should fall entirely below the criterion curve. [From Cohen, 10, fig. 5, based on several sources.]

Intermittent noise. In the case of intermittent noise, the tolerance limits depend on a trade-off between intensity and the relationship between the duration of exposure and the duration of subsequent nonexposure. An example of such trade-off is given in Figure 16-8. The curve for a given intensity indicates the period of nonexposure (the *off time*) which must follow noise exposures of the durations shown along the horizontal scale (the *on time*) to avoid temporary threshold shifts greater than 12 dB at 2000 Hz. (The 2000-Hz value is used since this is within the major range of frequencies that are important in understanding speech.)

THE HANDLING OF NOISE PROBLEMS

When a noise problem exists, or might be anticipated, there is no substitute for having good, solid information to bring to bear and for attacking the problem in a systematic manner. The manner of working out such problems is, of course, for the acoustical engineers, and we shall not go into this topic except in a very cursory way.[5]

Data Collection

There are two primary types of information that one should have in attacking noise problems. In the first place, the noise in suspected areas should be measured or estimated, especially in locations where persons

---

[5] Interested readers are referred to such sources as Beranek [1], Peterson and Gross [34], *Foundry noise manual* [40], and *Industrial noise manual* [43].

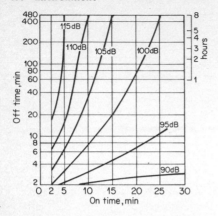

Figure 16-8 Guide to allowable exposure times for intermittent noise. (Each curve is labeled with the average value for octave bands with center line frequencies of 500, 1000, and 2000 Hz.) The vertical scale shows the off time which must follow noise exposure of the duration shown on the horizontal scale (on time) to avoid temporary threshold shifts greater than 12 dB at 2000 Hz. [From *Guide for conservation of hearing in noise,* 41.]

are normally exposed. The overall noise level can be measured with a sound-level meter. If the overall intensity approaches undesirable levels, a spectrum of the noise should be developed with an octave-band analyzer. Where the noise problem is one which is anticipated in connection with a system that is being designed, the overall intensity and the nature of the spectrum can only be estimated. This might be done from information about earlier versions of similar equipment (such as aircraft, production machines, etc.) or by engineering analyses. In the second place, individuals who have been exposed to high levels of noise should be given audiometric tests. In fact, this should be standard practice for all employees, new and old.

Defining the Noise Problem

If the results of noise measurement or of audiometric tests point up a noise problem in hearing loss, annoyance, communications, etc., it is then in order to define the problem clearly. This consists essentially of two phases: The first of these is the measurement of the noise as mentioned earlier. The second is to determine what noise level would be

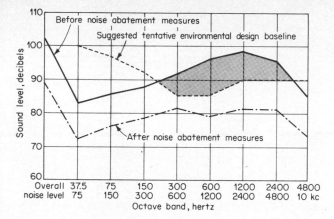

Figure 16-9 Spectrum of noise of foundry cleaning room before abatement, the base line that represents a desired upper ceiling, and the spectrum after abatement. The shaded area represents the desired reduction. The abatement consisted primarily in spraying a heavy coat of *deadener* on the tumbling barrels and surfaces of tote boxes. [From *Foundry Noise Manual*, 40, p. 52.]

acceptable in attempting to solve the problem. Such limits normally would be those adapted from relevant criteria such as discussed above. An example is shown in Figure 16-9. This figure shows the spectrum of the original noise of a foundry cleaning room, and a tentative design base line that was derived from a set of relevant noise standards. Incidentally, part of the original spectrum was below this level, but the high frequencies were not; the difference represents the amount of reduction that should be achieved (in this case the shaded area). The third line shows the noise level after abatement.

Noise Control

The general approaches to the control of noise include:

- Control at the source, such as proper design of machines, proper maintenance and lubrication, use of rubber mountings for machines, and use of vehicle mufflers
- Isolation of noise, as the use of enclosures, rooms, and other barriers; the closing of windows in a home typically reduces intensity by about 10 dB
- Use of baffles and sound absorbers
- Use of acoustical treatment
- Proper layout

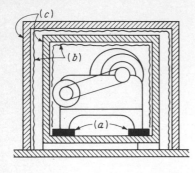 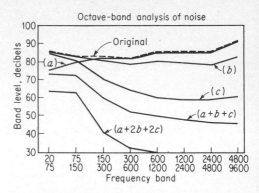

Figure 16-10 Illustrations of the possible effects of some noise-control measures. The lines on the graph show the possible reductions in noise (from the original level) that might be expected by vibration insulation a; an enclosure of acoustical absorbing material b; a rigid, sealed enclosure c; a single combined enclosure plus vibration insulation, $a + b + c$; and a double combined enclosure plus vibration insulation, $a + 2b + 2c$. [Adapted from Peterson and Gross, 34.]

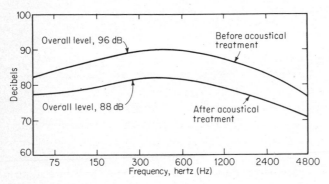

Figure 16-11 Spectrum of noise before and after acoustical treatment. [Adapted from Bonvallet, in 45, p. 41.]

Actually, there are many ingenious variations of these and other means that have aided in noise reduction. Although we shall not discuss the effects of noise control, Figure 16-10 illustrates the possible effect of various noise-control measures. As another example, Figure 16-11 shows the spectrum of noise in one situation before and after the use of acoustical treatment.

Ear Protection

Where the noise level cannot reasonably be reduced to "safe" limits, some form of ear protection should be considered for those people who are exposed to the noise. The most conventional form is some type of ear-protection device; these are of two general types, namely, earplugs that fit into the canal of the outer ear, and devices that fit over or surround the entire ear like muffs.

Effectiveness of ear-protection devices.   The effectiveness of ear-protection devices is somewhat variable, depending on the nature of the noise, the duration of exposure, the fit of the device, its attenuating characteristics, and possibly other variables. By and large, however, such devices provide a worthwhile degree of protection from high noise levels. An example of such protection is shown in Figure 16-12 by the use of ear-

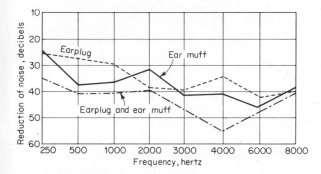

Figure 16-12   Protection provided by the use of earplugs, ear muffs, and a combination of the two. [USAF, Wright-Patterson AFB.]

plugs, ear muffs, and a combination. Although studies have shown that the attenuation of noise with ear-protection devices can be as high as 30 or 40 dB, or even more (especially in the higher frequencies), their typical attenuation is substantially lower. The substantially lower levels argue for the selection of appropriate devices, their proper fitting and of course, their consistent use when exposed to noise.

Acoustic Reflex

In recent years, some attention has been given to the possibility of taking advantage of the acoustic reflex as a means of providing some protection from the impulse type of noise. The acoustic reflex is a reflex contraction of the two intra-aural (middle-ear) muscles that occurs spontaneously when certain tones above about 70 dB occur. The contraction of these muscles serves to reduce the transmission of vibrations through the middle ear for a short time. The possibility of capitalizing on the acoustic reflex to provide protection against impulse noise would depend upon the feasibility of inducing the reflex just before the occurrence of the noise. In the case of such noises as from gunfire, drop hammers, and dynamite blasts, it presumably would be possible to cause a tone to be sounded at some predetermined brief interval before the impulse noise.

The possible use of the acoustic reflex with machine gunfire was investigated by Fletcher [17] and Fletcher and Riopelle [18]. A 1000-Hz tone of 98 dB was presented 200 msec (⅕ sec) before each shot. A comparison of the temporary threshold shift with and without the acoustic reflex was made, as well as with subjects using earplugs. The overall average threshold shifts were as follows [Fletcher, 17]:

No protection	19.23 dB
Acoustic reflex	6.27 dB
Earplugs	2.50 dB

While earplugs provided more overall protection, the acoustic reflex gave better protection up to and including frequencies of 1000 Hz but inferior protection at and above 2000 Hz.

In another study a comparison was made of protection by the acoustic reflex against the (recorded) sound of a drop hammer [Chisman and Simon, 9]. The acoustic reflex tone was increased from 70 to 100 dB over 250 msec, and was presented approximately 400 to 600 msec before each impact. Tones of 250 and 1000 Hz were used in different phases of the study. The mean temporary threshold shifts of the different conditions are shown in Figure 16-13. It can be seen that the acoustic reflex

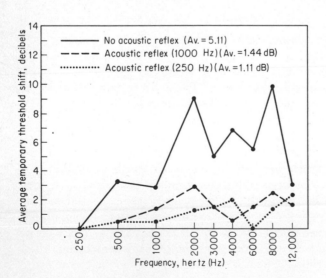

Figure 16-13 Average temporary threshold shift produced with and without acoustic-reflex-inducing tones. [From Chisman and Simon, 9.]

provided considerable protection (resulted in less threshold shift) than the control condition did (with no preceding tone), especially in frequencies of 2000 Hz and above; there was no appreciable difference between the 250- and 1000-Hz tones.

While it is probable that earplugs could provide as much, or more, protection than the acoustic reflex could, nonetheless there probably would be some circumstances involving impulse noise where this method of protection could be advantageous. As pointed out by Chisman and Simon, the acoustic reflex has such advantages as operating only when necessary, of not interfering with normal conversation, of not producing discomfort or unpleasantness, of not requiring the cooperation of the worker (as required in wearing earplugs), and of serving as a preimpact warning device for workers in the area.

SUMMARY

1. Noise can be characterized as *unwanted* sound set in an information-theory context, in which the unwanted sound (noise) is characterized as that auditory stimulus or stimuli bearing no informational relationship to the presence or completion of the immediate task.
2. There are two types of deafness, as follows: (a) *nerve deafness*, which is caused by impairment of the nerve cells; and (b) *conduction deafness*, which is caused by some conditions of the outer or middle ear that affect transmission of sound waves to the inner ear. Hearing loss through age and through continuous exposure to high noise levels usually is of the nerve type.
3. Hearing and hearing loss are measured by the following two methods: (a) *simple hearing tests* (whisper tests, voice tests, coin-click tests, etc.) and (b) *audiometer tests* (audiometers are instruments that reproduce sounds of speech with controlled intensity).
4. Hearing loss of individuals is typically plotted on an *audiogram*, showing the threshold of hearing at various frequencies.
5. Speech-intelligibility tests, given with audiometers, usually consist of words or other speech materials which have been carefully selected to cover the ranges of normal speech sounds.
6. Increasing age usually is accompanied by impairment in hearing, the loss generally being greater in the higher frequencies. Women generally have greater hearing loss in the lower frequencies than men, and men in turn have greater loss in the higher frequencies than women.
7. There is abundant evidence that continuous exposure to high noise levels contributes to hearing loss, although there is not much known about the degree and character of hearing loss caused by different

noise conditions. There is evidence, however, to support certain general conclusions regarding degree of hearing loss, as follows: (a) hearing loss is related to level of noise to which exposed; (b) hearing loss is related to exposure time for high exposure intensities, though to a limited extent, or not at all, for low exposure intensities; and (c) hearing loss usually is greater in the 4000-Hz range than in the 1000- and 2000-Hz ranges.

8.   Continuous and extensive exposure to noise levels above 80, 85, or 90 dB is generally considered to bring about hearing loss.

9.   *Impact* noise, such as that from drop forges, and *impulsive* noise, such as that from gun blasts, generally bring about hearing loss more quickly than exposure to continuous noise.

10.   While hearing loss may be temporary (after noise exposure), there is evidence that, with increasing duration of exposure, there is less and less recovery (temporary hearing loss) and increasing permanent loss.

11.   Initial exposure to loud noise usually causes certain physiological reactions, but adaptation is such that these processes settle back to normal, or nearly normal, levels. Whether there are any residual physiological effects (or other hearing loss) from continuous exposure is not yet known.

12.   While there is no evidence that noise *generally* brings about degradation of work performance, there is accumulating evidence that it affects performance on certain kinds of tasks, such as vigilance tasks, complex mental tasks, tasks requiring skill and speed, and those that demand a high level of perceptual capacity.

13.   People have considerable capacity to adapt to the annoying characteristics of noise, but this adaptation probably is not complete.

14.   The characteristics of noises that cause them to be annoying seem to be high intensities, high frequencies, intermittency, and reverberation effects.

15.   Procedures have been developed that make it possible to predict the probable neighborhood reaction to any noise condition in the area.

16.   An organization which knows, or suspects, that it has a noise problem can take certain actions. These include the following: (a) Measure the noise to determine its overall intensity and spectral characteristics; (b) give audiometric tests to personnel, and arrange for medical attention for those with noticeable hearing loss; (c) define the noise problem by comparing present noise levels with the level that it is desired to achieve; and (d) where excessive noise conditions exist, carry out noise control.

17.   Noise control can be accomplished in various ways, depending upon the circumstance. Some methods are (a) control of noise at the

source, such as through machine design, proper maintenance and lubrication, or mounting equipment on rubber; (b) isolation of noise with enclosures, rooms, barriers, etc.; (c) use of baffles and sound absorbers; (d) acoustical treatment; and (e) proper layout.

18. Appropriate ear protective devices can provide substantial protection from noise.

REFERENCES

1. Beranek, L. L. (ed.): *Noise reduction*, McGraw-Hill Book Company, New York, 1960.
2. Berrien, F. K.: The effects of noise, *Psychological Bulletin*, 1946, vol. 43, pp. 141–161.
3. Boggs, D. H., and J. R. Simon: Differential effect of noise on tasks of varying complexity, *Journal of Applied Psychology*, 1968, vol. 52, no. 2, pp. 148–153.
4. Botsford, J. H.: A new method for rating noise exposures, *American Industrial Hygiene Association Journal*, September–October, 1967, vol. 28, pp. 431–446.
5. Broadbent, D. E.: Effect of noise on an "intellectual" task, *Journal of the Acoustical Society of America*, 1958, vol. 30, pp. 824–827.
6. Broadbent, D. E., and E. A. J. Little: Effects of noise reduction in a work situation, *Occupational Psychology*, 1960, vol. 34, pp. 133–140.
7. Burrows, A. A.: Acoustic noise, an informational definition, *Human Factors*, August, 1960, vol. 2, no. 3, pp. 163–168.
8. Carhart, R.: The ears of industry, A.M.A. *Archives of Industrial Hygiene and Occupational Medicine*, 1950, vol. 2, pp. 534–541.
9. Chisman, J. A., and J. R. Simon: Protection against impulse-type industrial noise by utilizing the acoustic reflex, *Journal of Applied Psychology*, 1961, vol. 45, pp. 402–407.
10. Cohen, A.: U.S. Public Health Service field work on the industrial noise hearing loss problem, *Occupational Health Review*, 1965, vol. 17, no. 3, pp. 3–10.
11. Cohen, A.: Noise effects on health, productivity, and well-being, *Transactions of the New York Academy of Sciences*, May, 1968, ser. 2, vol. 30, no. 7, pp. 910–918.
12. Cohen, A.: *Noise and psychological state*, USPHS, National Center for Urban and Industrial Health, RR–9, July, 1968.
13. Coles, R. R. A., G. R. Garinther, D. C. Hodge, and C. C. Rice: Hazardous exposure to impulse noise, *Journal of the Acoustical Society of America*, 1968, vol. 43, no. 2, pp. 336–343.
14. Culbert, S. S., and M. I. Posner: Human habituation to an acoustical energy distribution spectrum, *Journal of Applied Psychology*, 1960, vol. 44, pp. 263–266.
15. Egan, J. P.: Articulation testing methods, *Laryngoscope*, 1948, vol. 58, pp. 955–991.

16. Fenwick, C. A.: *Judged annoyance of differing noise spectra*, unpublished master's thesis, Purdue University, Lafayette, Ind., August, 1959.
17. Fletcher, J. L.: *Comparison of attenuation characteristics of the acoustic reflex and the V-51R earplug*, USA Medical Research Laboratory, Report 397, 1959.
18. Fletcher, J. L., and A. J. Riopelle: *The protective effect of acoustic reflex for impulsive noises*, USA Medical Research Laboratory, Report 396, 1959.
19. Glorig, A., W. D. Ward, and C. W. Nixon: Damage risk criteria for noise exposure, *Archives of Otolaryngology*, 1961, vol. 74, pp. 413–423.
20. Harmon, F. L.: The effects of noise upon certain psychological and physiological processes, *Archives of Psychology*, 1933, no. 147, pp. 1–81.
21. Hudgins, C. V., J. E. Hawkins, J. E. Karlin, and S. S. Stevens: The development of recorded auditory tests for measuring hearing loss for speech, *Laryngoscope*, 1947, vol. 57, pp. 57–89.
22. Jerison, H. J.: Effects of noise on human performance, *Journal of Applied Psychology*, 1959, vol. 43, pp. 96–101.
23. Jerison, H. J., and R. A. Wallis: *Experiments on vigilance: one-clock and three-clock monitoring*, USAF, WADC, TR 57–206, April, 1957.
24. Jerison, H. J., and R. A. Wallis: *Experiments on vigilance: performance on a simple vigilance task in noise and in quiet*, USAF, WADC, TR 57–318, June, 1957.
25. Jerison, H. J., and S. Wing: *Effects of noise and fatigue on a complex vigilance task*, USAF, WADC, TR 57–15, January, 1957.
26. Jones, H. H., and A. Cohen: Noise as a health hazard at work, in the community, and in the home, USPHS, *Public Health Reports*, July, 1968, vol. 83, no. 7, pp. 533–536.
27. Kryter, K. D.: The effects of noise on man, *Journal of Speech and Hearing Disorders, Monograph Supplement* 1, 1950, pp. 1–95.
28. Kryter, K. D.: Damage risk criterion and contours based on permanent and temporary hearing loss data, *American Industrial Hygiene Association Journal*, 1965, vol. 26, no. 1, pp. 34–44.
29. Kryter, K. D.: Concepts of perceived noisiness, their implementation and application, *Journal of the Acoustical Society of America*, 1968, vol. 43, no. 2, pp. 344–361.
30. Kryter, K. D., W. D. Ward, J. D. Miller, and D. H. Eldredge: Hazardous exposure to intermittent and steady-state noise, *Journal of the Acoustical Society of America*, 1966, vol. 39, pp. 451–463.
31. LaBenz, P., A. Cohen, and B. Pearson: A noise and hearing survey of earthmoving equipment operators, *American Industrial Hygiene Association Journal*, March–April, 1967, vol. 28, pp. 117–128.
32. Machle, W.: The effect of gun blast on hearing, *Archives of Otolaryngology*, 1945, vol. 42, pp. 164–168.
33. MacLaren, W. P., and A. L. Chaney: An evaluation of some factors in the development of occupational deafness, *Industrial Medicine*, 1947, vol. 16, pp. 109–115.
34. Peterson, A. P. G., and E. E. Gross, Jr.: *Handbook of noise measurement*, 6th ed., General Radio Co., New Concord, Mass., 1967.

35. Plutchik, R.: The effects of high intensity intermittent sound on performance, feeling and physiology, *Psychological Bulletin,* 1959, vol. 56, pp. 133–151.
36. Rosenblith, W. A., and K. N. Stevens: *Handbook of acoustic noise control: Vol. II. Noise and man,* USAF, WADC, TR 52–204, June, 1953; Report PB 111, 274, U.S. Department of Commerce, Office of Technical Services.
37. Roth, E. M. (ed.): Compendium of human responses to the aerospace environment, NASA CR–1205 (5 vols.), November, 1968.
38. Teichner, W. H., E. Arees, and R. Reilly: Noise and human performance, a psychophysiological approach, *Ergonomics,* 1963, vol. 6, no. 1, pp. 83–97.
39. AF Regulation 160–3, USAF, October, 1956.
40. *Foundry noise manual,* 2d ed., American Foundryman's Society, Des Plaines, Ill., 1966.
41. *Guide for conservation of hearing in noise,* prepared by Subcommittee on Noise, American Academy of Opthalmology and Otolaryngology, revised 1964.
42. Guidelines for noise exposure control, *American Industrial Hygiene Association Journal,* September–October, 1967, vol. 28, pp. 418–424.
43. *Industrial noise manual,* 2d ed., American Industrial Hygiene Association, Detroit, 1966.
44. *Land use planning with respect to aircraft noise,* prepared by Bolt, Beranek, and Newman, Inc., Los Angeles, published as AFM 86–5, TM 5–365, Navdocks P–98 (Joint manual of the U.S. Departments of the Air Force, Army, and Navy); also published as FAA TR, October, 1965.
45. *Noise,* lectures presented at the Inservice Training Course on the Acoustical Spectrum, Feb. 5–8, 1952, sponsored by the University of Michigan, School of Public Health and Institute of Industrial Health, University of Michigan Press, Ann Arbor.
46. *Noise environment of urban and suburban areas,* prepared by Bolt, Beranek, and Newman, Inc., Los Angeles, Federal Housing Administration, Superintendent of Documents, Washington, D.C., January, 1967.
47. *Rating noise with respect to hearing conservation, speech communication, and annoyance,* ISO, Technical Committee 43, Acoustics, Secretariat–139, August, 1961.
48. *The relations of hearing loss to noise exposure,* USASI, New York, 1954.

As man has developed his technological competence, he has used this competence, in part, to increase his own mobility. The combination of the wheel with power in the last century was a major stride. The combination of power with an aerodynamic surface (an aircraft wing) was another major advance, to be followed by space capsules, zero-ground pressure vehicles, rockets strapped to one's back, etc. These mechanisms have made it possible for man to move at speeds and in environments that he had never experienced before and to which he is not biologically adapted.

These greater speeds, and new environments, impose human factors problems in the design of the gadgets that man has developed to increase his mobility. While the manned space ship is the most spectacular of these developments (and has a whole host of special human factors problems associated with it), jet and conventional aircraft, and even the earth-bound automobile, have created circumstances to which man is not by nature entirely adapted. The disparity between man's biological nature and that which is required for him to exist and perform in a different mobility environment specifies the domain within which human factors adaptations must be made.

The variables imposed by his increased mobility include vibration, acceleration and deceleration, weightlessness, and an assortment of more strictly psychological phenomena associated with his mobility, such as disorientation and other illusions. It is to some of these that we shall now turn our attention.[1]

VIBRATION

The earlier discussions of sound and noise have dealt with the physical parameters of vibrations that are audible. The same parameters, of course, apply to our present topic, except that here we are more concerned with

[1] An extensive bibliography has been prepared by Buckhout [7]. A compendium of information related to acceleration and vibration is presented by Roth [41]. See also von Gierke [47].

vibrations of lower frequencies, generally below about 100 Hz, most of which are not audible. As applied to vibration, these parameters and their derivatives are (1) *frequency* (the frequencies of vibration can be sinusoidal, complex, or random); (2) *displacement amplitude* (inches or centimeters of displacement, either a half-wave single-amplitude or a full-wave double-amplitude displacement); (3) *velocity* (first derivative of displacement, usually in inches, centimeters, or minutes per second, but not often used); (4) *acceleration* (second time derivative of displacement, usually in inches, centimeters, or minutes per second per second, and sometimes expressed as peak or maximum *g*; recommended international practice is in minutes per second per second (sec²); and (5) *jerk* (third time derivative of displacement, in inches or minutes per second per second per second (sec³), especially used with low-frequency vibrations); and (6) *duration* (in seconds or hours).

Displacement and maximum acceleration (in *g* units) are used most commonly in characterizing intensity of sinusoidal vibration. Random vibrations typically are characterized in terms of a power spectral-density curve (PSD).[2]

It should be pointed out that physical objects have their individual resonant frequencies that are in large part a function of their mass. For example, rubber-tired earth-moving vehicles have resonant frequencies primarily in the range from 1 to 4 Hz. The predominant frequencies for farm tractors and highway trucks are from about 2 to 7 Hz, and for track types of vehicles from about 4 Hz up [Radke, 38]. Even different parts of the body, such as the head, have their own resonant frequencies.

Whole Body Vibration

As the entire body is caused to vibrate, the various parts thereof tend not to vibrate in unison. Rather, the effect is the alternating displacements of body parts and organs and their supporting structures in relation to each other. The resulting tensions and deformations can be the source of localized pain. In addition, other more general discomfort can occur, along with certain subjective sensations. These symptoms can include abdominal, chest, and testicular pain, head symptoms, localized symptoms of other types, general discomfort, and anxiety [Magid, Coermann, and Ziegenruecker, 34]. There is, however, a fair amount of variation in the frequencies at which these various symptoms occur.

Amplification and attenuation effects.   In the process of vibrating the whole body, the vibration, as it is transmitted to the body, can be amplified, or attenuated, as the consequence of body posture (whether standing or sitting), the type of seating, and the frequency of vibrations.

[2] The PSD is derived from the squares of the rms of the frequency bands that characterize the random vibrations.

Such effects are interrelated with the resonance characteristics of the body and of its parts, which are subject to the same laws of vibration as other physical objects are. Man on a seat can be likened to a sprung mass, without damping, something like that shown in Figure 17-1. If

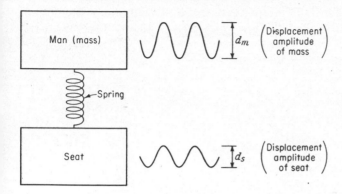

Figure 17-1 Analogy of a man (as a simple sprung mass) on a vibrating seat, without damping.

the seat is caused to vibrate with a frequency f_s and a displacement amplitude d_s, the mass will vibrate with that frequency f_s and displacement amplitude d_m (after equilibrium has been reached), where d_m is defined by the following equation (which applies to an undamped mass):

$$d_m = d_y \left[\frac{1}{1 - (f_s/f_r)^2} \right]$$

in which f_r is the resonant frequency of the sprung mass. In this equation if the sign of the resulting d_m value is minus, that sign should be disregarded. Where the vibrating frequency of the seat f_s is high, relative to the resonant frequency of the mass f_r, displacement amplitude of the mass d_m is high, which means that the vibration of the mass is amplified. Such amplification occurs when the ratio of the vibration frequency to the resonant frequency f_s/f_r is less than the square root of 2, or 1.414 [Radke, 38]. When that ratio is more than 1.414, there is an attenuation, which means that the displacement amplitude of the mass is less than that of the seat. The amplification becomes greatest where this ratio is around 1.0 (where the seat is vibrating at, or near, the resonant frequency of the body). With particular reference to the head, as frequency increases above resonance, the displacement amplitude decreases, and at around 10 Hz the head and seat vibrate at about equal amplitudes [Cope, 15]. At higher vibrating frequencies, the transmission to the head is decreased, until at about 70 Hz only about 10 percent of the amplitude of the seat is transmitted to the head [Coermann, 14, p. 36].

The amplification and attenuation that would be derived from the formula above, however, would be expected with no damping. In practice, some damping usually would be provided, either by intent or otherwise. Examples of the actual amplification and attenuation of the human body while standing and seated are illustrated in Figure 17-2. With the

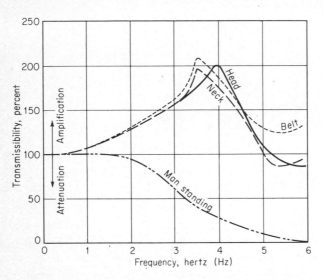

Figure 17-2 Mechanical response of man to vertical vibration, showing amplification and attenuation, by frequency, for seated and standing positions. [From Radke, 38.]

vibration of the vibrating table used as a base (expressed as 100 percent), the amplification, or attenuation, is measured by the use of accelerometers positioned at different locations, such as at the belt, the neck, and head. For a person standing, there is a typical attenuation effect; this is because the legs serve to absorb the effects as the individual bends and straightens his legs in response to the movement. In the case of a seated individual, however, there is an amplification effect (at the various body locations) especially in the range of frequencies of about 3½ to 4 Hz. At frequencies of 5 to 6 Hz the transmissibility is reduced to about the amplitude of the vibrating source (in experimental situations the vibrating table); the difference between these empirical results and possible theoretical (undamped) values probably can be attributed to some form of damping in the system, including the damping effects of the body itself, especially the buttocks.

Damping can be controlled to some degree by appropriate seat design, spring design, cushioning, etc. Without going into details, Figure 17-3 illustrates the amplification and attenuation effects of various seats. It can be seen that, in this particular comparison, the suspension seat resulted in substantial attenuation over the critical frequencies of 3 to 6 Hz.

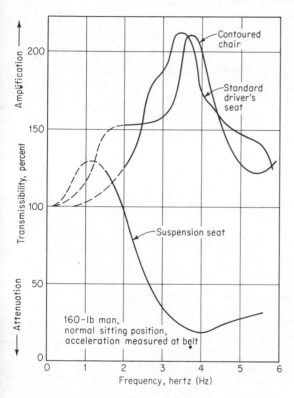

Figure 17-3   Mechanical response of a man's body to vibrations, when seated in different seats. [From Simons, Radke, and Oswald, 42.]

With respect to the attenuation of vibration in vehicles, however, it may be unwise to introduce complete attenuation, even if one could do so. This would have the effect of completely isolating the individual from the vibration effects and would keep him on an even keel (in space) with the vehicle moving about him. Such complete isolation could interfere with his ability to read the instruments or operate the controls.

Subjective responses to whole body vibration.   From the above discussion we can see that the physical effects of vibration are influenced by such factors as body posture, and any materials interposed between the body and the source of vibration (such as the nature of the seat, padding, and springs). As one would expect, these and other factors also influence the subjective reaction of people to vibration. Since we cannot probe all of these as they relate to subjective responses (even if data were available on each), we shall select a couple of examples for illustrations.

As one example, Chaney [11] shows comparative data on subjective reactions of standing and seated subjects when subjected to vibration of various frequencies on a shake table. Three tolerance curves are shown in Figure 17-4, these being the mean levels that were reported by subjects

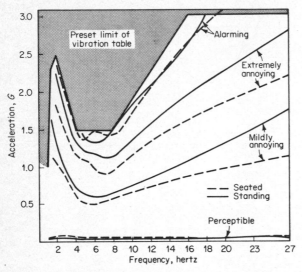

Figure 17-4 Subjective reactions of subjects, seated and standing, when subjected to whole body vibration. Each curve represents the combinations of frequencies and amplitudes that were judged by the subjects to produce the level of sensation in question (such as mildly annoying). [From Chaney, 11.]

to be mildly annoying, extremely annoying, and alarming. The greater tolerance of the standing subjects is undoubtedly the consequence of the damping effects of leg flexion as illustrated above in Figure 17-2. The lowest tolerance level is in the range from about 4 to 6 or 8 Hz.

As you would expect, people can tolerate mild vibration longer than really rough jarring. Some indication of this is shown in Figure 17-5, each line representing an estimate of the upper limit of the amplitudes of vibrations of various frequencies that people generally can tolerate for the specified time limits before fatigue effects would catch up with them; amplitudes of about 10 dB less than those indicated tend to characterize the upper

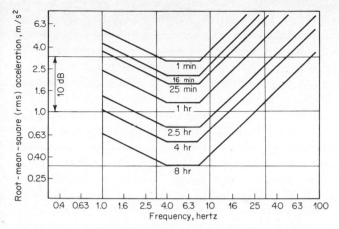

Figure 17-5   Tentative fatigue-decreased proficiency boundary for vertical vibration proposed by the International Organization for Standardization (ISO). The following weighting factors are proposed for the specified conditions: fore-and-aft or side-to-side vibration, subtract 3 dB from values; boundary of reduced comfort, subtract 10 dB; and safe exposure limits, add 6 dB. [From *Revised proposal to the secretariat: guide for the evaluation of human exposure to whole-body vibration,* 49.]

boundary of comfort, and the safe exposure limits are about 6 dB higher than the values shown. The particular set of curves is based on a draft proposal to the secretariat of the ISO, as prepared by a working group [49], and even though it is still tentative, it probably reflects the best available vibration-exposure criteria as a function of frequency.

Effects of whole body vibration on performance.   The effects of whole body vibration on performance can arise directly from the mechanical interference with the activity, or indirectly from the physiological alterations that are induced [Roth, 41, pp. 8–67]. We shall not attempt to pull together and summarize the results of the research relating to the effects of vibration on performance,[3] but a couple of reflections about such research are in order. Most of this research has dealt with sinusoidal vibration, usually for short exposure periods. Such research has indicated that sinusoidal vibration can affect performance on certain types of visual tasks (such as dial reading and visual scanning), simple motor tasks that require precision and coordination (but usually not those tasks that require the maintenance of firm muscular control), and usually more complex psychomotor tasks, including tracking tasks (although the effects depend in part on the nature of the task).

[3] See Roth [41, vol. 3, sec. 8] for such a review.

However, the shaking people have to take is sometimes more random than sinusoidal (although it may have dominant frequency components) and sometimes continues over a period of time, such as several hours. Two studies have exposed subjects to such vibration while performing tracking tasks. One of these involved 5 hr of vibration such as might be experienced in low-altitude, high-speed aircraft [Hornick and Lefritz, 27], and the other involved 6 hr of vibration also characteristic of aircraft, but with peak power at 2 or 5 Hz [Holland, 26]. Allowing for certain inconsistencies, the results of both studies showed degradation in tracking during such prolonged vibration, (but it should be noted that, despite some degradation in performance, the subjects were able to stand the gaff and to maintain at least a moderate level of performance for such periods). However, there is still some question as to whether the *degree* of degradation is associated with the *intensity* of vibration and as to what the possible differential effects are from various types of vibration spectrums.

Localized Vibration

In work activities of some types a part of the body is subject to vibration, especially the hands, as in handling or holding vibrating objects. Theoretically, there probably is no such thing as "localized" vibration, since vibrations applied to one area of the body are transmitted, to some degree, to other parts through bone structure and body tissue. In a practical sense this transmission probably is negligible in some circumstances; but in other situations it could be of considerable consequence. This would be true, for example, with jackhammer operators; the intensity of the vibrations in the hands is such that part of it is transmitted to the shoulders and upper trunk.

There is little data available regarding the tolerance to, or effects of, vibration that is localized, but there are some data on sensitivity thresholds of touch [Fibikar, 20]. Some of these data are given in Figure 17-6. In particular this figure shows the average and maximum sensation level (the minimum intensity) that can be detected by people, from vibrations of varying frequencies, with their fingertips and with a full handgrip (such as when gripping portable hand tools). The "maximum" is the lowest intensity that was detected by any individual in the group of 25 subjects. The general patterns in both cases are fairly similar in that sensitivity is greatest (meaning that smaller amplitudes can be perceived) in the range of frequencies around 300 Hz. While there were marked individual differences, the pattern of the greatest sensitivity (finger or hand) follows generally the pattern of the respective mean. While these data do not reflect tolerance amplitudes, it is probable that they would follow similar patterns.

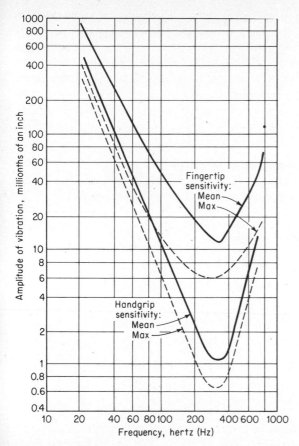

Figure 17-6   Mean and maximum sensitivity level
of vibration when object is touched with finger tips
and when gripped by hand, by frequency. [Adapted
from Fibikar, 20.]

ACCELERATION AND DECELERATION

If the human body is caused forcibly to change its velocity (such as in
a vehicle that is changing its velocity), the change in velocity causes some
displacement of body tissue, organs, and blood, since these body components are not rigid. Thus, as an automobile accelerates rapidly, the
internal organs and the blood tend to "lag behind" the structure of the
body as the body is propelled forward at increasing velocity. In normal
automobile driving this internal shift is not of any appreciable concern,
but for a man being projected into space the effect may be one of some
consequence.

Terminology

The basic unit of measure of acceleration is that of the pull of gravity g (32.16 ft/sec^2). Acceleration of the body in one direction, however, produces a physiological *reactive* force that is opposite that of the direction of motion; this physiologically descriptive force is abbreviated as capital G, and is expressed in multiples of the unit of acceleration. These, and other, systems of nomenclature are given in Table 17-1. This table

Table 17-1 Equivalents for Acceleration Terminology

Direction of motion	Direction of acceleration		Physiological reaction		
	Aircraft computer standard	Acceleration descriptive	Physio-logical computer standard	Physiological descriptive	Vernacular description
	(1)	(2)	(3)	(4)	(5)
Linear motion					
Forward	$+a_x$	Forward	$+G_x$	Transverse (A-P)* G Supine G Chest to back G	Eyeballs in
Backward	$-a_x$	Backward	$-G_x$	Transverse (P-A)* G Prone G Back to chest G	Eyeballs out
Upward	$-a_z$	Headward	$+G_z$	Positive G	Eyeballs down
Downward	$+a_z$	Footward	$-G_z$	Negative G	Eyeballs up
To right	$+a_y$	Right lateral	$+G_y$	Left lateral G	Eyeballs left
To left	$-a_y$	Left lateral	$-G_y$	Right lateral G	Eyeballs right
Angular motion					
Roll right	$+\dot{p}$		$-\dot{R}_x$	Roll	
Roll left	$-\dot{p}$		$+\dot{R}_x$		
Pitch up	$+\dot{q}$		$-\dot{R}_y$	Pitch	
Pitch down	$-\dot{q}$		$+\dot{R}_y$		
Yaw right	$+\dot{r}$		$+\dot{R}_z$	Yaw	
Yaw left	$-\dot{r}$		$-\dot{R}_z$		

* A-P = anterior-posterior; P-A = posterior-anterior.
SOURCE: Adapted from Gell [22] in *Aerospace Medicine*.

shows, for various directions of motion, the symbols and terms used for the various systems, certain ones of which are predicated on the conventional x (forward-backward), y (lateral), and z (vertical) coordinates used in specifying aircraft orientation. Systems 1 and 2 describe the direction of acceleration of, say, a vehicle, system 1 being the *aircraft computer* standard and system 2 characterizing acceleration in terms of the direction of motion of a human body. Systems 3 and 4, on the other hand, specify the effects of acceleration forces in terms of physiological reactions, specifically the direction of the displacement of the internal organs and blood. System 3 consists of engineering notations that relate physiological displacement to the *xyz* coordinate system. System 4 describes physiological displacement in terms that have commonly been used, especially negative acceleration (blood flow toward head from footward movement), positive acceleration (the opposite), and transverse (perpendicular to body). System 5 consists of vernacular terms that characterize displacement in terms of the movement of the eyeballs. Thus, an "eyeballs down" condition is one in which the acceleration forces tend to cause the eyeballs to be displaced downward; this would occur during headward motion. Systems 2 and 4 are illustrated in Figure 17-7. Whatever the direction of motion—linear or angular—the physio-

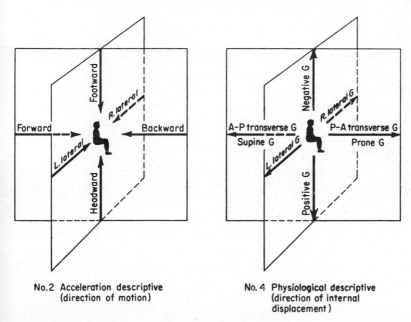

No. 2 Acceleration descriptive
(direction of motion)

No. 4 Physiological descriptive
(direction of internal
displacement)

Figure 17-7   Illustrations of two acceleration-terminology systems, acceleration descriptive, 2, and physiological descriptive, 4, as given in Table 17-1. [Adapted from Gell, 22, by permission of *Aerospace Medicine*.]

logical reaction is opposite to the direction of acceleration. Thus, for example, the directions of displacement of system 4 are opposite to directions of motion of system 2.

Physiological Effects of Acceleration

The nature and intensity of the physiological effects of acceleration depend on quite an assortment of factors such as the direction of the acceleration with respect to the axes of the body (and whether it is linear or angular), its magnitude and duration, its rate of onset and offset, and body posture. As illustrations of the physiological effects, let us consider certain aspects of linear acceleration. Such effects are most accentuated in the case of footward acceleration $-G_z$, which causes blood to flow toward the upper extremities and brings on initial sensations of facial suffusion and cranial fullness; at about 2 to $3G$, facial congestion and headache typically occur, and if such levels continue for about 5 sec, visual blurring and sometimes *redout* (a sensation of redness of the visual field) occur [Roth, 41, chap. 7]. With higher levels or continued exposure, unconsciousness usually occurs.

In the case of headward acceleration $+G_z$, there is an initial sensation of drooping of the facial and body tissues; by about $2\frac{1}{2}G$ it is nearly impossible to raise oneself from a seated position [Roth, 41, chap. 7]. A very few seconds of exposure at about 3 to $4G$ induce dimming of peripheral vision (*grayout*); and visual tunneling and complete loss of vision (*blackout*) occur with exposure of about 5 sec at these levels or up to about $4\frac{1}{2}$ to $5G$. A dose of a few more seconds usually brings about unconsciousness, although some people are much more durable than others. The time required to bring about grayout or unconsciousness depends in part on the rate of onset (with longer onset rates being related to shorter times to grayout or unconsciousness).

The effects of linear acceleration in the forward-backward directions $+G_x$ and $-G_x$ and lateral directions $+G_y$ and $-G_y$ are not as accentuated as in the vertical direction. The relatively higher G load that people can tolerate especially in the forward direction $+G_x$ has been the basis for the use of this posture in space vehicles. Although symptoms of discomfort and pressure and breathing difficulty occur, G loads of 12 to $15G$ can be tolerated by some subjects, in some instances over 100 sec [Chambers and Hitchcock, 10].

Voluntary Tolerance to Acceleration

The physiological symptoms brought about by acceleration of course have a pretty direct relationship to the levels that people are willing to tolerate voluntarily. The physiological effects of differential levels of linear acceleration in various directions mentioned above is paralleled by somewhat corresponding differences in average voluntary tolerance levels, such as shown in Figure 17-8. This figure shows the average levels for the various directions that can be tolerated for specified times. It can be seen that

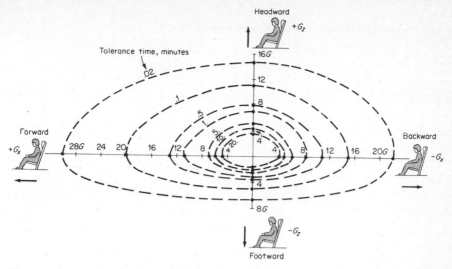

Figure 17-8 Average levels of linear acceleration, in different directions, that can be tolerated on a voluntary basis for specified periods of time. Each curve shows the average G load that can be tolerated for the time indicated. The data points obtained were actually those on the axes; the lines as such are extrapolated from the data points to form the concentric figures. [Adapted from Chambers, 8, fig. 6.]

tolerance to footward acceleration $-G_z$ is least, followed by headward acceleration $+G_z$, with forward acceleration $+G_x$ being the most tolerable. Individual differences are of considerable magnitude; trained and highly motivated personnel frequently can endure substantially higher levels than the average.

Further indications of the differential effects on tolerance of acceleration forces operating in different directions is shown in Figure 17-9. Bypassing the water-immersion condition a for the moment, that figure shows the tolerance time for various G loads when people are oriented differently with respect to the direction of the acceleration forces (along with a line showing the duration required to reach the escape velocity of 18,000 mph in the case of space ships). Here, again, we see evidence of the limited tolerance to footward acceleration $-G_z$ as reflected by condition F. By comparison, we can see that water immersion (with welcomed breathing arrangements, of course) increases one's tolerance, particularly in a supine position with about a 35° angle of the trunk to the line of force. Under immersion conditions, the same acceleration forces affect the water as well as the person immersed in it. This has the effect of equalizing the forces over the parts of the body. Individuals so immersed can even

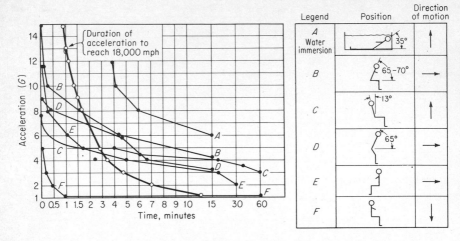

Figure 17-9 Comparison of acceleration tolerance of immersed and nonimmersed subjects in different postures. The angles indicated are not "back angles" as such, but rather are angles of the trunk relative to the direction of the motion that gives rise to the acceleration. [From Bondurant et al., 2, as adapted by Stapp, 44.]

move their body members under acceleration and can tolerate more acceleration than under any other known condition.

Performance Effects of Acceleration

Within the limits of physiological safety, acceleration can cause degradation in performance of certain kinds of human activities that might be required under acceleration conditions, such as in aircraft or space ships.[4]

Acceleration and gross body movement. As implied above, acceleration can make some simple physical movements difficult or impossible (and this effect, of course, carries over to the performance of certain physical tasks). The effects of vehicular (forward) acceleration $+G_x$ on gross body movements are shown in Figure 17-10. These effects are akin to increasing the *effective weight* of body members and thus increasing the muscular force required to execute the motion in question. This figure shows the typical levels of G that are near the threshold of ability to perform the acts indicated. Some of the implications for design decisions are fairly obvious; it would seem unwise, for example, to place the ejection-control level of an airplane over the pilot's head.

Acceleration and specific activities. Acceleration in its several variations (intensity, direction, etc.) has been found to be significantly

---

[4] For reviews of the effects of acceleration on human performance, see Brown [4], Eiband [19], Rogers [40], Roth [41, chap. 7], or Stapp [44].

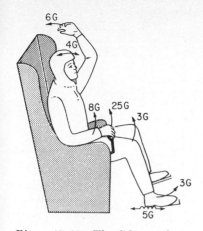

Figure 17-10   The *G* forces that are near the threshold of various body movements. For any given motion indicated, the movement is just possible at the *G* forces indicated; greater *G* forces usually would make it impossible to perform the act. [From Chambers and Brown, 9.]

related to performance on a varied spectrum of human activities and responses, such as reaction time, visual acuity, peripheral acuity, time estimation, tracking performance, and time to activate aircraft-ejection mechanisms, to mention a few [Chambers and Brown, 9, Brown, 4]. The general pattern of such performance degradation is fairly similar for different aspects of performance. Figure 17-11 gives a generalized impression of such degradation, showing average estimated impairment for several aspects of performance.

But while some types of human performance deteriorate under acceleration conditions, this is not necessarily true of all human performance under conditions of moderate acceleration. (Under near-tolerance or beyond-tolerance levels, all performance would deteriorate, of course.) For example, in a study by Brown et al. [6] it was found that the ability of Navy carrier pilots to stabilize disturbances in pitch and roll immediately after catapult takeoff was not affected by acceleration—even after exposure to nearly 12*G*. In still a different study, subjects performed on a tracking task in a centrifuge during exposure to accelerations which they could control [Brown and Collins, 5]. While their average error scores on horizontal and

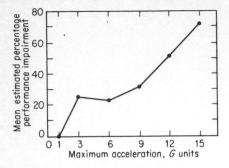

Figure 17-11 Average estimated performance impairment (across several aspects of performance) associated with acceleration forces. [From Chambers and Brown, 9.]

vertical tracking deteriorated under acceleration conditions, their flight coordination actually improved as indicated below:

Task	*Average error score*	
	No acceleration	*With acceleration*
Horizontal tracking	50	53
Vertical tracking	39	45
Flight coordination	52	37

It seems apparent, however, that the requirement under acceleration for additional attention to flight coordination (imposed by the fluctuating acceleration pattern) would appear to be met at the expense of some other aspect of performance (in this case, tracking performance).

Protection from Acceleration Effects

Although minor doses of acceleration (as in most land vehicles and most commercial airplanes) pose no serious problem either in safety and welfare or in performance, the effects of higher levels (especially with long exposures) usually require some protective measures. As indicated above, and as illustrated in Figure 17-9, certain postures make it possible to tolerate higher *G* loads than other postures. Immersion, although providing considerable protection, is generally not considered feasible in space flight because of the problem of lugging around a private swimming pool. Aside from the use of optimum postures, the presently available protection schemes include restraining devices, anti-G suits, and special body supports such as contour and net couches.

For high, sustained G loads in a forward direction $+G_x$, the use of contour couches (sometimes molded for the individual) are reasonably satisfactory; the specific posture that is provided, however, can affect the tolerance level, as illustrated in Figure 17-12. Such couches, however, do

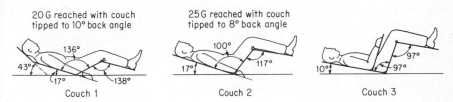

Figure 17-12   Illustration of three NASA-AMAL couch designs for space capsules. Design 2 permitted greatest tolerance, 25G, in a forward direction with reference to the subject (in this case up). [From Chambers and Brown, 9.]

not provide adequate support for the rearward direction $-G_x$. For such purposes additional contraptions are required to provide adequate restraint, such as a restraint helmet and supporting face and chin pieces, and frontal supports (such as nylon netting) for the chest, torso, and body members; to further hem in the individual, an anti-G suit may be in order [Roth, 41, chap. 7]. Anti-G suits, however, generally have their greatest utility is connection with headward $+G_z$ acceleration, with some designs being more effective than others [Nicholson and Franks, 36]. There is no particularly effective scheme for providing protection against footward acceleration $-G_z$, so, if you insist on being shot out of a cannon, have them shoot you out headward.

Deceleration and Impact

The normal deceleration of a vehicle imposes forces upon people that are essentially the same as those of actual acceleration, but in reverse. Thus, the effects of forward acceleration of a vehicle for a person facing forward would be the same as for a person facing backward in a vehicle that would be decelerating at an equal rate; both conditions would produce an eyeballs-in effect with the soft tissue, internal organs, and the blood being forced toward the back of the body. But acceleration usually is a gradual affair (except in unusual instances, such as with astronauts during blast off), whereas deceleration can be extremely abrupt, especially in the case of vehicle accidents.

Effects of vehicular impact.   When a vehicle hits a solid object or another vehicle head on, an unrestrained occupant will continue forward at his initial velocity until the body strikes some part of the interior of the vehicle. This is sometimes called the *second collision*. As pointed out by Cichowski and Silver [12], even though the forward portion of the vehicle collapses by, say, 30 in. (which absorbs some of the energy), the deceleration of the occupant is a function of the deformation of the part of the interior he hits, which might be only 3 in.; the occupant in this case would undergo the same velocity change as that of the vehicle in only one-tenth the distance. Thus, the occupant benefits only from the crush of the vehicular interior and not from the absorption of energy by the car itself as its front end caves in. To be protected from this second collision, the occupant must be tied to the vehicle and decelerate with it (such as through, say, a distance of 30 in.).

Restraining devices. An occupant can be "tied" to the vehicle by quite a few types of restraining devices, in automobiles and commercial aircraft these nearly always being seat (lap) belts. In some automobiles shoulder harnesses are used by themselves or in combination with seat belts, and in military aircraft more complex restraints are used, individually or in combination. A major deficiency of automobile seat belts is, of course, the possibility of head damage caused by the torso and head being propelled forward with the body hinged at the hips. Considerably greater protection is provided by lap-shoulder belts, which aid the occupant in "riding down" the vehicle as it impacts. A general comparison of some of these effects is shown in Figure 17-13. (However, there is some

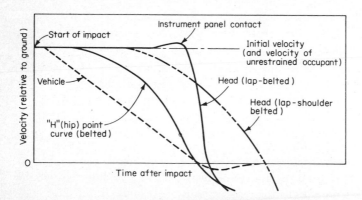

Figure 17-13 Velocity of vehicle occupants following impact with lap belt and shoulder-lap belt. The shoulder-lap restraint causes the upper torso and head to follow more closely the vehicle velocity change than does the seat belt alone. On impact, an unrestrained occupant would be propelled forward at the impact velocity until he hit some part of the interior. [From Cichowski and Silver, 12, fig. 1.]

suspicion that, under impact conditions, shoulder belts might apply excessive force to the spine.)

Additional protection to the head is provided by *pre-positioning* the head when impact seems imminent, for example, by placing the arms on the steering wheel or dashboard, and cradling the head in the arms. The relative probabilities of head injuries under various conditions are given below [Cichowski and Silver, 12]:

	Not pre-positioned	Pre-positioned
Unrestrained	100 (baseline index)	
Lap belt	80	15
Lap-shoulder belt	35	20
Lap belt plus energy-absorbing steering column		10

Further indication of the protection provided by seat belts comes from a survey in California; the use of seat belts reduced the incidence of fatalities by 35 percent, largely because of the prevention of ejection [Tourin and Garrett, 46]. Although the seat-belt wearers were injured with the same frequency as nonwearers, the degree of injury was lower for those who wore belts.

However, as pointed out by Muller [35], there is much to be done in developing vehicles and nuisance-free restraint systems that will provide reasonable protection against the effects of impact. To avoid the second collision the combination of vehicle and restraint systems should *minimize* that portion of the deceleration travel of the occupant which must occur *within* the vehicle's rigid body (by use of restraints) and should *maximize* the travel of the occupant allowed by the "deforming" of the vehicle structure [Muller, 35]. The minimum portion which must occur within the vehicle's rigid body is based on certain physiological assumptions of the deceleration forces people can tolerate, and is a function of vehicle impact velocity [Muller, 35].

Other protection procedures. Although we have become almost completely adapted to forward-facing seats in most passenger vehicles, aft-facing seats have the edge over forward-facing seats because of the extent to which they can absorb energy in the case of head-on impact; it has been estimated that properly designed aft-facing seats could protect occupants against as much as 40-G impacts with little or no injury [Eiband, 19]. Such seats would be particularly feasible in aircraft, trains, etc., but are not completely unrealistic for use in automobiles (if we could adapt ourselves to them). In the case of vehicles with forward-facing seats involved in rear-end crashes, the use of a full back support with headrest minimizes the possibility of whiplash.

Weightlessness

Weightlessness is a condition that is characterized by the absence of gravity. For clarity, it has been proposed that the terms *zero g* or *null g* be used to refer to the physical state of an object or body and that the term *weightlessness* be used to refer to the physiological and psychological experience of living organisms under zero *g* conditions [Ritter and Gerathewohl, 39]. In our earthbound experiences weightlessness is seldom encountered, and when it is, it is for only a brief time period. It is only in outer space that zero gravity conditions are possible for an extended period.

Since zero *g* conditions are not very common in our earthbound environment, the creation of such conditions for research purposes is a bit of a problem. Among the methods employed is the use of aircraft following a Kaplarian trajectory, which produces, at the peak of the trajectory, a zero *g* condition for up to half a minute or so. Most other methods, such as water immersion, bed rest for extended periods of time, etc., do not really produce zero *g* conditions, but only simulate certain features of them. Certain methods (such as immersion), however, may at least reduce the physical requirements of activity to what presumably is quite comparable to true zero *g* conditions [Knight, 31].

Physiological effects of weightlessness. Among the various physiological effects of short-term weightlessness are nausea, vestibular disturbances, blood pooling, slight drop in blood pressure, and variable effects on the heart rate. The evidence to date, however, suggests that the physiological effects of short-term weightlessness are not of major consequence. The possible effects of long-term exposure, however, may be another matter. The evidence to date on prolonged exposure is generally based on the United States and Russian space flights, all of which are limited to several days. These, and some simulated earthbound studies, have indicated that during extended zero *g* conditions most heart and vital body functions remain within reasonable bounds, but that some cardiovascular adaptation results in lowered blood volume (about 7 to 15 percent) with decreases in both the plasma and red cell fractions of the blood [Roth, 41, chap. 7]. There also is sometimes a loss of total body weight. These changes sometimes result in orthostatic intolerance (a tendency toward postural disorientation).

Body movement under zero *g*. Under zero *g*, human motor coordination and movement are very different from those under normal gravity conditions. The (otherwise) simple process of locomotion is, of course, affected, in part because there are no sensory gravity cues to indicate which way is down, and more importantly because locomotion produces no friction. However, subjects become "oriented" in space as soon as any body mobility ceases; "down" is the direction in which his feet are [Simons, 43]. The lack of friction between the feet and the contact plane makes walking virtually impossible without some method of overcoming the frictionless condition.

With other body movements, the law of conservation of momentum produces a very different, unreal world. For example, the force a weightless man exerts on a fixed object causes him to move away from the object; if he applies a torque to some object, the force will cause him to rotate around the object.

Performance under weightlessness. The markedly different "rules of the game" in doing manual work under weightlessness would cause one to suspect that manual performance on at least some types of tasks would be adversely affected. These suspicions have been generally confirmed in many experiments [Dzendolet, 17; Dzendolet and Rievley, 18; Kama, 30, and others; see especially reviews by Hammer, 25, and by Loftus and Hammer, 32]. Some moderate effects on visual acuity have also been reported [Pigg and Kama, 37]. These studies will not be summarized here. Some of the more general findings and implications of this research, however, seem to be in order. While man's performance on many tasks when weightless is adversely affected, it is probable that man can perform adequately on certain types of tasks under such conditions. It seems generally evident that the free-floating orbital worker may be overwhelmed with problems of orientation, inadvertent tumbling, locomotion, stabilization, and material handling. The effect of free-floating weightlessness on performance on various manual tasks is well documented. Such effects include increase in time required to perform certain operations; for example, the time required to use push buttons, toggle switches, and rotary switches increased significantly when in a condition of weightlessness; the increase in time averaged 15 percent [Wade, 48].

But while weightlessness may affect manual performances somewhat, the effects on some activities can be minimized by the use of certain aids. Some suggestions along these lines are given below [Dzendolet and Rievley, 18]:

1. If a man is to perform a turning or tightening task by hand, he can exert his maximum torque with a minimum of resultant body movement by positioning himself so that his body is at right angles to the axis of rotation of his turning task.
2. A handle should be available within the work area, so that the man can maintain himself in position.
3. The handhold angle should be as nearly parallel as possible to the imaginary line joining the handhold midpoint and the point of the man's force application.

4. Units or push buttons should not require a torque of greater than 2.5 lb-ft if the torque is to be exerted for longer than 3 sec.

5. A large-amplitude, short-time force or torque can be used in many cases to perform a task with less body reaction than that resulting from a smaller, long-time force.

6. Special tools which fasten to the work and do not require continual application of contact force are best for many maintenance tasks. These tools can simultaneously function as handholds.

ILLUSIONS DURING MOTION

When man is in motion, he receives cues regarding his whereabouts and motion from sense organs, especially the semicircular canals, the vestibular sacs (the utricle and saccule), the eyes, the kinesthetic receptors, and the cutaneous senses. The interactions among these senses are quite intricate, and under some circumstances the sensory cues received are misleading and result in errors in perception and illusions, especially when certain cues are at odds with each other. Such conditions are perhaps most common (and most critical) in aircraft and space flight because of the unique factors and special circumstances that occur under such conditions [Clark and Graybiel, 13].

Disorientation (vertigo)

A rather common perceptual phenomenon in flight is associated with the sensation of orientation in space (such as sensing which direction is up). There are several manifestations of disorientation (vertigo), these generally resulting from stimulation of the vestibular and kinesthetic sense organs from acceleration and angular changes in direction. For example, especially during changes in direction and in velocity the sensation of the vertical is felt more in relationship to the aircraft than to the earth because of the manner in which the acceleration forces act on the vestibular organs. This effect is sometimes so strong that the sensation of tilt is grossly underestimated. The effect is illustrated by Figure 17-14, which shows, for blindfolded pilots during aircraft flight, the average judgments of angle of tilt (bank) for various actual angles of aircraft tilt. The errors in judgments are of major proportions. It was also found that the perception of any tilt lagged behind the actual turn by an average of over 9 sec and that the turns were judged to be much shorter in time than they really were. In some cases the recovery from turn was accompanied by an impression of turning in the opposite direction, although the plane had straightened out. The tendency to underestimate the amount of bank misleads the pilot into "perceiving" the plane as being ori-

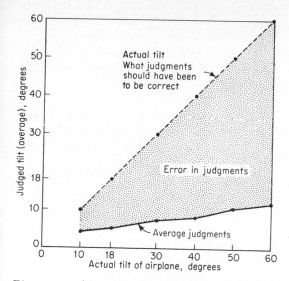

Figure 17-14 Averages of judgments of tilt of blindfolded observers in comparison with actual tilt of airplane. [Adapted from MacCorquodale, 33, by permission of the *Journal of Aviation Medicine*.]

ented more vertically than it really is in such situations. If in such a case there is no external visual frame of reference (because of fog, darkness, clouds, etc.), the pilot also tends to perceive the ceiling of the plane as being vertical. Thus, both gravitational cues and visual cues can operate to "confirm" each other—and both can be wrong!

In another related study the subjects, when facing forward in an airplane, reported sensations of backward tilt (or going into a climb) under conditions of forward acceleration, and of forward tilt (going into a dive) under conditions of forward deceleration. One can readily see how easy it would be under conditions of poor visibility for a pilot decelerating when landing to get the impression of going into a dive, and "correct" for it by overshooting the landing field. This emphasizes the importance, in building landing strips, of providing reliable visual cues (such as lights) to help pilots orient themselves visually when visibility is poor.

Other variations of the vertigo theme include sensations of climbing while turning, sensations of diving when recovering from a turn or following pull-out from a dive, the coriolis phenomenon (a loss of equilibrium that occurs when the pilot is rotating with the aircraft and moves his head out of the plane of rotation), and a sensation of reversed rotation that sometimes occurs after spinning [Roth, 41, chap. 7].

Visual Illusions

Certain illusions that sometimes occur during flight are essentially visual. One of these is autokinesis which is manifested by a confusion with lights at night in which a fixed light may appear to move against its dark background. It has been reported that pilots have attempted to "join up" in formations with stars, buoys, and street lights which appeared to be moving [Clark and Graybiel, 13]. Some pilots have experienced illusory horizons. This can be brought about, especially under conditions of poor visibility (such as over water, snow, and desert), when there is no clearly defined landmark. Such illusions can bring about errors in the estimation of altitude, distances, or directions. The oculogyral illusion is character-ized by apparent movements of objects that actually are stationary. They occur following angular acceleration which stimulates the semicircular canals; the illusion is something of an aftereffect of their stimulation. Another type is the oculogravic illusion, which is induced by conflicting cues from the eyes and the vestibular sacs. It is characterized by the apparent displacement of objects in space as well as body displacement, the direction of the apparent displacement depending on the posture of the individual relative to the line of the force to which he is subject.

Discussion

The reduction of disorientation and illusions generally depends more on procedural practices and training than upon the engineering design of aircraft. Among such practices are the following [Clark and Graybiel, 13]: understanding the nature of various illusions and the circumstances under which they tend to occur, maintaining either instrument *or* contact flight, avoiding night aerobatics, shifting attention to different features of the environment, learning to depend upon the correct cue to orientation (such as using visual cues when possible and otherwise using instruments), avoiding sudden accelerations and decelerations at night, and avoiding prolonged constant speed turns at night.

VEHICLE DRIVING

Very few among us will ever have the opportunity to experience the sensa-tions of acceleration and of weightlessness such as those of the astronauts in spacecraft. And on the other hand, there are few among us for whom mobility in the family automobile is not part of our daily lives. The "system" of which a vehicle driver is a part consists of himself, his vehicle, the road, the signs, the surrounding traffic, and the physical environment, each of which abounds with human factors problems. Although previous reference has been made to certain human factors aspects of vehicles (such as rapid deceleration and some aspects of road signs), we shall here mention very briefly a few other aspects.

Sensory Input in Driving

Although a number of senses are sources for information input in driving, vision is undoubtedly the most important. However, there has been more speculation than factual data on the extent to which information from the features of the visual environment is used.

Visual search and scan patterns in driving. Some inklings of the visual search and scan patterns of drivers come from a study by the Systems Research Group, The Ohio State University [50], in which eight drivers followed a specified route, their eye movements being recorded en route with an eye camera. Some indication of the results are shown in Figure 17-15, which gives the percentages of time during which the eyes

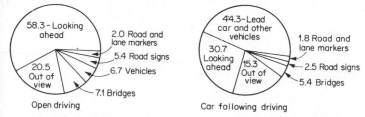

Figure 17-15 Percentages of time spent by automobile drivers viewing different features of the environment. The figures are for the third of three trials, or runs, for "open" driving and for "car following" driving. ("Out of view" represents time during which eye position could not be picked up by the eye camera.) [From *Specifications for partially automated control systems for the driver*, 50.]

were viewing environmental features (actually for "open" driving and for "car following" driving).

In the open driving, the primary visual fixation when *looking ahead* tends to be at a point located slightly above the horizon, with the line of sight passing over the right edge marker of the highway, about 200 ft ahead of the driver. Contrary to popular opinion, the drivers seldom fixated on the center of the lane in which they were driving. These results jibe quite well with those obtained by Gordon [23], who used a somewhat different camera for locating the point of visual regard of drivers. He found that drivers tended to use the road edge, plus the dividing center line, in keeping the car within bounds. Because of the visual dependence of drivers on the *edges* of his driving domain, it is obvious that the edges of this domain should be made visible to drivers at all times. Harking back to the first study [50], we can see in Figure 17-15 that the visual fixation of the car-following driving task markedly affects the fixation pattern, with the dominant point being the lead car and other vehicles. It might also be added that, with both types of driving, experience with the route to be followed (each route being driven three times) causes moderate changes in visual fixations.

Response to road signs. One important source of input to drivers consists of road signs. Actually, there probably are two basic aspects in the effective use of such signs. In the first place, they must be perceived by drivers, and in the second place, their meaning must be clear. As to the first, some rather discouraging results of a survey are reported by Johansson and Rumar [28], who stopped and interviewed 1000 drivers to ascertain how many recalled having seen five specified types of road signs in a previous stretch of road. On the average, the signs were "recalled" by the motorists only 47 percent of the time; however, there were differences in the percents for the individual signs, as follows: speed limit (78 percent), "police control" sign (63 percent), road condition sign (55 percent), general warning (18 percent), and pedestrian crossing (17 percent). It is probably speculative to consider these results as being representative of drivers generally, but they do suggest that relevant road signs should be so designed and positioned as to stand a reasonable chance of being perceived by passing motorists.

As a sample of the second aspect (the meaning of signs), in one study, different symbols were investigated to identify the optimum symbols for lane control purposes such as on multilane highways or bridges [Forbes, 21]. It was desirable, for example, to identify a symbol that would say, in effect, "do not drive in this lane." Various designs were used (such as red and yellow bull's eyes and X's, arrows pointing up and down, and diagonal slashes). It was found that a red X had the strongest association with the intended meaning and the weakest association with "stop;" (a red bull's eye had a fairly strong association with "stop," because of its common use in traffic signs). The red X sign, and an accompanying green arrow pointing up to indicate "go in this lane," are now used on the Mackinac Bridge in Michigan, and elsewhere.

Judgments of Vehicle Drivers

Once the information (really, the stimuli) is received by a vehicle driver, the driver needs to make judgments on certain choices that he then feeds into his decision-making process.

Psychophysical judgments of velocity changes. As an example, drivers have to make psychophysical judgments relating to the changes in velocity of other vehicles (e.g., the acceleration and deceleration of vehicles). In this connection Braunstein and Laughery [3] carried out a study with two cars, a lead car and a following car. The subjects, in the second (following) car, were asked to judge the points in time when the lead car underwent predetermined velocity changes. The detection times of the subjects (e.g., the time required to detect such changes) varied as follows: (1) Detection time *increased* when the distance between the vehicles was increased and (2) detection time *decreased* when the rate of change of velocity of the lead car (e.g., its acceleration) was increased. The latter point was generally confirmed by Torf and Duckstein [45], but they also found that the detection of deceleration took longer than the detection of acceleration.

Judgments in passing other vehicles. Another type of judgment that vehicle drivers have to make is that in passing other vehicles. The human error in making such critical judgments was reflected by the results of a study by Jones and Heimstra [29] in which 19 subjects, serving as drivers, were asked to indicate the last possible moment they felt they could pass a lead car that was driving at a specified speed, when another car was approaching. It was possible for the experimenters to determine whether the driver *could* have overtaken the lead car (had he actually attempted to do so), taking into account recorded speeds, distances, etc. Of 190 such judgments (10 for each subject), 93 were *underestimates* of time; such an underestimate would, in an actual situation, result in a potentially dangerous passing situation. To add to the potential danger of passing other cars, Crawford [16] reports that, based on an observation study of actual drivers, judgments of the available time interval are less realistic in passing vehicles traveling at high speeds than at low speeds. As far as advice to the driver is concerned, he urges that a driver should *not* decide to pass a car ahead if there is *any doubt* about being able to make it, and that he should be especially wary in deciding to pass cars traveling over about 40 mph.

Discussion

Probably one of the major problems of present day living is vehicular travel, particularly reflected by the number of deaths, injuries, and economic costs involved in accidents. The human factors problems associated with conventional motor vehicles cover the spectrum of highway and road design, the design and location of road signs, the many aspects of vehicular design including vehicle dynamics and handling qualities, occupant restraining devices, and (in the words of the king in "The King and I") et cetera, et cetera, et cetera. Although we cannot (at least not yet) fully protect ourselves from the risks that other drivers take with our lives, attention to some of these problems will make the open road at least a bit safer, and the driving more comfortable, but (if you will pardon a pun) there is still a long road ahead. In the meantime some people are scratching their heads about new modes of transportation, such as with external automatic control on superhighways. And maybe our descendants will each have his own private flying saucer to use in going to his plant, office, or grocery store.

SUMMARY

1. The discomfort and physiological damage that can occur under conditions of excessive whole body vibration are caused by displacements and deformations of body organs or parts (these effects generally being due to the different resonant frequencies of the organs and parts).

2. Tolerance of the whole body to sinusoidal vibration is least when in the range of frequencies from about 4 to 8 Hz.

3. Vibration transmitted to the body can be amplified, or attenuated, as the consequence of body posture (standing versus sitting), the type of seating, and the frequency of the vibration. Amplification is greatest when the seat is vibrating at, or near, the resonant frequency of the body.

4. The effects of vibration of the whole body on performance depend on the nature of the vibration (its frequencies and amplitudes) and the activity being performed. For example, sinusoidal vibration can affect performance on certain visual tasks, simple motor tasks that require precision and coordination (but usually not those that require firm muscular control), and complex motor tasks.

5. Sensitivity of the hands to sinusoidal vibration is greatest around 300 Hz.

6. Acceleration of the human body causes displacement of body tissues, organs, and blood; these displacements are in the direction opposite to the direction of body movement.

7. Human beings can tolerate more acceleration when moving in a forward or backward direction than upward or downward; they can tolerate least in a downward direction.

8. Within the bounds of tolerance, acceleration affects different human performances, such as reaction time, visual acuity, time estimation, tracking, and time to activate control mechanisms. While some aspects of human performance are not affected under moderate levels of acceleration, virtually all aspects are affected at near-tolerance levels.

9. If a body member is being moved against an acceleration force, the amount of force required to overcome this is proportional to the number of G units affecting the body.

10. Protection from the effects of acceleration includes appropriate posture (including, where pertinent, seats and couches), immersion in water (which has been used experimentally), the use of pressure

suits, and the use of properly designed body restraints. (Body restraints such as seat belts are used most commonly, however, for protection from the body being propelled forward during rapid deceleration.)

11. The effects of deceleration are most critical in the event of impact. In vehicular impact, the most effective protection is the use of full-body restraining devices (such as lap and shoulder belts) that permit the individual to "ride down" the vehicle as it impacts (rather than being propelled forward as the vehicle is brought to a stop).

12. Under conditions of weightlessness, coordination and body movement are very different from those under gravity conditions. Possible methods of control of body movement include the use of special shoes (magnetic, adhesive, and with suction cups), gyros, rockets, hand and tethering lines, and handholds.

13. Manipulative tasks are also affected by weightlessness (such as turning a wrench, which causes the body to rotate). Such effects can be minimized, however, by the use of special equipment such as handholds and special tools and by appropriate body posture.

14. While there are some physiological effects that are brought on by weightlessness of short duration, these generally are not of major consequence; the effects of long exposure to weightlessness are well known, but experience and research to date indicate that most heart and lung functions would remain within reasonable bounds, but that certain physiological changes would occur, such as lowered blood volume and decreases in blood-plasma and red-cell fractions.

15. When man is in motion, as in aircraft, he may experience various types of illusions, especially when sensory cues are in conflict with each other. Under conditions of acceleration and angular changes in direction, the stimulation of the vestibular and kinesthetic sense organs can give rise to disorientation effects.

16. The human factors aspects of vehicle driving include those associated with the road system; the signs, and other visual inputs such as traffic; the dynamics of the vehicle; and the judgments needed in driving.

REFERENCES

1. Biggs, N. L.: Directional guidance of motor vehicles—a preliminary survey and analysis, *Ergonomics*, 1966, vol. 9, no. 3, pp. 193–202.

2. Bondurant, S., N. P. Clark, W. G. Blanchard, et al.: *Human tolerance to some of the accelerations anticipated in space flight*, WADC, TR 58–156, 1958.

3. Braunstein, M. L., and K. R. Laughery: Detection of vehicle velocity changes during expressway driving, *Human Factors*, 1964, vol. 6, no. 4, pp. 327–331.

4. Brown, J. L.: Acceleration and motor performance, *Human Factors*, November, 1960, vol. 2, no. 4, pp. 175–185.
5. Brown, J. L., and C. C. Collins: Air-to-air tracking during closed-loop centrifuge operation, *Journal of Aviation Medicine*, 1958, vol. 29, p. 795.
6. Brown, J. L., W. H. B. Ellis, M. G. Webb, and R. F. Gray: *The effect of simulated catapult launching on pilot performance*, USN Air Development Center, Report NADC-MA-5719, Dec. 31, 1957.
7. Buckhout, R.: *A working bibliography on the effects of motion on human performance*, USAF, MRL, TDR 62–77, July, 1962.
8. Chambers, R. M.: "Operator performance in acceleration environments," in N. M. Burns, R. M. Chambers, and E. Hendler (eds.), *Unusual environments and human behavior*, The Free Press, New York, 1963, pp. 193–320.
9. Chambers, R. M., and J. L. Brown: *Acceleration*, paper presented at Symposium on Environmental Stress and Human Performance, American Psychological Association, September, 1959.
10. Chambers, R. M., and L. Hitchcock: "Effects of high G conditions on pilot performance," in *Proceedings of the National Meeting of Manned Space Flight*, Apr. 30–May 2, 1962, Institute of the Aerospace Sciences, New York, 1962, pp. 204–227.
11. Chaney, R. E.: *Whole body vibration of standing subjects*, Boeing Company, Wichita, Kans., BOE–D3–6779, August, 1965.
12. Cichowski, W. G., and J. N. Silver: *Effective use of restraint systems in passenger cars*, paper presented at Automotive Engineering Congress, Detroit, Michigan, Jan. 8–12, 1968, SAE 680032.
13. Clark, B., and A. Graybiel: *Disorientation: a cause of pilot error*, USN School of Aviation Medicine, Research Project NM 001 110 100.39, Mar. 2, 1955.
14. Coermann, R. R.: Effect of vibration and noise on the human body, *Ringbuch der Luftfahrttechnic*, 1939, vol. 5, no. 1.
15. Cope, F. W.: Problems in human vibration engineering, *Ergonomics*, 1960, vol. 3, pp. 35–43.
16. Crawford, A.: The overtaking driver, *Ergonomics*, 1963, vol. 6, no. 2, pp. 153–170.
17. Dzendolet, E.: Manual application of impulses while tractionless, *Human Factors*, November, 1960, vol. 2, no. 4, pp. 221–227.
18. Dzendolet, E., and J. Rievley: *Man's ability to apply certain torques while weightless*, USAF, WADC, TR 59–94, April, 1959.
19. Eiband, A. M.: *Human tolerance to rapidly applied acceleration*, NASA Memo 5–19–59E, 1959.
20. Fibikar, R. J.: Touch and vibration sensitivity, *Product Engineering*, November, 1956, vol. 27, no. 12, pp. 177–179.
21. Forbes, T. W.: Human factors in highway design, operation and safety problems, *Human Factors*, 1960, vol. 2, no. 1, pp. 1–8.
22. Gell, C. F.: Table of equivalents for acceleration terminology, *Aerospace Medicine*, 1961, vol. 32, no. 12, p. 1109.
23. Gordon, D. A.: Experimental isolation of the driver's visual input, *Human Factors*, 1966, vol. 8, no. 2, pp. 129–137.

24. Guignard, J. C.: *The physical response of seated men to low-frequency vertical vibration*, FPRC–1062, Flying Personnel Research Committee, Air Ministry, London, April, 1959.

25. Hammer, L. R.: Aeronautical Systems Division studies in weightlessness: 1959–1960, USAF, WADD, TR 60–715, December, 1961.

26. Holland, C. L.: Performance effects of long-term random vertical vibration, *Human Factors*, 1967, vol. 9, no. 2, pp. 93–104.

27. Hornick, R. J., and N. M. Lefritz: A study and review of human response to prolonged random vibration, *Human Factors*, 1966, vol. 8, no. 6, pp. 481–492.

28. Johansson, G., and Käre Rumar: Drivers and road signs: a preliminary investigation of the capacity of car drivers to get information from road signs, *Ergonomics*, 1966, vol. 9, no. 1, pp. 57–62.

29. Jones, H. V., and N. M. Heimstra: Ability of drivers to make critical passing judgments, *Journal of Engineering Psychology*, 1964, vol. 3, no. 4, pp. 117–122.

30. Kama, W. N.: *Speed and accuracy of positioning weightless objects as a function of mass, distance, and direction*, USAF, WADD, TR 61–182, March, 1961.

31. Knight, L. A.: An approach to the physiological simulation of the null-gravity state, *Journal of Aviation Medicine*, 1958, vol. 29, pp. 283–286.

32. Loftus, J. P., and L. R. Hammer: *Weightlessness and performance: a review of the literature*, USAF, ASD, TR 61–166, June, 1961.

33. MacCorquodale, K.: Effects of angular acceleration and centrifugal force on nonvisual space orientation during flight, *Journal of Aviation Medicine*, 1948, vol. 19, pp. 146–157.

34. Magid, E. B., R. R. Coermann, and G. H. Ziegenruecker: Human tolerance to whole body sinusoidal vibration, *Aerospace Medicine*, November, 1960, vol. 31, pp. 915–924.

35. Muller, G.: "Why a second collision?" in M. L. Salzer et al. (eds.), *Prevention of highway injury*, Highway Safety Research Institute, The University of Michigan, Ann Arbor, 1967.

36. Nicholson, A. N., and W. R. Franks: "Devices for protection against positive (long axis) acceleration," in P. I. Altman and D. S. Dittmer (eds.), *Environmental biology*, AMRL, TR 66–194, November, 1966, pp. 259–260.

37. Pigg, L. D., and W. N. Kama: *The effect of transient weightlessness on visual acuity*, USAF, WADD, TR 61–184, March, 1961.

38. Radke, A. O.: *Vehicle vibration: man's new environment*, ASME, paper 57–A–54, Dec. 3, 1957.

39. Ritter, O. L., and S. J. Gerathewohl: *The concept of weight and stress in human flight*, USAF School of Aviation Medicine, Report 58–154, 1959.

40. Rogers, T. A.: The physiological effects of acceleration, *Scientific American*, February, 1962, vol. 206, no. 2, pp. 60–70.

41. Roth, E. M. (ed.): *Compendium of human responses to the aerospace environment*, NASA CR–1205, vol. 2 and 3, November, 1968.

42. Simons, A. K., A. O. Radke, and W. C. Oswald: *A study of truck ride characteristics in military vehicles*, Bostrom Research Laboratories, Milwaukee, Report 118, Mar. 15, 1956.

43. Simons, J. C.: *Walking under zero-gravity conditions*, USAF, WADC, Technical Note 59–327, October, 1959.
44. Stapp, J. P.: Acceleration: how great a problem? *Astronautics*, February, 1959, vol. 4, no. 2, pp. 38–39, 98–100.
45. Torf, A. S., and L. Duckstein: A methodology for the determination of driver perceptual latency in car following, *Human Factors*, 1966, vol. 8, no. 5, pp. 441–447.
46. Tourin, B., and J. W. Garrett: *Safety belt effectiveness in rural California automobile accidents*, Automotive Crash Injury Research of Cornell University, February, 1960.
47. von Gierke, H. E.: "Response of the body to mechanical forces, an overview," in *Lectures in Aerospace Medicine*, 6th series, Feb. 6–9, 1967, School of Aerospace Medicine, Brooks AFB, Tex., pp. 325–344.
48. Wade, J. E.: *Psychomotor performance under conditions of weightlessness*, USAF, MRL, TDR 62–73, June, 1962.
49. *Revised proposal to the secretariat: guide for the evaluation of human exposure to whole-body vibration*, ISO, Technical Committee 108 (mechanical vibration and shock), working group 7, ISO TC 108/WG 7, December, 1968.
50. *Specifications for partially automated control systems for the driver*, Systems Research Group, Department of Industrial Engineering, The Ohio State University, Columbus, Report EES 277B–3, March, 1969.

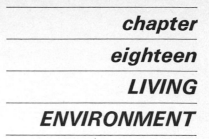

To date the field of human factors probably has been largely concerned with the design of specific, reasonably definitive, circumscribed systems, perhaps of a predominantly "hardware" nature. People usually are involved with such items of equipment either in earning their daily bread (such as in using industrial machines, tractors, tools, and computers in their jobs) or in using specific consumer items (such as passenger automobiles or dishwashers).

Aside from using such specific products of our civilization, however, we also serve as users of a broad and ill-defined spectrum of features or aspects of our total living environment. Such daily involvement includes our use of: buildings and their related facilities (houses, apartments, schools, offices, factories, stores, theaters, houseboats, etc.); the community proper, of whatever size and arrangement (including its aesthetic character, its facilities for entertainment, culture, and recreation); the transportation facilities; and the physical (ambient) environment (including the natural environment itself plus whatever we mortals do to it, such as polluting it and adding to its noise level). As Wells [25] points out, the environmental "unit" of an individual is never as simple as a single room or building, but rather is the composite of many interacting features of our total environment, such as depicted in Figure 18-1. All of these aspects of the living environment have an impact, good or bad as the case may be, on human beings. One would hope that these features of our civilization would become the new frontiers of human factors attention. There is, indeed, considerable current interest in, and talk about, some of these problems; but to date this concern has perhaps not resulted in a systematic attack on their human factors aspects (as has been the case in the more conventional human factors problems), or the development of a broad body of substantive research that would support such a systematic attack.

To contemplate the human factors aspects of this living environment requires a significant jump or two from the conventional human factors context of airplanes, industrial machines, and automobiles. In the first place, the systems of concern (the community, for example) are amorphous and less well defined than are pieces of hardware. And, in the second

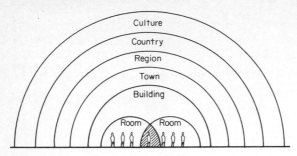

Figure 18-1 Representation of the different features of our total environment. The living environment of an individual is never as simple as a single room or building. (The shaded portion represents passage space between rooms.) [From Wells, 25, fig. 1, p. 680.]

place, the relevant criteria for evaluation tend to be rather different. Instead of being primarily interested in system performance criteria (such as for a computer system) or physiological criteria (as in the case of astronauts), perhaps the dominant considerations in living environments tend to be more subjective, with a major focus on the fulfillment of human values and on the achievement of human satisfactions; such considerations, of course, have significant overtones in the social interactions among people.

DEPENDENT AND INDEPENDENT VARIABLES

It is not feasible within the remaining pages to delve into the living environment with its many human factors ramifications. But it may be useful to touch on some of the dimensions of the problem, especially the dependent (i.e., criterion) variables and the independent (i.e., situational) variables. Beyond that, we shall bring in the results of a few relevant studies, but more for the purpose of illustration than for their substantive content as such.

Dependent Variables (Criteria)

Let us suppose someone asked you, and me, and others, to specify the standards (i.e., the criteria) by which we would evaluate any given living environment (but let us restrict this poser to the criteria relevant to those features of the environment that man can do something about). There would, indeed, be wide individual differences, but such an assortment might generally fall into the following categories:

· Performance of activities (work in offices, factories, hospitals, etc.; preparing meals at home; working in home workshop; engaging in sports and games; etc.)

- Physical convenience (convenience of things that people use and proximity to places people want to go to)
- Convenient mobility (efficient mobility in going from one place to another by public or private transportation or by foot)
- Physical welfare (maintenance of physical welfare, such as health and personal safety)
- Physical comfort (temperature, seating comfort, avoidance of noise, etc.)
- Emotional health and welfare
- Personal space (adequate personal space relevant to the situation, such as at work, at home, or in travel; opportunity for privacy)
- Social interaction (opportunity for desired social contacts and interchange; individual and group interaction)
- Aesthetic values and preferences
- Fulfillment of personal values (opportunity for selection of activities and situations that fulfill one's own, individual values, such as recreation, entertainment, and culture)
- Financial considerations

These categories generally cover the same types of criteria as discussed by Ittleson [10], namely, goal directed behavior, health and safety, emotional responses, social responses, and perceptual and cognitive responses.

Independent Variables

On the other side of the coin, one might likewise categorize the features of the total environment (i.e., the independent variables) that might impinge upon one or more of the above criteria (i.e., dependent variables). Such categories might include the following:

- Building design characteristics (structural characteristics such as size and arrangement of rooms, number and size of windows and doors, halls and passageways, and architectural style)
- Physical environment (nature and arrangement of furniture and other facilities, decor, etc.)
- Ambient environment (ambient outdoor environment, interior illumination, temperature control, noise control, etc.)
- Community (layout, arrangement, size, recreational and cultural facilities, shopping facilities, beauty and other aesthetic aspects, etc.)
- Transportation facilities (location, speed, etc., of public transportation; and streets, highways, parking facilities, etc., for private transportation)

To put these two together, one could construct something of a matrix of the independent and dependent variables, such as in Table 18-1. In viewing this matrix it would be apparent that only certain of the cells would have meaningful implications, specifically those in which a particular characteristic of an independent variable (such as, say, room arrangement) would have an *effect* on a specific criterion (such as convenient mobility).

Table 18-1 Matrix of Possible Independent and Dependent Variables Relevant to Human Factors Aspects of Living Environment

	Independent variables														
	Building characteristics			Physical environment characteristics			Ambient environment variables			Community characteristics			Characteristics of transportation facilities		
Dependent variables (criteria)	1	2	etc.	1	2	etc.	1	2	etc.	1	2	etc.	1	2	etc.
Performance of activities															
Physical convenience															
Convenient mobility															
Physical welfare															
Physical comfort															
Emotional health and welfare															
Personal space															
Social interactions															
Aesthetic values															
Personal values															
Financial considerations															

Design Criteria

One could envision the possibility umpteen years from now when research might make it possible to set forth, for at least certain of the individual cells of this matrix, the specifications of that independent variable parameter that should be fulfilled in order to achieve a desired criterion value. These specifications would then become *design* criteria as mentioned in Chapter 2. Some such criteria, of course, are already available, as in the ambient environmental variables of illumination and temperature (discussed in earlier chapters). But, in the case of at least some of the cells, the current design criteria (if, in fact, any exist) are based on outmoded regulations (such as city building codes), on habit or rules of thumb that have somehow evolved, and on beliefs or hypotheses that may or may not have intrinsic validity.

Some of the illustrative examples discussed below may suggest the way in which research can contribute to the development of such design criteria. However, in viewing the role of the behavioral scientists in this general domain, we should be aware of Sommer's reflections in which he expresses misgivings about such scientists becoming involved in the actual design of buildings [17, p. 157]. Rather, he argues that their greatest potential contribution would be as data gatherers (that is, as researchers). This jibes somewhat with Wells' observation [22] that the most useful contribution that psychologists could make at this stage of knowledge is to describe the psychological implications of a variety of conditions and to educate architects, clients, and others concerned with buildings, about the possible human consequences of their technical decisions.

To complicate the data-gathering process (and the subsequent use of such data in establishing design criteria) Sommer [17] points out the sobering fact that research in somewhat similar circumstances frequently results in substantially different findings; for example, the students in one college might tend to prefer one type of dormitory and those in another college, a different type. To add further complications, since much research in this area relates to human values and opinions, one would of course expect marked differences in the subjective reactions, and the consequent behavior, of different individuals. People differ, for example, in their preferences for degree of privacy, for group interaction at work or elsewhere, for different decor, for use of leisure time, and for ways of using their abode (house, room, penthouse, or igloo). Wells [25] acknowledges this when he observes that there is no reason to expect that a given subjective rating made by one individual on a given occasion should have any *necessary* validity for another individual.

Since we are here dealing primarily with human values and assuming that such values are relevant in their own right, we might well expect that design criteria relating to some features of our living environment should not be standardized for all individuals; rather, they should provide choice (if appropriate) for the fulfillment of individual values. There are also other complications in behavioral research in this area. The behavioral scientists, however, should not veer away from conducting behavioral research in these environmental areas simply because there are rocks in the road. We are already paying a heavy price for many past failures, in the development of buildings and communities, to design them adequately for human use.

ARCHITECTURAL DESIGN

Many architects presumably have been cognizant of the possible behavioral effects of the environments they design; Noble [14], for example, wrote, in part ". . . as architects we help to shape people's future behaviour by the environment we create." But it is only in recent years that psychologists and other behavioral scientists have started to get curious about the implications of architectural design for human behavior—in the home, schools, offices, hospitals, etc. These stirrings of interest have resulted in the coinage of the term *architectural psychology* by Taylor at the University of Utah [in Taylor, Bailey, and Branch, 20, pp. 2-1 to 2-21]. Our British cousins have also started to poke around in this area, such as at the University of Strathclyde [Canter, 4, 5, 6; Wells, 24, 25]. As a starting point in illustrating some of the probing efforts in this area, let us first introduce the concept of personal space.

Personal Space

Both research and observation confirm the notion that individuals have what has been termed *personal space*, that is, some area surrounding the individual, usually with invisible boundaries, within which "intruders" may not come [Sommer, 17, p. 26]. Such space is, in a sense, portable, in that the individual takes it with him wherever he goes. However, it has some of the flexibility of an accordion in that we will tolerate being closer to others in some circumstances (such as in a subway or at a football game) than in other circumstances (such as in a conversation at home or in an office). The amount of space also shrinks or expands depending on whether an invader is a close acquaintance or a stranger, and on the differential status of individuals (for example, a lowly subordinate of a high-ranking executive typically will remain farther away from the executive in the executive's offices than, let us say, an individual with more equal status). The amount of personal space is also a function of cultural background.

There are many manifestations of the existence of personal space, such as in seating oneself in a conversational group; in looking for a seat in a library, on an airplane, or in a theater (when it is usual for one to try to find a seat separated from others); in "keeping one's distance" when walking along a sidewalk; in city gangs that have their own "territory"; and in "staking out one's claim" on territory in a picnic ground. People use various schemes (some rather ingenious) to define their territory, such as glaring at a stranger who sits too close on a park bench (or even getting up and leaving, taking one's portable space with him), or by putting a coat or brief case on a chair beside oneself in a public place, or by building a fence around one's yard.

The matter of status sometimes adds an additional twist, such as in some old people's homes in which some individuals "own" certain chairs in lounge rooms, the ownership being a function of relative status, with the higher-status individuals having preferred territories. We can see the obvious parallel between such human behavior and that of certain animal species in which, for example, individual animals have their own territory, or in which high density results in increased aggression, or in which there is a rank order of status (as the pecking order of chickens) that defines the relative prerogatives of individuals, as at the feed trough.

In all of this, we probably should grant that some individuals exceed normally acceptable bounds in exercising their own prerogatives about their own personal space, or in invading the personal space of others. Aside from such excesses, however, the relevance of this concept to our present concern with living environment probably can be viewed in this frame of reference: Some aspects of architectural design and arrangement of living space do in fact influence (for better or worse) the extent to which the personal space of individuals can be preserved; personal space is a reasonable human value that should be respected; therefore, when relevant, the features of architectural design and the arrangement of living space should provide reasonable opportunity for such space.

Room Usage

Let us now look at some examples of human research studies relating to building environments, perhaps with more of a focus on how people have gone about such research than on the results as such. To begin with, let us consider room usage.

Room usage in private dwellings.   In some housing studies, indexes are used that relate to the spatial adequacy of dwelling units. One of these is the number of *persons per room* (PPR), which is the simple ratio of the total number of occupants divided by the number of rooms. As pointed out by Black [2, p. 58], the upper limits of what are usually considered acceptable PPRs are somewhere around 1.00 or 1.20, with a national average of about 0.69. Another index that is sometimes used is the *square feet per person* (SFPP). As pointed out by Black, there are no United States norms, or standards, for the SFPP, although Chombart de Lauwe [7] has proposed the categories given below; the last column shows the percent in each category resulting from a survey of 121 houses in Salt Lake City [Black, 2, pp. 62–63].

SFPP	Category	Salt Lake City survey, percent
Less than 130	Poor housing	2
131–215	Adequate	21
Over 215	Very good	77

It might be added that, of the home owners surveyed, 7 out of 10 were satisfied with their present houses, and in the case of those who were not, house size had no apparent relation to their dissatisfaction.

Although Black's survey resulted in a correlation of −.77 between the PPR and SFPP values for the 121 houses,[1] he expresses the general opinion that the SFPP is the more discriminating of the two indexes (except in the case of overcrowded houses).

But, aside from deriving these gross indexes of spatial adequacy of the houses in the survey, Black was more concerned with usage of rooms, and in this connection used a questionnaire in which home owners reported the frequency of use of various rooms for such purposes as studying, TV viewing, family activities, and seeking privacy. An example of the responses to three such questions are given below, specifically questions regarding rooms in which reading usually was done.

	Type of reading		
Space	Books, percent	Magazines, percent	Newspapers, percent
Living room	50	62	50
Recreation room	24	25	21
Bedroom	24	12	2
Kitchen-dining area	5	7	32
Other (and "never")	6	3	2

As another example of a survey of room usage, Burnham [3] used a questionnaire to elicit responses of students relating to their use of various areas of a new campus union building. The frequencies of use of certain areas are given below, these values being percentages:

	Frequency of use, percent				
Area	Very often	Often	Sometimes	Seldom	Never
Main lobby	22.4	27.8	22.2	14.0	2.5
Lounge areas (in corridors)	13.8	20.9	33.9	21.3	9.8
Cafeteria	8.8	12.3	22.6	32.4	24.3
Browsing room	8.6	9.2	22.6	26.4	38.3
Den	0.2	1.3	6.9	21.5	69.9

Data collected by such methods can, of course, serve as useful input in the design of related facilities.

[1] The negative correlation is brought about by the fact that PPR values *decrease* with house size, whereas the SFPP values *increase*.

Large versus Small Offices

As another example of the effects of architectural features on people, let us consider the differences between large and small offices.

Office size and social behavior. Wells conducted a sociometric study of 295 office personnel, 214 of whom worked in a large open area, and 81 in surrounding smaller enclosed areas [24]. A sociometric study is one in which individuals are asked to indicate their choices of others to be associated with in some context; in this study each person was asked to indicate the individuals beside whom he (she) would like to work. The results, when analyzed, tend to reveal the affinities of groups of individuals. Some of the data from this study are summarized below, in particular the percents of the choices made by individuals (in open and small areas) of other persons who were in their own section or department.

	Percent of choices to members of own section	Percent of reciprocal choices within own section
Open areas	64	38
Small areas	81	66

These and other related results show that there was a much greater degree of internal cohesion among those members of sections working in the smaller areas, than among those in the larger areas. It should also be added that there were more isolates (individuals not chosen by anyone) in the small areas than in the open areas.

Office size and expressed preference. The greater degree of group cohesiveness of personnel in small areas (rather than open areas) is paralleled by a tendency for office personnel to prefer small areas [Manning, 12, p. 129]. These preferences were elicited [Wells, 22] from the same office personnel who responded to the sociometric questionnaires discussed above, and comprised an additional phase of that research program. The procedures involved, in part, the presentation to the office personnel of illustrations of all possible pairs of five different office arrangements (ten pairs in all). The illustrations were pictures of three-dimensional models of the five layouts, but simplified versions of the layouts are shown in Figure 18-2. By summarizing the numbers of choices for each layout, it was possible to determine the order of preference of the personnel surveyed. Such rank orders were derived separately from the preferences expressed by clerks, by supervisors, and by office managers, all of these being shown in Figure 18-2. We can see that clerks and supervisors ranked the layouts strictly in terms of degree of openness, with the most closed

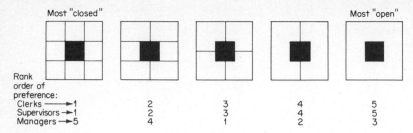

Most "closed"   Most "open"

Rank
order of
preference:
Clerks ——————►1   2   3   4   5
Supervisors ——►1   2   3   4   5
Managers ——►5   4   1   2   3

Figure 18-2   Rank order of preferences expressed for different office layouts by clerks, supervisors, and managers (1 = most preferred; 5 = least
preferred). The center represents the elevator and hall area, and the lines
represent partitions of the office space. [Adapted from Manning, 12, pp.
113, 123, and 126.]

being most preferred and the most open being least preferred. The managers, however, tended to favor the more open layouts than the more
closed layouts, which reflects a difference from the preferences of the
clerks and supervisors.

Responses to associated questions shed some additional light on the
nature of the differences between the clerks and managers. Higher percentages of clerks than of managers, for example, agreed with statements
such as: ". . . larger offices make one feel relatively unimportant" and
". . . there is an uncomfortable feeling of being watched all the time
in a large office." On the other hand, managers tended to discount any
disadvantages of larger (open) offices, and posed a number of advantages
of open planning, such as easier overall supervision and greater economy
and flexibility of space.

In discussing some of these results Wells [24] offers these reflections:
"The internal cohesiveness of the sections in the small areas (closed plan)
meet many of the criteria for small group effectiveness. On the other hand,
the sections in the open areas have better interconnections with other
parts of the department and may, therefore, constitute a more effective
total working unit than the sum of the individual sections."

We should add, however, as an aside, that previous experience usually
has at least a modest effect on the expressed preferences of people; in the
above study it was found, for example, that people who had actually
worked in large offices tended to be somewhat more favorably disposed
toward them than those who had not been so exposed.

Office size and performance.   There seems not to be much hard
evidence available regarding office size as it might relate to work performance, but one straw in the wind comes from a study by Canter [5]
in which office personnel were administered two tests (a clerical and an
analogy test) in offices of various sizes. One person would take the tests
while everyone else worked as usual. Canter found correlations between

room size and test performance of $-.47$ and $-.28$ for the two tests, the test performance being *poorer* in the *larger* office areas. The investigator ruled out any room distraction or personality variables as possible factors, which leaves the strong implication that the performance was influenced by some motivational or attitudinal factors that stem from being in large versus small offices. Granting that tests are, of course, not like usual office activities, the results do at least suggest the possibility of room size affecting actual work performance.

Discussion. It is not up to us, here, in reflecting upon the results of such surveys, to make a pronouncement that offices should be small, or large, or in between. But it is relevant to our interests to note that the use of office space apparently does affect the social behavior of people, that people do have opinions and preferences about such space (most people preferring smaller offices), and that the size of offices may influence work performance. The implications of all of this to architects, managers, etc., are that the manipulation of the various aspects of work space may very well have some effects on the behavior and attitudes of people.

Windows or No Windows?

It is probable that the primary bases for the use of windows have been those of providing light and ventilation. Since these can now be provided artificially, however, the question has arisen as to the value of windows in certain kinds of facilities such as offices and factories. On one side of this fence certain hypotheses have been proposed to support the use of windows, such as the belief that people have a "psychological need" for some "contact with the outside world" [Manning, 13] or for actual daylight.

There are not many data to bring to bear on this general question, but one study by Wells [23] sheds a little light on this topic. He was interested in obtaining opinions from clerical personnel in offices about certain aspects of daylight versus artificial illumination; but since this was done with people in large offices, it was possible to relate the responses to the distances people were from the windows. In one aspect of the survey the people were asked to estimate, for the total illumination at their work stations, the percentage of that illumination that was daylight (from the windows). The means of these estimates are shown in Figure 18-3 in relation to distance from the nearest daylight source, along with the actual (measured) amount. We can see there was a marked overestimation at the farther distances (compared with the actual values, which were quite low at those distances). Replies of 2500 people to two questions included in a prior survey also are interesting in that they reveal the attitudes (perhaps the stereotypes) of people about daylight. One question dealt with opinions on whether it was important to be able to see out of the office even if there were plenty of artificial light; an overwhelming 89 percent felt this to be the case. In reply to a question as to whether it was as

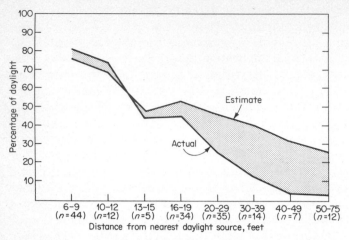

Figure 18-3   Actual percentages of daylight, and estimates of percentages of daylight made by office personnel, at various distances from daylight source. [Adapted from Wells, 23, fig. 3, p. 61.]

"good" for the eyes to work by artificial light as by daylight, the following responses were received:

Daylight better	No difference	Artificial as good as daylight
69%	18%	13%

These responses imply opinions that are generally held, and are not related to the actual distance from windows at which people work.

Discussion.   These results do not provide a firm yes or no regarding the use of windows (as in offices). However, a couple of relevant points can be made. In the first place, one needs to take cognizance of the fact that most people prefer to be able to "see out," which strongly suggests the desirability of windows. However, it seems evident that a subjectively satisfying environment does not critically depend upon the actual amount of daylight available (in fact, people some distance from windows generally overestimated the proportion of such light). This tends to support the concept of deep buildings with windows, granting that some people will be some distance from the windows. But this type of construction (although it is more economical) implies the use of more open space than closed space (which will possibly influence the social interactions of people).

Movement within Buildings

As an illustration of the effects of another architectural feature on the behavior of people, let us consider the movement of people as influenced by the physical arrangement of space. As an example, Whitehead and

Eldars [26] report from a survey that the movement of personnel in a typical hospital operating theater, and into and out of various related areas, occupied 38 percent of the working time, and they estimate that a more optimum arrangement would reduce this time by at least one-fourth. If this time saving could be reflected in reduced numbers of personnel or of hours worked, the salary saving would be over 8 percent.

As a method of analysis of human movements in such situations, they drew upon industrial engineering methods of observing movements, followed by a rearrangement (in new layouts) that would minimize movement time of personnel. As one phase of such an activity analysis, they prepared an *association chart* that shows the number of movements of personnel between the rooms and stations. An example of part of such a chart is shown in Figure 18-4. (This is really a form of sequential link,

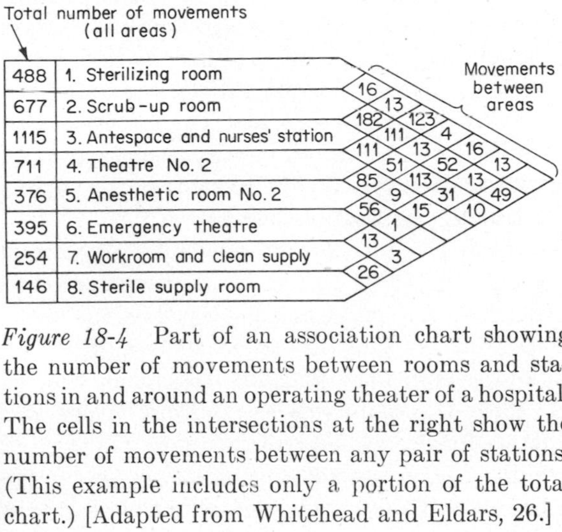

Figure 18-4 Part of an association chart showing the number of movements between rooms and stations in and around an operating theater of a hospital. The cells in the intersections at the right show the number of movements between any pair of stations. (This example includes only a portion of the total chart.) [Adapted from Whitehead and Eldars, 26.]

discussed in Chapter 13.) In this particular case a computer was used to derive a diagrammatic layout that would minimize travel movements.

Mental Health Facilities

There are various types of cues about the possible behavioral effects of the nature of the facilities for mental health patients. For example, Lipman [11] observed that in conventional large dayrooms the patients tended to occupy their "own" chairs in groupings that formed social cliques, with some accompanying hostility between the groups. In smaller rooms, however, the cliques could not be formed as readily since people were in closer proximity to each other and the groups could not separate. On the other hand, in the larger rooms the patients seemed to be more mentally alert and tended to participate more in social interaction, even though some such interaction tended to be more of an aggressive, hostile nature.

In connection with the accouterments of space (as contrasted with space itself), Taylor [19] relates some observations regarding the use of carpets on the floors of mental health centers. In one center, with carpeted floors, the patients tended to interpret their situation as one reflecting the feeling of "society" that they (the patients) were "worthwhile people." And in still another mental health facility (a psychiatric ward) Ittleson [10] was interested in observing the behavior of patients in certain areas. In a solarium, for instance, he found only "isolated" withdrawal types of behavior. The room had little furniture, and no drapes, and was hot even in winter. The addition of drapes and furniture resulted in increased use of the room and, more importantly, use for more constructive activities.

Room Color

There have been speculations about the emotional associations of color and about the effects of color on human reactions. In this connection Birren [1] has made some generalizations regarding the emotional and behavioral effects of different hues, along with suggestions regarding the appropriate hues for particular types of environments. The empirical evidence about the behavioral effects of color in our environment, however, is relatively meager. Its paucity may in part be due to some of the experimental problems involved in tracing the possible effects.

One interesting example of research on color is reported by Srivastava and Peel [18]. The setting of the study was a room in a museum of art at the University of Kansas, with Japanese paintings displayed on the walls. Paid (volunteer) subjects were informed that the experiment was intended to determine their reactions to the paintings (and this was done). But for the first half of the subjects (301) the room was painted light beige and had a corresponding rug, whereas for the second half the room was painted dark brown and had a matching rug. A special sensing device under the rug (a *hodometer*) made it possible to record each square foot of area over which the subjects walked, and a watch was used to measure the time spent in the room. Some of the comparisons of movements of the two groups are given below:

Room	No. of steps	Area covered, ft^2	Time spent, sec
Beige	42.7	9.0	38.1
Brown	46.2	17.9	26.4

It can be seen that the subjects in the brown room took more steps, covered more area (almost twice as much), and spent less time than those in the beige room. Although one might be hard pressed to explain why the differences in color affected the movements of people so markedly, one must conclude that the effect must be the indirect consequence of some subjective reaction to the environmental color. Granting that this study dealt with environmental color in a single situation, the results tend to support the suspicions that color can, under some circumstances, influence human reactions and behavior.

Psychological Dimensions of Buildings

As people perceive virtually any stimulus object (people, institutions, news media, and, in our present context, rooms and buildings), it is their tendency to have reactions in each of several perceived *dimensions*. For example, we might "peg" individuals at positions along different dimensions such as "trustworthiness," or "friendliness." So, we might have reactions to any room or building along each of various subjective dimensions. The identification of such relatively independent dimensions is a bit tricky, but one procedure that has been used is the *semantic differential* [Osgood et al., 15]. This method is based on judgments made by individuals along scales for each of several, or many, pairs of opposing adjectives, such as pleasant-unpleasant, and the subsequent identification of groups of such scales for which the responses tend to be correlated.

Using this approach, Canter [6] has identified about 10 different dimensions of the reactions of people to buildings, the 4 primary ones being pleasantness, comfort, friendliness, and coherence (this dimension being associated with words such as stable, tidy, and harmonious). Such a scheme makes it possible to *quantify* human reactions to buildings and rooms (including their furniture and decor), and this, in turn, could lead toward a couple of other objectives. In the first place, one could systematically study the relationship between characteristics of rooms or buildings and the reactions of people to them in terms of such dimensions, to answer such questions as, What features of rooms are conducive to their being perceived as *pleasant?* or *comfortable?* This approach was followed by Canter [6], albeit with descriptions of rooms in terms of lighting, heating, furniture, and noise level rather than with real rooms or pictures of them. The relative importance of these four room features to the judged pleasantness and comfortableness of the rooms (as described) are given below:

Dimension	Relative importance, percent				
	Lighting	Heating	Furniture	Noise	Other
Pleasant	29	27	21	6	17
Comfort	29	41	21	0.3	3

The second possible objective could be that of determining the personality characteristics of people as related to their own preferences for various *types* of rooms or buildings.

Discussion

These have been something of a sample of the behavioral and attitudinal effects of types of building space and related accouterments. Although the results of studies such as the ones cited probably cannot be considered to provide very much of a substantive base for specifying design criteria, they may at least point up the potential effects of architectural design upon human beings and suggest approaches to the further study of such effects.

THE NATURE OF COMMUNITIES

The burgeoning problems of urban centers undoubedly comprise one of the major challenges of current life. The many facets of these problems leave few inhabitants unscathed. A partial inventory would include problems associated with health, recreation, mobility, segregation, education, congestion, physical housing, crime, and loss of individuality; one could certainly add to these. The current manifestation of these problems lends some validity to the forebodings of Ralph Waldo Emerson and Henry Thoreau, who viewed with deep misgivings the encroachment of civilization on human life, especially in the form of large population centers. The tremendous population growth, however, makes it inevitable that many people must live in close proximity to others (and thus the existence of urban centers); accepting this inevitability, however, it is proposed that one should operate on the hypothesis that, by proper *design*, urban centers *can* be created which might make it possible to achieve the fulfillment of a wide spectrum of reasonable human values—perhaps even those that Emerson and Thoreau, and maybe those that you and I, might esteem.

Following the discussion at the beginning of this chapter, we shall here try to view this problem in something of a human factors frame of reference, with the particular intent of trying to see how different community characteristics (i.e., independent variables) influence human behaviors and reactions (i.e., dependent variables, or criteria). For this purpose, the basic interrelationships of Table 18-1 might be relevant.

Criteria

For city design, Lynch [in Holland, 9, pp. 120–171] proposes what he refers to as desirable formal goals or performance characteristics of cities, as follows: (1) accessibility, (2) adequacy of facilities (houses, schools, recreation, etc.), (3) congruence (the fit of the system, the coordination of the parts in operation), (4) diversity, (5) adaptability, (6) legibility (a term something akin to the "image" of a city), (7) safety, (8) stress (achieving a balance in physiological and psychological stress, neither too much nor too little); and (9) efficiency. One can recognize most of these as variants of the criteria given in Table 18-1.

Urban Patterns

Whether by intentional design or (more frequently) by fortuitous development, urban areas have some *pattern*. Forgetting for the moment the reality of existing patterns (and the almost insurmountable problem of major overhaul), let us consider the implications of some patterns for the nature of human life.

Aspects of urban patterns. The patterns of urban areas are formed by many elements, such as the physical structures, the facilities for the circulation of people (roads and transit systems), and the fixed facilities that draw upon or serve the population (stores, factories, hospitals, parks, theaters, (etc.). With reference to these elements of spatial pattern, Lynch [in Rodwin, 16, chap. 6] points out that the most significant features of such patterns are the *grain* (the degree of "intimacy" with which houses,

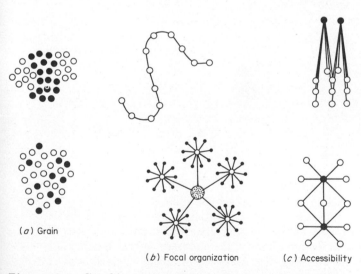

(*a*) Grain

(*b*) Focal organization   (*c*) Accessibility

Figure 18-5 Graphic representation of three aspects of the spatial patterns of urban areas, namely, grain, focal organization, and accessibility; see text for description. [From Lynch in Rodwin, 16, p. 106; reproduced with the permission of the publisher, George Braziller, Inc. Copyright © 1960 by the American Academy of Arts and Sciences.]

stores, and other facilities are intermixed), the *focal organization* (the interrelation of the "nodes" of concentration and interchange contrasted with the general background), and the *accessibility* (the general proximity, in terms of time, of all points in a region to a given kind of activity or facility). All three of these aspects of spatial pattern can be depicted graphically, as illustrated in Figure 18-5.

Variations in patterns.   Given these three parameters of the patterns of urban areas, any given area can be depicted graphically in such a manner as to reflect these features. And, in general terms, there are different types of patterns that can be identified (in the case of existing urban areas) or conceptualized (for planning). Lynch [in Rodwin, 16] presents five such variations, which are illustrated graphically in Figure 18-6.

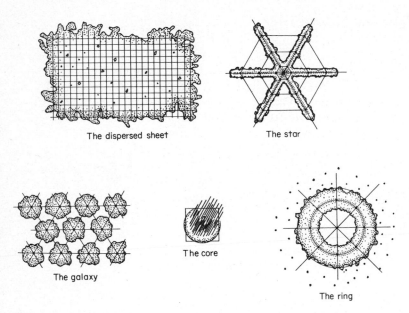

Figure 18-6   General representations of five different spatial patterns of urban areas, reflecting grain, focal organization, and accessibility. Each pattern has different effects on the nature and quality of the living of the inhabitants. [Adapted from Lynch, in Rodwin, 16, figs. 4–8; reproduced with the permission of the publisher, George Braziller, Inc. Copyright © 1960 by the American Academy of Arts and Sciences.]

The *dispersed sheet* is characterized by low densities with substantial interstices of open land kept in reserve, with dispersion of virtually all activities over a very broad area; there would be no outstanding nodal points or terminals. The *galaxy* consists of numerous small community units, each with an internal peak of density and each separated from the next by a zone of low or zero structural density. The *core city* consists of a highly compact community, literally three-dimensional. The *urban star* is formed of a dominant core with tongues of land radiating from the center, these having moderate density levels with secondary centers distributed along the main radials and transportation systems along the radials and between them. The *ring* is like a doughnut, the "hole" having low density, with a surrounding high-density ring, with nodes of concentration for various community activities and functions; channels of flow systems would be series of annular rings plus feeder radial lines.

Effects of urban spatial patterns. The central point that we are leading up to is that differences in such urban patterns can have direct effects on the nature and quality of the living of people in urban areas of such patterns. The total spectrum of these effects probably is not now known. However, certain effects probably can be inferred from the representation of such structures, although we need to recognize such inferences as being substantially subjective. In presenting these particular patterns, Lynch suggests some such possible effects, as summarized below:

· *Dispersed sheet:* Convenient circulation of individual vehicles, dispersion of traffic loads, good accessibility, personal comfort, encouragement of local participation, negation of metropolitan character.

· *Galaxy:* Providing centers of activity, sharpened image of local community, accessibility to open country, concentration of city-wide activities, possible monotonous similarity among communities but otherwise effects generally similar to dispersed sheet.

· *Core city:* Dependence upon public transport, good accessibility, effects of high-density housing, spontaneous communication, high but possibly restricted privacy, possible discomfort from noise and poor climate, restriction of choice of habitat, strong sense of community identification, possible need for second (weekend) house in countryside.

· *Urban star:* Maintenance of metropolitan character with central-city facilities, opportunity for choice of habitat, varied access to services depending on individual location, possible transportation congestion along radials, and possible transportation problems in circumferential movements.

· *Ring:* High accessibility to services and to open land, wide choice of habitat, variety and strong character of specialized centers, identity with individual centers, possible problem in controlling growth and maintaining open spaces.

Discussion

We can see that certain criteria probably are intricately influenced (one way or the other) by the spatial pattern of the urban area, such as relative accessibility to various services, the type(s) of habitat available, availability of open spaces, identity with community, metropolitan character, etc. The patterns differ in the degree to which these and other criteria might be fulfilled. However, other kinds of criteria probably are unrelated to the urban pattern and depend more on other features of the community. Although we may not now have very many guidelines to follow, it would seem that the community should provide for the fulfillment of the several criteria mentioned earlier, such as physical welfare and aesthetic values, but in particular the opportunity for choice in terms of social interchange and privacy and in terms of goods, services, facilities, habitat, recreation, entertainment, culture, etc. In all of this there must, of course, be compromises, but insofar as possible the nature of the compromises also should be a matter of personal choice.

As to mankind's ever striving for some undefined goal of satisfaction in life, we probably cannot now say very much about the extent to which man's progress toward that goal is influenced by the features of the community that he *can* control. This, of course, specifies the direction of (we hope) the behavioral research that would provide the basis for such guidelines.

MOBILITY

As implied in the above discussion, a dominant aspect of living environment is its facilities for movement of people either by public transportation system or private vehicles. At various stages of development (from ideas, to drawing boards, prototypes, and experimental versions, to actual operation) is an assortment of new transportation systems and equipment including steam and electric cars, minicars, automated highways, rapid-transit systems (even one with passenger capsules), high-speed railroads, vertical takeoff and landing aircraft (VTOLs), short-takeoff and landing aircraft (STOLs), hydrofoil boats, air cushion vehicles (ACVs), amphibious cars, and more.

The technology for these and other new ways of getting from here to there is now available or on the threshold. The implementation of such technology consists in the combination of development plus the horrendous financial, political, land-procurement, and administrative aspects of such systems (in some cases by major changes of existing communities).

Human Requirements for Mobility Systems

It is not for us here to explore the human factors aspects of all these systems, but it may be useful for our present purpose to speculate about certain basic human factors requirements of such systems. The dominant requirements seem to be convenience (being able to go where one wants to go when he wants to go), speed, physical safety, and comfort. For convenience and speed of mobility, it would seem reasonable to specify something of a door-to-door feasibility. In broad terms this implies the need for individual (personal) vehicles and for public transportation facilities.

Two examples. Many notions are being bounced around regarding the development of transportation systems to fulfill these two requirements, but for illustration, let us toss in some of the concepts that have evolved in connection with Project Metran [Hanson, 8]. The ideas in this project were generated by students and staff as a group effort in a systems engineering course at the Massachusetts Institute of Technology, the focus being to develop ideas for an integrated transportation system for the Boston metropolitan area. Although the project was essentially a class project (and not a real system design), it may have some useful implications for real systems. The complete integrated system used four components, namely, automobiles (because of their flexibility), a BOS (bus only street) system, the PERC (signifying PERsonalized Capsule), and an automated guideway. For our purpose the last two are of particular interest.

The PERC system was designed to provide convenient transportation for highly dense core areas, and would consist of two-man capsules operating along a track, in some places at grade level (with limited access) and in other places at second-floor level (either on a free-standing elevated track or attached to the facade of structures along the street). Frequent, readily accessible stations would make it possible for passengers to get on and off at convenient locations. A sketch of this system is shown in Figure 18-7.

Figure 18-7 Sketch of the PERC system (PERsonalized Capsule) proposed for personal transportation in high-density center-city areas. It could operate at ground or second-floor level, in some areas against the facade of buildings. [From Hanson, 8, fig. 10-13.]

   The second component of particular interest in this proposed system is an automated guideway. This system would be designed so that individual vehicles could operate on regular streets and highways and also move onto an automated guideway with automatic (computerized) control of vehicles. A sketch of this concept is shown in Figure 18-8. It is

Figure 18-8 Sketch of the entrance-way of a possible automated-guideway transportation system, with vehicles that could be operated in the conventional manner or moved onto the guideway where they could be moved by automatic control. [From Hanson, 8, fig. 2-12.]

estimated that the guideway capacity, for travel at 60 mph, would be about 21,000 vehicles per hour, as contrasted with 12,600 with 10-ft headways, or 7050 with much safer 30-ft headways.

DISCUSSION

The problems that would be involved in a rather thorough rehabilitation of many urban areas could easily cause one to throw up his hands in utter despair, especially because of the economic and political aspects. In this connection it is the contention of many knowledgeable people that the magnitude of the problem would require a major national effort [Myrdal, Dyckman, and others, in Warner, 21].

   But aside from the rehabilitation of parts of existing communities, every day additional elements of the fabric of communities are being constructed, such as new houses, schools, public and private buildings, streets, and highways. Probably a substantial portion of such construction perpetuates the errors of the past and will only amplify the total problem of developing communities that would be fairly optimum in the human values that community living should fulfill. But as one looks at either side of the coin—rehabilitation or further development—it probably must be said that our available guidelines for achieving our goals are still fragmentary. The challenge to the behavioral scientists is to carry out the behavioral research that will help to fill in the missing guidelines.

REFERENCES

1. Birren, F.: *Color psychology and color therapy*, University Books, Inc., New Hyde Park, N.Y., 1965.
2. Black, J. C.: *Uses made of spaces in owner-occupied houses*, Ph.D. thesis, University of Utah, Salt Lake City, April, 1968.
3. Burnham, T. L.: *Behavioral criteria in an analysis of the University of Utah Union Building*, M.S. thesis, University of Utah, Salt Lake City, June, 1968.
4. Canter, D.: The psychological study of office size: an example of psychological research in architecture, *The Architects' Journal*, March, 1968.
5. Canter, D.: Office size: an example of psychological research in architecture, *The Architects' Journal*, Apr. 24, 1968, SfB (92):Aa3:UDC 725.23:301.151, pp. 881–888.
6. Canter, D.: *The study of meaning in architecture*, Building Performance Research Unit, University of Strathclyde, Glasgow, Scotland, GD/16/DC/A, Apr. 25, 1968.
7. Chombart de Lauwe, P.: The sociology of housing methods and prospects of research, *International Journal of Comparative Sociology*, March, 1961, vol. 2, no. 1, pp. 23–41.
8. Hanson, M. E. (ed.): *Project Metran: an integrated, evolutionary transportation system for urban areas*, The M.I.T. Press, Cambridge, Mass., 1966.
9. Holland, L. B. (ed.): *Who designs America?* Anchor Books, Doubleday & Company, Inc., Garden City, N.Y., 1965.
10. Ittleson, W. H.: "Environmental psychology of the psychiatric ward," in C. W. Taylor, R. Bailey, and C. H. W. Branch (eds.), *The second national conference on architectural psychology, May 26–28, 1966, Park City, Utah*, University of Utah, Salt Lake City, September, 1967.
11. Lipman, A.: Building design and social interaction, *The Architects' Journal*, Jan. 3, 1968, SfB (94):Aa3:UDC 725.56:301, pp. 23–30.
12. Manning, P. (ed.): *Office design: a study of environment*, Pilkington Research Unit, Department of Building Science, University of Liverpool, Liverpool, England, SfB (92):UDC 725.23, 1965.
13. Manning, P.: Windows, environment and people, *Interbuild/Arena*, October, 1967.
14. Noble, J.: The how and why of behavior: social psychology for the architect, *The Architects' Journal*, Mar. 6, 1963, pp. 531–546.
15. Osgood, C. E., G. J. Suci, and P. H. Tannenbaum: *Measurement of meaning*, The University of Illinois Press, Urbana, 1957.
16. Rodwin, L. (ed.): *The future metropolis*, George Braziller, Inc., New York, 1960.
17. Sommer, R.: *Personal space: the behavioral basis of design*, Prentice-Hall, Inc., Englewood Cliffs, N.J., 1969.
18. Srivastava, R. K., and T. S. Peel: *Human movement as a function of color stimulation*, Environmental Research Foundation, Topeka, Kan., April, 1968.
19. Taylor, C. W.: *Architectural psychology: a pioneering program*, mimeographed paper, Department of Psychology, University of Utah, Salt Lake City, 1968.

20. Taylor, C. W., R. Bailey, and C. H. W. Branch (eds.): *The second national conference on architectural psychology*, *May 26–28, 1966, Park City, Utah*, University of Utah, Salt Lake City, September, 1967.

21. Warner, S. B., Jr. (ed.): *Planning for a nation of cities*, The M.I.T. Press, Cambridge, Mass., 1966.

22. Wells, B. W. P.: A psychological study with office design implications, *The Architects' Journal*, Oct. 14, 1964, SfB (92):Ba 4:UDC 725.23, pp. 877–882.

23. Wells, B. W. P.: Subjective responses to the lighting installation in a modern office building and their design implications, *Building Science*, 1965, vol. 1, SfB:Ab 7:UDC 628.9777, pp. 57–68.

24. Wells, B. W. P.: The psycho-social influence of building environment: sociometric findings in large and small office spaces, *Building Science*, 1965, vol. 1, SfB (92):UDC 301.151, pp. 153–165.

25. Wells, B. W. P.: Towards a definition of environmental studies: a psychologist's contribution, *The Architects' Journal*, Sept. 22, 1965, SfB:Ab1: UDC 61, pp. 677–683.

26. Whitehead, B., and M. Z. Eldars: An approach to the optimum layout of single-storey buildings, *The Architects' Journal*, June 17, 1964, SfB:Ba4: UDC 721.011, pp. 1373–1380.

part
seven
OVERVIEW

As an identifiable part of the process of developing systems and devices, human factors engineering is, of course, still relatively young, stemming from about the 1940s. As in the case of people, cheese, and wine, however, it has indeed matured with the years.

HUMAN FACTORS PHILOSOPHIES

The rather brief history of human factors in the design and development of systems has been characterized by certain shiftings in points of view. Van Cott and Altman [10], for example, refer to three such changes, starting from a philosophy of machine-oriented design (essentially a pre-human factors philosophy), with primary emphasis on the machine and the individual being adapted to the machine through selection and training. This philosophy was somewhat replaced by a man-oriented design point of view, with the reverse twist of the adaptation of the machine to the man. A third philosophy (and the one that is probably most current) is more system oriented; its emphasis is on designing a system in which there is something of an optimum integration of men with machine components, toward the end of the most effective achievement of the system objectives. This concept might be thought of as one in which the relative "talents" of both the physical components and of human beings are utilized most effectively, but the primary control of the system is essentially the responsibility of the human beings in it. More recently, however, there has emerged an inkling of a new philosophy that is focused more on basic human values and the consideration of individuals in the creation of the systems, facilities, and environments [Christensen, 2; Lyman, 6].

But man, in a sense, is his own worst enemy. Equipped with the knowledge and skills resulting from the scientific and technological achievements, he can modify much of his environment. Granting that such modifications have often contributed to the fulfillment of human values, the unhappy fact remains that in many instances these achievements have in part been to his own detriment, to wit, air and water pollution, urban congestion, highway injuries and deaths, the loss of "identity" (as reflected in a minor way by individuals becoming numbers in punch cards), and the increased dependence upon *being* entertained by others (such as by TV) as contrasted with dependence on oneself for overt, active recreational and cultural enrichment. These and other negative by-products of our scientific and technological achievements offer challenges to our society in general, but they should have special relevance to those concerned directly or indirectly with the human factors aspects of systems, facilities, and environments of our civilization. It is not enough to make pronouncements in favor of human values and individuality; one must actively seek to promote the creation of the elements of our total environment (including the many and varied physical accouterments of it) that are conducive to the fulfillment of such human values and to the preservation of individuality. Christensen, in harping on this theme [2], reinforces his point with some of the reflections of Gardner [5], as follows: "Too often in the past we have designed systems to meet all kinds of exacting requirements except the requirement that they contribute to the fulfillment and growth of the participants"; and "We must learn to make technology serve man not only in the end product *but in the doing.*" As implied by Christensen, this concern for humans does not mean that we cannot take advantage of technology, but rather that we should adapt technology toward the fulfillment of individual human values. As a couple of specific examples he refers to the use of adaptive control systems and adaptive teaching machines that do take intelligent account of individual differences among the users. And, in a broader frame of reference, we certainly have the technology to minimize air and water pollution, to develop improved transportation systems, to build better housing facilities, to create better working conditions, to improve health conditions and medical services, and to provide prosthetic devices for the physically handicapped and better communication devices for the blind and deaf. The stumbling blocks in the way of these and other objectives include economic considerations (which we sometimes cannot do anything about) and our collective motivation (which we can do something about).

It is, of course, much easier to attack the specific human factors aspects of a washing machine, an automobile, a computer, a jackhammer, or even an airline-reservation system, than to attack the broad and ill-defined spectrum of human problems arising from the involvement of people in everyday living. But it would seem that these are human factors problems in a broad sense, and that the various disciplines that have to do with some segment of human factors need to be addressing themselves to such problems as well as to the human factors features of the more hardware-oriented types of equipment and facilities that comprise their current bill of fare.

HUMAN FACTORS ASPECTS OF SYSTEM DEVELOPMENT

Some of the basic processes in the human factors aspects of system development were discussed in Chapter 2. It might be useful, here, however, to discuss further a few of these aspects. (We should point out, however, that most of these would be relevant only to certain types of systems, not across the board, and in some circumstances would take on quite a different hue.)

Consideration of System Interactions

The process of system design generally requires sequences of decisions, each one having some subsequent effect on other decisions. It is sometimes desirable to be able to trace these effects in a systematic manner, so the right hand knows what the left hand is doing.

System analysis and integration model (SAIM). One procedure that has been proposed for this purpose is the System Analysis and Integration Model, SAIM [Shapero and Bates, 9]. While the total analysis procedures are too extensive to elaborate here, its central rationale is predicated on a model in which the *effects* of various system *elements* are identified. In this formulation, three different types of elements are characterized: system *determinants* (elements that determine the nature, form, and limits of a system, such as its mission, performance requirements, inputs, and constraint), system *components* (mechanisms, men, facilities, etc., that are internal to the system proper), and system *integrators* (operational sequences, communications, organization, dual decision structure). Actually, the analytical method used in the SAIM includes the use of a checkerboard matrix of all combinations of all elements with provision for identifying interacting elements; this includes the elements that *are affected by* any given element (and in what way) and, in turn, those elements that *affect* any given element.

Of the interaction effects, those of machine-component elements upon human elements influence in particular the decision processes in designing or selecting machine-component elements. The following (hypothetical) example illustrates the sequential decisions that might be involved in deciding how to present some item of information; for each one, the *affecting element* is indicated (this provides some form of *machine output*, which, in turn, affects the *human input*) [adapted from Shapero and Bates, 9].

· *Decision 1:* It is decided to have an operator receive information from a console. Affecting element: display (all kinds are available at this point, visual, audio, tactile, proprioceptive).

- *Decision 2:* It is next decided that the information will be presented visually. Affecting element: visual display (possible types are label, indicator, light, scope).
- *Decision 3:* It is next decided that the visual display will be a light. Affecting element: visual light display.
- *Decision 4:* Light display is completely specified. Affecting element: visual light display specified in terms of color, brightness, contrast, configuration, temporal characteristics, etc.

At each decision point, one would, of course, have to take into account any pertinent data, such as the relative advantages of different sensory modalities and different types of displays.

Function and Task Analysis

In fairly complex systems, the decisions made during the successive stages of design and development have extensive ramifications for the human functions and tasks that will be required in the operation and maintenance of the final system. In such cases some systematic procedure may be in order to develop and maintain current information regarding the functions and tasks which (at any given stage) are tentatively implied. Among the purposes of such analyses are the following: identifying the functions or tasks that individually or in combination are incompatible with human abilities; the development of training programs for personnel who will later be involved with the system, including the development of training materials and training aids; and personnel procurement and associated manpower planning.

While there is considerable variation in techniques and forms used for function and task analysis, they typically result in an organized presentation of the tasks that are to be carried out in the use or maintenance of the system. Depending on the particular method being used and the system in question, supplementary information related to the tasks may also be provided. Although there are variations among the methods and the specific information provided for, a basic process is common to all, namely, that of inferring, from the design of the equipment, what human tasks will be required when the system is completed; this is a process of *predicting* the nature of future tasks. Some such inferences are quite evident; others are less so. It should also be added, however, that some tasks are not necessarily implied by the equipment characteristics, since some tasks come into being by reason of other considerations, such as decisions regarding safety practices and philosophies regarding maintenance procedures.

As an illustration, an excerpt from a task analysis of a missile system is given below; in actual practice such data would be displayed in appropriate columns or forms provided for such purposes [adapted from Rabideau, Cooper, and Bates, 8]:

- *Function:* Activate test instruments.
- *Human functional element:* Provide corrective inputs and monitoring when self-test yields a "no go."
- *Task:* Isolate and replace malfunctioning analyzer module.
- *Subtasks:*
  (*a*) Check code displayed in malfunctioning window against modular code table.
  (*b*) Remove module from analyzer and take to storage.
  (*c*) Obtain replacement module from storage, take to analyzer, and insert it.
  Human output:
  (1) Obtain module from bin.
  (2) Carry module to analyzer.
  (3) Insert module
  (4) Close module latches.
  Decision: Decide which bin replacement module is stored in.

Since the above excerpts are removed from their context, their contextual meaningfulness is limited. (The complete example would include several functions and functional elements, several tasks, numerous subtasks, and the elaboration of subtask analysis, as illustrated above for only the last subtask.) Nonetheless, this may illustrate the general nature of such procedures. Subsequent analyses provide for determining (through estimates) the time required for various tasks and subtasks; the extent to which such tasks would occur sequentially, simultaneously, or otherwise; the difficulty of the tasks; and ultimately the various personnel positions (jobs) that should be planned for in the completed system.

Another task analysis scheme, while differing in details, provides a format (actually a form with the headings given below) as follows [Gael and Stackfleth, 4]:

Task number	Performance Criteria
Task name	Time
Indicator (usually some type of display)	Accuracy
Response (behavior)	Location of task
Feedback	Criticality
Incidence	Newness
Time	
Frequency	

In at least the U.S. Air Force, plans provide for keeping this task information up to date, and changes are made as they are brought about by equipment modifications or other considerations. This program has become known as the QQPRI (quantitative and qualitative personnel requirements information).

Operational Procedures

The use of most systems involves certain procedures. These can consist of formally established policies, rules, and specified routines established by the organization, or they may be informal practices, or ways of doing things, that have evolved in the organization. For some systems these procedures are established during the process of development of the system, especially when personnel are to be trained during the development. In fact, in the case of some systems (especially in the military services) the specific procedures to be used can influence the actual system design. For example, the design of a system might be influenced by the maintenance "philosophy" that would ultimately be adopted—whether, for example, maintenance of components would be done on the spot or by the replacement of components (the "black-box" concept).

Job Design

Task analysis in effect characterizes what work activities are (or will be) performed. Within reasonable limits, however, those work activities that are to be performed can be combined in different ways, each such combination comprising a separate job (or possible job). The formation of jobs for systems that are under development requires the use of estimates of various kinds, such as the time that would be required to perform tasks and the frequency with which certain functions will have to be performed. It should be noted that information about the time required for different tasks, when added together, gives some impression of the number of individuals who would be required to perform the activities in question, and it can be used for developing quantitative manpower requirements.

Where the circumstances suggest a systematic and analytical approach to the organization of activities into jobs, certain types of data and specific procedures can facilitate the process.[1] Any specific procedures, however, should be predicated on guiding principles. In this connection, certain guidelines have been set forth to assist in making decisions relative to job design. It should be kept in mind that these would not apply to every system and should therefore be used selectively. These principles are given below [Folley, 3]:

1. Principle of supervisory structure. Tasks requiring responsibility or
    authority for assigning personnel, for distributing work, for evaluating

[1] For a discussion of this the reader is referred to Folley [3, chap. 9] or *Personnel Subsystems* [11].

work, and for making decisions in emergencies should be assigned to the appropriate level of supervisory jobs.

2. Principle of time utilization. Tasks assigned to a given job should keep the man busy for a typical work period without excessive repetitions.

3. Principle of knowledge relationship. The tasks assigned to one job should require similar and related background knowledge and information.

4. Principle of job progression. Each job should be related to a group of other jobs where the primary difference is that of level of advancement, so that a job progression is possible from relatively routine jobs to jobs with substantial responsibility and authority involving similar tasks or types of tasks.

5. Principle of location relationship. The tasks assigned to one job should all be performed in the same job location.

6. Principle of function relationship. A series of tasks closely related in function and in time should be assigned to the same job.

7. Principle of aptitude relationship. The tasks assigned to one job should require the same or similar aptitudes.

8. Principle of equipment relationship. The tasks assigned to one job should usually all relate to the same equipment or subsystem.

Obviously there would be some conflicts among these principles. Where conflicts do occur, the ones earlier in the list usually should take precedence.

Selection and Training

Where a job exists or has been designed for the future, the selection of personnel becomes an early item on the agenda if the users are to be selected (as opposed to being self-selected). The sequel is of course training of the individuals who are to use the system. While the principles and procedures involved in these processes will not be discussed here, there are certain facets of these processes that should be mentioned briefly.

Relationship to system development.   The processes of selection and training should not be viewed as separate, independent chores that have to be carried out after the physical system is in being. Rather, they should preferably be considered as part and parcel of the total system; in fact, their consideration early in the game can have a bearing upon the system development process. The system should be developed with a view toward its use by the *kinds* of people who will use it (machine operators, soldiers, housewives, farmers, engineers, doctors, or bird watchers). It also should be developed with an eye to the type of training or instructions that would likely be provided; where the training might have to consist of, say, sets of instructions that accompany a consumer item, the product design might appropriately have to be modified more (to make it more foolproof) than if the opportunity should exist for more organized training. Another possible influence of training considerations upon design is in the possible provisions for feedback information. Since feedback (knowledge of the consequences of one's actions) usually contributes to learning, a system that provides for feedback (through its design features) probably would facilitate the learning process.

Personnel Subsystem

We can see that, for at least some systems, one has to start worrying about *personnel* affairs during most of the design and development process, such affairs including task and function analysis, job design, operating procedures, personnel selection, training, the development of training aids, and manpower planning. Undoubtedly the most comprehensive program of this type is that of the U.S. Air Force [11] which is referred to as the *personnel subsystem* (PSS). (In fact, the term is also being used somewhat outside the Air Force.) As developed by the Air Force, this program embodies a systematic set of procedures and guidelines to be followed during the development of at least major systems, for which trained personnel must be available at the time the final system is produced. Although many (perhaps most) systems do not require an elaborate personnel subsystem in this sense, there may be somewhat corresponding functions that require attention during system development for the benefit of the ultimate user of the system. Such functions might include actual training of people (such as mechanics for a new model of automobile), and more likely the preparation of instructional manuals for self-training of users (such as how the householder can sharpen the blade of his new lawnmower).

Human Factors Evaluation

Evaluation in the context of system development has been defined as the measurement of system-development products (hardware, procedures, and personnel) to verify that they will do what they are supposed to do [Meister and Rabideau, 7, p. 13], and, in turn, human factors evaluation is the examination of these products to ensure the adequacy of attributes that have implications for human performance. Actually, almost every decision during the design of a system includes some evaluation, such as deciding whether to use a visual signal or an auditory signal. Although many such evaluative decisions need to be made as part of the on-going development cycle, for most systems certain systematic procedures should be followed to evaluate the system-development products (hardware, procedures, and personnel). Our interests are, of course, focused on the human factors evaluation. We shall mention at least a few such evaluation processes.[2]

Experimental procedures. The testing of a system or component (either actual or simulated) is essentially an experiment and therefore requires the use of procedures in which appropriate experimental practices should be followed. While it is not feasible here to describe or discuss such practices, various texts deal with this topic.[3] While conventional experiments involve the manipulation of experimental conditions (and while some tests also involve such manipulation), in other test situations there is no variation in such conditions; in such cases the purpose may be to evaluate performance under a single specified condition, usually to compare with some predetermined performance requirement standard. It should be pointed out, however, that in any event some measure of performance (a criterion) is essential.

Test conditions. The conditions under which the evaluation test is carried out should usually simulate as closely as possible those in which the system is ultimately to be used. This includes physical environment, procedures, numbers of personnel, and where possible, even conditions of stress.

Subjects. The subjects used in the tests should be the same types of individuals that are expected to use the system, taking into account considerations of aptitude and training.

Adequate number of repetitions. As in any experiment, repeated observations, or trials, are required in order to give a reliable indication of performance. Where feasible, there should be enough replications to provide the basis for appropriate tests of statistical significance.

HUMAN FACTORS DESIGN CONSIDERATIONS

It is probably fairly obvious that despite the heavy reliance of human factors engineering on research data, the application of human factors data to system-development processes does not (at least yet) lend itself to the formulation of a completely routine, objective set of procedures and solutions. However, a systematic consideration of the human factors aspects of a system usually would at least focus attention on features that should be designed with human beings in mind. In this connection, therefore, it might be useful to characterize at least some of the considerations that are in order in approaching a system design problem. These

[2] For a thorough treatment of human factors evaluation in system development, the reader is referred to Meister and Rabideau [7].
[3] A particularly useful source is Chapanis [1]. There are also several experimental psychology texts available.

considerations will be presented in the form of a series of questions (with occasional supplementary comments). These are intended as possible reminders of desirable objectives in the development of man-machine systems. The implementation of some of these in specific problems would require the application of appropriate information or principles, such as presented in earlier chapters.

Two or three points should be made about the implication of these considerations. In the first place, some of these would not be pertinent in some systems; in turn, this is not intended as an all-inclusive assortment of considerations. Further (and as indicated frequently before), the fulfillment of one objective may of necessity be at the cost of another. Perhaps in a general sense, then, these may serve to summarize some of the human factors considerations that should be taken into account in system development.

1. What are the functions that need to be carried out to fulfill the system objective?
2. If there are any reasonable options available, which of these should be performed by human beings?
3. For a given function, what information external to the individual is required? Of such information, what information can be adequately received directly from the environment, and what information should be presented through the use of displays?
4. For information to be presented by displays, what sensory modality should be used? Consideration should be given to the relative advantages and disadvantages of the various sensory modalities for receiving the type of information in question.
5. For any given type of information, what type of display should be used? The display generally should provide the information when and where it is needed, in a manner that will ensure reception. These considerations may take into account the general type of display, the stimulus dimension and codes to be used, and the specific features of the display. The display should provide for adequate sensory discrimination of the minimum differences that are required.
6. Are the various visual displays arranged for optimum use?
7. Are the information inputs collectively within the reasonable bounds of human information-receiving capacities?
8. Do the various information sources avoid excessive time sharing? If two or more displays would require very much time sharing, information reception might be degraded.
9. Are the decision-making and adaptive abilities of human beings appropriately utilized? Although human decisions usually are more rapid and decisive when the decisions under specified conditions are

predetermined, such decisions sometimes can be (and preferably should be) programmed for automatic execution by the system. When such automatic programming of decisions is not possible, the flexible, adaptive decision-making abilities of human beings should be utilized to the degree that is commensurate with system objectives.

10. Are the decisions to be made at any given time within the reasonable capability limits of human beings?

11. Granting that aspects of some systems will be automated, is the basic *control* of the system that of the individual?

12. When physical control is to be exercised by an individual, what type of control device should be used?

13. Is each control device easily identifiable?

14. Is the operation of each control device compatible with any corresponding display, and with common human response tendencies?

15. Are the operational requirements of any given control (as well as of the controls generally) within reasonable bounds? The requirements for force, speed, precision, etc., should be within limits of virtually all persons who are to use the system. The man-machine dynamics should so capitalize on human abilities that, in operation, the devices meet the specified system requirements.

16. Are the control devices arranged conveniently and for reasonably optimum use?

17. If there is a communication network, will the communication flow avoid overburdening the individuals involved?

18. Are the various tasks to be done grouped appropriately into *jobs*? Each job should consist of a compatible combination of tasks.

19. Do the tasks which require time sharing avoid overburdening any individual or the system? Particular attention needs to be given to the possibility of overburdening in emergencies.

20. Is there provision for adequate redundancy in the system, especially of critical functions? Redundancy can be provided in the form of backup or parallel components (either men or machines).

21. Are the jobs of such a nature that the personnel to perform them can be trained to do them?

22. If so, is the training period expected to be within reasonable time limits?

23. Do the work aids and training complement each other?

24. If training simulators are used, do they achieve a reasonable balance between transfer of training and costs?

25. Is the work space suitable for use by the range of individuals who will use the facility?

26. Are the environmental conditions (temperature, illumination, noise,

etc.) such that they permit satisfactory levels of human performance and provide for the physical well-being of individuals?

27. In any evaluation or test of the system (or components) does the system performance meet the desired performance requirements?

28. Does the system in its entirety provide reasonable opportunity for the individual(s) involved to experience some form and degree of self-fulfillment and to fulfill some of the human values that we should all like to have the opportunity to fulfill in our daily lives?

29. Does the system in its entirety contribute generally to the fulfillment of reasonable human values? In the case of systems with identifiable outputs of goods and services, this consideration would apply to those goods and services. In the case of systems that relate to our life space and everyday living, this consideration would apply to the potential fulfillment of those human values that are within the reasonable bounds of our civilization.

In the resolution of these and other kinds of human factors considerations, one should draw upon whatever relevant information is available. This information can be of different types, including principles that have been developed through experience or research, sets of normative data (such as frequency distributions of, say, body size), sets of factual data of a probability nature (such as percentage of signals that are detected under specified conditions), mathematical formulas, tentative theories of behavior, hypotheses that have been suggested by research investigations, and even the general knowledge acquired through everyday experience.

With respect to information that would have to be generated through research (as opposed to experience), while there is very comprehensive information available in certain areas of knowledge, in others it is pretty skimpy, and there are some areas in which one draws virtually a complete blank. The paucity of information relating to human factors in some areas probably actually serves as a block to the development of some types of systems; the manned-space programs, for example, depended on the resolution of certain human factors aspects.

In other types of systems, however (especially those in which the human factors aspects are less critical), it may not be feasible to defer the development of a system until all the human factors aspects have been resolved to a perfectionist's standards. In other words, design decisions relating to human factors aspects of some systems need to be made (and will be made) with, or without, the benefit of human factors information. Where adequate information is not available, the opinion is expressed that considered judgments based on partial information will, in the long run, result in better design decisions than those that are pulled out of a hat. But let us reinforce the point that such judgments usually should be made by those whose professional training and experience put them in the class of *experts*, whether in the field of night vision, physical anthropometry, hearing disorders, perception, heat stress, acceleration, learning, decision making, or otherwise.

We should here, again, mention the almost inevitability of having to trade off certain advantages for others. The balancing out of advantages and disadvantages generally needs to take into account various types of considerations—engineering feasibility, human considerations, economic considerations, and others. Granting that there probably are few guidelines to follow, nonetheless the general objective of this horse trading is fairly clear. This basically goes back to the stated or implied system objectives and the accompanying performance requirements. In other words, any trade-offs should be made on the basis of the considerations of their relative effects in terms of system objectives.

A FINAL WORD

The rapid scientific and technological developments of modern life make possible the development of many new and improved items of equipment and systems. Especially in some of the complex systems that are made possible by recent technology, it is of critical importance that they be designed with human considerations in mind. And even where the stakes are less than critical, those facilities that are to be used by human beings usually can be improved by systematic consideration of their human factors aspects. Such improvements usually can be characterized in terms of systems performance and human benefits such as reduction of stress, increased convenience, safety, health, and comfort.

The many facets of human considerations that can arise in the development of systems range over virtually the entire gamut of human attributes and behavior. It has not been the presumptuous intent in this volume to compile and organize a tremendous amount of available pertinent information relating to human attributes and behavior. Rather, the intent has focused around certain more modest objectives. These include the presentation of at least some information about certain human characteristics (such as sensory and motor processes) that might contribute to greater understanding of human performance and behavior. In addition, efforts have been made to illustrate the ways in which human abilities and other characteristics are related to their performance in various aspects of human work and to the performance of systems of which they are a part. In this process, examples have been presented that deal with the relationship between design features on the one hand and performance on the other hand. While these examples are intended to be only illustrative, in many instances they deal with relationships that may have fairly general utility. It has also been the intent to present discussions of at least some techniques and methods, such as those relating to the conduct of human factors studies, and to discuss some aspects of the application of human factors data and principles to practical problems.

But perhaps the more important underlying objective has been to develop an increased sensitivity to, or awareness of, the many human aspects of the systems and situations that abound in our current civilization. Such awareness is at least the first prerequisite to the subsequent processes of creating those systems and situations in which human talents can be most effectively utilized in the furtherance of human welfare.

REFERENCES

1. Chapanis, A.: *Research techniques in human engineering*, The Johns Hopkins Press, Baltimore, 1959.
2. Christensen, J. M.: Individuals and us, *Human Factors*, 1966, vol. 8, no. 1, pp. 1–6.
3. Folley, J. D., Jr. (ed.): *Human factors methods for system design*, The American Institute for Research, Pittsburgh, 1960.
4. Gael, S., and E. D. Stackfleth: *A data reduction technique applied to the development of qualitative personnel requirements information (QPRI)—Keysort card system*, USAF, WADD, Technical Note 60–133, 1960.
5. Gardner, J. W.: *Self-renewal, the individual and the innovative society*, Harper and Row, Publishers, Incorporated, New York, 1963.
6. Lyman, J.: Measuring metafacts, *Human Factors*, 1969, vol. 11, no. 1, pp. 3–8.
7. Meister, D., and G. F. Rabideau: *Human factors evaluation in system development*, John Wiley & Sons, Inc., New York, 1965.
8. Rabideau, G. F., J. I. Cooper, and C. J. Bates, Jr.: *A guide to the use of function and task analysis as a weapon system development tool*, Northrop Corp., Norair Division, Preliminary final draft of USAF, WADC, TR (undated).
9. Shapero, A., and C. J. Bates, Jr.: *A method for performing human engineering analysis of weapon systems*, USAF, WADC, TR 59–784, September, 1959.
10. Van Cott, H. P., and J. W. Altman: *Procedures for including human engineering factors in the development of weapon systems*, USAF, WADC, TR 56–488, October, 1956.
11. *Personnel subsystems*, USAF, AFSC design handbook, ser. 1–0, General, AFSC DH 1–3, 1st ed., Jan. 1, 1969, Headquarters, AFSC.

AFB	Air Force Base
AFHRL	Air Force Human Resources Laboratory
AFSC	Air Force Systems Command
AMD	Aerospace Medical Division, Air Force Systems Command
AMRL (of USAF)	Aerospace Medical Research Laboratory, Aerospace Medical Division, Air Force Systems Command
ASD	Aeronautical Systems Division, Air Force Systems Command
ASHRAE	American Society of Heating, Refrigerating, and Air-conditioning Engineers
ASHVE	American Society of Heating and Ventilating Engineers (now ASHRAE)
ASME	American Society of Mechanical Engineers
FAA	Federal Aviation Administration
HFRB	Human Factors Research Branch, Adjutant Generals Research and Development Command, US Army
IEEE	Institute of Electrical and Electronics Engineers, Inc.
IES	Illuminating Engineering Society
IRE	Institute of Radio Engineers
ISO	International Organization for Standardization
MRL (of USAF)	Medical Research Laboratory (see AMRL)
NAS	National Academy of Sciences
NASA	National Aeronautics and Space Administration
NAVTRADEVCEN	United States Naval Training Device Center (formerly Special Devices Center)
NEL	Navy Electronics Laboratory (USN)
NRC	National Research Council

NRL	Naval Research Laboratory (USN)
ONR	Office of Naval Research (USN)
SAE	Society of Automotive Engineers
SDC	Special Devices Center (USN) (See NAVTRADEVCEN)
TDR	Technical Documentary Report (AMRL)
TR	Technical Report (term used by various organizations)
USA	United States Army
USAF	United States Air Force
USASI	United States of America Standards Institute
USN	United States Navy
USPHS	United States Public Health Service
WADC	Wright Air Development Center, USAF (see AMRL and AFHRL)
WADD	Wright Air Development Division, USAF (see AMRL and AFHRL)

Table B-1 presents a brief evaluation of the operational characteristics of certain types of control devices.[1] Table B-2 presents a summary of recommendations regarding certain features of these types of control devices.[2] In the use of these and other recommendations, it should be kept in mind that the unique situation in which a control device is to be used and the purposes for which it is to be used can affect materially the appropriateness of a given type of control and can justify (or virtually require) variations from a set of general recommendations or from general practice based on research or experience. For further information regarding these, refer to the original sources given in the reports from which these are drawn.

COMMENTS REGARDING CONTROLS[3]

- *Hand push button:* Surface concave, or provide friction. Preferably audible clock when activated. Elastic resistance plus slight sliding friction, starting low, building up rapidly, sudden drop. Minimize viscous damping and inertial resistance.
- *Foot push button:* Use elastic resistance, aided by static friction, to support foot. Resistance to start low, build up rapidly, drop suddenly. Minimize viscous damping and inertial resistance. Hinged pedal preferable if space permits.
- *Toggle switch:* Use elastic resistance which builds up and then decreases as position is approached. Minimize frictional and inertial resistance.
- *Rotary selector switch:* Provide detent for each control position (setting). Use elastic resistance which builds up, then decreases as detent is approached. Minimum fraction and inertial resistance. Separation of detents should be at least $\frac{1}{4}$ in.
- *Knob:* Preferably code by shape if used without vision. Kind of desirable resistance depends on performance requirements. Inertial resistance

---

[1] Adapted largely from J. H. Ely, R. M. Thomson, and J. Orlansky, "Design of controls," chap. 6 in C. T. Morgan, J. S. Cook, III, A. Chapanis, and M. W. Lund (eds.), *Human engineering guide to equipment design,* McGraw-Hill Book Company, New York, 1963.

[2] Adapted largely from *ibid.* and from *Personnel subsystems,* USAF, AFSC design handbook, ser. 1–0, General, AFSC DH 1–3, 1st ed., Jan. 1, 1969, Headquarters, AFSC.

[3] Adapted from Ely, Thomson, and Orlansky, *op. cit.,* chap. 6.

Table B-1 Comparison of the Characteristics of Common Controls

Characteristic	Hand push button	Foot push button	Toggle switch	Rotary switch	Knob	Crank	Lever	Hand-wheel	Pedal
Space required	Small	Large	Small	Medium	Small–medium	Medium–large	Medium–large	Large	Large
Effectiveness of coding	Fair–good	Poor	Fair	Good	Good	Fair	Good	Fair	Poor
Ease of visual identification of control position	Poor*	Poor	Fair–good	Fair–good	Fair–good†	Poor‡	Fair–good	Poor–fair	Poor
Ease of non-visual identification of control position	Fair	Fair	Good	Fair–good	Poor–good	Poor‡	Poor–fair	Poor–fair	Poor–fair
Ease of check reading in array of like controls	Poor*	Poor	Good	Good	Good†	Poor‡	Good	Poor	Poor
Ease of operation in array of like controls	Good	Poor	Good	Poor	Poor	Poor	Good	Poor	Poor
Effectiveness in combined control	Good	Poor	Good	Fair	Good§	Poor	Good	Good	Poor

* Except when control is backlighted and light comes on when control is activated.
† Applicable only when control makes less than one rotation and when round knobs have pointer attached.
‡ Assumes control makes more than one rotation.
§ Effective primarily when mounted concentrically on one axis with other knobs.

Table B-2 Summary of Selected Data Regarding Design Recommendations for Control Devices

	Size, in.		Displacement		Resistance	
	Mini-mum	Maxi-mum	Mini-mum	Maxi-mum	Mini-mum	Maxi-mum
Hand push button						
Fingertip operation	1/2	None	1/8 in.	1 in.	10 oz	40 oz
Foot push button	1/2	None				
Normal operation			1/2 in.			
Wearing boots			1 in.			
Ankle flexion only				2 1/2 in.		
Leg movement				4 in.		
Will *not* rest on control					4 lb	20 lb
May rest on control					10 lb	20 lb
Toggle switch			30°	120°	10 oz	40 oz
Control tip diameter	1/8	1				
Lever arm length	1/2	2				
Rotary selector switch					1 in./lb	6 in./lb
Length	1					
Width		1				
Depth	5/8					
Visual positioning			15°	40°*		
Nonvisual positioning			30°	40°*		
Knob, continuous adjustment†						
Finger-thumb						4 1/2–6 in./oz
Depth	1/2	1				
Diameter	3/8	4				
Hand/palm						
Depth	3/4					
Diameter	1 1/2	3				
Crank†						
For rate, radius	1/2	4 1/2				
For force, radius	1/2	20				
Rapid, steady turning						
<3-in. radius					2 lb	5 lb
5–8 in. radius					5 lb	10 lb
>8-in. radius					?	?
For precise settings					2 1/2 lb	8 lb
Lever†§						
Fore-aft (one hand)				14 in.	2 lb	30–50 lb
Lateral (one hand)				38 in.	4 lb	20 lb
Finger, diameter	1/2	3				
Handwheel†				90°–120°		
Diameter	7	21				
Rim thickness	3/4	2			5 lb	30 lb‡
Pedal						
Length	3 1/2					
Width	1					
Normal use			1/2 in.			
Heavy boots			1 in.			
Ankle flexion				2 1/2 in.		10 lb
Leg movement				7 in.		180 lb
Will *not* rest on control					4 lb	
May rest on control					10 lb	

* When special requirements demand large separations, maximum should be 90°.

† Displacement of knobs, cranks, and handwheels should be determined by desired control-display ratio.

‡ For two-handed operation, maximum resistance of handwheel can be up to 50 lb.

§ Length depends on situation, including mechanical advantage required. For long movements, longer levers are desirable (so movement is more linear).

of little practical consequence, but may counteract harmful effects of friction and vice versa.

· *Crank:* Use when task involves two rotations or more. Friction (2 to 5 lb) reduces effects of jolting, but degrades constant-speed rotation at slow or moderate speeds. Inertial resistance aids performance for small cranks and low rates. Grip handle should rotate.

· *Lever:* Provide elbow support for large adjustment, forearm support for small hand movements, wrist support for finger movements. Elastic resistance may improve "feel."

· *Handwheel:* For small movements, minimize inertia. Indentations in grip rim to aid holding. Displacement usually should not exceed $\pm 60°$ from normal. For displacements less than 120°, only two sections need be provided, each of which is at least 6 in. long.

· *Pedal:* Pedal should return to null position when force is removed; hence, elastic resistance should be provided. Pedals operated by entire leg should have 2 to 4 in. displacement, except for automobile-brake type for which 2 to 3 in. of travel may be added. Displacement of 3 to 4 in. or more should have resistance of 10 lb or more. Pedals operated by ankle action should have maximum travel of $2\frac{1}{2}$ in.

appendix C
SELECTED
REFERENCES

This appendix includes a list of selected books and journals that deal with human factors engineering. With a few exceptions, the books listed are general references; for references dealing with topics that are specific to the topics of the various chapters, the reader is referred to the lists following the individual chapters.

BOOKS

1. Bennett, E., J. Degan, and J. Spiegel (eds.): *Human factors in technology*, McGraw-Hill Book Company, New York, 1963.
2. Bilodeau, E. A., and Ina M. Bilodeau (eds.): *Principles of skill acquisition*, Academic Press, Inc., New York, 1969.
3. Burns, N., R. Chambers, and E. Hendler: *Unusual environments and human behavior*, The Free Press, New York, 1963.
4. Chapanis, A.: *Research techniques in human engineering*, The Johns Hopkins Press, Baltimore, 1959.
5. Chapanis, A.: *Man-machine engineering*, Wadsworth Publishing Company, Inc., Belmont, Calif., 1965.
6. Damon, A., H. W. Stoudt, and R. A. McFarland: *The human body in equipment design*, Harvard University Press, Cambridge, Mass., 1966.
7. Fogel, L. J.: *Biotechnology: concepts and applications*, Prentice-Hall, Inc., Englewood Cliffs, N.J., 1963.
8. Forbes, T. W., and M. S. Katz: *Summary of human engineering research data and principles related to highway design and traffic engineering problems*, The American Institute for Research, Pittsburgh, 1957.
9. Gagné, R. M. (ed.): *Psychological principles in system development*, Holt, Rinehart and Winston, Inc., New York, 1962.
10. Hanrahan, J. S., and D. Bushnell: *Space biology*, Science Editions, John Wiley & Sons, Inc., New York, 1961.
11. Harris, D. H., and F. B. Chaney: *Human factors in quality assurance*, John Wiley & Sons, Inc., New York, 1969.
12. Jennings, B. H., and J. E. Murphy: *Interactions of man and his environment*, Plenum Press, Plenum Publishing Corporation, New York, 1966.
13. Kelley, C. R.: *Manual and automatic control*, John Wiley & Sons, Inc., New York, 1968.
14. McFarland, R. A., et al.: *Human factors in the design of highway transport equipment*, Harvard School of Public Health, Boston, Mass., 1953.
15. McFarland, R. A., R. C. Moore, and A. B. Warren: *Human variables in motor vehicle accidents*, Harvard School of Public Health, Boston, Mass., 1955.

16. McFarland, R. A., and A. L. Moseley: *Human factors in highway transportation safety*, Harvard School of Public Health, Boston, Mass., 1954.

17. Meister, D., and G. F. Rabideau: *Human factors evaluation in system development*, John Wiley & Sons, Inc., New York, 1965.

18. Morgan, C. T., J. S. Cook, III, A. Chapanis, and M. W. Lund (eds.): *Human engineering guide to equipment design*, McGraw-Hill Book Company, New York, 1963.

19. Murrell, K. F. H.: *Human performance in industry*, Reinhold Publishing Corporation, New York, 1965.

20. Nadler, G.: *Work design*, Richard D. Irwin, Inc., Homewood, Ill., 1963.

21. Roth, E. M. (ed.): *Compendium of human responses to the aerospace environment*, NASA, Washington, D.C., NASA CR–1205(1), vols. 1–5, November, 1968.

22. Sells, S. B., and C. A. Berry (eds.): *Human factors in jet and space travel: a medical psychological analysis*, The Ronald Press Company, New York, 1961.

23. Sinaiko, H. W.: *Selected papers on human factors in the design and use of control systems*, Dover Publications, Inc., New York, 1961.

24. Spector, W. S.: *Handbook of biological data*, USAF, WADC, TR 56–273, October, 1956.

25. Stevens, S. S. (ed.): *Handbook of experimental psychology*, John Wiley & Sons, Inc., New York, 1951.

26. Webb, P. (ed.): *Bioastronautics data book*, NASA, Washington, D.C., NASA SP–3006, 1964.

27. Woodson, W. E. and D. W. Conover: *Human engineering guide for equipment designers*, (2d ed.), University of California Press, Berkeley, 1964.

28. *Personnel subsystems*, USAF, AFSC design handbook, ser. 1–0, General, AFSC DH 1–3, Jan. 1, 1969, Headquarters, AFSC.

29. *Studies of human vigilance*, Human Factors Research, Inc., Goleta, Calif., January, 1968.

JOURNALS

Ergonomics, Taylor & Francis, Ltd., London.
Human Factors, Pergamon Press, New York.
Journal of Safety Research, National Safety Council, Chicago.

NAME INDEX

SUBJECT INDEX